Resource Mobilization for Drinking Water and Sanitation in Developing Nations

Proceedings of the International Conference sponsored by the Water Resources Planning and Management Division and the Environmental Engineering Division of the American Society of Civil Engineers

San Juan, Puerto Rico: May 26–29, 1987

Editors:
F.W. Montanari
Terrence P. Thompson
Terence P. Curran
Walter Saukin

Foreword by Dr. Abel Wolman

A contribution to the
International Drinking Water Supply and
Sanitation Decade

Published by the
American Society of Civil Engineers
345 East 47th Street
New York, New York 10017-2398

ABSTRACT

The ASCE International Conference on Resource Mobilization for Drinking Water Supply and Sanitation in Developing Countries was held at the Caribe Hilton Hotel in San Juan, Puerto Rico, May 26-29, 1987. The papers presented at the conference and included in this book focus on successful implementation and sustainability of projects related to the International Drinking Water Supply and Sanitation Decade (IDWSSD). Four major components are addressed. In the area of human resources development, community education and training of managers, technicians and professional engineers are discussed. Financial resources for water and sanitation projects, actions needed to improve financial performance, and economic policies followed by international agencies are reviewed. Papers on appropriate technology and technology transfer applicable to planning and research in water supply and sanitary engineering in developing countries are included. Discussions on strategic planning for operation and maintenance of water supply and sewerage systems include means of upgrading plant performance, evaluating O&M, and developing design concepts aimed at reducing operational problems.

Library of Congress Cataloging-in-Publication Data

Resource mobilization for drinking water and sanitation in developing nations: proceedings of the international conference/sponsored by the Water Resources Planning and Management Division and the Environmental Engineering Division of the American Society of Civil Engineers, San Juan, Puerto Rico, May 26-29, 1987; editors, F.W. Montenari . . . [et al.]: foreword by Abel Wolman.

p. cm.

"A contribution [sic] to the International Drinking Water Supply and Sanitation Decade"—T.p.

Includes index.

ISBN 0-87262-629-6

1. Water-supply—Developing countries—Congresses. 2. Sanitation engineering—Developing countries—Congresses. 3. International Drinking Water Supply and Sanitation Decade. 1981-1990—Developing countries—Congresses. I. Montenari, F.W. II. American Society of Civil Engineers. Water Resources Planning and Management Division. III. American Society of Civil Engineers. Environmental Engineering Division. IV. ASCE International Conference on Resource Mobilization for Drinking Water Supply and Sanitation in Developing Nations (1987: San Juan, P.R.) V. International Drinking Water Supply and Sanitation Decade (1981-1990)

TD353.R44 1987 87-33366
363.7′2′091724—dc19 CIP

Library of Congress Catalog Card No.: 87-33366
ISBN 0-87262-629-6
Manufactured in the United States of America.

Voluntary cash donations from the Conference participants to UNICEF made possible the purchase and installation of a TARA intermediate-lift plastic handpump, similar to the one pictured above, in the Village of Horipur in Paba Upazila, Bangladesh. The well serves approximately 200 villagers and provides easy access to clean water, especially in the dry season when the water table drops below the range of suction pumps.

FOREWORD

Abel Wolman

The Proceedings of the International Conference represent a continuing global task directed to two important decisions unanimously fostered at Alma Ata, USSR, in the late 1970s. One was the courageous decision to provide life-giving amenities more rapidly to rural and urban people throughout the world. The second was to develop public and private understanding and acceptance of the means by which implementation of these desires and hopes would be fulfilled.

Unfortunately, the goals for the Decade have not been realized. They remain as the guides for the future. The delayed pace must be accelerated, and can be, given the will to do so. Undue pessimism is unwarranted, since many, if not all of the facilities, have reached millions of people since 1980. Reasons for revived hope are amply demonstrated in the sessions here recorded.

The International Conference is aptly named as an arena for Resource Mobilization. Within that rubric, one finds the familiar ingredients of human resources, finance and economics, technology and engineering, and operation and maintenance. The four days offered participants the opportunity to reassess change, failure, timing, and new approaches. Courage and enthusiasm still prevail in the deliberations to rescue some 2.8 billion people from unwarranted early sickness and death. Of even greater importance is the significance of the effort to provide the under-pinning for industrial and agricultural development.

Since the future mirrors the past, some closing comments are warranted. The world population in 1936 was about 2.1 billion. Today, it is 5 billion. It is interesting to note that the U.S. population was 6.4 percent of the world in 1936. It is 4.8 percent now, with the prospect in post-2000 it will drop to less than 3.0 percent. The issues which these figures portent are great, complex, and challenging—they pervade proposals and analyses throughout the Proceedings!

Danny Kaye, in 1980, said it all:

"The greatest natural resource that any country can have is its children. They are more powerful than oil, they are more beautiful than rivers, and far more determined that the world shall exist."

ACKNOWLEDGEMENTS

While the conference arrangements and environment contribute to successful ventures, these facets are often forgotten in the "getting out of the proceedings." The Steering Committee gives special credit to the sponsors' representatives, the advisory committee and most importantly to the local committee under the chairmanship of Max Figueroa D. of the Colegio de Ingenieros y Agremensores de Puerto Rico. The technical program committee developed a four track program of high quality providing the substance of the conference, making these proceedings valuable to a wide range of practitioners of the art and science of providing water and sanitation throughout the world. ASCE staff members Harry Tuvel and Elizabeth Yee along with Denise Nurse supported the members and associates making possible the delivery of a most successful meeting as these proceedings testify.

SPONSOR

American Society of Civil Engineers
Water Resources Planning & Mgmt. Div.
Environmental Engineering Div.

CO-SPONSORS

American Academy of Environmental Engineers
American Public Works Association
American Water Resources Association
American Water Works Association
Asociacion Interamericana de Ingenieria Sanitaria y Ambiental
Colegio de Ingenieros y Agrimensores de Puerto Rico
Water Pollution Control Federation

PARTICIPATING ORGANIZATIONS

Interamerican Development Bank
Organization of American States
Pan American Health Organization
United Nations Development Programme
United Nations Environment Programme
United Nations Department of Technical Cooperation for Development
UNICEF (United Nations Children's Fund)
U.S. Coordinator for the Water Decade
U.S. Army Corps of Engineers
World Bank
World Health Organization

CO-SPONSOR'S REPRESENTATIVES

William C. Anderson, *American Academy of Environmental Engineers*
Robert D. Bugher, *American Public Works Association*
Eugene DeMichele, *Water Pollution Control Federation*
Max Figueroa Dominguez, *Colegio de Ingenieros y Agrimensores de Puerto Rico*

Maria Flores de Otero, *Asociacion Interamericana de Ingenieria Sanitaria y Ambiental*
Kurt Keeley, Dale Kratzer, *American Water Works Association*
Kenneth Reid, *American Water Resources Association*

ADVISORS
Juan Alfaro, *Interamerican Development Bank*
Steven Bender, *Organization of American States*
Martin Beyer, *UNICEF*
Marcia Brewster, Enzo Fano, *U.N. Department of Technical Cooperation for Development*
Noel Brown, *U.N. Environment Programme*
Bernd Dieterich, Odyer Sperandio, *World Health Organization*
J. Randall Hanchey, *U.S. Corps of Engineers*
Franciscus Hartvelt, *U.N. Development Programme*
John W. McDonald, Jr., *U.S. State Department*
Horst Otterstetter, *Pan American Health Organization*
Michael Potashnik, Alfonzo Zavala, *World Bank*

STEERING COMMITTEE
Honorary Chairman, Abel Wolman
General Chairman, F.W. Montanari
Vice-Chairman/Secretary, Terrence P. Thompson
Treasurer, Harry Tuvel
Technical Program Co-Chairman, Walter Saukin, Terence P. Curran
Chairman, Publicity and Promotion, Mark Rothenberg
Exhibits Chairman, Arun Deb
Chairman, Financial Sponsors, Albert Machlin
Local Committee Chairman, Max Figueroa Dominguez

TECHNICAL PROGRAM COMMITTEE
Carl Bartone
Terence P. Curran, *Co-Chairman*
James Jordan
John Kalbermatten
Charles Morse
Horst Otterstetter
Michael Potashnik
Walter Saukin, *Co-Chairman*
Harold Shipman
Dennis Warner

LOCAL COMMITTEE
Juan A. Bonnet
Rafael Cruz-Perez
Benigno Despiau
Max Figueroa Dominquez, *Local Chairman*
Juan L. Figueroa Laugier
Maria Flores de Otero
Rafael Miranda Franco
Orlando Guihurt
Jairo Fco. Lascarro
Alejandro Palau
Carl-Axel Soderberg

All papers included in the proceedings were accepted for publication based on abstract review by the conference program committee. All papers are eligible for discussion in the appropriate ASCE Journal. In addition, these papers are eligible for ASCE awards.

CONTENTS

*Manuscript not received in time for publication.

Closing Plenary Session

Role of Professional and Technical Associations in International Water Supply and Sanitation

Preconference panels are usually mundane gatherings brought together to solve some problem associated with the conference. This panel was anything but mundane. It addressed a principal area of considerations and contributed significantly to the objective of the conference—a real contribution by the professional associations to the International Drinking Water and Sanitation Decade.

For the first time responsible officers of the principal U.S. professional associations gathered around a table and discussed what is probably the major world problem—water and sanitation. The scene was ably and clearly set by Ambassador John McDonald, Jr., formerly U.S. Coordinator for the Water Decade (albeit justifiably critical of the U.S. participation in support of the Decade to date). Each association representative addressed the Decade and what his organization was contributing. Time ran out long before the subject of the Decade and U.S. professional associations could be adequately explored. Beth Turner, President-Elect of the Water Pollution Control Federation, provides the high point from the association side of the ledger. She suggested that the U.S. associations should join together, through a committee or some device, to pool resources and efforts.

This panel discussion was indeed in tune with the entire conference and an appropriate prelude for the opening session and Dr. Abel Wolman's keynote.

The conference steering committee congratulates the following on their contributions:

Ambassador John W. McDonald, U.S. Coordinator Water Decade, Moderator
Juan A. Bonnet, Colegio de Ingenieros y Agramensores de Puerto Rico (CIAPR)
Paul L. Busch, President-Elect, American Academy of Environmental Engineers (AAEE)
Robert T. Chuck, President American Water Works Association (AWWA)
Phillip E. Greeson, Past President, American Water Resources Association (AWRA)
Luis Jauregui, Past President, Asociacion Interamericana de Ingenieria Sanitaria y Ambiental (AIDIS)
Glenn Stout, Secretary General, International Water Resources Association (IWRA)
Beth Turner, President-Elect, Water Pollution Control Federation (WPCF)
William Zoino, Vice President, Zone 1, American Society of Civil Engineers (ASCE)

Ambassador McDonald and Beth Turner have framed the challenge and an important response. Are the U.S. professional organizations prepared to move? Read Ambassador McDonald's presentation to decide what you should do, and then do it!

THE UNITED NATIONS WATER DECADE
BY
AMBASSADOR JOHN W. MCDONALD

I. INTRODUCTION

Good Morning! I am John McDonald-your moderator for this morning's Pre-Conference Panel. It is great to be in this beautiful city of San Juan and in this lovely hotel. I want to thank Monty and Terry Thompson and our Puerto Rican hosts and so many others for making this all possible!

Some of you are probably wondering what a United States Diplomat is doing up here, especially since he is a lawyer and not even an engineer. Let me answer the question by saying that we are going to be talking about the United Nations Water Decade this week and I was named the U.S. Coordinator for the "Decade" by the State Department in 1979, because of my United Nations expertise. So I have some knowledge of the subject. But the main reason I am up here is, because Monty asked me. I believe it is fair to say that the last reason is why all of us on the panel are here.

I will be the first speaker this morning, giving you an overview of the Water Decade: how and why it happened and where it is going, with some recommendations for future actions. And I will then call on this very distinguished panel as follows:

1. Robert Chuck, President of AWWA
2. Phil Greeson, Past President of AWRA
3. Paul Busch, President-elect of AAEE
4. Juan Bonnet, Member of CIAPR
5. Glenn Stout, Secretary-General of IWRA
6. Beth Turner, President-elect of WPFC
7. Bill Zoino, Regional Vice-President and Representative of ASCE and finally,
8. Luis Jauregui, Argentina, Past President of AIDIS and President of the United Nations Water Conference.

II. THE DECADE-ITS BEGINNINGS

I have often been asked, "Why did the United Nations get involved in drinking water as an issue in the first Place?" Let me add that it is usually North Americans who ask me that question, not realizing what we take totally for granted in our society-putting a glass under a faucet in the kitchen and getting back a glass of cool, clear, clean and safe drinking water-is, in most countries of the world, a miracle of life!

The United Nations got involved because:
- over half of the people in the 127 countries called the Developing World, or the Third World, do not have easy access to safe drinking water;
-three quarters of the people in the Third World have no sanitation facilities at all;
-80% of all sickness and disease in the world comes from unsafe water and sanitation;
-50,000 people a day- 40,000of whom are children under five- die every day from unclean water.

I could go on, but you get the picture. The experts realized in the mid-70's that the technology was there to stop this continuing disaster. What was needed was the political will and the structure and organization at the national and international level, to change this picture.

The United Nations General Assembly acted and in March 1977 the first World Conference of governments on water took place at Mar Del Plata, Argentina. The assembled governments agreed that all peoples of the world should have access to safe drinking water and sanitation, and urged the United Nations to launch a Decade, with the slogan "Safe Water and Sanitation for the World by 1990".

The World heard briefly- and then forgot, and for two years nothing happened. In March of 1979 the first follow-up meeting was called by the United Nations. I headed the U.S. Delegation to a special session of the Natural Resources Committee of the Economic and Social Council of the U.N., and proposed five ideas to make something happen:

1. there should be a one-day special session of the United Nations General Assembly in November 1980 to officially launch the Decade·
2. every nation should develop a National Plan for the Decade and present it to the special session;
3. the United Nations Development Program (UNDP)-Resident Representatives should coordinate United Nations efforts in Third World countries and help the country coordinate their efforts;
4. a United Nations Steering Committee should be formally established under the UNDP (United Nations Development Program), not the WHO (World Health Organization) since this is broader than health, to coordinate the United Nations' action·
5. an inter-governmental donor group should be formed to talk about money.

These recommendations were all adopted by the Natural Resources Committee, ECOSOC, the UNDP, WHO and finally the

United Nations itself. The Decade was launched on November 10, 1980 and forty ministers of Development spoke in support of the Decade.

In fact, the United Nations system has played a major, positive role in the 6½ years of the Decade.

1. the UNDP and its Resident Representatives have done an outstanding job of national and international coordination;
2. the WHO has assumed an impressive, substantive
3. UNICEF allocates 28 % of its budget into water and sanitation projects;
4. the World Bank has allocated from 500 million dollars to 1 billion dollars a year for water and sanitation projects·
5. the Inter-American Bank puts 14% of its budget into water and sanitation projects:
6. PAHO(Pan American Health Organization) has established a major program in this field.

The United Nations system has made a critically important contribution to the success of the Decade.

III. DONOR GOVERNMENTS

How have donor governments done? Not very well overall. One donor-conference on water was held- no financial pledges were made: just talk, and then all began to cut back on their aid by Mid-Decade. Donor aid has been uncoordinated, with each government pushing its own pumps and other hardware. Governments seemed to take little interest in training the villagers how to maintain the pumps, or to recognize the value of health education and the necessary participation of women in the whole process.

The one governmental exception to this statement is the German Agency for Technical Cooperation (GTC). Over the last few years they have done some out-standing work on their own and with WHO, developing guidelines for project preparation and funding and organizing regional, subregional and national conferences to stimulate the development of water and sanitation projects which donor countries could fund, if they so desired.

The United States-AID has focused mostly on urban water systems and ignored the rural water problems, where 80 % of the world's population lives. However,there are two bright spots on the U.S. scene:

1. The Peace Corps has over 500 volunteers in the field in over 30 countries in water and sanitation, working on village-level projects
2. WASH (Water and Sanitation for Health), funded by AID, is a fantastic contractor, providing technical assistance, help and guidance in water and sanitation, in a fast and efficient manner.

IV. PRIVATE SECTOR

What about the private sector? On the business side, the British, French, Germans and Italians are far ahead. The U.S. Department of Commerce says the U.S. has 5 % of the $ 50 Billion a year business in world-wide water and sanitation hardware. However, there are three bright spots on the not-for-profit side of the private sector:

1. The International Reference Center at the Hague is doing a great job of information exchange:
2. ZONTA International, a global women's group-raised $1 million and put in 4,000 wells and pumps in Sri Lanka over a five year period:
3. Global Water, an NGO which Dr. Peter Bourne and I founded to push the goals of the Decade, has small-scale projects in half a dozen countries and is lobbying in the United States for the Decade. Dr. Bourne testified last month before Congress, as did President Carter, in support of the Water Decade and urged AID to do more in this field.

Professional associations have been supportive with resolutions, but not much action.

V. THIRD WORLD

Well, where is the action then? It is in the Third World, and it is exciting and far-reaching, wherever the political will is stimulated. Over 35 countries, at the head of state level, have publicly accepted the goals of the Water Decade. Over 20 others will meet 80 to 90 % of the goals of the Decade. Over 80 countries have national action committees and over 100 have national plans or strategies for the Decade.

Let me focus on the two largest countries and what they are doing. CHINA: China's seventh 5-year plan (1986-1990) states that 80 % of China's 800 million rural population will have easy access to safe drinking water by 1990. That means adding 200 million people to the list in four years, but the political structure has pledged to this. I think they will achieve it. They hope to reach the Decade goal by the year 2000. INDIA: When the Decade started, 30 % of the country had ready access to clean water. By 1990 they estimate 70 % of the total population will have access to clean water. The present 5-year plan has allocated $ 5 billion to water and sanitation. India's private sector is now manufacturing 100,000 hand pumps a year.

China, India, Pakistan, Bangladesh and Indonesia represent over half the world's population, and more than 2/3 of all the unserved people in the world, and all of these five governments have embraced the goals of the Water Decade and are doing a good deal to make it happen.

So, even though the United States is on the sidelines-exercising no leadership in this field-amazing progress is being made on a global scale.

VI. CONCLUSIONS

Where do we go from here? I, for one, would like to see the United States get back into the picture and make an impact, for business reasons and for humanitarian reasons.

I believe this is possible and that this conference can start the process by taking the following actions:

1. Every U.S. based organization attending this conference has a newsletter or a professional journal. You should pledge to use this vehicle to educate your membership about the goals and objectives of the Water Decade and what has happened to date
2. Describe the business opportunities that exist and are ignored because many U.S. companies are afraid and do not know how to compete abroad in this field;
3. Explain that political risk is covered by a U.S. Government Agency (OPIC). It issues insurance policies to U.S. firms investing in Third World countries. And the World Bank also issues political risk insurance policies to encourage investment;
4. Push the human resource development angle- the "software" - part of the development process, not just the easy part, the "hardware", and
   a) publicize, support and help expand the world-wide training program in water and sanitation that our own John Kalbermatten is building in Third World countries;
   b) publicize, support and help to expand the teachings and recommendations of our own Mary Elmendorff on the role of women and water. Dr. Elmendorff is a universally recognized expert in this field. Her properly executed recommendations for action in this area of "software" will double the impact of any "hardware" project.
5. Urge all U.S. members to lobby the U.S. Congress on the matter of drinking water and sanitation and the United Nations Decade goals. Urge that U.S.-AID be restructured so that it can perform a major service in this area and urge that more funds be provided so that U.S.-AID can make a difference!
6. Finally, in your writings and in your conversations, support the idea of the United Nations extending the Water Decade concept, structures and goals for another Decade, to the year 2000. I am convinced, in a second ten-year period, the goals of the original Decade "safe water and sanitation for the world" can be achieved!

In conclusion, let us make the last quarter of this decade the most productive quarter of all-it is not too late! Thank you.

STATUS OF THE INTERNATIONAL DRINKING WATER SUPPLY AND SANITATION DECADE

Alexander H. Rotival*

The best introduction to a report on the status of the decade is to provide the latest data on where we stand on meeting objectives as we move into the second half of the IDWSSD. The World Health Organization (WHO) in Geneva has recently compiled a statistical review of progress in meeting decade quantitative coverage of populations as of 1985 at mid-Decade point. The report, containing a wealth of data, is still unpublished but statistical data provided by governments reveal the following with reference to world population served (excluding The Peoples Republic of China) by water and sanitation facilities as of December 31, 1985:

	Population Served	
	1980	1985
Urban Water	509.0 (72%)	670.0 (77%)
Urban Sanitation	395.0 (54%)	558.0 (62%)
Rural Water	472.0 (32%)	679.0 (42%)
Rural Sanitation	207.0 (14%)	297.0 (18%)

These statistics indicate that worldwide 350 million additional people in 1985 have access to an adequate safe water supply and an additional 230 million to appropriate sanitation; however, remaining unserved, and once again excluding The Peoples Republic of China, at the end of 1985 are 1,200 million persons without adequate and safe water supply and 1,630 million without sanitation.

Breaking these global statistics on population coverage down on a regional basis we find the following:

In Africa - a reported 16 to 19 percent coverage with respect to urban water supply and sanitation. This important increase in coverage in the five years from 1980-1985 should be treated with some caution since it includes two extremely populous African countries which did not report in the Africa Region previously.

Statistics concerning coverage of rural populations are more sobering. With respect to rural water supply, there has only been a three percent increase in coverage between 1980-1985 representing from 58 to 73 million

---

*UNDP/WHO Coordinator, IDWSSD; Geneva, Switzerland

persons or a percentile change from 22 to 25 percent. The situation is comparable for rural sanitation in Africa.

In Latin America and Caribbean Region, there has been, with respect to urban water supply, a reported six percent increase in coverage with an 84 percent coverage of overall urban population or 233 million persons. It appears reasonable to expect that more than 90 percent of the population will have adequate and safe water supply at the end of the Decade. For urban sanitation, while the percentile increase of 11 percent from 1980 to 1985 is impressive, coverage of population at 219 million or 79 percent is less than that of urban water supply. Coverage levels of rural population are less impressive. There has been a five percent increase in rural water supply with a coverage in 1985 of 60 million persons or 47 percent of the rural population. It should be underlined that a matter of particular concern is the dispersed rural population. There has been a 10 percent increase in rural sanitation services representing 34 million inhabitants or 27 percent coverage.

Looking at South and South-East Asia Regions, the statistics on the evolution of water coverage from 1980 to 1985 on urban populations are preoccupying. There has only been a one percent reported increase in the first five years of the Decade resulting in a percentile coverage of 65 percent or 196 million people in 1985. The urban sanitation situation is more encouraging since there has been a net increase of three percent in coverage with 33 percent of the population or 99 million persons covered by appropriate sanitation facilities. It should be noted that urban population increased by an estimated 20 percent from 1980 to 1985 and that the growth of urban populations in the developing regions throughout the world represents two to three times the growth of rural-based populations. This is a phenomenon that requires increasing attention in the setting of Decade priorities.

On coverage of rural populations in South and South-East Asia, there has been a rather remarkable 16 percent increase in coverage of populations in rural water supply to 419 million population or 47 percent of total rural population. It would be, I believe, accurate to state that this achievement is due in no small part to the handpumps program sponsored by the United Nations Development System and in particular to the action of UNICEF in certain key countries. The rural sanitation situation is far less encouraging with only a two percent increase in coverage in the first five years of the decade to a very low total coverage of nine percent of the population or 22 million additional people covered in the five years. This must be contrasted with the total of 700 million inhabitants, excluding the Peoples Republic of China, not covered.

The quantitative data provided on certain developing regions as they stand in 1985 compared to 1980 - the initiation of the IDWSSD - are revealing but need to be treated with some caution. Governments, who prepared and submitted the statistics to WHO, have been refining their statistics since the initial data provided in 1980. As an example, one populous country in Asia lowered their estimate on coverage in urban water supply from 65 percent in 1980 to 49 percent in 1985. Governments are making attempts to ensure more accurate statistical data, and not incidentally, more reasonable targets for coverage levels.

The statistics also underline the necessity in the last five years of the Decade to give higher priority to rural as well as peri-urban populations to meet the Mar del Plata Plan of Action "...giving priority attention to the segments of the population in greatest need."

The Mar del Plata Plan of Action emphasized as well the importance of accompanying new systems with programs of health and domestic hygiene education. This came to be what is known as "The Decade Approach" including community participation at all levels, institutional development and inter-sectoral coordination in general. It is therefore heartening that 80 percent of reporting countries indicate that water supply and sanitation programs are being implemented within the context of an overall primary health care strategy.

With respect to coordination both at a global and country level, the IDWSSD compares very favorably with other development sectors. At a global level, with the support of the Development Assistance Committee of the OECD which held a High-Level Meeting on "The Water Decade" in 1985, ever increasing mutual consultations and cooperation with the objective of harmonization of efforts is taking place between bilateral and multilateral donors and concerned organizations of the United Nations Development System. The United Nations System itself through the Steering Committee of the IDWSSD plus an informal consultative mechanism between WHO, The World Bank, UNICEF and UNDP is ensuring an increasingly coordinated "Decade Approach" to activities and programs.

At a country level, the UNDP Resident Representatives has been designated as focal point for the decade and is responsible in close cooperation with the WHO Country Representative and other external donors to assist the government with coordination at the country level. A prime objective is fostering the preparation of Decade Sectoral Plans by government which can then lead under government leadership to the holding of Decade Consultative Meetings (DCMs) between government and donors on the mobilization of financing for programs an projects identified within a clearly defined national strategy. Since 1986, the UNDP/World Bank Sector Development Teams established at Abidjan, Cote d'Invoire for West and Central Africa and Nairobi, Kenya for East and Southern Africa as well as in New Delhi, India are paying an important role in assisting governments in defining national decade strategies and investment programs. As of 1985, 71 percent of the developing countries reported that Decade Sectoral Plans existed and within this group 84 percent of the Least Developed Countries (LDCs) reported the preparation of Sectoral Plans. This is an encouraging achievement.

It is also encouraging to be able to report that the "Decade Approach" is increasingly being applied at country level as well as by donors. The second half of the Decade will concentrate increasingly on software issues such as those identified in the Mar del Plata Plan of Action as well as the interaction between software and hardware. Important results have been achieved during the first half of the Decade particularly through the joint UNDP/World Bank Program in developing low-cost and adapted technologies for both rural water supply and rural and peri-urban sanitation systems. It must be stated, however, that more strides have to be made to change mentalities towards accepting the low-cost technology option by policy-makers at the developing country and donor levels as well as professional engineers in both developing and developed countries.

Another issue which must be addressed, and in this case by the developing countries themselves, is cost recovery. The WHO mid-Decade assessment which I have referred to previously indicates that an estimated 83 billion US dollars are needed in capital investment to meet Decade targets in developing countries during the period 1986-1990. Of this total, it is anticipated that while developing countries financing will be at a level of 50 billion dollars, 42 percent of capital investment on average will have to come from external sources while in sub-Saharan Africa the percentage of external vis-a-vis total anticipated investment will be at a level of 75 percent.

The level of financing required both generated internally by developing countries and anticipated and required from external sources of financing is sobering particularly when one takes into consideration the phenomenon of "donor fatigue", the investment priorities of other development sectors such as agriculture, not to mention net capital outflow from developing countries and the decreasing availability of external investment capital.

It is clear that developing countries must maximize the generation of additional resources by giving increased attention to the issue of cost recovery in their drinking water and sanitation programs.

In conclusion, while much has been achieved during the first half of the Decade, both quantitatively and qualitatively, much more remains to be done before the end of the Decade and clearly beyond. More emphasis is required in the real delivery of an integrated and appropriate technology approach wedding hardware to an increasing emphasis on software issues thus providing further guarantees that systems will be both sustainable and replicable. Donors and the professional organizations in developed countries must play their part.

I am convinced that this International Conference will play a critical part in getting this message across, and on behalf of UNDP and the United Nations Development System, I offer our hand in partnership to all non-governmental organizations and associations present at this conference so that together we can achieve these objectives.

THE CHALLENGE AND YOUR OPPORTUNITY

F. W. Montanari, P. E. *

Abstract

The International Drinking Water and Sanitation Decade (DECADE) is a unique and important world-wide effort to provide basic services of clean water and sanitation to much of the world's population. The ASCE Puerto Rico Conference, involving a number of other professional organizations, is designed to contribute evaluation, state-of-the-art information and, above all, inspiration and support to the DECADE. It offers an opportunity for exchange of world-wide experience and scientific information and a challenge to make these efforts a success of Civil Engineering -- a people-serving profession.

Introduction

The International Drinking Water and Sanitation Decade (DECADE) continues to be a most wonderful opportunity to help the world. Obviously, there will be other occasions and programs, but it is difficult to imagine them having quite the same stimulating purpose and universal spirit. What can be more effective in helping people than to provide that essential first step up the long, long road to a healthy, satisfying life? Most of us have never faced -- nor will we -- existence without available water and sanitation, and it takes little imagination to appreciate why it was the developing countries themselves who proposed the DECADE.

The ASCE, and principally those who dreamed up this Puerto Rico Conference, are to be highly congratulated. Against the backdrop of "Water and Sanitation for ALL by 1990", caring professionals made the May 1981 International Conference in New York City a success. That General Session, devoted to the DECADE, was followed by a whole afternoon packed with experiences and information exposing past successes and failures in meeting this age-old challenge. Impressed with the opportunity of contributing from the store of civil engineering resources, and bitten by the bug "I can make a difference", a small group of professionals invested their imagination, time and effort to get this current conference underway. They were inspired and motivated by what they heard at that International Convention in 1981.

*Consulting Engineer 307 Yoakum Parkway #912 Alexandria, VA. 22304

Terrence Thompson, and Terence Curran (T^2) were ably supported by Harry Tuvel of ASCE staff, who has long contributed to international interests and concerns.

The success of this week's venture will be measured not only by what you take from this conference but then apply to solving the problem further. We challenge each and every one to "squeeze the most riches" out of the sessions, the experiences presented and, above all, the associations and fine international friendships which are generated here, (much as the wine-maker squeezes his grapes to produce a better wine). Here, our objective is more clean water!

Conference Goals

This conference and the effort it supports (The DECADE) have an impeccable pedigree. The start at the New York City International Conference was excellent, a follow-up to the United Nations General Assembly's only all-day session devoted to a single subject of such world-wide consequence. Thus, in November 1980 was launched the DECADE. This went unnoticed, sadly, by many, but it was of great importance to those who lacked these necessities for a decent life. It inspired hope, and it caught the attention of eager sanitary engineers. Many governments around the world refocused their efforts on this long-recognized need, and attempted new activities toward this goal. Too soon, however, some faced the unexpected world recession, which necessarily dampened and prolonged the accomplishments, although it did not quench the desire.

With the three-quarter -- yes, far past the half-way -- mark reached the United States has really missed the departure of this special train, and therefore an important, rewarding part of the journey. There are many excuses ("reasons," some claim)." The goals were too high, beyond accomplishment," it is said. But is this true? To get the world's largest public-works project off the ground imaginations needed to be fired and fueled with energetic inspiration. Many view this significant aspect as crucial. The world-wide effort is well started, and with further support will successfully continue. Perhaps this is where we can make up for a late start. We hope this will be considered here at the conference.

Our meetings offer a rich array of subjects, people and experiences, as the program outlines. Even before this opening session, officers and key personnel of the cooperating organizations have reviewed their roles and contributions to the DECADE, under the able chairmanship of Ambassador John McDonald. Can, and more importantly, will more now be accomplished by these associations in international endeavors?

DECADE, under the able chairmanship of Ambassador John McDonald. Can, and more importantly, will more now be accomplished by these associations in international endeavors?

Your organizing committee and the advisors worked hard (at least long!), fashioning a program which could be of significance to the DECADE. Under the strong leadership of the Track Coordinators, four streams of technical presentations, papers, and panels were developed:

1) Human Resources

2) Finance and Economics

3) Technology and Engineering

4) Operations and Maintenance

The selected speakers range from individual "practitioners" -- including engineers, physical and social scientists, and professors -- to representatives of international organizations, such as the multinational banks and members of prestigious international engineering firms. Discussion is not only provided for, but ardently hoped for. It is in these unexpected fora that the exciting ideas frequently evolve. We challenge you to provide discussion -- yes, arguments -- and give birth to new, useful ideas.

The most productive attendance methodology,though at times arduous and even boring (perish the thought!) is to remain with a specific track or stream. Then, intellectually,you will be in a position to contribute from your experience to the ultimate results of the deliberations. The final proceedings of the conference will stand as a useful reference document relating the up-to-date experiences and knowledge to serve prosessionals and practitioners all over the world. Your individual contributions shape and refine the presentations, improving the final product through the discussions.

The objectives are simple in expression, and at the same time a challenge -- a challenge that can be satisfied only by a personal role. By your very presence here you, as a caring professional, a sincere human being, indicate you are willing and eager to assist those less fortunate.

The DECADE, launched as the result of understanding of universal need and a deep concern for fellow-men and women, seeks a worthy goal -- water and sanitation facilities for all. This conference is designed -- yes, destined -- to make a significant contribution to this crusade. I repeat, you, each one, have the unique opportunity, using your individual experience and intelligence, combined with those of your working colleagues, to give direction and inspiration to the task yet ahead. It is immense, and asks the best that we have to offer.

An obvious response to "What do we do now?" is "We push on -- harder." Simply, the job must be carried on with greater emphasis, and the services provided to those who need them. But How, is the question and thus the real challenge to us here.

The remuneration will be priceless -- satisfaction on both personal and professional grounds. It is our conscience that enjoys having shared individual talents, each in his own way, in helping to solve the most pressing and personal of problems. Many practitioners have paid their dues over the years, the centuries. Let us join them. The success of this gathering, with all that has gone into its preparation, rests in your hands, -- but even more in your hearts.

The Decade Program Is Still Alive!

Abel Wolman

The Johns Hopkins University
Baltimore, MD

When the Biosphere Conference was convened in Paris in 1968, few predicted that it would ultimately generate the demand for primary health care for everyone, on a foundation of safe water reasonably available to all to drink, to bathe, to cook, and to remove human excreta from within every house to a safe disposal.

This intention was ratified in Stockholm, Alma Ata, and Mar del Plata, by a unanimity extraordinary in global decision-making. Ever since, the goals have been diligently pursued. The results have been less than satisfying. Deadlines for 1990 will not be met, although significant amenities have been provided for millions of people previously unserved in both urban and rural communities.

1987 is now late in the decade course. Why has the hope expressed in late 1970s failed to be fulfilled. Without priority of causes, a few are listed. Renewed efforts at implementation must go forward. A "Crowded Earth" is undoubtedly one of the restraining phenomena. Growth of population was dramatic and outstripped the new facilities provided.

In 1950, world population aggregated 2.564 billion. By 1985, it had reached 4.865, and now, 5 billion struggle for existence! Even though some countries are reducing birth rates, the prospect for the year 2000, just around the corner, intimidates with 6.159 billion people.

Of greater importance, the world economic recession has set back individual incomes at least ten years. The poor are poorer today than they were then. Their recovery is likely to be very slow. Poverty and hunger became more extensive than the provision of water and sanitation.

The consequences are reflected in two important indices of the ability of people to survive. The World Bank, not given to hyperbole in its reports, uses a phrase descriptive of the fate of those who have "brief and painful lives". In some 40 countries, the birth rate per 1000 in 1983, varied from a low of 25 to a high of 55. Most had rates between 40 and 50. The corresponding figure in the United States was 16.

Life expectancy at birth, in years, in these same 40 countries, is dismal. Only one has reached 67 years. The remainder have a range of 36 to 63, with most struggling to reach 40 to 50. The U. S. records 75 years.

At the risk of drowning you in statistics, one more figure is descriptive of the world in which billions of people live deprived and die too early. This last index, per capita annual income, in U. S. dollars, rarely reaches $300. I conjure up a modified index, "a Gross National Production of Disease." GNPD makes its first public appearance here today. It should be included in all global accounts!

What Now?

The global situation, presented before a similar conference in Atlanta, Georgia on November 10, 1986, has now been published.* It supplements today's discussion and need not be repeated.

We now seek new or modified approaches to circumvent familiar constraints. These were lack of motivation by governments, lack of

*Abel Wolman. International Conference on Water and Human Health. American Water Resources Association, Water Resources Bulletin, Vol. 22, No. 6 December 1986.

skilled manpower, absence of managerial competence, and deficiency of money. The decision to retain the goals of the Decade as originally posed is a wise one. Our chore becomes one of discovering new means of implementation.

Alternative Approaches

The setting, in which approaches must be devised, is provided by specific data collected by Dr. Bernd Dieterich, on his retirement in 1986 as Director of the Division of Environmental Health of the World Health Organization. The status is reasonably accurate for the period ending October 31, 1986.

> For urban dwellers, 77 percent have access to safe water, compared to 72% in 1980. Only 60% have appropriate sanitation, compared to 54% five years ago. 36% of people in rural areas have safe water, compared to 31% in 1980. Appropriate sanitation reached only 16%--with no increase since the beginning of the Decade.

Some good news should be recorded. Within the last five years, 270 million people were provided with water supply and 180 million with sanitation. If present rate of progress is maintained, and there is no guarantee that it will be, approximately 542 million more people would be supplied by 1990 and 357 million more would have sanitation. This would leave us with 1,236 billion still unserved with water, and 1,809 billion without sanitation. The message is most sobering!

(a) Some practitioners, with decades of experience in less fortunate countries, draw upon the century for lessons of water and sanitation progress in western industrialized countries. They emphasize that facilities were built without interminable discussions with communities.

When built, they came into use with reasonable rapidity, as consumers realized the safety, comfort, and health benefits they afforded. For the most part, they were locally sponsored.

These workers view with jaundiced eyes the present time consuming efforts with public participation in which most public agencies and private groups are now assiduously engaged. Educating the public is attempted via diligent sessions with local leaders and their constituents, and the distribution of pamphlets, cartoons, and other published materials. These not only take time, but have not been distinguished by too many successes! The rapid increase in the use of television, even in the tribal villages, promises a better return in educating the public in personal hygiene, particularly mothers and children.

Fascination with desire to have public understanding of science-technology is not new. One hundred and fifty years ago the British Association for the Advancement of Science was already engaged in this effort.

The B. A. President, in 1986, Sir George Porter, picks up the gauntlet in these startling terms:

> "Should we force our science down the throats of those who have no taste for it? Is it our duty to drag them kicking and screaming into the 21st century? I am afraid that it is."*

At the annual meeting in 1986, Dr. Bernard Dixon, former editor of "New Science" argued for less science news reporting. "The received wisdom is that more science in the mass media would be good for science" but he argued that "interminable sequences of claim and counterclaim by 'an expert' and 'another expert' did nothing for the public's

---

*Chemistry in Britain, Volume 22, No. 10, October, 1986

understanding of debates such as those over nuclear power or food additives. Rather, they were likely to leave the public confused and misinformed. Too many scientists," Dixon said, "believe that journalists should simply report what they are told, without digging below the surface, putting the story into context, or informing the public".*

The participants this week may well debate the pros and cons of this approach. It runs counter to the majority of activities now underway throughout the less favored world.

(b) The advocates of detailed public participation are at the helm trying to speed up the provision of services. They insist that public perception and participation, at every step from initial concept to ultimate maintenance and operation, must be pursued. Unless this route is followed zealously, failures will dominate the scene.

(c) We are now flooded with documents directed toward the means of improving the depth and permanence of community understanding, responsibility and development.

The most elaborate of these aids to community action is by Anne Whyte** of the University of Toronto, Canada. At the request of WHO, she chaired a conference at the Hague, Netherlands, on the community participation process. The sessions resulted in a book of 53 pages. The author describes the volume as presenting "in a simple and readily understandable form to lead the planner through the 'what, when, where, why, how and who' questions associated with the participation process". The audience will make the ultimate decision as to how well her promise may be fulfilled in field practice.

---

*Chemistry in Britain, Vol. 22, No. 10, October, 1986

**Anne Whyte. Guide Lines for Planning Community Participation Activities in Water Supply and Sanitation Projects, WHO Offset Publication No. 96, Geneva, 1986.

Dr. Whyte is fully aware that the objective of the Decade can be successful only if such participation is assured. She recognizes that "the switch from centrally managed to community-based projects will not happen overnight".

The document may even teach us not to move thousands of men, women, and children from farming in their birthplace in the cool highlands to the hot lowlands where irrigation projects are a must--and are a completely unknown way of farming to the newcomers. And these disasters are being repeated!

A similar effort, of some 67 pages, was developed by the South-East Regional Office of WHO in 1985. The nine countries agreed upon a document* after two regional workshops in 1983 and 1985. It carries the reader step by step toward a presumed successful community participation and permanent responsibility.

(d) In the meantime, some successful undertakings are being reported. They demonstrate that approaches vary, but may be accomplished even under most difficult and diverse cultural, religious, and socio-economic-political circumstances. Inquiring members of this conference may be interested in these examples. None of them are characterized by simplicity of action. Decisions are arrived at through a thicket of national governments, ministries of health and public works, the private sector, external aid donors, internal agencies, and local governments. Historically, local governments have really had a minimum of local governance. Behavioral change in a community, whether tribal, rural or urban, will require ample time to build links with the people to fathom their feelings and understand their hopes and desires.

---

*Achieving Success in Community Water Supply and Sanitation Projects. WHO Regional Office for South-East Asia, New Delhi, India, Regional Health Papers, No. 9, 1985.

No small part of the problem stems from the realization that the spearhead of action is WHO and its international and local leaders. These invariably are medical in origin. Their training and education are usually devoid of communication skills as well as understanding of engineering disciplines and practice.

One of the most difficult demonstrations of proven success of people participation is directed at vicious diseases associated with water as well as directly water borne. The projects are in the Bandiagara Plateau in Mali, east of the Niger River. The main parasitic diseases with vectors or intermediate hosts are schistosomiasis, dracunculiasis (guinea worm disease), onchocerciasis, and malaria. Control has been successful. Most importantly is the fact that the measures can be mastered by rural communities, who took almost complete responsibility for the required detailed field operations and evaluations. Reference to the complex activities are noted* for those who wish to examine them more fully.

An equally complex undertaking is described in a group of villages in Thailand. The project** was directed toward the improvement of drinking water supplies and sanitation facilities. Much of its success was due to the fact that the communities were properly informed about what was being attempted. The villagers trusted each other, their local leaders, and the government officials assigned to the project. The latter worked on it without remuneration!

A completely different approach to the provision of services is in

---

*M. Kassambara, P. Poudougo, B. Philippon, E.M. Samba, and D.G. Zerbo. Village Community Participation in Onchocerciasis Vector Control, World Health Public Health Forum, Vol. 7, No. 1, P. 57, 1986, Geneva, Switzerland

**Anant Menaruchi. Drinking-Water and Sanitation: A Village in Action World Health Forum, Vol. 7, No. 3, 1986, P. 303, Geneva, Switzerland

the rural sanitation program* in Sierra Leone. This has special importance because of the high infant mortality rate of 200 deaths per 1000 births and an average life expectancy of 34 years, one of the lowest in the world.

In 1981, a small non-governmental organization in Sierra Leone, called the Community Development Council (CDC), developed plans to determine whether community approach for water and sanitation was possible and whether it would result in health improvement of village people. The settlements chosen, with controls, were remote and difficult of access, but selected by the "Paramount Chief" of the regions. Prospective beneficiaries took charge and determined whether real impact upon their own wellbeing was accomplished. The results were gratifyingly positive.

Fortunately, evaluation of these effects was undertaken by Roderic Beaujot, Associate Professor, Department of Sociology, University of Western Ontario, London, Ontario, Canada. He is likewise the author of the Document* containing the details of the project. Much of the success is attributed to S. Kabbah, the local project leader.

(e) Economists and bankers have pursued for years the "god of cost-effectiveness" in place of large capital investments and consequent maintenance and operation costs, inherent in water supply and sanitation. Some six years ago, Hopkins workers in Bangladesh re-discovered oral rehydration therapy and its value in recovery from diarrhoeal diseases.

The underlying principle had been enunciated some 150 years ago by a physician from Scotland who came to London during a devastating cholera epidemic. He attempted, via intravenous injection, to demonstrate the

---

*Roderic Beaujot. Rural Sanitation in Sierra Leone: With Our Own Hands, Research for Third World Development, The International Development Research Center of Canada, Ontario, Canada, 1986, p. 103.

validity of his hypothesis. He lost his license and the medical profession prevented his practice in London. He returned to Scotland. His principle was recorded in the British Lancet journal, where it rested until now.

ORT has now become the major weapon in the diarrhoeal diseases battle. It is undoubtedly valuable in saving lives, particularly with young children. Studies are being carried out in various regions of the world, in order to arrive at the best logistics for its world-wide use. The field studies disclose that universal use is not easy to attain. In-house use poses problems, because mothers are not familiar with even prepared materials for application. Some religious groups refuse to use it, because it runs counter to their ancient custom of managing diarrhoeas. The benefits of ORT are great when wisely used. Costs are low. The disabilities are obvious. ORT is repetitive, non-preventive of disease, saves lives of those who return to environments where personal hygiene is unknown--and they die. The theoretical cost-effective feature is a delusion.

It may come as a surprise that, even in the United States, studies have shown that, in major American cities, dehydration caused by diarrhea is among the top five causes of hospitalization for children. Only recently has it been recognized that dehydration is also a major problem with the elderly, AIDS patients, and, of course, travelers.

The formulas generally used abroad, consisting of glucose, salt, and a few other common substances mixed with water, may give way to better and more familiar mixtures. The one now being favored is mother's familiar and tasty chicken soup and rice. They provide excellent ingredients to cope with dehydration. Dr. William B. Greenough, of Johns Hopkins

University, indicates that this mix is more effective than present ORT formulas.

A similar panacea is considered by some who wish to escape the money necessary for permanent preventatives. Their hope is pinned to vaccines, so far unavailable, but promised before too long. History has amply shown that vaccines alone, again indubitably valuable, have not been successful in preventing or eliminating disease.

(f) Enough has been said to illustrate the baffling obligation we have assumed to soften the disease load imposed upon people of the third world. I leave this arena with one more intimidating statistic. Now that China has reappeared in the world, with its one billion one hundred million souls, we must avoid the footnote in all international documents of the last decades, eg. "exclusive of mainland China".

Carl Taylor, recently retired from the Johns Hopkins University School of Hygiene and Public Health, is the head of the UNICEF programs in China. His timely comments, on his work in 28 provinces, are especially valuable:

> "China's great efforts over the past three decades have reduced the infant mortality-----But many serious child health problems remain.
>
> "Parasitic diseases, especially worms, are found in 80 to 90 percent of children in many rural areas.
>
> "Some health problems result from economic pressures associated with the modernization of China, such as increasing air and water pollution in both urban and rural areas, a marked decline in breastfeeding in urban areas, and a reversal in improving sanitation because of the use of human excreta for fertilization in rural areas."

How does one recapture the health implications of the Decade objectives? An observer in another field of global activity provides a lovely phrase descriptive of what is happening in "divestitures". He says they represent "sweepstakes for futility"--strange in a 1987 world with an infant mortality, the highest in the world, now affecting Africa.

(g) A Codicil: Day by day, investigations are continuing to determine whether the amenities here discussed actually prevent disease. Highly competent workers in Bangladesh have now carried out such studies, with an introductory criticism of prior studies loosely and erroneously recording of field experience. Their summary* of present findings is important:

> "An educational intervention was designed to improve three water-sanitation behaviors empirically shown to be associated with high rates of childhood diarrhea in Dhaka, Bangladesh: lack of handwashing before preparing food, open defecation by children in the family compound, and inattention to proper disposal of garbage and feces, increasing the opportunity for young children to place waste products in their mouths. Fifty-one communities, each comprising 38 families, were randomized either to receive (n = 25) or not to receive (n = 26) the intervention. During the six months after the intervention, the rate of diarrhea (per 100 person-weeks) in children under six years of age was 4.3 in the intervention communities and

---

*An Educational Intervention for Altering Water-Sanitation Behaviors to Reduce Childhood Diarrhoea in Urban Bangladesh. I. John A. Clemens, et al; II. Bonita F. Stanton, et al. American Journal of Epidemiology, Vol. 125, No. 2, February, 1987, pp. 284 to 301.

5.8 in the control communities (26% protective efficacy; $p < 0.0001$). A corresponding improvement in handwashing practices before preparing food was noted, although no improvement was observed for defecation and waste disposal practices. These data suggest that educational interventions for water-sanitation practices can have an important beneficial effect upon childhood diarrhea in developing countries, particularly when the interventions are designed in a simple way to promote naturally occuring salutory behaviors that are empirically associated with lower rates of childhood diarrhea."

U. S. Practice Abroad

We pride ourselves on the high levels attained by our science and technology accomplishments. We wish to share these riches with less favored countries, by transfer of our skills. The results of such efforts are not always salutary, often inappropriate, sometimes too sophisticated, frequently too big, and too costly in dollars.

These disabilities are surfacing in many countries. They are turning away from U. S. consultants, standards, and designs, and moving to indigenous competent consultants and advisors from Japan, Italy, Great Britain, West Germany and the Far East.

We operate on a two-way street. The other world develops solutions useful to us, because necessity generates innovation! Modifications in design and consequent reductions in unit costs are surprising. "Example after example in which academic, highly paid advisers were wrong and peasant farmers were right."*

---

*Paul Richards. Indigenous Agriculture Revolution. Westview Press, Boulder, Colorado, 1985.

We are still confronted with the ancient myth that water supply should be provided free as a "social good". Some fifteen years ago, I appeared in this same building to address public works engineers from most of the Latin American countries. The subject was a familiar one--namely why you should pay for both capital and maintenance and operating costs. Virtually none of those present appeared to buy the principle! The issue remains less than universally resolved!

The Prospect

The 1986 Report of the United Nations Environment Programme* provides a succinct and sad prophesy for hundreds of millions of people. Its views are a warning that the task ahead is great. If well-done and more speedily accomplished, the well-being of many millions will be greatly advanced.

With 4.6 million children under 5 years of age dying each year of diarrhoea, and almost as many of respiratory diseases, with 100 million acute cases of malaria and 2 million deaths, with 200 million cases of schistosomiasis, and with cases of onchocerciasis and the various trypanosomiases------the toll of the environment on human health in developing countries is extremely high. It is still higher if one adds the host of other communicable diseases linked to the environment endangering those living in tropical and sub-tropical regions.

Natural and man-made disasters claim an increasing number of victims. These increased many times between 1960s and today. Fatalities average in the thousands, concentrated largely in the most crowded and lowest income countries.

---

*United Nations Environment Programme: The State of the Environment. Environment and Health, June 10, 1986. Nairobi, Kenya.

In developed and developing countries, operational releases of pollutants to the environment, particularly into water supplies, have created and will continue to create, major problems for their control. Their visible health costs, however, are far less than that of communicable diseases in warm countries. The major cases involving fatalities have been episodic, due to a concurrence of causes. They resulted in mortality figures four orders of magnitude lower than the annual numbers of deaths from some tropical communicable diseases.

The working environment has in general become safer in the last decade. It should be less and less hazardous with the applications of known technology.

It is important to note, where adequate water supply and sanitation services are provided in one area, a major reduction in diarrhoeal diseases would be expected, even if the same measures were not taken in a neighboring town. These effects are in sharp contrast, for example, in the case of air pollution where control in some, but not all regions, does affect distant areas.

Most important of all is the recognition that the effects of water-sanitation services are not confined to human health. They are essential for the development of most resources which create socio-economic progress.

On this 40 year celebration of the life of UNICEF, I close with a comment that tells all, by Danny Kaye in 1980.

> "The greatest natural resource that any country can have is its children. They are more powerful than oil, they are more beautiful than rivers, and far more determined that the world shall exist."*

---

*B'nai B'rith International, Washington, D. C., 1980

Introduction to Human Resources Development

Horst Ottenstetter and Michael Potashnik

Specialists from national and international agencies presented methodologies and experiences in the development of human resources in the area of water and sanitation.

The training of professionals and technicians and community education is discussed along with the production of training material. Special attention is given to the organization and implementation of national and international training projects and networks.

The importance of management training is recognized and special sessions deal with this matter. In the same context the contribution of training to the overall performance of the water agencies is analyzed and discussed.

As part of this track, institutional performance problems are discussed. The role of the community is also considered.

The Common Denominator - Management

Martin Lang*

Abstract

Despite the wide diversity in geography, culture, demographics, water resources, and political and fiscal structures, the common attribute of a productive, self-supporting, and reliable water supply system is proper management. Hence, success in this area is not a purely engineering undertaking, but requires a high degree of managerial skill to cope with the personnel, political, fiscal, and sociological challenges in a developing country.

A pertinent and relevant case history is presented, of the past five years of the National Water Commission of the Government of Jamaica. A firm commitment, in 1982, to managerial improvement, by the Prime Minister and the Minister of Public Utilities and Transport, has transformed the Commission, by 1987, from a fragmented and unprofitable agency, into an increasingly robust and self-sustaining utility now serving the entire country.

To accelerate the process, a six-man interim team of Camp Dresser & McKee, specialists, designated "Operation Backstop," assisted the National Water Commission, beginning in 1983. Within a year, the enhanced credibility of the NWC won World Bank support. Tattered ledgers were replaced by minicomputers. By the middle of 1986, a newly indoctrinated cadre of Jamaican managers was in control, and Backstop was successfully completed by December 1986.

This "Backstop" approach, aimed at managerial enhancement through a multidisciplinary total assistance effort, is deemed to be widely applicable.

Introduction

The four post-war decades have been marked by extensive and expensive transfer of water technology, and provision of money, talent, and equipment -- to developing countries.

Have these programs been successful? A lot of wells have been drilled, reservoirs have been built, pipe has been laid, and an array of western technology has been installed. If we define "success" only as the construction of facilities and the delivery of hardware, then these programs have largely met the criterion.

But, of course, the bilateral and multilateral agencies which spent billions on these programs had higher goals. Their intent, through these programs, was to catalyze the creation of viable water utilities.

*Senior Consultant, Camp Dresser & McKee International Inc., 40 Rector Street, New York, New York 10006, Member ASCE.

What is a successful water utility? By U.S. and U.K. standards, it is an entrepreneurial organization which reliably delivers an adequate and supply of pure water, and does so efficiently profitably; is self-sustaining from its own revenues for operation and maintenance, renewal and replacement, debt service for programs for growth, and a rate of return on investment. For developing countries there is a reasonable consensus that a viable water utility is one that could become self-sustaining, from its own revenues, for recurrent operation and maintenance costs.

If we review the record of development of water utilities in our time, can we identify some common denominator generally applicable to those systems which evolved into successful utilities?

It is my thesis that, although under the rubric of "developing countries" there is a tremendous diversity of geography, culture, demographics, water resources, and political and fiscal structures, the common attribute of <u>any</u> productive, reliable, and self-supporting water system is proper management.

Therefore the conception and delivery of a viable, self-perpetuating water supply program is not the sole turf of the sanitary engineer, since the <u>managerial, fiscal, political, and sociological challenges to such a program in a developing country are usually much more difficult that the technical aspects</u>.

For example, it is easy to generate national (and international) recognition of a highly visible new water project. But, after the ribbon cutting, competent management should already be in place, with the resources to staff, motivate, train, retain, and supply the requisite human resources to prudently operate and maintain that facility forever - and do it with local money.

Program Assessment

The recurrent paradox has been ample outside millions for new design and construction - and scarce local pennies to protect and preserve a huge capital investment already in place.

It is in that context that we attempt to assess these programs after almost 40 years of substantial investment. These findings are not specifically applicable to any one city or country, but they do apply, in part or in whole, to many situations -- situations, incidentally, that can also be found in the United States.

1. In many cases, management could not even adequately maintain or operate its pre-existing facilities.

2. In some cases, management could not cope with proper administrative, technical, and financial control of major new construction.

3. In many cases, aid programs did not compel infrastructure maintenance, renewal, and rehabilitation, and management "mined out" new facilities by running them to failure after the newly-built "honeymoon" phase.

4. In some cases, management made no effort to effectively deploy usually over-manned staffs, or to control overtime. Sometimes, there was a prevailing philosophy that the organization should act as a socially desirable mass employer.

5. In some cases, management kept putting off the tough social and political decisions for properly billing and collecting for water.

6. In some cases, the continuity of water supply service was predicated on the indefinite continuation of central government subsidies.

7. In some cases, management was not adaptable to accelerating urbanization and the need to extend service to new, sprawling conurbations.

8. In many cases, management in developing countries kept putting off the inevitable confrontation with the problems of industrial wastes control, ignoring the particular vulnerability of the water supplies of those countries to toxic and hazardous substances.

9. In many cases, water supply augmentation was by ad hoc improvisation, rather than by a phased, long range master plan preceded by an inventory of water resources.

Many of these are symptoms of a more profound disease, the lack of utility-oriented aggressive management. The ultimate solution will not be found in hardware, computers, or technical procedures, but in people.

One obvious remedial action is restructuring the organization to conform with goals of productivity, reliability, efficiency, and profitability, with concentration on metering, billing, collection, and managing existing assets.

This action must, however, be accompanied by the recruitment, motivation, training, and retention of managerial staff - solving the "people problem."

Case Study - Jamaica

A recent and continuing case history is cited to illustrate a pragmatic remedial approach -- a support program for the National Water Commission (NWC) of the Government of Jamaica, enabling it to effect change and move rapidly to effective in-country management through a three-year program of focused management and functional support.

Since the 18th century, Kingston, the capital city of Jamaica, has been served by some form of public water supply. In this century, this system was owned and operated by the Kingston - St. Andrews Water Commission (KSAWC), which served the individual consumers in Kingston and its environs.

The National Water Authority (NWA), an arm of central government, supplied bulk water to the individual Parish Councils throughout the island of Jamaica for sale and distribution to their individual consumers. The NWA owned about 90 individual underground and surface sources. Some of the Parish Councils augmented the NWA water from their own sources.

The KSAWC, with two large reservoirs, an aqueduct, four water treatment plants, a river intake, about 20 wells and pumping stations, some wastewater facilities, a metered distribution system, and a reasonably proficient operation and maintenance staff, had been deemed, about 20 years ago, as a model Caribbean water system. However, it suffered a continuing attrition of professional and technical staff, and dwindling resources for maintenance, parts, supplies, and rehabilitation. The deterioration accelerated during the 1970s, at the very time that the service population almost doubled. In this same period, the reservoir capacity was diminished by siltation, and the yield of some wells was curtailed by excessive drawdown and chloride intrusion. By 1980, Kingston was subject to intermittent water lock-offs during the dry seasons, and the situation was exacerbated by unprecedented drought in 1983, continuing to 1985.

The NWA billed the Parish Councils for the bulk water delivered, but only collected about 15% of its billing. In effect, it operated by an annual subsidy by government. It could just about maintain the status quo, and could not generate the investment required to rehabilitate or expand its services. At the same time, the rural areas were inadequately served by the Parish Councils.

The government of Jamaica, and particularly its Ministry of Finance, had a sophisticated insight into the advantage of a utility approach. It took a courageous first step in 1980 by legislation to consolidate the KSAWC and NWA into a National Water Commission (NWC). The intent was clear:

1) Bring into a countrywide organization the capabilities formerly demonstrated by KSAWC.

2) Bill and collect for the water sold to the Parish Councils.

3) Eventually absorb the Parish Council Systems into NWC.

4) Reduce or eliminate central government subsidy.

It became apparent that combining two agencies, each in a deficit mode, one primarily wholesale and one retail, each with differing administrative, accounting, personnel, and commercial practices, into a viable self-sustaining utility could not be accomplished just by legislative fiat.

The Prime Minister of Jamaica, the Honorable Edward Seaga, requested engineering, managerial, and transport assessments of the NWC. These tasks were delegated to the Jamaica National Investment Company, Inc.

(JNIC), a parastatal semi-autonomous arm of government. JNIC selected Camp Dresser & McKee Inc., (CDM) of Boston, Massachusetts to assist in the three-month engineering assessment. JNIC obtained funding from USAID, for this study.

The NWC is administered by the Ministry of Public Utilities and Transport (MPUT). When the CDM team assembled in Jamaica, they received from the Honorable Pearnel Charles, Minister of MPUT, a clear and classic directive that has already become a legend:

> "We don't need any more damned reports! Just tell us what is wrong and what to do about it, and don't worry about making friends in Jamaica!"

The report developed into an "Engineering and Operational Assessment." It was unsparing in detailing the cumulative effect of what appeared to be a decade of deterioration, it linked operational problems to managerial deficiency, but concluded that there was the intrinsic potential for development into a viable utility. It recommended a number of tough and unprecedented remedial actions.

It was successively reviewed and discussed in detail with JNIC, with the top management of NWC, and finally by the Honorable Prime Minister.

The meeting with NWC was important. The candid discussion that ensued marked the beginning of a continuing effort to establish a degree of rapport between the brash newcomers to Jamaica and the harried career team that had been wrestling with the problems of dwindling resources for years.

The Prime Minister estimated that it would take several years to achieve utility status. There was no prospect for international funding. Jamaica's cherished foreign exchange could only be doled out sparingly. With extraordinary prescience, the government of Jamaica undertook to fund this effort for one year, with the intent of restoring the credibility of its water programs with international funding agencies.

Essentially, the catalyst for change was to be the presence of a six-man on-site resident team of specialists, each a counterpart to a top manager. They were not to displace or replace top management, but to assist and enhance managerial, technical, fiscal, and administrative performance. The government of Jamaica contracted with CDM to deliver these services, beginning March 1983. CDM designated the project "Operation Backstop," to emphasize its support function.

There was a unique aspect to this contract. Although the services were to be provided to NWC, the contractual client was an autonomous arm of central government, the JNIC. It was inevitable that initially some incumbents could view this team warily and apprehensively.

The Backstop director told the team "local credibility and acceptance have to be earned by performance, not just expected as a matter of course. The burden is properly on us to establish the mutual esteem and respect essential to a joint enterprise."

It was good the team started on that prudent premise. Within a few months, the concerns about Backstop, either candidly expressed or plainly implied, were largely dispelled, as the Backstop team established their presence and their role.

The six-man team provided personal counterpart support to the Managing Director, Deputy Managing Director, Director of Engineering, Director of Operation and Maintenance (Kingston), Director of Operation and Maintenance (Rural), and Director of Human Resources and Development.

These assignments became broader during the year to ensure sustained productivity of Backstop, even when some of the counterpart incumbents were absent.

The year was marked by a ferment of activity. These actions are properly attributable to the increasing acceptance by NWC of the attempt to approximate a profitable utility, rather than continue as a dependent ward of central government.

Some of these actions were:

- By June, 1983, an administrative reorganization was effected with clear lines of accountability and responsibility in specific functional areas.
- A three-year financial plan was generated in October, 1983.
- The Backstop computer-generated series of rate and revenue options became a major input to the historic rate increases of November 1983 and August 1984.
- A uniform accounting system was established by April, 1984.
- Controls were imposed on hiring practices.
- The proliferation of overtime came under control.
- A credible budget was structured with a monitoring procedure to ensure adherence, and was cited as a model for other agencies.
- A personnel policies manual was developed, and a major segment adopted by the Board.
- In-house formal training was revived, and 700 personnel received instructions.
- Dispersed operation-and-maintenance units were centralized.
- A continuing effort was mounted to ensure some persistent chlorine residual at the ends of each system.
- The conversion of the manual billing system to computerization was under way by 1984.

- o A central emergency control room was established.
- o The laboratory became an operational O&M arm, instead of a data repository.
- o NWC collaborated with PAHO in developing the first water quality standards for Jamaica.

The year should not be described in effusive, glowing terms. The cumulative problems of decades cannot be solved in one year. Nevertheless, there was enough significant change to encourage intensive on-site assessment by the World Bank, which sent several observations teams to Jamaica, beginning in June 1983. The intent of the government of Jamaica to strengthen the emerging utility status of NWC was signaled by the promulgation of a 50% increase in water and wastewater tariffs in November 1983. At that time, the Bank requested the submission of a Technical Assistance Programme proposal.

Perhaps the best indication of the extent to which Backstop had been integrated into the NWC management was their successful joint production of a comprehensive document to buttress a proposed loan within one month.

This Programme was to provide specific, detailed resources in equipment and services to enhance water supply and distribution, rehabilitate wastewater facilities, upgrade and standardize pumps, centralize and modernize a maintenance facility, augment transportation, improve chlorination, institute control of unaccounted-for water, microcomputerize key activities, provide some parts and supplies, and, importantly, strengthen all aspects of fiscal control and commercial operations. All this was projected within a two-year timetable.

The Bank's normal lead time for the consideration and processing of such a loan would usually extend for a year or more. In this instance, the Bank team, under the direction of a knowledgeable engineering and fiscal specialist, and aided by an economist, compressed these actions into six months. This unprecedented acceleration reflected the Bank's realistic judgment that the momentum of change must be sustained.

With the Bank's support, Operation Backstop was eventually extended until December 1986. By that time, virtually an entirely new Jamaican in-house management cadre was in firm control of NWC. NWC was now delivering retail water service to the entire country, an increase from 100,000 accounts to 200,000 between 1983 and 1986. It was now self-sustaining for operation and maintenance from its own revenues. The IBRD Technical Assistance Programme has physically transformed NWC in terms of vehicles, radios, microcomputers, maintenance equipment, chlorination equipment, pumps, meters, and the new pipeline delivering treated Yallahs River water from the New Hope Plant to Kingston. $4.5 million (U.S.) have been prudently expended from the Bank loan, and another $4.5 million (U.S.) is scheduled for commitment in the next 18 months, largely for vehicles, billing system improvements, and sewerage upgrading.

Typical of the new resilience of NWC was its performance in 1985. In just two months, it administratively incorporated 900 new systems and 100,000 new accounts by taking over the Parish Council systems, and then, in the same year, reducing the manning of the expanded NWC by 30%. All this was done without the administrative chaos that had been predicted. Another evidence of adaptability to change was the smooth transfer of a substantial part of the Kingston billing operation to the private sector, with significant savings.

In 1982, Kingston was subject to almost daily prolonged lock-offs of water. In 1987, Kingston was receiving a reliable daily supply, with a volume almost 50% greater than in 1982. Most of this is attributable to the new Yallahs River supply, but a substantial increase is attributable to improvements in the distribution system and greatly improved response to, and elimination of, leaks.

NWC designed, obtained rights of way, and constructed the new pipe line from the New Hope Plant to Kingston, within two years, and synchronized with the plant construction. This is indicative of a new style in prompt decision-making and execution.

The lifeblood of NWC is not water, it is money. By now the government of Jamaica is only contributing to a portion of the debt service. The commercial operation of NWC will require two more years to shake down to a smoothly computerized and uniform island-wide system, but it already has boosted rural collections from 15% to 60%. Of course, significant revenue enhancement followed unprecedented rate increases in three successive years. Backstop's computers generated the various options and consequences, to assist in fixing the rates. In 1986, CDM submitted a Tariff Study as a basis for future rate changes.

Summary

A diagnostic preliminary audit of engineering and operation distinguished between the obvious operational deficiencies and the underlying cause of managerial shortfall.

A Backstop team of water supply, financial, administrative and commercial specialists, assisting, at first, specific counterpart managers, and then, functional areas, helped catalyze profound changes in all functions of NWC. NWC still will require years of improvement, but has already been converted to a much more robust organization, virtually self-sufficient in revenues, and, with its international credibility restored, eligible for further aid. It has already met the basic goals of the 1980 legislation forming NWC.

Based on over four years experience, the approach used by the CDM Backstop team merits consideration for application in other countries. It is essentially a multidisciplinary total assistance effort, instead of a narrow technical contribution. An absolute prerequisite for such joint effort is the establishment of mutual rapport between the expatriates and the in-house managers.

The problems encountered were certainly not unique, but rather typify many similar situations. The team members could cite closely parallel situations in the U.S., South America, Puerto Rico, Trinidad, Sri Lanka, Lebanon, Egypt, and Yemen.

Assessment of Effective Water and Sanitation Institutions

David Laredo *

1.0 Introduction

Over the past several years an abundant body of literature has accumulated citing the critical need to strengthen organizations involved in many nations' water and sanitation sectors. Echoing themes have reminded us that investments in infrastructure and programs activities should be made only if sector organizations have the capacity to effectively implement programs and manage facilities. In other words, investments should be made to benefit only those water and sanitation sector organizations which are able to focus adequate financial and human resources to insure the fulfillment of their intended missions.

However, at the same time, rising needs and expectations for the basic human services provided by water and sanitation sector projects have continued to expand. This combination of forces has created many situations in which governments, faced with diminishing sources of external aid, are finding the indigenous human and financial resources of their water and sanitation sectors strained almost to the breaking point.

Notwithstanding the crescendos of rhetoric, often rising to thunderous heights, it is still rare to find instances in which program (and/or project) planning (including those efforts funded via external sources of aid) has considered the long term effect on sector investments if the institutions responsible for managing and operating systems and programs are not strengthened. In fact, the dilemma faced by many institutions is to identify their problem areas, or assess their institutional capacity.

This paper focuses on problem identification. As such, it presents a framework for executing an institutional assessment, and reviews the key issues involved in conducting an assessment and implementing changes.

1.1 "Complexity Dimension" In The Analysis of Public Sector Institutions

Institutional analyses as discussed herein fall under the broad category of public sector management. An inherent problem in the analyses of public sector institutions, especially those in lesser developed countries (LDC's) is that they do not lend themselves to the typical "management study" models which are so often cited in the literature of management science.

*Camp Dresser McKee Inc., Boston, MA

Public sector institutions have no "bottom line" or "profit motive" with which to measure results, as exists in the private sector. Public sector management, more often than not, is conducted within environments characterized by scarce resources and expanding service populations. Further, the political/social dimensions of public sector activities often play a crucial role in how institutions operate and must be factored into any analysis. (This is especially true for the institutions in the water and sanitation sectors of LDC's, where the emotional factors aligned with the provision of basic (often life sustaining) services are involved.)

Thus, this "complexity dimension" inherent in the study of any public sector institution requires that the analysis be framed to characterize the entire breadth of operations so that an institution's "capacity" can be properly illustrated. Institutional capacity can be defined as its <u>success level attained through the exertion of its combined skills and measured throughout all aspects of its operations</u>. An institutional analysis therefore, must first focus on identifying broad areas of operational performance. The analysis, per se, must concentrate upon determining the effect of the <u>inter-relationships</u> of these <u>broad activities</u> to an institution's pursuit of its mission.

Thus, institutional analysis is "cross-cutting" in nature. A proper assessment requires the recognition and analysis of all institutional activities related to its mission, and the determination of how these activities and procedures impinge upon one another.

2.0 Approach

The approach presented herein has been adopted from the research accomplished by the United States Agency for International Development (USAID) in its Water and Sanitation for Health (WASH) Project.* The basic elements of the analysis include: data collection, identification of strengths and weaknesses through diagnostic analysis, and recommendations for improvement. These are similar for any "management study". However, the approach is unique in that the analysis concentrates on particular "performance categories" rather than the more limited procedural activities and/or systems. Performance categories as defined will encompass wide ranging sets of related skills, systems, procedures, and capabilities which "cross-cut" through an organization, defining the institution's performance categorically rather than by specific functions or systems. (The application to a specific institution may necessitate varying degrees of emphasis on particular categories.)

---

* The research is described in detail in the WASH publication: Field Report No. 37 - Guidelines for Institutional Assessment of Water and Wastewater Institutions, prepared by CDM and Associates for U.S. Agency for International Development. (WASH Project, 1611 North Kent Street, Room 1002, Arlington, Virginia, 22209, USA.)

The set of performance categories, discussed in further detail in the next section of this paper, are general enough to be applied to any type of organization, and include:

1. Organizational autonomy
2. Leadership
3. Management and administration
4. Commercial orientation
5. Consumer orientation
6. Technical capability
7. Develop/maintain staff
8. Organizational culture
9. Interactions with other institutions

Performance categories should be analyzed in part, based upon the data gathered in the sub-analyses of "output measures" or "indicators of (high) performance". The measures or indicators should be directly related to the organization's mission and/or to the performance category being evaluated. Some of these may be numerical measures and thus easy to quantify (i.e. volume of water supplied or new customers per year); others more difficult (i.e. customer satisfaction or quality of leadership). However, notwithstanding the possible difficulties in quantification, it is important to recognize that the process of selecting proper performance categories for a specific institution will be greatly enhanced by the assignment of indicators or output measures. Further, classifying the performance categories using output measures or indicators allows an institution to be examined in terms of the qualitative and quantitative results of the product produced by the institution and the results of delivery of the product.

Thus, the output of the institutional assessment, through conscious design of the process, will be a compilation of the institution's successes by performance category. This compilation allows a comprehensive profile to be drawn, which illustrates the strengths and weaknesses of each of the performance categories. Since the performance categories can be prioritized in terms of their importance to an institution's fulfillment of its mission, a prioritized set of implementation requirements can be established. Implementation requirements can then be assessed in terms of financial and human resource needs, and these requirements can be factored into the institution's planning process. Thus, the process will allow the effects of delayed implementation to be assessed, and could lead to the setting of new goals and objectives for the institution, or could form the basis for changing priorities for resource allocation to obviate the effects of delayed implementation.

3.0 Analysis of Performance Categories

Performance categories should be defined in a general manner. Table One defines the nine categories referenced in the previous section. This set of categories appears to be almost generic. In fact, the categories were chosen for application to a wide range of water and sanitation institutions. The detail level of the investigations for any of these categories will vary depending on the specific institution being studied. In fact, the categories themselves could be different than those shown depending upon the analysis required.

Table One

Definition of Performance Categories

1. Organizational Autonomy - The level or degree of freedom an institution has in pursuing its mission without undue regulatory, bureaucratic and political constraints. Thus, autonomy measures the power of the institution in making critical decisions necessary to achieve its objectives.

2. Leadership - Leadership is the ingredient which "energizes" organizations to pursue its objectives. Leaders - at all levels in institutions - as positive role models, inspiring high commitment to objectives by all staff. Leaders recognize needs compared to responsibilities, and continually work to improve the status and resources of the institution in order to achieve its goals.

3. Management and Administration - Management is the organization of all resources to accomplish objectives. Good managers optimize available resources, and foster cooperation, teamwork, and communications. Administration is the combined body of policy procedures and rules which guide management. These refer to the specific systems which direct daily activities.

4. Commercial Orientation - Measures the optimum use of financial resources and the degree by which institutions are driven by cost effective considerations.

5. Consumer Orientation - Measures the degree to which activities/systems serve and educate the users of their service, and the level of responsiveness in acting on complaints, and/or other feedback directly provided by users.

6. Technical Capability - Measures the level of in-house competence in performing all technical tasks required, and the ability to transfer technical knowledge throughout the organization.

7. Develop/Maintain Staff - Measures the ability to retain and upgrade adequate staff at all levels, paying competitive salaries, providing a good working environment, training and continued education.

8. Organization Culture - Measures the values, attitudes, and norms which guide activities. A positive culture is indicated by evidence of good sense of purpose and mission, and pride of institution's history.

9. Interactions with External Institutions - Measures ability to effect positive coordination with all outside institutions impinging upon its mission.

The performance categories will be further defined by the selection of output measures or indicators. If carefully selected, these will act to guide the analyses of the performance categories, and will almost force their inter-relationships to be illustrated. The output measures or indicators can be numeric, quantitative, or both. The key to their selection is to recognize how they will better define the cross-cutting performance categories under study.

A sample set of output measures/indicators related to the nine performance categories is presented in Table Two. It is well to note that the indicators in Table Two are presented as a list of topics or areas selected to focus on the analysis of the performance categories by their component parts.

Once analyzed, the performance categories can be rated on either an analytical or very simple scale. A simple rating is recommended (3 level - good, average, poor, or 5 level - excellent, above average, average, below average, poor). Whatever method of rating is used, the rating system should be one which can easily be justified by the analysis. (Ratings justified by "gut feelings" alone may be excellent to provide guidance to analysts. However, recommendations for improvements should have a wider justification.)

An overall profile of each performance category can be drawn and by comparing the analysis of each to one another, the areas of strength and weakness throughout the organization can be identified. A plan can then be formulated, for the institution as a whole, to correct the apparent weaknesses.

(The basis for the correction plan will be the data analyzed in preparing the individual profiles. Thus it must be recognized that if only a general analysis is performed to generate the profiles, a more detailed analysis may be required for the implementation plan.)

The flexibility in this approach is worth noting: The process itself requires that the institution be examined through the use of generally drawn performance categories which are better defined by the use of indicators pertinent to the institution under study. In effect the process forces an examination of the institution so that the inter-related problem areas of the institution can be clearly identified. No corrective measures are formulated without first examining the institution as a whole.

4.0 Proposed Analysis Framework

A simple framework for applying the approach discussed above is as follows:

- The institution under study should be examined in terms of the nine performance categories presented (or revised as required, to insure that the institution's responsibilities (mission) can adequately be described within the major areas chosen).

Table Two

Performance Category	Sample Performance Indicators
1) Organizational Autonomy	- Power/ability to set and revise policy - Establish (and successfully obtain) adequate capital and operating budgets - Maintain adequate staff at all levels - Work without undue regulatory or bureaucratic constraints
2) Leadership	- Leaders act as role models and are respected throughout organization - Motivation in pursuit of goals apparent - Integrity and clear performance standards are evident at all levels - Continually work to improve organization while planning for future
3) Management/ Administration	- Clear sense of responsibilities by group, and high interaction within and amongst groups - Problems solved at lowest level - Managers formulate budgets; plan and monitor operations. Timely information passed upward and routinely utilized and in actions of senior managers - Proper systems in place allowing smooth operations. Timely management information is provided, adding to successful operations.
4) Commercial Orientation	- Expenditures are balanced against budgets - Operations cost effective and very efficient - Financial/accounting systems adequate and provide timely information
5) Consumer Orientation	- Activities/procedures geared to consumer acceptance - Consumer "orientation" maintained throughout organization - Proper emergency and complaint-response mechanisms in place
6) Technical	- Technical staffs are competent for responsibilities - Standards/procedures updated - Technical data produced as aid in operations - Technical information shared throughout organization

Performance Category	Sample Performance Indicators
7) Staff Development/ Maintenance	- Systems to increase staff knowledge, and training in place - Active skill transfer programs - formal and informal - in place - Adequate in-house promotions policy in place - Staff adequately compensated with decent fringe benefits - Adequate personnel systems in place
8) Organizational Culture	- Morale high; staff active in promoting organization - Staff believes power shared and organization's success reflects on them - Informal alliances amongst staff act to complement overall mission of organization - Staff positive as to individual roles and their fit into overall institution
9) Interactions with External Organizations	- "Turf problems" mitigated as managers work to foster inter-organization cooperation - Managers act to develop formal/informal extra-organizational ties to pursue mission - Managers informed on a sector-wide basis, to overcome obstacles to performing mission

- Indicators of performance (quantitative and qualitative as the case dictates) should be determined to act as criteria for analysis of the performance categories. (The performance categories can be revised again at this point, based upon the selected indicators.)

- Data should be gathered and analyzed. The analyst team should use a multi-disciplinary approach to ensure that technical, financial and managerial areas are all covered. (It may be well to utilize individuals from outside the institution being studied to obviate the natural bias of in-house managers. However, an initial in-house analysis should not be discouraged.) Further, the analysts, especially if from outside the institution, should be well briefed in the cultural, social and political environment in which the institution operates (the briefing could easily be part of the data gathering procedure). A list of indicators should be agreed on by the analysts and at least the top managers of the institution being analyzed. The preliminary list of indicators should be drawn up by the top managers of the institution.

- The analyst team should perform the analysis making sure all components of the analysis are compared to one another by the total team (rather than the individual team members analyzing them on a piece meal basis). Preliminary findings should be reviewed with the institution's management to insure that the analysts have correctly interpreted all the data and information.

- Final recommendations should be formulated together with an implementation plan. The plan should indicate strengths and weaknesses, with recommendations for a specific prioritized schedule of improvements (indications as to the difficulty of implementing the various improvements should also be presented).

- The final point in the analysis should be an assessment of how delayed implementation will effect the institutions' responsibilities, especially its goals and objectives.

5.0 Conclusion

Self examination is never easily accomplished. However, in an environment characterized by scarce human and financial resources, public sector institutions would be shirking their responsibilities if such self examinations were not undertaken.

The procedures outlined above are not difficult to implement. They can be carried out in-house but perhaps are best conducted by outside parties. The cost to the institution is a pittance when compared to the public expenditures involved, and more often than not they pay for themselves manyfold in very short periods of time.

It is no coincidence that public institutions with the highest reputations are those which welcome the type of analysis outlined in this paper. The best institutions remain best through their continued quest for excellence.

GUATEMALA NATIONAL PLAN

Paul C. Dreyer and David Laredo*

INTRODUCTION

Background

The Republic of Guatemala has embarked on a wide ranging National Plan to upgrade water and wastewater service in the nation's 328 urban areas (Guatemala City is not included in this study as it is administered through its own agency). Development of the National Plan for Water and Wastewater for the Urban Centers in the interior of the Republic is the responsibility of the Institute for Municipal Development (INFOM) of the Republic of Guatemala. Funding for the National Plan development is a joint effort of the Government of Guatemala and the Inter-American Development Bank (IDB). INFOM selected the association of Camp Dresser and McKee International Inc. (CDM) of Boston, Massachusetts, and Cordon y Merida, Ings. (CyM) of Guatemala City to provide professional engineering services required for the development of the National Plan. INFOM issued the notice to proceed in November 1986, and the work is progressing with a scheduled completion date of December 1987.

INFOM is an executive agency founded in early 1957 as the national institution with responsibilities to aid local governments for municipal development. INFOM provides support and technical assistance in planning and implementation of a variety of projects, and a large part of INFOM's work is in development of water and wastewater projects. Over the last ten years INFOM has invested US $50 million in over 600 water supply and sanitation projects throughout Guatemala's twenty-two Departments. INFOM's main purpose in developing a National Plan at this time was to streamline the process of selection of future projects for implementation.

Thus a key element of the project was to develop a database that provided reliable information that was easily accessible and updated. As such, the database could be used by INFOM for planning future projects and programs. Training of INFOM staff prior to completion of this program will insure the continued use of this planning tool is essential to the continued success of future programs.

*Associates, Camp Dresser & McKee International Inc., 2001 NW 62nd Street, Fort Lauderdale, FL 33310; and One Center Plaza, Boston, MA 02108, respectively.

Work Plan

The scope of work for the development of the National Plan requires the identification and prioritization of water and sanitation projects in each of Guatemala's 328 municipalities. The task of the consultant is to develop the National Plan using a planning process that is comprehensive enough to allow long and short term national needs to be determined, and flexible enough to allow for continual updating by INFOM to accomodate future conditions and changes.

The core of the process was the compilation of a detailed database or inventory which fully described the national urban water and sanitation sector conditions (outside of Guatemala City). The database was to be designed so that it was suitable for use in the study, and at the same time, appropriate for INFOM's use in future activities.

To accommodate these needs, the project work plan divided the overall effort into three phases as follows:

- Inventory - The purpose of the inventory phase was to compile the database on the conditions in the 328 municipalities throughout the country. (Information on file was thought to be incomplete and there was some concern as to the consistency of the data due to the number of years over which it has been collected.)

- National Plan Development - The second phase of the project referred to the prioritization program to determine the projects to be implemented within the short term, and to determine an investment program for the next 15 to 20 years.

- Prefeasibility Studies - The third phase of the project includes the preparation of prefeasiblity studies for the short term projects (i.e., immediate needs) selected as part of the National Plan.

The timing of the notice to proceed in late 1986 required the questionnaires be devised, the effort be staffed, and all data collected within a four-to five-month period to obviate any difficulties which might be caused by the rainy season (May to October.) The program is scheduled to be completed in a twelve-month period. A schematic diagram of the work plan showing the interrelationship of work tasks is presented in Figure 1.

DATA COLLECTION

Questionnaire Design

Three questionnaires were used in the field surveys, one each for the water supply system, sewerage/sanitation system, and

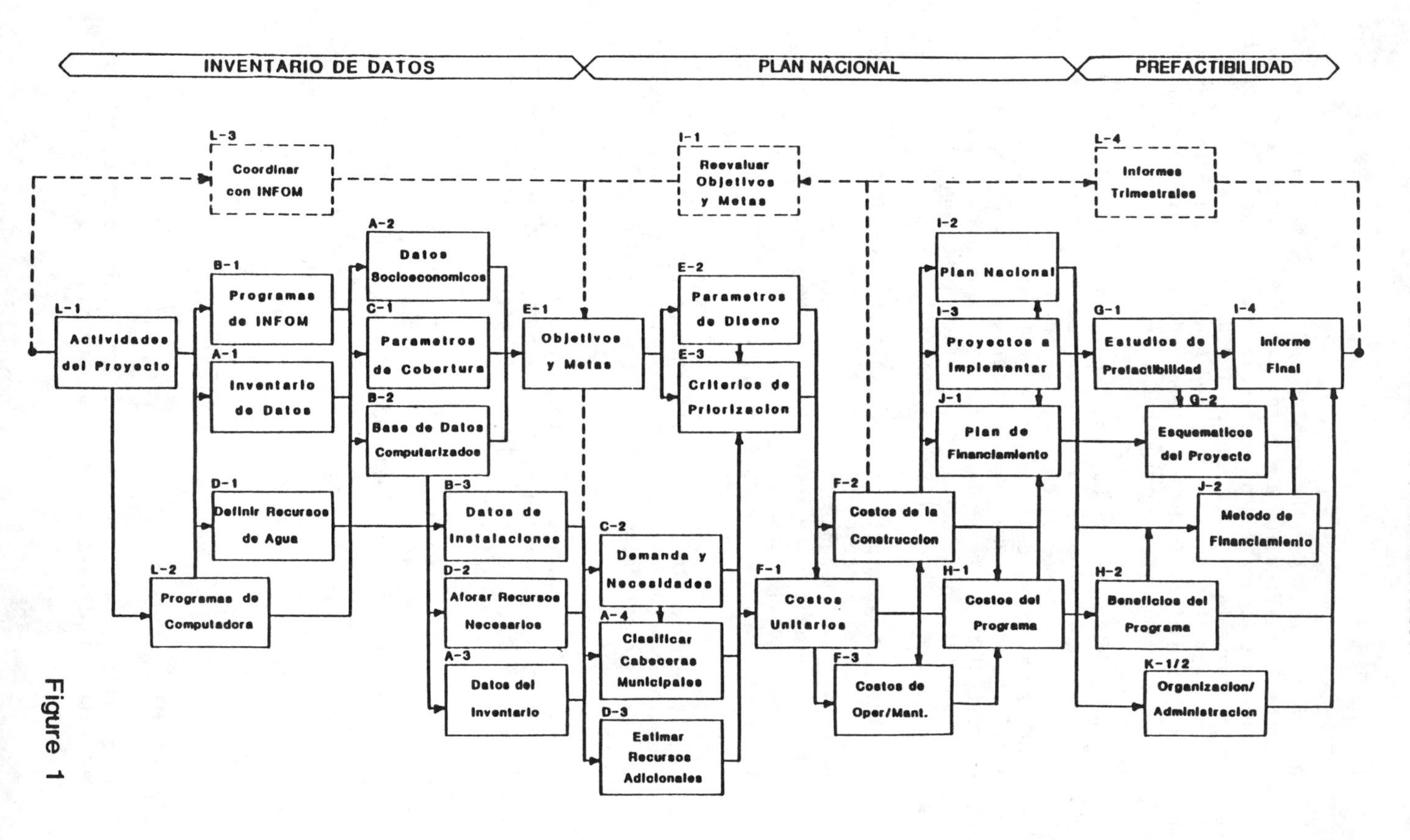

INVENTARIO DE DATOS
PLAN NACIONAL
PREFACTIBILIDAD
L-3 Coordinar con INFOM
I-1 Reevaluar Objetivos y Metas
L-4 Informes Trimestrales
L-1 Actividades del Proyecto
B-1 Programas de INFOM
A-1 Inventario de Datos
D-1 Definir Recursos de Agua
L-2 Programas de Computadora
A-2 Datos Socioeconomicos
C-1 Parametros de Cobertura
B-2 Base de Datos Computarizados
B-3 Datos de Instalaciones
D-2 Aforar Recursos Necesarios
A-3 Datos del Inventario
E-1 Objetivos y Metas
E-2 Parametros de Diseno
E-3 Criterios de Priorizacion
C-2 Demanda y Necesidades
A-4 Clasificar Cabeceras Municipales
D-3 Estimar Recursos Adicionales
F-1 Costos Unitarios
F-2 Costos de la Construccion
F-3 Costos de Oper/Mant.
H-1 Costos del Programa
I-2 Plan Nacional
I-3 Proyectos a Implementar
J-1 Plan de Financiamiento
G-1 Estudios de Prefactibilidad
G-2 Esquematicos del Proyecto
J-2 Metodo de Financiamiento
H-2 Beneficios del Programa
K-1/2 Organizacion/ Administracion
I-4 Informe Final

Figure 1

household survey. Great care was taken in the design of these questionnaires. The water and sewerage questionnaires were prepared using maps of the municipalities supplied by INFOM. The maps indicated housing densities, roadways, and topographic features. The questionnaires were formulated considering how the data was to be utilized and to assure that the formats provided allowed the field teams to collect the data in an easy manner. All questionnaires were tested in the field and revised as necessary.

The facilities questionnaires were designed to be completed in one or two days for the smaller communities and three or four days for larger communities. Much of the data came from discussion with various officials of the municipalities and the system operators. The remainder of the data was obtained by field measurements. The maps were revised in the field to reflect the new roadways and buildings constructed since the maps were prepared and additional information (e.g., number of buildings served, extent of the water and sewer systems installed, and the length of pipelines) were entered into the questionnaires. Sketches of facilities (i.e., type, size, location, and condition) and system schematics were drawn in the field.

The household survey questionnaires were designed to be completed in a fifteen minute interview. The number of houses to be surveyed was designated. (These ranged from ten percent of the houses for communities having populations of 500 to two percent for the larger communities. The houses surveyed were also numbered on the maps. Some 13,000 household surveys, representing about 70,000 persons will have been completed in this fashion. This survey population is about 8 percent of the total 1985 population in all 328 municipalities.

Quality Control of Field Work

A set of detailed instructions was prepared for the field teams' use. The instructions included how the data was to be utilized, the manner in which it should be collected, and the manner in which questions should be worded. A detailed briefing was given to each field team supervisor prior to the commencement of any field work. Further, the field supervisors were briefed one to two weeks after commencement of the work and the completed questionnaires were reviewed. Instructions were issued based upon these reviews. Spot checks of the field-work were also made as a further measure to assure the quality of the data being collected.

Another key element in the quality control of the field-work was the selection of the supervisors and assignment of their areas of work, which was greatly assisted by the local firm, CyM. The twenty-two departments in the country were divided into fourteen areas for purposes of data collection required to complete the inventory phase of the project. Field work for data collection was accomplished through the work of fourteen survey teams whose supervisors were selected for their knowledge and experience in these regions, familiarity with the indigenous population, and the ability to communicate in the local native languages. (There are about two dozen linguistically distinct groups in Guatemala.) A fifteenth group was assigned a few more complicated communities and served to field check the results obtained in other areas. Teams consisted of two to four individuals who visited each community for a period of two to several days. Detailed data and information was collected based on procedures and questionnaires discussed previously.

Computerized Data Base

The information collected in the field was entered into a computerized database. The software used (Knowledgeman - 2 program) provided the flexibility of database management and analytical analysis in one package.

An IBM-AT personal computer (to be provided INFOM at the end of the project) was utilized to store and process the data. The database management is shown schematically below.

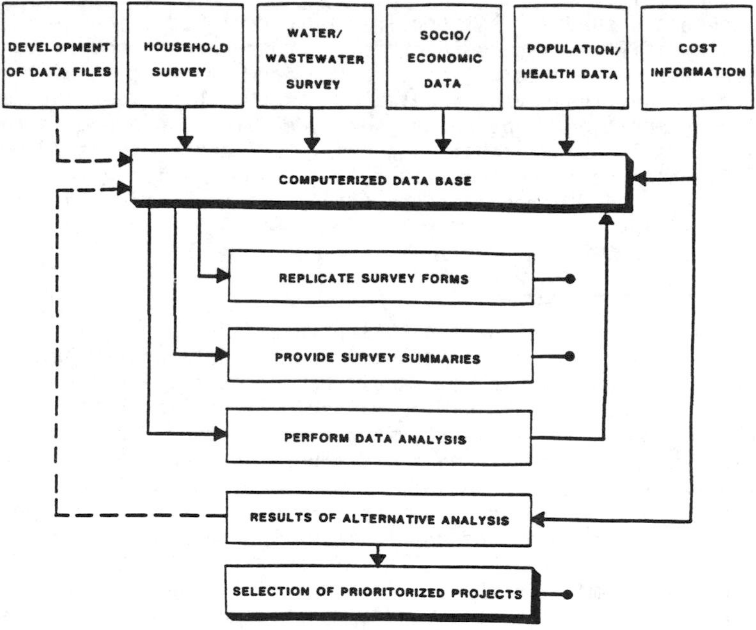

As shown, the data from the surveys will be stored together with other parameters and statistics for use in prioritizing the projects. The computer program will have the ability to directly reproduce the data collected in the water and wastewater surveys, produce various summaries of this data, and generate summaries of data created through computations and analysis (using the base data and other parameters as input.)

For instance, the program can compute future populations, populations served, maximum day demand at future target years, facilities requirements, annual cost of improvements (operating and capital), and cost per customer or cost per unit volume of water supplied. Inputs for these analyses would be questionnaire data and parameters supplied (e.g., future target year, rate of population growth, target percentage served, maximum-day to average-day factor, costs of facilities and improvements, and financing data) The impact on INFOM'S overall improvements program, in terms of project costs and annual investment levels, can be examined under various policy alternatives. (For instance, a level of 90 percent served for water supply for year the 2000 can be compared to any other percentage costs for the entire list of projects, and cost per customer can be computed and compared to INFOM's expected financing levels and the ability to pay of customers.)

The program can also be utilized in the prioritization analysis. For a given set of policy alternatives, a list of project costs and other parameters can be generated. The projects can then be "sorted" by any of several combinations of parameters (or indices) developed. The actual prioritization procedure will depend upon the policy set adoped by INFOM.

Methodology

The methodology finally adopted for use in prioritization projects will depend upon the "policy set" adopted by INFOM. Development of this kind of policy is difficult for any agency. The procedure is often subject to natural political pressure exerted in almost all cases of resource allocation for public services. The overall issue facing INFOM is that of determining a "global" approach. For example, should the main guideline for investment planning be one of coverage? (i.e., providing the greastest coverage for the funds available), or should the municipalities with the poorest service receive the most investments? The "greatest coverage" option leads to providing extentions in coverage on the basis of lowest unit cost (i.e., lowest cost per person). This generally leads to investments in the larger systems as they benefit from a scale factor. The "worst-first" option leads to providing investments in smaller systems, but providing greater incremental benefits in each system as service levels increase. The actual investment pattern will probably fall within the range described.

At the time of this writing, the prioritization methodology has not been finalized and reviewed by INFOM; however, an amendment describing the actual methodology wil be distributed at the conference.

Criteria

The criteria for prioritization was evaluated in detail, and under consideration include the following:

- Project Cost - Total project cost (individual municipality) and cost per capita are included. Inherent in this paramater is the projected population growth rate and the capacity of the existing source. The problem of fast-growing municipalities with limited existing sources are apparent, if INFOM adopts a policy of maintaining at least the existing level of coverage (i.e., percent served).

- Ability to Pay - This parameter (including social rate of return on the new investment) is related to service level and project cost (the higher the level usually the higher the cost.) If a community has the ability to finance the project, its likelihood will be higher since the risk of investing limited resources will be low.

- Quality of Existing Service - This parameter is related to the management and operation of the existing system. The higher the existing quality, the lower the risk in terms of new facilities being effectively operated and maintained.

- "Importance" Factor - This parameter is related to the municipality's "place" in Guatemala's overall development plans. A community making important economic contributions (i.e., existing industrial developments, large agricultural centers, or tourist centers) will be strong candidates for improved infrastructure to insure the continuance of the economic contribution.

- Health/Socioeconomic/Environmental Factors - These parameters are obvious indicators for water supply and sanitation projects (several analyses are underway to formulate these in parameters in terms of indices.) One such index is the potential sewerage generated and the disposal methods in use. Another is the environmental impact of the disposal alternatives on the land and/or water body acting as the disposal mechanism.

The methodology being considered is determining a score for several parameters derived in accordance with this discussion and weighting these parameters to determine an overall ranking for each municipality. An alternative ranking system being considered is to determine three or four levels of need and classifying all municipalities by "need level." The prioritization could be finally determined by applying cost per capita, ability to pay, and existing service quality parameters. An example of the alternative of selection criteria is as follows:

Criteria	Factor	Percent Served		
		Alt. 1	Alt. 2	Alt. 3
Cost per capita	10			
Ability to pay	15			
Quality of existing service	20	(to be determined based on actual community need)		
Importance factor	15			
Economic factor	10			
Health/ socioeconomic	15			
Environmental	15	____	____	____
TOTALS	100			

Whatever option is finally decided will reflect INFOM policy and the level of funds available for investment.

DISCUSSION

Conclusions

INFOM's national planning project presents many lessons which can be applied to other programs. First and foremost, INFOM has decided to establish a long term process for use in their future planning. The process developed will allow for continual updating as conditions change, and will allow INFOM to make firm investment decisions on a logical basis which are adaptable to the technical, economical, and political pressures any such agency is forced to work under.

The project also recognized the need to establish a comprehensive database, and much time and effort has been allocated to this task. Further, the use of a computerized database and analysis program (using "off-the-shelf" software) gives strong indication that INFOM recognizes the need to use new and sophisticated planning mechanisms for their future investments.

The database design will allow INFOM to analyze "real world" conditions, from budget constraints to changing demands for service, on a strategic basis. Thus, INFOM can react quickly to changed conditions and effectively allocate their ever pressurized, and in some cases, falling capital resources.

Acknowledgement

The authors appreciate the support and collaboration received from their colleagues at CDM and CyM, especially Ing. R. Octavio Cordon M of CyM.

RURAL WATER SUPPLY AND SANITATION SYSTEM DEVELOPMENT

BANGLADESH EXPERIENCE

Ali Basaran *

ABSTRACT

Water supply and sanitation programme (CWSS) development in Bangladesh; and through its implementation, the achievements and failures of the environmental health promotion programme is described. By the end of 1985 rural water supply coverage exceeded 46% while the coverage in shallow tubewell areas reached 67%, both based on one operable tubewell for every 75 persons, a remarkable achievement in such as short time.

The present sanitation coverage is far behind that of safe water. Although the rural sanitation programme (VSS) progressed admirably since its start in 1974 the coverage reached an insignificant 3% by the end of 1985.

It is concluded that the lack of complimentarity between water supply and sanitation in the national programme should be reduced/eliminated by further intensifying the sanitation programme, using an integrated approach for which some guidelines are given.

INTRODUCTION

Estimated total population of Bangladesh is 100.6 million, the population growth rate is 2.6% per year, and density is 695 persons per sq.km. (1). About 87% of the total population live in rural areas under the greatest poverty levels and under-developed state(1).

The status of health in Bangladesh reflects to a great extent the status of water supply, sanitation and personal hygiene. At present, although cholera no longer appears in epidemic form, other enteric diseases related to water supply and sanitation such as dysentery, and diarrhoeal attacks continue to be a national concern, resulting in a high rate of infant and child morbidity and mortality. Infant and child mortalities are 136 and 21 per 1000 respectively, both among the highest in the world (2).

Government's national socio-economic policy is implemented under five year development plans (FYDP). At present, the Third FYDP (1985-90), is being executed (3). Primary objective of the national socio-economic policy is to improve the health and well being of the

*World Health Organization, Dhaka, BANGLADESH

Table 1: IDWSS DECADE TARGETS FOR BANGLADESH

Year	RWS	VSS	Remarks
1985	46	3	Present coverage
1990	77	13	Initial targets
1990	68	11	Revised targets

population by increasing the rate of development expenditure and ensuring inter-sectoral co-operation and co-ordination in education, housing, potable water supply and environmental sanitation (3).

The Ministry of Local Government Rural Development and Cooperatives (MLGRDC), through its executing agency, Department of Public Health Engineering (DPHE), is solely responsible for the water supply and sanitation (CWSS) sector in Bangladesh. Therefore, the scope here will be limited to the development programmes undertaken and perceived by DPHE in its past, on-going, and future programmes(4,5,6).

As a part of national committment to the International Drinking Water Supply and Sanitation (IDWSS) Decade and considering the existing socio-economic condition of the country plus the fact that the affordability and willingness of the population is relatively low, realistic national IDWSS Decade targets have been fixed as shown in Table 1 and the Government is actively pursuing it in a sustained manner to achieve the targets with extensive technical and financial support from UNICEF, WHO and other bilateral donor agencies (4).

APPROACHES & STRATEGIES IN SYSTEM DEVELOPMENT

Traditionally, because of the riverine nature of the country, people naturally resorted to surface water for all their needs. Then, the shallow handpump tubewell (STW) system of water supply was first introduced some 50 yrs ago, more as a consequence of repeated outbreaks of cholera (4).

Like many developing countries Bangladesh undertook a relatively large rural CWSS programme since its independence in 1971 with substantial external, technical and financial assistance. The immediate objective was to provide new STW and to rehabilitate the non-functioning ones in order to increase the availability of safe water among the rural people. At the on-set, one public STW was available for an average of 400 people which was far too thin a coverage. On the sanitation side, the prevailing system of excreta disposal in the country could best be described as very primitive. Open and random defecation had been the prevailing practice and only a negligible number of farm houses had traditional "kutcha (pit)" latrines (3,4). Developments in both rural water supply (VSS) and sanitation (VSS) are elaborated in following paragraphs.

Rural Water Supply (RWS) Programme

The hydrogeological conditions in almost all of Bangladesh is favourable for groundwater extraction. In general, the groundwater table is within 4m of the surface, but may drop during the dry season to depths of more than 15m in some areas. Fresh water aquifer is available between 30m to 300m in most places(4).

The programme progressed in phases and the first RWS programme was launched in 1973, shortly after the liberation of Bangladesh (3). It provided 100,000 new STWs and resinking of 60,000 chocked-up STWs by 1976, raising the STW coverage from the initial 400 to 250 rural people on average (3). During the second phase, 155,000 STWs and 5,000 deep handpump tubewells (DTW) were installed between 1976 and 1980. The rural population coverage by the end of 1980 was 36%, based on one operable tubewell for every 75 persons. During the third phase a total of 250,000 STWs, 10,000 DTWs and 8000 a new generation of deep-set handpump tubewells (DSP), also known as TARA pump, were installed between 1980 and 1985. Thus, as of the end of 1985, the total coverage exceeded 46% while in STW areas the coverage reached 67% a remarkable achievement in such a short time (3). A year by year account of these from 1972-85 are presented in Table 2.

Rural Sanitation Programme (VSP):

The first rural sanitation programme (VSP) started as a DPHE and WHO collaboration in 1950 (4). It was primarily an applied research and demonstration of appropriate technology, and its effectiveness in cholera control. In 1962, an experimental pilot project provided, free of cost, a number of households with low-cost sanitary latrines (LCSL) of direct pit, pour-flush water seal (WS) latrines (WSL)

The first comprehensive rural sanitation programme (VSS) was launched in 1974. The programme provided WSLs free of cost to 8,000 household and 75 schools by 1978 (4,5). This programme was revised in 1979 and incorporated sale of WSLs to the beneficiaries through (DPHE) production and selling centers (VSPC). 256 VSPCs were established all over rural Bangladesh by 1982 and 135,000 WSLs were produced and sold at subsidised prices (4,5).

The second phase of the programme started in mid-1982 and by the end of this phase in 1985 a total of 460 VSPCs were established and a target number of 340,000 WSLs were produced (4,5). The sanitation programme, though progressed admirably, the coverage reached an insignificant 3% by 1986 (4,5). As a consequence of such a large gap between coverages by water supply and sanitation in the national programme, the net health impact has not been very discernible

At present, the third phase of the VSS is being executed which will run until mid-1990s (1,5,6).

Table 2: PHYSICAL ACHIEVEMENTS OF RWS AND VSS

	STW		DTW		DSP		RESINKG - STW		VSS/LCSL	
July-June	Plan'd	Achv'd	Plan'd	Achv'd	Plan'd	Achv'd	Plan'd	Achv'd	Plan'd	Achv'd
1972-73	20,000	15,000	1,000	900	-	-	30,000	24,000	-	-
1973-74	20,000	17,800	200	190	300	200	25,000	20,000	-	-
1974-75	25,000	23,000	-	-	500	300	20,000	16,000	-	-
1975-76	45,000	42,110	-	-	500	500	-	-	-	-
1976-77	25,000	30,900	500	400	-	-	-	-	10,000	5,331
1977-78	69,000	42,690	1,500	600	-	-	-	-	10,000	7,694
1978-79	59,000	38,500	1,200	1,450	1,000	690	-	-	10,000	8,965
1979-80	38,000	30,700	1,600	1,710	2,000	1,480	-	-	25,000	18,306
1980-81	44,000	37,940	1,000	1,070	2,500	1,300	15,000	4,950	41,630	35,860
1981-82	31,000	24,850	900	940	500	490	15,000	7,170	83,840	59,606
1982-83	47,000	42,330	1,000	810	1,000	480	18,000	9,460	50,000	52,595
1983-84	34,000	30,846	1,000	813	500	520	15,000	9,657	101,550	84,300
1984-85	41,000	31,230	400	495	400	430	20,000	12,861	93,000	89,268
TOTAL	498,000	407,896	10,300	9,378	9,200	6,390	158,000	104,098	425,020	361,925

As per the national IDWSS Decade target, by the end of 1990, the coverage by sanitation is expected to be increased to 11%. Therefore, the number of DPHE VSPCs will be increased from the present 460 to 1000 in 1990 (4,5). In addition to the government's efforts, many non government organisations (NGO) and private enterprises have also come forward with their own programmes of sanitation.

RESULTS AND DISCUSSION

Rural Water Supply (RWS) Programme

In the first programme, there was no community participation except that one volunteer caretaker was selected for each tubewell to look after the well's security. The Government employed contractors to sink new wells or to replace the choked-up ones at her cost while UNICEF provided the supplies. During this programme, improvement in STW and handpump designs, including introduction of PVC pipe and screen, were brought about . The choice of STW location and the platform design used were not suitable for all purpose use of the STW water. At a later stage, people were tried to be motivated to use HTW water for all purposes and to get rid of all water and excreta related diseases. As a result, STW water use for drinking increased because people started to percieve it to be good for their health. However, this situation did not improve much even though the gross coverage per STW increased to 138 rural people as of 1985 (3). It is estimated that only 10% of the rural population use STW water for all purposes.

A pilot project was also carried out during this phase to study the extent of peoples' participation and it was found that people were willing to share the cost of sinking (4). However, the main objective to participate in obtaining a STW by giving contribution by the entire prospective beneficiaries did not materialise. This was mainly because of two reasons: a) the poor families were not mobilised adequately to participate and the relatively better-off took this opportunity to contribute the entire sum; and b) the local body which had a vital role in STW allocation treated the richer people more favourably due to political and other reasons. As a consequence, the STWs could not be located in a manner to insure equal accessibility. However, as a result of contribution in the sinking of STWs, the sense of ownership developed in the community and the operational status of the STWs improved. For small repairs or replacement, the community, in most cases one or two families, did not wait for DPHE mechanics and took care of the work. In other words, both DPHE and the caretakers became partners in STW maintenance.

Based on findings of the studies conducted during the second programme and also on some new studies conducted in the early part of the third programme some changes and additions were made in programme delivery (5). Besides, a number of studies were undertaken to find alternate sources of water in STW-unsuccessful areas, to develop community type iron removal plants for iron problem areas, to improve components of the handpumps used, to develop a revival process of choked-up wells by desanding, to determine extent of pollution of groundwater from LCSLs and safe distance of a tubewell from a LCSL, to develop a low-cost DSP to economically cope with the depleting water table and to make a comprehensive socio-economic study in order to develop a strategy to achieve a significant impact on health of water and sanitation interventions.

During the third phase, the contributory choked-up STW replacement programme was converted into a self-help resinking programme by the community. Similarly, the contribution of the community for the STW sinking was first raised from 50% to 75% and finally, in early 1985, it was also converted into a self-help sinking programme (6). Sinking of DTW and DSP wells however, remained as a contributory system with DPHE sinking the tubewells through the contractors because, the sinking of these wells were not as simple as that of the STWs. To involve the caretakers directly in the maintenance of the STWs, a training programme was designed and started for them, where the necessary hand tools were also provided; and progressed satisfactorily. But its impact is yet to be seen as the caretakers have not yet gained access to the spare parts needed. The positive findings of the pilot project of selling spares among the caretakers is planned to be started in 1987 (5). Once it starts it will not only be a self-help system, but also it will save a considerably large amount of government expense too.

During the Third FYDP the programme will give more attention to underserved areas such as DTW areas, DSP areas, and STW-unsuccessful areas (3). The coverage by tubewells in these areas is very poor compared to the coverage by STWs in similar tubewell-feasible areas. Another major thrust would be to augment physical facilities at pump site in order to encourage the people to use tubewell water as much as possible. The most crucial activity would be to introduce a system whereby: a) the poorer section of the people also will be able to obtain tubewells according to their requirements; b) the siting of the tubewells will be such that the women who collect most of the water for families can have easy access and also can play an important role in maintenance; and c) an effective programme of water and sanitation related health and hygiene education will be undertaken.

Rural Sanitation Programme (VSP)

An evaluation of the LCSL programme revealed that the poorer section of the people still are not getting the benefit of this programme (1,3). It further showed that some improvements took place in the use and maintenance of WSLs including non-breaking of WS which was very common in the early part of the programme. Provision for higher subsidy for a minimum basic unit of WSL has been kept for 1985-90 period to attract the poorer section of the people. But the actual success of VSP will depend on intensity and effectiveness of a public sanitation education and motivation programme to encourage the people to have their own sanitary latrines built. DPHE perceives its programme as a means of creating awareness among the people and to provide various alternative solutions to the sanitation problems.

Several types of LCSLs were tried in two villages to test their applicability and public acceptance in rural Bangladesh (1,3). But the on-going WSL was again confirmed by the people as their primary choice. An interesting finding was that the women used the latrine more, irrespective of the type of the latrine and its odour problem if the superstructure was intact. On the other hand, the children were found to use the latrine less when the superstructure was intact.

CONCLUSIONS AND RECOMMENDATIONS

Needless to say that the success of the water and sanitation programmes as health intervention is largely dependent on how people will perceive them, how easily they will get access to these facilities and how convenient these facilities will be. Under the present socio-economic situation it seems to be a very difficult task, especially when the women are difficult to reach with the existing Government programme.

Evaluation of both self-help STW sinking and resinking programmes revealed that these were viable community based activities needing proper supervision and communication support. One of the important features of these systems is that, except for establishment, carriage, etc., they are now independent of Government budget allocation, which has been a severe constraint over the last few years.

For each cluster of houses, known as "Bari", there should be at least two public tubewells, and one of these should be located at a common place within the inner compound so that the women can use it for all purposes including bathing which they will not generally do in public places. The other should be located outside so that the male members of one, two or more Baris can use them for all purposes. Therefore, each tubewell should have adequate size platform.

The day-to-day maintenance and minor repair and replacement of the parts of the handpumps should be done by the beneficiaries themselves and at their costs.

People will continue to use the competitive, convenient and unprotected sources for washing and bathing. It is necessary to establish the health problems caused by this practice and its extent, to be more convincing to people, so that they may be expected to change their habits.

The collaboration between DPHE and Directorate General of Health Services (DGHS) at District level and below should continue and be further promoted for the successful formulation and implementation of the National Sanitation Programme.

More emphasis should be given to health and personal hygiene education of school children in their curricula. Furthermore, the school teachers should be trained to carry out this task both in the classroom and in the community.

Alongside the Government programmes of water and sanitation a self-supporting door-to-door health and hygiene education needs to be undertaken by the community in a sustained manner. The aim being to achieve minimum hygiene and sanitation practices especially, when it is now clear that diarrhoea is also caused by poor hygiene practice and poor food sanitation.

Religious and community leaders should also be trained in health and personal hygiene topics so that they can effectively contribute to public education and awareness in the community.

NGOs and private sector should be encouraged to get more involved in water and particularly sanitation programme to supplement government's efforts to increase production.

REFERENCES

1) Anonymous, 1985, Area and Population, Household and Housing Charact.; and Educ., Health, Fam. Plann. Soc. Welfare and Sports, in "Stat. Pocket Book of Bangladesh 1984-85," Bangladesh Bureau of Statistics, Stat. Div., Min. of Plann., Dhaka, Aug. 18, 1985.

2) Anonymous, 1985, Country Report on Primary Health Care in "The Pyongyang Conference: Primary Health Care in Action," WHO , SEARO Regional Health Papers No. 6, New Delhi, 1985.

3) Anonymous, 1985, Phys. Plann. Housing and Water Supp. in "The 3rd 5-yr Plann, 1985-90," Plann. Comm. Min. of Planning, Dhaka, 1985

4) Anonymous, 1950-1987, DPHE Records and files in "Rural Water Supply and Sanitation Programme & Annual Development Programmes/Plans for 1950-1987", Dept. of Public Health Engineering, Min. of LGRDC, People's Rep. of Bangladesh, Dhaka, 1986.

5) Anonymous,1985, Rev. Report of the Working Group on Rural Water Supply and Sanitation in "Planning Towards the Third Five Year Plan (1985-90)," Dept. of Public Health Engr., Min. of LGRDC, Dhaka, 1985.

6) Anonymous, 1986, Guidelines for Implementation of Rural Water Supply and Sanitation Programme, in "Annual Development Programme for Rural Water Supply and Sanitation ," DPHE, Min. of LGRDC, Dhaka, 1987.

7) Anonymous, 1986,Country Situation Presentation in "Draft Proc. of Inter Country Workshop on Accel. of Nat'l Progr. on Sanitary Disposal of Human Excreta," WHO SEARO, New Delhi 27-31 Oct. 1986

8) Basaran, A., 1985, WHO Experiences on Water Supp. and San. Progr. in Bangladesh, in "Proc. of Workshop on Water & Sanitation Interv. Related to Diarrhoeal Disease in Bangladesh, Dec. 17-19, 1985," A. Basaran and R. Islam ed., Int'l Centre for Diarrhoeal Disease Res., Bangladesh, Dhaka, 1987

9) Ko Ko, U., 1985, Promotion of Envr. Health in "The Work of WHO in the South East Asia Region: 37th Ann. Rept. of the Reg. Dir.," WHO Reg. Off. for S. E. Asia, New Delhi, 1985.

THE MANAGEMENT OF WATER WORKS IN TAIWAN PROVINCE AND KAOHSIUNG CITY, ROC.

Lien-chuan Chen*

I. Prelude

Taiwan was called "Ilha Fromosa" (means beautiful island) by Portuguese a few hundred years ago. It lies at 119 to 122 degree East Longitude, 21 to 25 degree North Latitude. The Tropic of Cancer passes through its southern part. Annual average temperature is 21.6°C in the north and 24.3°C in the south. The island is strategically located between the Ryukyu Islands and the Philippines, 128 miles east of China mainland. The total area of Taiwan is 35,981 square kilometers with the Central Range as its backbone running from north to south. The mountains are sprawling gradually down the vast plain in the western part. In Summer and Autumn, there is a lot of heavy storms and typhons which produce about 80% of 2,400 mm annual average rainfall. Dozens of dams have been built for multipurpose along the rivers in the mountainous area of the west. The eastern part of the mountain is steep and has only a very narrow plain along the coast. Additionally, the river beds are nearly dry throughout the year except in the typhoon season; therefore the development of the water resources in the east is not so satisfactary as it is in the west.

By the end of 1986, the population of Taiwan is about 19.45 million.

Now, there are two water utilities in Taiwan area, one is Taipei Water Department under the Taipei Municipal Government, and the other one is the Taiwan Water Supply Corporation (TWSC) owned by the Taiwan Province and its local governments and Kaohsiung city. They supplyed 7.12 million m^3/day of water for 15.5 million persons in 1986. The percentage of population served was about 88%.

II. History of water supplies in Taiwan

Before 1941, the Taiwan water utilities consisted of 118 water supply systems in total. Besides those in major cities and towns, most of them were simple water supply facilities, and could only produce a few hundred cubic meters of water per day. Moreover, most of them were damaged during World War II, with no more than 20% left that could be operated to supply water regularly.

*Deputy Commissioner, Reconstruction Dept. of Taiwan Provincial Government,
Mantou, Taiwan

The percentage of population served was only 17% in 1945. Our government then made efforts to rehabilitate the water supply facilities and undertook a long-range water supply development plan.

Yet owing to the rapid population growth, when the first phase development projects were completed in 1968, the percentage of population served only rose to 35%. By the end of 1973, when the second phase development projects were completed the percentage reached 41%.

Since our government tried to operate the regional consolidation of water supply in 1954, due to objections raised by some of the city and township councilmen under various pretexts, this policy has not been fulfilled effectually for the past 18 years.

There were 128 water works in Taiwan Province and Kaohsiung city in 1974. These 128 independent water utilities were managed separately by the province, different cities, countys and townships. There were 7 manged by the province, 8 by the cities and countys, 7 by associated townships, and 106 by different townships. The situation is shown in Table 1.

Table 1: The Situation of Water Works in Taiwan before Taiwan Water Supply Corporation Was Established.

Managed by	Waterworks		Output Capacity	
	No.	%	CMD	%
Province	7	5.5	347,000	25.4
Hsiens & cities	8	6.2	477,000	35.0
Partnerships of towns	7	5.5	93,000	6.8
Townships	106	82.2	448,000	32.8
total	128	100	1,365,000	100

Owing to the original separatly managed waterworks had so many defects as follows:

1. It was difficult to raise enough expenditures for construction and expansion of installations because of low water rates.

2. Because water of streams are not rich even drought expect rainy season. Therefore, some of water works used the same water resource so as to invest duplicate money in development and facilities.
3. Owing to the underpayment and less of the chance for promotion, most of the excellent staffs of water works try to look for other job with high salary.
4. It was uneasy to renew the facilities, even could not to maintain the equipment properly because of low water rates (NT$ 2.00 = US$ 0.05 per cubic meter in 1973).
5. Unable to take a scientific management by establishing a sound system, to run the business to increase the efficience since the limitation of manpower and finance.
6. The five defects were mentioned above so as to cause worse business.

In the spring of 1973, the Taiwan Provincial Government, at the instruction of Premier Chiang, formulated the "Plan for Centralized Management of Provincial Water Utilities". As Premier chiang put it in his instruction, " With a view to do developing public water supply efficiently, it is imperative to set up a province-wide water supply company with centralized management so as to reduce ooperating cost and raise investment returns." Accordingly,, the Taiwan Water Supply Corporation (TWSC) was formally established on January 1, 1974. The original 128 waterworks were merged into the Corporation to become the shareholders.

III. Organization and Management of Taiwan Water Supply Corporation (TWSC)

1. Organization

The organization of TWSC is shown in Fig. 1 .

The business affairs of the corporation is centralized management, but proceeded separately operation. The head office is located at Taichung city to take charge of overall planning and evaluation, and to coordinate manpower and purse. Under the head office, there are 11 district offices in charge of production, operation, maintenance, business and service to customers, and three district engineering departments in charge of expansion projects. Besides, there are a staff training center and two water meter repair shops in charge of training course and repairing water meters respectively.

Up today, there are 23 water works, 94 operation stations and 39 service centers scatted in the province and Kaohsiung City.

The organization of the original waterworks is as follows,

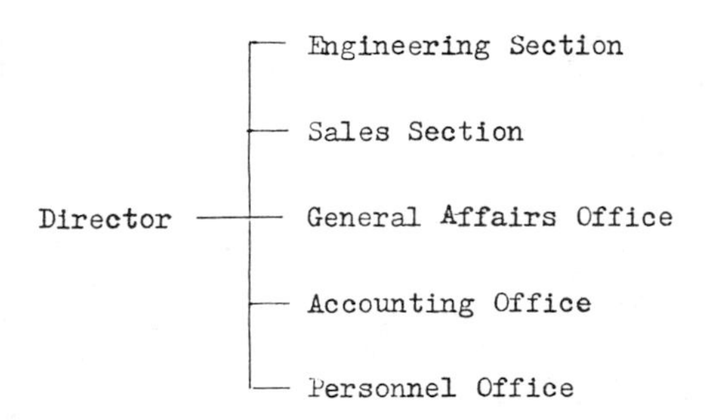

2. The Manpower Utilization

The existing TWSC formed by merger of 128 individual water treatment plants at counties is now manpowered with 4,432 persons. As the quality of some personnel from county water treatment plants was much to be desired an on-job training program was initiated on active personnel after the activation of this company. For production of better operational and business function, and uplifting the service quality, an employee training office was established to meet the requirement. So far, 8851 persons/times have been trained until 1987; of which, 69.17% were managing persons; 19.55% were technicians; and 11.28% were administrative personnel. The current project is that each employee is schedualed to receive refreshment training once every two years. As employee's apprehensions, concepts, and stand-points can be integrated and unified, after the training, their method, and techniques are improved, and in turn, their working effectiveness is uplifted year by year. In addition, some specialists are sometimes employed to meet the casual requirement. It is hereby to illustrate the water supplying performance for the recent five years in table 2.

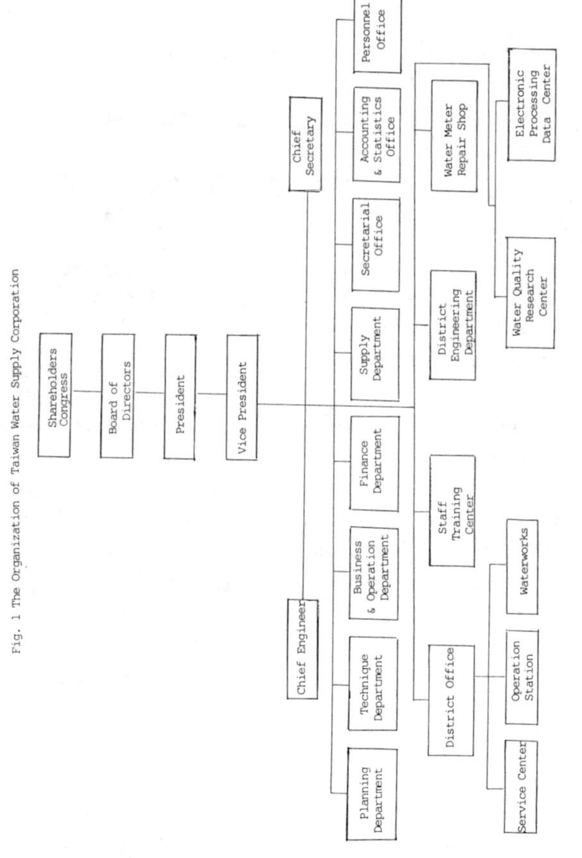

Fig. 1 The Organization of Taiwan Water Supply Corporation

- 5 -

Table 2 : The water Supply Performance of TWSC for the Recent Five Years

Item / Performance / Year	Customers			Water Sale			Water Consumption						Employees			House Served / Employee		
	Total, household	Increasing Rate, %		Total, 10^3 m^3/day	Increasing Rate, %		Total $\times 10^3$ m^3/day	Increasing Rate, %		Per Capita /day	Increasing Rate, %		Total Capita	Increasing Rate, %		Houses / Capita	Increasing Rate, %	
		Annual	Total		Annual	Total		Annual	Total		Annual	Total		Annual	Total		Annual	Total
1980*	1,619,689	—	—	583,342	—	—	472,557	—	—	150	—	—	5951	—	—	272	—	—
1981	1,828,230	12.9	12.9	644,563	10.5	10.5	531,389	12.5	12.5	157	4.7	4.7	6512	9.4	9.4	281	3.3	3.2
1982	2,044,528	11.8	26.2	718,505	11.5	23.2	607,851	14.4	28.6	168	7.0	12.0	6756	3.8	13.5	303	7.8	11.4
1983	2,201,946	7.7	36.0	792,987	10.4	35.9	672,182	10.6	42.2	172	2.4	14.7	6751	−0.1	13.4	326	7.6	19.9
1984	2,354,013	6.9	45.3	844,893	6.6	44.8	714,182	6.3	51.1	179	4.1	19.3	6815	0.9	14.5	345	5.8	26.8
1985	2,489,354	5.8	53.7	905,691	7.2	55.3	772,045	8.1	63.4	186	3.9	24.0	6971	2.3	17.1	357	3.5	31.3
Average	—	9.0	10.6	—	9.2	11.1	—	10.4	12.7	—	4.4	4.8	—	3.3	3.4	—	5.6	6.3

* Basic year

7a

3. The Financial Condition

In accordance with the water supply law, TWSC shall be operated in business manner, and on self-sufficient basis. As we know that the water supply is not popularized yet throughout Taiwan, and in the served area, follow-on investment is required in water supply facilities; even the funds required for some new construsction is made available through loan. As the business is brought under integration, and all the funds can be consolidated for effective use, the company can maintain its business, and operation. At present, the averaged water tariff is about N.T.$ 6.60 (U.S.$ 0.20) per cubic meter; the individual's annual water bill would take only 0.32% of per capita income. The proceeds of water fee could leave some balance as profit after it is used to support the general business expense and to pay the interest for the loan. It is herby to illustrate the recent three year's operating income and expense of the company in the table 3.

Table 3: The Operating Income and Expense of TWSC for Recent Three Years

Year / Money (thousand NT$) / Items	1984	1985	1986
Operating Income	6,228,317	6,591,390	6,870,111
Other Income	86,324	88,110	70,019
Subtotal Income	6,314,641	6,679,500	6,940,130
Operating Expense	4,868,490	5,234,706	5,734,649
Other Expense	952,917	897,941	816,161
Subtotal Expense	5,821,407	6,132,647	6,550,810
Net Income (or Loss)	493,234	546,853	389,320
Project Investment	2,600,591	2,534,760	2,432,902

Remark: 1 US$ = 30 NT$

4. The Water Rate

With a view to let the people in remote area or high cost area use supplied water at cheaper rate, and to facilitate the popularization of water supply, TWSC has consolidated and standardized more than 60 different water rates; however, the active water rate fails to reflect the due cost in time because the adjustment of water rate is subject to the obstructions of native councils. For instance of the recent five year's comparason, the averaged water supply pipe lines per thousand persons for the remote area is increased 55.1% in length as compared with that in the dense population area, and the horse power of water supply is increased 22.5%. The company's profit objective can be still achieved neverthless impossible to raise the water rate; this achievement should be credited to the effective utilization of manpower, and the decent control on operational/business cost. Table 4 illustrates the increasing ratio of cost for some major subordinate units in recent decade. Fig. 2 illustrates comparason of water rate with cost for the recent decade. In accordance with the water supply law, the investing reward rate in Taiwan water supply enterprise should be between 5 and 8 percent; actually, the current investing reward rate lies only between 1 and 4 percent on account of increasing cost in remote area facilities, and the fact that the water rate failed to reflect the due cost in time.

Table 4: The Increasing Ratio of Costs for Some Major Subordinate Units in Recent Decade

Year \ Ratio \ Cost due to	Manpower		Power		Raw Water Purchase		Operating		Interest		Total		Depreciation	
	Link	Base (1976)	Link	Base (1976)	Link	Base (1976)	Link	Base (1976)	Link	Base (1976)	Link	Base (1976)	Link	Base (1976)
1976	100	100	100	100	100	100	100	100	100	100	100	100	100	100
1977	94.6	94.6	100	100	144.4	144.4	89.3	89.3	238.5	238.5	107.9	107.9	92.3	92.3
1978	91.4	85.6	94.9	94.9	100	144.4	100	89.3	103.2	246.2	98.4	106.1	104.2	96.2
1979	88.4	75.7	83.8	79.5	107.6	155.6	84.0	75.0	89.1	219.2	91.0	96.6	110.0	105.8
1980	96.4	73.0	109.7	87.2	128.6	200	102.4	76.8	112.3	238.5	104.2	100.7	100.0	105.8
1981	103.7	75.7	120.6	105.1	88	177.8	106.0	82.1	101.6	250	103.7	104.4	100.0	105.8
1982	106.0	80.2	107.3	112.8	106.3	188.9	104.4	85.7	86.2	215.4	102.0	106.5	105.5	111.5
1983	105.6	84.7	95.5	107.7	152.9	288.9	100.0	85.7	94.6	203.9	105.8	112.6	115.5	128.9
1984	101.1	85.6	95.2	102.6	107.7	311.1	102.8	87.5	117.0	238.5	104.2	117.4	104.5	134.6
1985	101.1	86.5	97.5	100.0	100.0	311.1	104.1	91.1	90.3	215.4	100.9	118.4	108.6	146.2
Average	98.7	-	100.5	-	115.2	-	99.2	-	114.7	-	102.0	-	104.5	-

9a

Fig. 2 The Comparasion of Unit Water Rate with Cost for the Recent Decade.

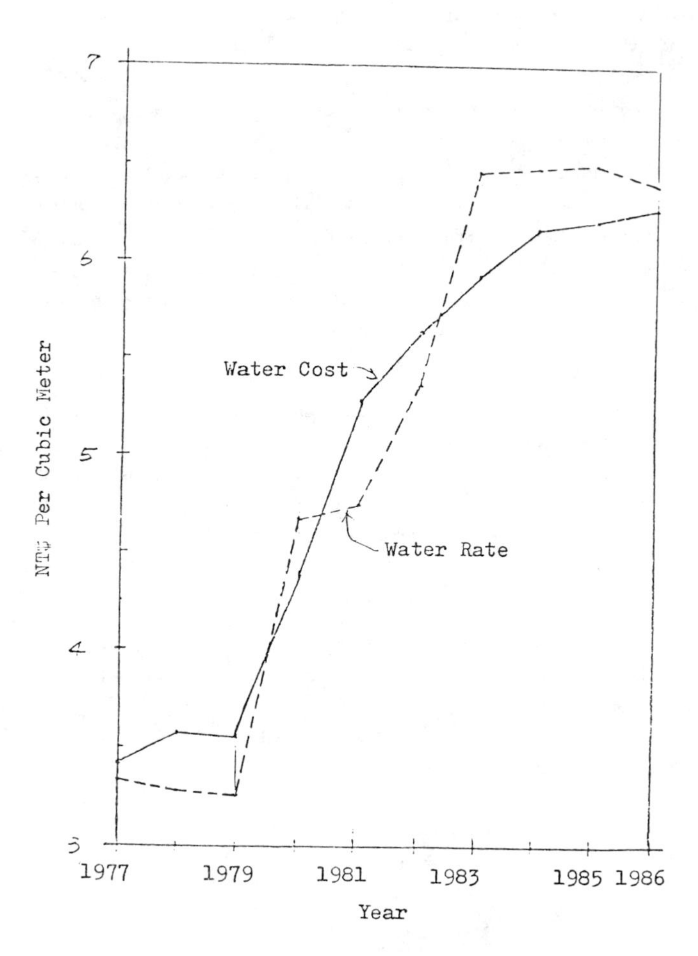

IV. The Performance (outturn) and Achievement of TWSC.

TWSC has been in operation for 13 years since its activation in January 1974; its morethan 13 year's effort has produced the following achievements:

(1) The swift growth of served population and its percentage. Fig. 3 and Fig. 4 respectively illustrate developing tendency of served population and its percentage from 1963 to 1983. In 1963, the served population was only 3,900,000 persons with 33%; until the activation of the company at the end of 1973, the served population amount to 5,800,000) persons with 42% in this decade, the served population was increased by 1.49 times with only 9% increased (the averaged annual increasing rate to be only 0.9%. Since the activation of the company, the served population from 5,800,000) persons was increased to 12,000,000 persons during the decade --- by 2.07 times; while the percentage of population served was increased from 42% to 77.5% --- by 35.5 percent more, averaged annual growth rate to be 3.6 percent. This growth rate is 4 times of that before the activation of the company.

(2) The uplifting of output capacity of water supply:

Fig. 5 illustrates the growing tendency of output capacity of Taiwan water supply from 1963 to 1983. In 1963, the total output of water supply for Taiwan water plants was 800,000 m^3/day, 1,364,000 m^3/day at the end of 1973; the output capacity was increased by 1.7 times within the decade. From 1974 to 1983, the output capacity was increased from 1,364,000 m^3/day to 4,600,000 m^3/day by 3.37 times.

(3) The other performances (outturn) and achievements:

(a) Till 1986, the company's performances and achievements in other fields as illustrated in table 5. It is noted as compared with that before the activation of the company in 1973, the growth rate in the 13 years is more than 200 percent. This is also indicating that the integrated business effectivenes has been uplifted a lot after the activation of a corporation for water supply operation.

Fig. 3: The Growth of Population Served before & after TWSC was established in 1974

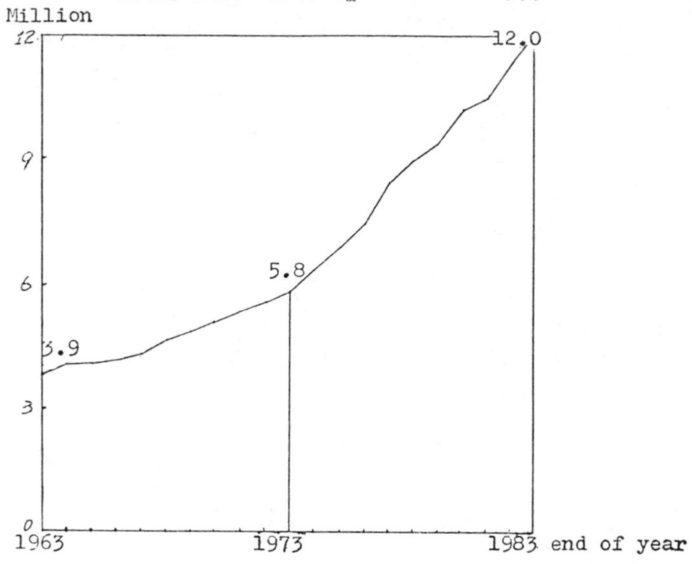

Fig. 4: The Percentage of Population Served before & after TWSC was established in 1974

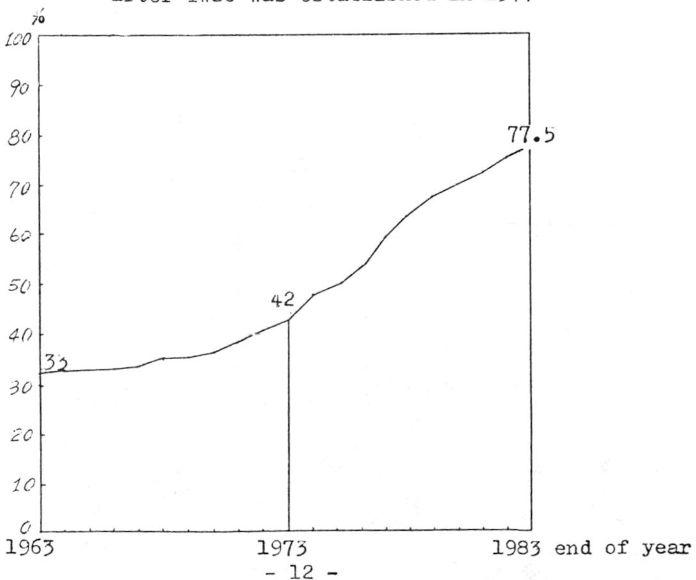

Fig. 5: The Trend of Output Capacity before & after TWSC was established in 1974

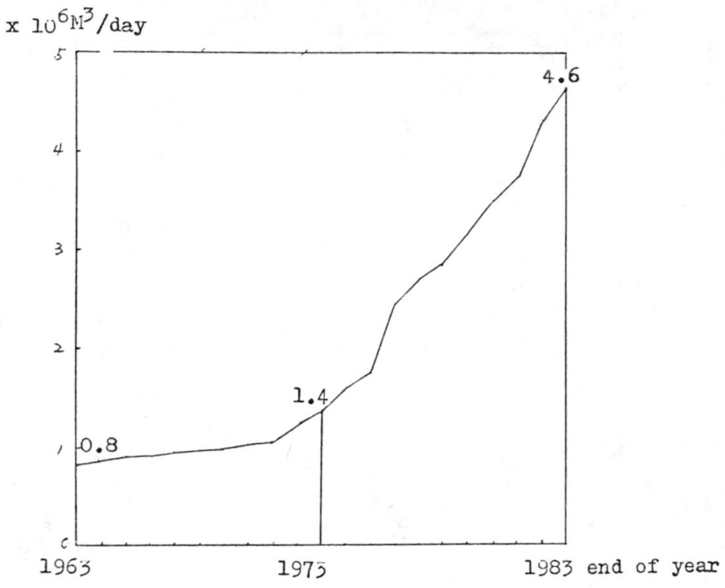

Table 5: A Comparison of Business Achievements between June 1986 and Jan. 1974.

Item \ Year or Growth	Jan. 1974	1986	Net Growth	Growth Rate %
Output Capacity, x 10^3 m^3/day	1,364	5,200	3,836	281
Water Sales per day, x 10^3 m^3/day	764	2,572	1,808	236
Sales Rate, %	68.1	78.1	10.0	-
Rate of Fee Collected, %	93.0	99.7	6.7	-
Customers, Household	661,081	2,550,713	1,889,632	286
percentage of Population Served, %	42.0	82.7	40.7	-

(b) About 2,130,000 persons who drank water contain arsenic have been served with safe drinking water.

(c) TWSC was able to mobilize a large number of its staffs, equipment and material to repair the facilities which were damaged by typhoons and heavy rains, and restored water supply in a very short time.

V. Conclusion

Thirteen years ago, the Government of of the Republic of China encouraged and supported the formation of the provincial water supply corporation. Today, we can see how our manpower and funds have been economically used and the accelerated water supply development program has been launched. These benefits would not come so soon without TWSC.

By June 1986, the output capacity of water supply in Taiwan province and Kaohsiung City increased 281% over the amount registered before the establishment of TWSC. The percentage of population served was also increased from 42 to 82.7% in the corresponding period.

Owing to most of water supply projects will have to take water from remote sources or polluted and more rural. TWSC is expected the invest NT$ 4,700,000,000 (US$ 156 million) each year in compliance with the 10-year development program.

To extend first class service to the customers, service center of each operation station has been set up since 1979. In addition, a public opinion survey is being conducted once every two years for the purpose of understanding comsumers' needs. TWSC always reminds its staff that one of the most important things is to satisfy consumers.

Cherokee Nation Self-Help Rural Water System

by
Owen W. Scott, M.ASCE [1]
and
Charlie Soap [2]

ABSTRACT

Certain predominantly Cherokee-speaking communities in northeastern Oklahoma lack adequate housing, a good water supply, and job opportunities and are losing their young people through out-migration. The prospect of its school's closing because of declining enrollment prompted the Bell community leaders to ask the Cherokee Nation of Oklahoma for help. The Cherokee Nation took this opportunity to try out a new approach of treating a whole community experiencing these common problems as a single client. By involving community members in defining their problems, proposing solutions and working toward those solutions, the tribal leaders hoped to build upon the "people power" of the community members. Two practical solutions evolved: (1) a mutual self-help housing program and (2) a self-help waterline construction program for residents of Bell and nearby Adair County communities. This paper describes the development of a self-help rural waterline construction program.

The people of the community served as elected neighborhood representatives and work crews that laid the waterline in their home neighborhoods. The self-help program developed for the Bell community was successful in quickly providing a water system at a lower cost and also in meeting other community needs, e.g., housing and health care. The Cherokee Nation has adopted the self-help method for development projects in other communities and the U.S. Bureau of Indian Affairs is considering its applicability as a model for other tribes throughout the country.

I. Background

a. Cherokee Nation of Oklahoma (1)

The Cherokee Nation, originally located in the Smoky Mountains of eastern Tennessee, the western Carolinas and upper Georgia and Alabama, was removed in 1838-39 on the "Trail of Tears" to its present location in northeastern Oklahoma, then known as Indian Territory.

[1] Hydraulic Engineer, U.S. Army Corps of Engineers, North Central Division, Chicago, IL

[2] Director, Community Development Department, Cherokee Nation of Oklahoma, Tahlequah, OK

The Cherokee people have a long tradition of helping one another as well as of adapting to progress and innovation. Aboriginal Cherokee organized informal communal work associations (known as "gadugi") within each township to tend crops raised in common and to help members in times of need. Today, structurally similar free labor companies exist among some conservative Cherokee groups.(3) With the invention of Sequoyah's alphabet (or more accurately called a syllabary) in 1822, the Cherokee quickly became a highly literate people. Sequoyah (in whose honor the giant Sequoia redwood trees of California were named) is the only person in history to develop, single-handedly and without previous knowledge of writing, a new system of writing. Within a short time, the Cherokee had a written constitution, their own tribal newspaper and an extensive public school system. At the time of their removal, the Cherokee had a higher literacy rate in their own language than Whites who displaced them had in theirs.(2)

After removal to Indian Territory, the Cherokee rebuilt their progressive lifestyle in the new location. Their new capital, Tahlequah, became a hub of business activity and a cultural oasis. Their bilingual newspaper and periodical became the first publications in the new territory. They strung the first telephone line west of the Mississippi River to Tahlequah. They reestablished an educational system; their 144 elementary schools and two higher education institutions rivaled all others at the time. However, Cherokee lands were gradually ceded to make room for others.

The Cherokee Nation does not occupy a reservation, as such. Tribal land was allocated to the individual members in severalty just prior to the organization of the State of Oklahoma and much of it soon passed into the hands of others. The tribal council and courts were dissolved in 1906 to make way for Oklahoma statehood in 1907.

Several decades after dissolution, the Cherokee Nation reorganized itself in accordance with the terms of the Indian Reorganization Act of 1934. The present Constitution of the Cherokee Nation of Oklahoma was ratified in 1975. The Cherokee tribal government is designed after the original democratic form of government of the Iroquois, as is that of the United States. In the eyes of the federal government, the Cherokee Nation of Oklahoma is a sovereign nation with the rights and privileges to govern itself and create laws within the framework of the U.S. Constitution.

The present-day Cherokee Nation encompasses all or part of fourteen northeastern Oklahoma counties, where 42,000 of the Cherokee Nation's 74,500 tribe members reside. The highest concentration of Cherokee population is in the four counties nearest the tribal capital in Tahlequah - Cherokee, Adair, Delaware and Sequoyah.

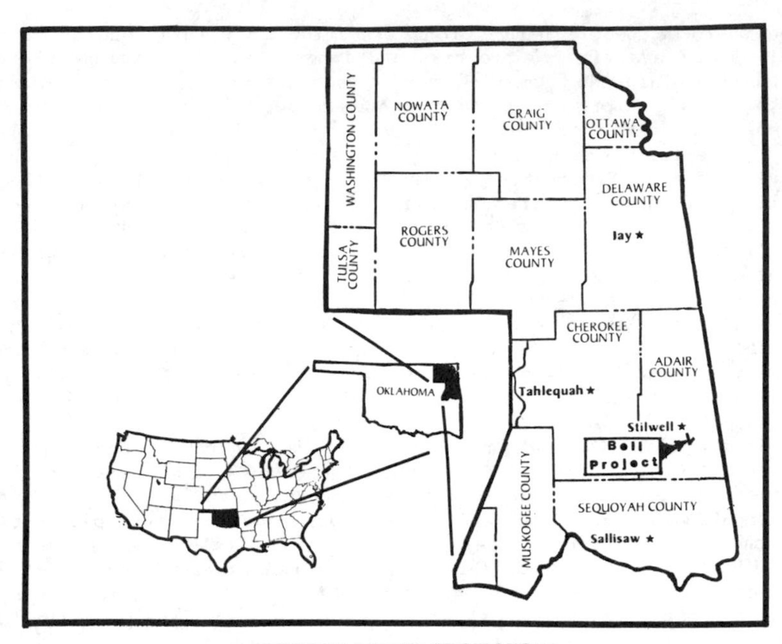

CHEROKEE NATION OF OKLAHOMA

b. Rural Community Situation

The rural Cherokee community of Bell is nestled among the Ozark foothills of southern Adair County. In 1981, this 104-resident community contained many indications of decline. More than half the houses were sub-standard. One quarter of the families hauled their domestic water, making gardening nearly impossible. The mean family income was $3,000 per year. For elders, it was less than $1,500. Unable to find jobs and facing a bleak future if they remained, many of the young people were leaving the community. The enrollment at Bell School had declined by twenty-two percent between 1976 and 1981.

As conservative Cherokees, the residents possess a strong sense of fellowship and a desire to help neighbors (a trait that is too often latent nowadays but found expression in "gadugi" organizations in former times) and a determination to hold onto these traditional values. Highly independent, the community in the past had not chosen to cooperate with the tribal government for services. The stereotype of Bell was of a violent, lazy people who would not work for pay, much less as volunteers. However, faced with the closing of the Bell School because of the declining enrollment, the Bell community leaders approached the Cherokee Nation for assistance in improving the community.

II Development of a Self-Help Program

a. Conceptualization

At the same time that the Bell community leaders were asking for assistance, the Cherokee Nation was searching for new ways of community development and self-help community renewal was being discussed. This situation provided the Cherokee Nation an opportunity to try out a new approach of treating a whole community, where residents experience common problems, as a single client. From experiences of the past and the "nature" of the Bell people, the community organizers from the Cherokee Nation realized that the building of trust was the key to a partnership.

The so-called experts and top-level policy makers have long determined the needs of Indian people who, in turn, are expected to sit passively as wards while "the government" delivers services through a system the tribal members are never a part of and do not understand. Most such programs perpetuate dependency. By involving community members in defining their problems, setting goals, proposing solutions to those problems and becoming involved in working toward the solutions, the Cherokee tribal leaders hoped to tap the enormous and generally unrecognized capacity of Cherokee people to help themselves. This would not only build on the Cherokee "gadugi" tradition of caring for one another but, hopefully, would restore the people's confidence in their own ability to solve problems.

b. Implementation Planning

From the onset, the Bell residents were expected to plan and implement community renewal, with the organizers from the Cherokee Nation acting as facilitators and funding brokers. Those terms were expressed in a partnership agreement between the Cherokee people of Bell and the Cherokee Nation. A primary concern was whether a community existed. This was determined by observations of the interactions of the people, of the community's natural rhythm, its history, the identity of its leaders and the basic values held (e.g., "Are they Cherokee?"). This information was then connected with the problems and solutions. Self-help community development was the strategy used to connect the Bell community with the outside resources. After the people of Bell and the Cherokee Nation both agreed to proceed with community work, the following goals were established:

1. Install a 27-kilometer (17-mile) rural water system using community volunteers.

2. Rehabilitate 20 old homes using homeowners as volunteer workers.

3. Construct 25 new energy-efficient houses using Indian Houses Authority resources.

4. Rehabilitate the dilapidated community center.

The underlying social objective was to bring the community together to collectively solve common problems. The residents of Bell clearly understood, from the beginning, that this would be their project and that they would be doing something entirely new.

III Implementation

a. Community Involvement, Organization

The organization techniques chosen involved having community people perform various pre-planning tasks, thereby building trust between partners as well as accomplishing tasks necessary for the success of the project. The first meeting consisted of a discussion of why the Cherokee Nation personnel were there. Very specific problems and concerns were identified. Small activities, such as posting notices of the next meeting and gathering information on housing, water bills, etc., which got as many people involved as possible, were identified. The first meetings were attended by few people. One task of these people was to bring other members. To create a comfortable atmosphere for individual participation, meetings were often in people's homes rather than the more formal setting of the community building or school. Much attention was given to assessing the audience to determine whether Cherokee or English would be used. Often, meetings were bilingual. In every case, clear and concise communication was paramount. Given the past experiences and "nature" of the Bell people, the building of trust was vital. The "method" used was to get to know the people personally and to keep them involved by assigning small projects to be completed by the community. Further, when people did show up, their input was welcomed. In a word, the people were treated as partners, not subjects.

Certain community residents were trained to conduct surveys of who wanted water and whether (and when) they were willing to work. Also, a survey was made to see what skills were possessed by the community members. As it became clear who were leaders, who would work and who merely talked, formal committees were selected by the community. To avoid factionalization of various sections of the community, enough members were needed to represent the entire community. Ten members were finally selected as the "Bell Water and Housing Committee" for decision making, idea formation and getting out those people who had signed up to work on a particular section of waterline.

The community was divided into eight neighborhoods of about ten families each. Neighborhood work crews were organized, each of which was responsible for laying about two miles of waterline pipe in its home neighborhood. Each committee member served as crew chief for one section of waterline. One committee member became coordinator, supply procurer, motivator and strategist for all the sections.

b. External Funding Assistance

Necessary funds and services came from a variety of federal and private sources, e.g., the Kerr Foundation of Oklahoma provided an organizing cost grant and the Indian Health Service designed the waterline and contributed septic systems and drainfields. The Cherokee Nation, itself, is dependent on categorical grants from the Federal government. Nearby construction companies leased or donated the use of equipment such as backhoes and cranes. As an enabling center to process the needed grant proposals, "broker" funds available to current needs, contract for outside services needed to supplement community capabilities and obtain easements, in addition to organizing community efforts, the Cherokee Nation in May, 1981, established its Community Development Department, with a staff of about sixty personnel. The director of that department was initially Wilma Mankiller (now Principal Chief of the Cherokee Nation), who was succeeded by Mr. Charlie Soap (co-author of this paper) in 1983 when Ms. Mankiller filed to run for Deputy Chief. The directorship recently passed to Ms. Bertha Alsenay.

An aspect peculiar to these Cherokee self-help projects is the desire to provide for non-Indian families living in the neighborhood, for total community involvement. Fund-raising events such as box suppers, etc., were held to pay for inclusion of non-Indians when they would have been excluded by categorical funding definitions.

c. Construction

The people of the community served as volunteer workers and as elected crew chiefs representing each neighborhood. Section-by-section completion (rather than hourly time) was used to establish each family's responsibility. Younger people and local church groups donated labor on behalf of the elderly and disabled. Work on the waterline included following the backhoe to keep the ditch clean, laying the pipe and backfilling the ditch. To provide the quality control required for a safe and reliable water system, a professional crew supervisor gave training sessions on trench digging, bedding and laying pipe, cutting and joining pipe, backfilling and meter construction. These sessions occurred immediately before construction, with "hands-on" practice and the instructions translated into Cherokee to ensure that all the community workers could easily understand the procedures. The same supervisor from the existing Cherry Tree Rural Water Distict (which would maintain the completed system) also inspected the work as it progressed, to ensure minimum standards. Technical work such as installing chlorination equipment and erecting standpipes was contracted to private firms. Work on the waterline commenced in August, 1982, and was completed in May, 1983.

IV Challenges Met and Overcome

In a sense, this project was not unprecedented, in that many other rural communities have undergone community development and the construction was not complicated, e.g., the waterline, itself, was ordinary plastic pipe. However, the Bell project presented certain significant challenges that were met and overcome: (1) the residents were generally bilingual though largely more fluent in the Cherokee language; (2) improvements were to be planned and implemented by community volunteers, themselves, and (3) the work was done in conjunction with an Indian tribe.

Past experiences of the community had fostered isolation and suspicion of outside entities, including the tribal government. The community was impoverished and declining. Barriers ranged from old patterns that were hard to break and a general sense of defeat, to the veiled threats of a local white rancher, to the institutional resistance to new procedures on the part of funding agencies. Key decisions by some administrators almost doomed the project when, for example, needed equipment was not provided at the proper time. A large threat to this type of project which required high motivation and long-term commitment was "down time" when, say, needed resources or approvals were not forthcoming, the backhoe broke down or bad weather occurred. The Bell project staff dealt with this down time in several ways: social functions such as community suppers, small projects such as proposal writing, media scheduling and constant communication through news updates about the "behind the scenes" work. Occassionally people ran into a barrier and retreated, saying, "Things will never change." The successful strategy for counteracting these problems was to continue to collectively deal with each problem in a solution-oriented manner and to continue to demonstrate the viability of the project. However, care was taken to avoid over-expectations, to assure that lack of immediate activity was not perceived as a failure of the project.

As they solved problem after problem and moved purposefully toward the goal, the people of Bell and the Cherokee Nation staff gained confidence in their own ability to shape and control community development while giving consideration to their own culture and lifestyle. Their success was demonstrated in the fact that, while the projected starting date was delayed from April to August, 1982, because of several problems, the waterline was completed four months ahead of schedule in mid-May, 1983, rather than the scheduled project completion date of September, 1983.

V. Conclusions

The method and ideas used to organize the Bell Community have been refined and used again and again to organize other Cherokee communities in Northeastern Oklahoma. Four other rural water systems have been built or are nearing completion, while others are under consideration.

Five points which are important in doing similar projects follow:

1. The community self-help concept stresses the development and use of the capacity of people to do things for themselves. This includes decision making. The organizers must be

sensitive enough to understand the strengths and history of a community, but not so presumptuous as to impose the "right way" of doing something on them.

2. There must exist a sense of solidarity by the group of people (the community). Various factors make a community. Mainly it is the result of many years of living together. If solidarity does not exist, it can sometimes be developed.
3. The partnership must be built on the strengths of the various groups. In many cases, outside groups want to "help" the Indian people. Their remedies are usually based on their values rather than any appreciation of the strengths (and weaknesses) of the community.
4. In case of lack of technical sophistication in isolated communities, an advocacy group must serve as intermediary to perform the procedural requirements of the "outside" people. The core group must focus on the project rather than the problems and frustrations caused by any type of organizational change.
5. There must be an enabling center to handle administrative tasks.

The Cherokee Nation Community Development Department has successfully served as the enabling center for several other Cherokee communities that have undertaken major self-help projects, though none quite as challenging as the initial one at Bell community. The people of Bell and the Cherokee Nation organizers risked a lot on their belief in the people's ability to solve problems and accomplish projects for themselves. The risk paid off beyond all expectations.

The U.S. Bureau of Indian Affairs is considering the community self-help method's applicability as a model for other tribes throughout the country. Perhaps others, working with developing countries in other parts of the world, may profit from the lessons learned in using this method to improve total communities as well as to construct rural water supply systems.

Acknowledgements:

Portions of a previous article written by one of the authors, Mr. Charlie Soap, was used in preparing this paper with the permission of the Seventh Generation Fund. The authors express their appreciation to Ms. Bertha Alsenay and Ms. Lil Perry for their assistance and to Ms. Lueretta Jones for typing this paper.

Appendix - References

1. Cherokee Nation of Oklahoma, Nation with Promise brochure, Tahlequah, Oklahoma.
2. Feeling, D., in a five-part series on Three Phases of Literary Among the Cherokee, Cherokee Advocate, Vol. IX, Nos. 4-9, May-September 1985.
3. Gulick, J., Cherokees at the Crossroads, Institute for Research in Social Science, University of North Carolina, Chapel Hill; Revised Ed. 1974, pp. 88-94.

Shelter and Sanitation for the Urban Poor

Dr. K.N. Ramamurthy*

Abstract

Slums and squatters exist in every country due to the acute shortage of housing for the people in the lower income group. It is almost becoming an impossible task for the government to provide housing and infrastructure at affordable cost to the urban poor and mainly due to the enormous land costs. As the formal land supply system cannot bring on to the market an adequate quantity of land at prices within their reach, the poor find their own solution by squatting. Such informal housing sectors must be recognised and the service facilities should be extended in a phased manner. It is seen that the community takes great care in maintaining and improving their shelter and environment when the views of the residents were given due consideration through participatory planning. The paper identifies the various design parameters covering the technical, functional, social and economic aspects of the shelter and sanitation for the urban poor.

Introduction

More than one billion people, a quarter of the World's population, are either literally homeless, or live in extremely poor housing and unhealthy environments. More than 100 million people have no shelter whatsoever. The problem of housing shortage is universal. It is common to industrialised and developing countries, to urban and rural areas. In the United States, the number of homeless is estimated at 2.5 million. In Canada, it is reported that about 5000 people live in streets. In the developing world, up to 50 percent of the urban population live in slums and squatter settlements. It is almost becoming an impossible task for the government to provide housing and infrastructure at affordable cost to the urban poor mainly because of the high land costs. As the formal land supply system cannot bring on to the market an adequate quantity of land at prices within their reach, the poor find their own solution by squatting. Sample survey shows that the squatter housing in Trinidad and Tobago constitutes about 25 percent of the building stock. Analysis of the housing situation in most of the developing countries, identifies that there is need for continuous maintenance and rehabilitation of the existing deteriorating structures and buildings, which presently constitute about 40 percent of the building stock in the Caribbean countries.

*Department of Civil Engineering, University of the West Indies, St. Augustine, Trinidad, West Indies.

The informal housing sector must be recognised as an integral part of the housing and the provision of service facilities should be extended in a phased manner.

Participatory Planning

The deficiency in housing are of two types, namely qualitative and quantitative. Merely building few million houses cannot solve the problem, since the qualitative aspect of the housing need is equally important. It is necessary to identify the link between the demography, sociology and housing. Formulating the housing policy and working out the house building programme involves basic social judgements on how the people want to live, how they live at present, the effect of environment on their lives and the social characteristics of the households. It is here, that the citizen participation or participatory planning could play a key role in providing information to the planners on the matters concerning that environment.

Case study analysis of a number of upgrading programmes carried out in the various developing nations show that the community takes great care in maintaining and improving their shelter and environment when the views of the residents were given due consideration at the early design phase. User participation is not aimed at to displace or override the responsibility of the planners or government agencies involved in promoting housing. It is to supplement and help professional planners in providing additional information on their realistic needs, wants and special characteristics of the community.

Slum Upgrading Programme - A Case Study[2]

Growth of slums is a phenomenon found in all metropolitan cities of the world. Slums in Madras city, India cover about 10 percent of the total area in the city and they contain more than 30 percent of the city's population. Under the Slum clearance scheme the slums were demolished and multi-storeyed tenements were erected for the slum dwellers as alternative housing on the same site. In many cases, huts used to reappear at the fringes of the tenements causing peripheral slums and were showing signs of rapid degradation. The reason for the failure of the slum clearance scheme was mainly due to rejection of the tenement housing by the community since it did not suit their technical, functional and social needs. There is lack of privacy and no facility to maintain their cattle. Functional analysis also shows that the slum house or a hut provides better thermal comfort than the concrete structure. The slum clearance programme will only benefit a small segment of the population, and hence is not workable.

The present phase in slum housing (Figure 1) accepts the formation of slums as demand based housing solution, and the environmental improvement is carried out with the help from the community through slum upgrading. It is a means by which the poorest segment of the urban population are provided with low cost shelter and with easy access to employment in addition to providing a means of retaining and improving the existing housing stock. The improvements envisaged in the upgrading programme are the following:(i) Provision of drinking

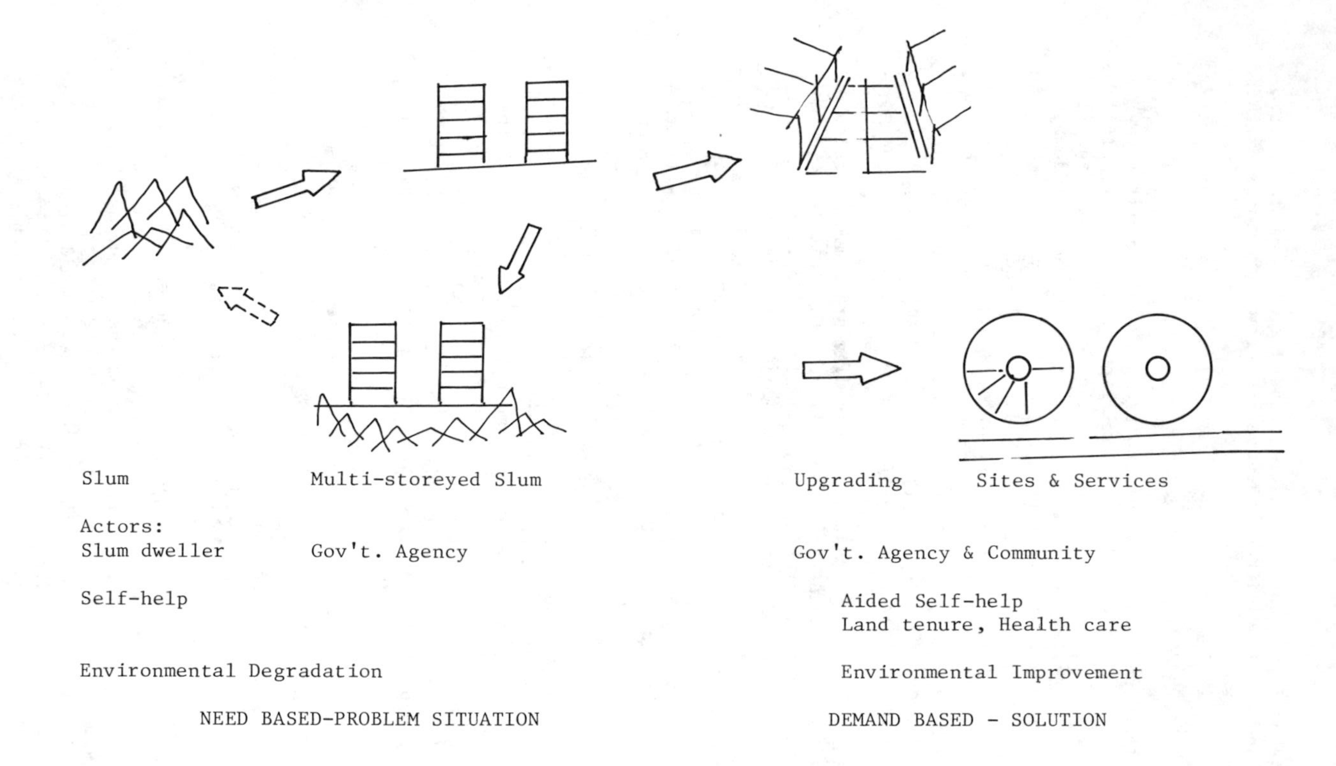

Figure 1 - Evolution Concept - Slum Housing

water supply with public stand pipes (One number for 10 houses), (ii) Community toilets (One number for 10 houses), (iii) New and improved roads, footpaths and drainage, (iv) Community facilities such as pre-schools (One for 200 houses), primary schools and clinics (One for 400 houses). An important feature of the scheme is the provision of land tenure to the slum dwellers. Once they are assured of their land titles, the community takes great care in improving their shelter and environment. The upgrading programme was successful mainly because it provided the residents - A share in decision making, The opportunity to build, and Experience of close cooperation with the city council and among themselves.

Sites and Services Programme

A radial layout pattern has been tried out for the various sites and services schemes in Tamilnadu, India. Sites and services programme envisages development of building sites with the provision of infrastructure including water supply and sanitary facilities. In the radial layout, the individual sanitary core units of a group of plots are arranged around a central point from where the single service connection is provided. This eliminates the necessity of carrying the network of utilities along the length of the plots. The utilities are made available to all the plots in the circular group through the central core. This brings about reasonable savings in the service connections. Similarly the distribution of water and electricity could also be facilitated through the central core area. In certain cases, a biogas digester is located at the centre which supplies gas for cooking and lighting thus minimising the fossil fuel energy use in low income housing.

Maintenance of Community Facilities

The maintenance of public facilities and services has usually taken a low profile particularly in the development of infrastructure. Since the resource allocation for maintenance and the upgrading operations are always limited, there is need for 'defensive design' through proper choice of equipments and fittings. Setting high metropolitan standards for the sanitation and service facilities for the low income housing is often inappropriate and cannot be afforded by the poor. It has been found through experience[3], that the pit latrine - though of a lesser grade, had much longer life and has provided better service than the standard water closet bowl, especially for the community latrines in the low income housing. Provision of high quality fittings require constant maintenance which is not feasible because of its increased cost. In addition these fittings are not tamper proof and as a result may suffer from theft and vandalism. It is necessary to make the community realise the importance of protection of local environment and proper use and maintenance of sanitary facilities. One of the easiest ways of achieving this is through participatory planning of involving the community while deciding about the location of such facilities and their maintenance.

Engineers are given the responsibility of providing the best quality service at the lowest life cycle cost. It is seen that for all sizes of the design population, the water supply service with hand pumps can be provided for approximately half the cost of the standard service with reticulation(1). Because of the shortage of funds for the recurrent costs, fuel and spare parts which are often in short supply, most of the times there is loss of pressure in the distribution network and subsequent ingress of polluted water. On the other hand, designing an elevated water tank is considered as a task well suited to an engineer's training, whereas installing several hundreds simple hand operated water pumps may not always be seen in the same light, even though hand pumps look to be the most suitable solution.

Self Help in Housing

In many respects, the squatters have shown their creative capabilities of inventiveness and self reliance in creating their own shelters at low costs and within their affordable limits, mostly through cooperative efforts and self help(4). A social survey carried out in Arima, Trinidad has shown that the residents are interested in working collectively towards effecting environmental improvements. As a community effort, the residents themselves provided the surfacing for the road and laid the water supply lines. The government should come forward to encourage such active involvement of the benefiting residents in the provision and maintenance of infrastructure and service facilities through collective self help schemes.

The potential of the aided self help approach can be seen as it enables an organised and controlled contribution, using the low cost unspecialised labour, and technologies appropriate to local conditions, without costly capital intensive construction. Self help approach also provides supplementary employment to the residents, which thereby results in effecting improvement in their affordability levels. Self help facilitates greater participation of the user in the planning and development phases of the residential layouts and will ease out any possible conflict which may arise out of possible relocation of housing sites while carrying out an upgrading operation.

Squatters have shown the possibility of building their own shelters by the use of locally available or used - second hand materials like automobile parts, wooden crates, tar drums, etc. It is necessary to identify and promote the application of recycling and reuse of waste materials in housing construction, as the large reduction in the cost of recycling process will itself make them more economical and competitive. Most of the construction techniques employing the recycled or domestic materials are relatively simple and easily adaptable to self built techniques thus offering enormous savings in the labour costs.

References

1. Franceys, R., "Hand Pumps or Reticulation Systems", Proceedings of the 12th WEDC Conference, Calcutta, 1986, pp. 17-20.

2. Ramamurthy, K.N., "Housing Policy and Programme", United Nations Conference on Housing for the Lower Income Population Groups, Bandung, Indonesia, 1981, pp. 1-15.

3. Ramamurthy, K.N., "Project Planning for Mass Housing", Proceedings of the 10th CIB Congress, Washington, D.C., 1986, pp. 1914-1921.

4. Ramamurthy, K.N., "Appropriate Technology for Housing in Developing Countries", IAHS World Congress on Housing, Miami, 1986, 6 pages.

SOCIO-POLITICAL IMPACTS ON POHNPEI'S WATER SYSTEM

By: Shahram Khosrowpanah[1]
Rosalind L. Hunter-Anderson[2]

ABSTRACT: In spite of adequate input to the modern water distribution system of Kolonia, Pohnpei, substandard quality water is available to consumers for only a few hours per day. Previous studies indicate certain technical flaws in the system, such as leakage and a lack of maintenance; however, field observations, consultations with local officials, and a reviews of the post-European contact history and socio- political background of the Pohnpeian people indicates that inappropriate consumer behavior and attitudes are partially responsible.

Post-European contact history has engendered a marked passivity toward the water system, just as it has toward other technologies introduced to the island largely for the benefit of foreigners. Substantial rate subsidies have precluded consumer awareness of the finite delivery capability of the system.

Recommendations include 1) public education about the water system 2) household metering and re-design of the rate structure 3) technical upgrades to improve water quality and reduce leakage 4) strong government support for effective implementation of these measures.

Introduction

The state of Pohnpei (Ponape), capital of the Federated States of Micronesia, is approximately 500 miles (805 km) north of the equator, 2,300 miles (3,700 km) southeast of Tokyo, Japan, and 3,100 miles (5,000 km) southwest of Honolulu, Hawaii. With a total land area of 110 square miles (334 km^2), Pohnpei is divided into five municipalities reflecting traditional political units. Kolonia, in Net municipality, is the seat of the state and national governments.

A high volcanic island, Pohnpei is surrounded by extensive fringing and barrier reefs. The weather is warm and humid throughout the year. Temperatures range around 80°F (27 °C), and the mean relative humidity is about 80 percent. The interior of the island is estimated to receive 330 inches (8,800 mm) of rain a year while coastal zones

[1] Asst. Prof., Water and Energy Research Institute, University of Guam, UOG Station, Mangilao, Guam 96923.

[2] Research Associate, Water and Energy Research Institute, University of Guam, UOG Station, Mangilao, Guam 96923.

receive an average of 190 inches (4,826 mm) a year (NOAA, 1983). Due to its volcanic character, Pohnpei has plenty of surface water which flows in a radial pattern following major drainages from the interior to the coast.

The present Kolonia water distribution system, encompassing some 2 square miles (5 km^2), was built in the early 1970s. It incorporates parts of the old Japanese system. In spite of an abundant supply of fresh water from the Nanpil River, the government continues to restrict water service to a few hours during the morning and evening.

The problems which have created the need for Pohnpei's inconvenient water hours are: 1) technical difficulties with the water distribution system, such as leakage, inadequate maintenance, and untrained personnel to operate the system (Khosrowpanah and Heitz, 1987) and 2) consumer attitudes and behavior, specifically, inappropriate water use habits and attitudes which result in excessive consumption. This second kind of problem is the focus of this paper.

Pohnpeian Socio-Cultural Background

No comprehensive study has been made of indigenous fresh water usage on Pohehpei but some inferences can be made on the basis of preliminary field and archives investigations (Hunter-Anderson, in preparation). Interviews with Pohnpeians about the Kolonia water system have provided additional facts about contemporary uses of and attitudes toward fresh water.

The high annual rainfall, lack of marked seasonality in rainfall, and a basaltic substrate produce abundant surface water year-round on Pohnpei. There are several large rivers, and numerous smaller streams, seeps, springs, water falls and pools throughout the island. Under these conditions, it is not surprising that fresh water has always been viewed by the Pohnpeians as an "unlimited free good". In contrast to islands with less surface water and lower annual rainfall distributed unevenly, there is no conservation ethic regarding fresh water on Pohnpei.

Archaeological surveys and oral histories indicate that favored prehistoric habitation sites were along narrow alluvial plains bordering rivers. The traditional settlement pattern was one of dispersed homesteads amongst garden areas (Streck, 1985). Residential siting along rivers implies a preference for performing key domestic activities near running water. In turn this suggests that people were not used to transporting fresh water over great distances but rather made use of it where it occurs naturally most abundantly. Recent fieldwork on Pohnpei documented no indigenous technology for storage or transport of quantities of water larger than about one quart. No rainfall catchment was practiced either, prior to the modern era.

Pohnpeians living in the Kolonia say they used to locate their homes near streams, as is indicated by the archaeological record. They cite frequent bathing, preferably three times a day, in streams, pools and water falls as one reason. Pohnpeians state that they do not feel clean unless the water flows abundantly on their bodies during bathing.

They usually remove the shower head in modern facilities, to avoid the fine spray. Another reason is to be near water for cooking, food preparation, and cleanup.

In Kolonia households, the outdoor water faucet is often left open. During much of the day no water flows through the faucet but during water hours it flows freely as waste after filling nearby containers. Some Pohnpeians say the faucet is left open in order to keep storage containers full; others see no reason to turn it off because 'there is so much water. The latter attitude apparently follows from ignorance of the design and capability of the Kolonia water system as well as from the absence of a need for conservation. As will be shown in the historical review below, since European contact and throughout the Colonial era to the present, there has been no pressure for a change in Pohnpeian water use habits. The Pohnpeian experience in Kolonia itself, until some years into the American period, has been one of exclusion and non-participation in the town's affairs, and municipal services existed only for the benefit of the town's foreign residents.

Water Use during the Spanish and German Administration

Europeans penetrated Micronesia as early as the 16th century but only established colonies during the late 19th century. The successive Spanish (1885-1899) and German (1899-1914) administrations in the Carolines invested minimally in the physical and monetary support of their Pohnpeian settlements. Apparently neither power built a water distribution system. According to available documents, the early European settlements consisted of a few buildings and a wharf area. Probably wells and nearby streams were used for drinking water; sanitation was undoubtedly minimal. These arrangements had no effect on the Pohnpeians, who continued to live elsewhere.

The Spanish settlement, called "La Colonia," was fortified and surrounded by a high stone wall. The governor rarely left the safety of the enclosure. The German colony ("Die Kolonie"), which was also limited to non-Pohnpeian residents, included the governor and his staff, a Melanesian military garrison, and a Japanese trading company. Upon their takeover of Pohnpei, the Germans immediately destroyed most of the Spanish wall and obliterated other vestiges of Spanish rule. They expanded the area of colonial occupation for the purposes of commerce. Much private land was expropriated by the government. Presumably Kolonia's water continued to be supplied by streams and wells. After a devastating typhoon in 1905, extensive repairs and roadbuilding within and outside the town, using forced labor, were undertaken.

Water Use during the Japanese Administration

During the 1920s and 1930s the Japanese colony on Pohnpei developed into a major center for trade and industry in the Eastern Carolines (Yanaihara, 1940). Like Belau and the Northern Marianas, Pohnpei was a major destination for thousands of immigrants from overcrowded Japan, Okinawa, and the Bonin Islands. The government attempted to create and maintain favorable conditions for the immigrants and for the commercial development of Pohnpei. The area under direct government control increased considerably over that of the earlier regimes. It included an airfield, a hospital, an agricultural experiment station and plantations, and outlying farming communities linked to the capital by roads, in addition to many small restaurants, hotels, shops, factories, and houses in Kolonia. The town had telephones and electricity generated by a small hydropower plant along the Nanpil River.

As before, the Pohnpeians were excluded from residence in bustling "Korronia" although some of them worked in Japanese-owned enterprises. Other Micronesians, such as outer islanders from Nukuoro and Pingelap, worked for the Japanese in Pohnpei and were allowed to settle in outlying parts of Kolonia. The Micronesians were assigned the lowest social status whereas native Japanese were at the apex of the rigidly hierarchical social system .

The Japanese built the first modern system of fresh water delivery on the island, centered on Kolonia. It included small dams and diversions on several streams and rivers; water catchment facilities at the airfield; a small treatment plant with raw water storage capacity, sand filters, and clear wells; several concrete storage tanks and an earth-covered storage facility; and a transmission network of mainly galvanized iron pipes which provided running water and sewerage (Barrett, Harris and Associates, Inc., 1983). Apparently the sewer system served only the industrial sector of the town.

Water Use during the American Administration

The American bombings of 1944 destroyed much of what the Japanese had built on Pohnpei. After the surrender, the American military occupation forces were poorly housed and equipped, and the restoration of Kolonia was slow and inefficient. In marked contrast to the previous foreign administrations, under the Americans Pohnpeians have been permitted to live, as well as to work, in Kolonia. Their residing in town was not common, however, until the 1960s, when government employment became a major source of income for local families.

Although civilian administration under the United Nations Trust Territory government was established by 1951, it was not until two decades later that an expanded and modernized water distribution system was built. Some of the Japanese facilities were integrated into the new system but neglect during the early post-war years precluded extensive reuse.

During the 1970s Kolonia became the seat of the national government and the capital of Pohnpei State. The population has increased tremendously in the last decade, straining the relatively small water system. The average use of water per person is 100 gallons a day (Khosrowpanah field observations, 1986).

Recently the local government has instituted a public education program aimed at making consumers more aware of the causes of the water shortage in Kolonia and encouraging water conservation.

Currently the monthly charge for water is a flat rate of $1.50 per household. Billing and collection is far from efficient.

Political Impacts on the Water System

After a hundred years of colonial history, of native non-involvement and exclusion based on social distinctions imposed by outsiders, Pohnpeians now receive and are responsible for municipal services in Kolonia. The local government, in trying to solve the problems associated with the water distribution system, is faced with consumer attitudes and behavior that have not significantly changed since the advent of the Europeans. A marked passivity and lack of involvement persist, in addition to a traditional cultural preference for abundantly flowing water near households.

During the Spanish and German administrations, the native population rapidly declined from European-introduced infectious diseases against which they had no immunity. However, settlement practices in most of Pohnpei remained relatively unchanged from precontact times until very recently. The Spanish did little to encourage commercial activities on the island, remaining primarily a military presence, whereas the German government actively promoted trade. Pohnpei was to provide raw materials, especially copra, to German industry and in turn to furnish a market for German products (Fletcher, 1920). The use of forced labor, land expropriation and assignment of individual titles, all tended to diminish the power of the chiefs. Government actions removed much land from the Pohnpeian land tenure system, and in Kolonia particularly, traditional attitudes of responsibility toward and personal involvement in community affairs never developed.

The political relations between Pohnpeians and the Japanese were characterized by a hierarchical social system. The Japanese assumed racial and cultural superiority over the Micronesians, and this attitude affected the quality of services provided to the native population. There was minimal investment by the Japanese government in directly improving the living conditions of the Pohnpeians, whereas the Japanese and Okinawan immigrants, particularly those living in Kolonia, enjoyed a material prosperity not seen in the island before. Pohnpeians participated in the colonial economy as low-wage laborers and as consumers with limited buying power. As under the Spanish and Germans, Pohnpeians' non-involvement in the affairs of the town continued.

Since the 1960s Pohnpeians have been encouraged to participate in American style democracy. As universal education was seen as essential to this form of government, the public education system on Pohnpei was extended to the junior college level. Generous financial assistance was made available to hundreds of post-secondary students for study in U.S. colleges. Returning to Pohnpei after U.S. schooling, these students have occupied elected office, government administrative positions, and private sector jobs.

Significantly, the emphasis on education for self-government did not include technical training to monitor and service the modern technological apparatus which has accompanied the American presence on Pohnpei. There has been virtually no Pohnpeian technical expertise, and no significant local participation in the planning, design, and implementation of major engineering projects, such as the Kolonia water distribution system.

Summary and Discussion

The inadequacy of Kolonia's water distribution system can be attributed to both technical and non-technical factors. The focus in this paper has been on non-technical factors such as inappropriate consumer attitudes and behavior toward the water system. These appear to derive from traditional Pohnpeian culture as well as from continuing colonial experiences since the last century.

Pohnpeian cultural traditions do not include a water conservation ethic, and there has been no pressure for change in this regard throughout the islanders' colonial history. Consumer attitudes toward the water system are characterized by a lack of personal involvement and even contempt due to its poor performance, especially in light of the abundant surface water on Pohnpei. Piped water is available only a few hours per day and the quality of the water is low.

At first excluded from Kolonia, Pohnpeians now live in the town. Many consumers let their outdoor water taps run continuously during water hours, whether containers are full or not, like a small stream next to the house. While living in town is far different from the traditional residence pattern, the cultural preference for nearby abundantly flowing water, as well as an ignorance of the capacity and design of the piped water system, are reflected in this practice.

Pohnpeians seem not to realize the consequences for the water system of leaving the tap open. Most do not know how the pipe system works and why there is not enough pressure in the pipes. Some people believe that when rainfall is heavy there is more water available to the system. Thus they are puzzled and angered when water hours remain unchanged regardless of rainfall. The lack of conservation is reinforced by the fact that the cost of water to the consumer is unrelated to the amount used, and there is no adverse result if the bill is not paid.

Adequate water delivery in Kolonia is seen by the Pohnpeians as the government's responsibility, not that of individual householders. This attitude, which separates the interests of the government and those of individuals, follows from colonial policies which prevented Pohnpeians from participating in the affairs of the town. These policies broke the traditonal ties to the land, which normally ensure close personal involvement in its management. Now, when a household's or neighborhood's water pipe breaks or starts to leak, it may go unreported for long periods, and no measures are taken by consumers to remedy the situation even temporarily. The government's response time is often longer than it should be, reinforcing the negative attitudes of consumers toward the water system and reflecting a technical inability to cope with or prevent such system breakdowns.

The colonial history of Pohnpei is one of attempted economic and political gain by occupying powers. The first three foreign regimes largely ignored Pohnpeian needs, focusing on commercial exploitation of the island's resources. During this time local population decline and massive immigration from outside combined to weaken traditional social and political structures. It has been argued (McHenry, 1975) that since the mid-1960s the American government deliberately cultivated economic dependency on the part of the native population through acculturative education and technological enhancements such as the modern water distribution system (see also Baron, 1973; Peterson, 1977).

The recent history of the Kolonia water distribution system may exemplify the American technique of creating dependency. The price of water paid by consumers is heavily subsidized by the Pohnpei government (Juan C. Tenorio, Inc., 1980), which in turn is solely financed by grants from the U.S. Treasury. Under this system of management, the people of Pohnpei can use as much water as they want.

Conclusions and Recommendations

Improvement of the Kolonia water system will come from technical upgrading, changes in system management, and changes in consumer attitudes and behavior toward the system. With technical and managerial improvements resulting in higher quality and more water available per day, Pohnpeian consumers should likely be more responsive to incentives for changing their water use habits and attitudes.

Given technical and managerial improvements of the system, a primary recommendation is that the low flat rate schedule be changed to one based on actual usage. The new rate schedule need not reflect the true cost of water delivery, yet it could be designed to reward significant reductions in household usage. Rewards could include lower rates or credits but need not be limited to these. The efficiency of fee collection should be increased as well. Household metering is essential.

To counteract public ignorance, the basic principles of the water system could be taught in the elementary and secondary schools as part of the science curriculum. The same information could be disseminated throughout the adult communities in Kolonia via community meetings and workshops, posters, direct mailings, and the radio.

Improvements in the Kolonia water system also require the strong government support. Leadership in the Department of Public Works is important, as is adequate funding for technical upgrading that includes physical improvements and repairs to the system, more effective operating procedures, and better trained Pohnpeian technicians. These measures can create the conditions under which rate structure incentives and a public education program can induce desirable changes in attitudes and behavior toward the water system.

References

1. Baron, Dona G. "Policy for Paradise." Ph.D. dissertation, Columbia University, New York, 1973.

2. Barrett, Harris, and Associates. "Leak Detection Services, Kolonia Water System, State of Ponape, Federated States of Micronesia." Report to Pohnpei State government, Honolulu, 1983.

3. Fletcher, C.B. "Stevenson's Germany, the Case against Germany in the Pacific." W. Heinemann, London, 1920.

4. Hunter-Anderson, Rosalind L. "Indigenous Fresh Water Management Technologies of Truk, Pohnpei, and Kosrae, Eastern Caroline Islands, and of Guam, Mariana Islands, Micronesia." Water and Energy Research Institute Technical Report, University of Guam, Guam, in preparation.

5. Juan C. Tenorio and Associates. "Water Resources Study Ponape, E. Caroline Islands." Tamuning, Guam, 1980.

6. Khosrowpanah, S., and L. Heitz. "Appropriate Design of Water Systems in Micronesia." Paper presented at ASCE International Conference on Resource Mobilization for Drinking Water Supply and Sanitation in Developing Nations, Sanjuan, Puerto Rico, May 1987.

7. McHenry, D. "Trust Betrayed." Carnegie Endowment for International Peace, New York, 1975.

8. NOAA. "Local Climatological Data, Annual Summary with Comparative Data, Ponape Island, Pacific." Asheville, NC., 1975.

9. Peterson, G. "Ponapean Agriculture and Economy: Politics Prestige, and Problems of Commercialization in the Eastern Caroline Islands." Ph.D. dissertation, Columbia University, New York, 1977.

10. Streck, C. "Intensive Archaeological Site Survey for the Proposed Nanpil River Hydropower Project, Net Municipality, Pohnpei, Federated States of Micronesia." Report to the US Army Corps of Engineers, Pacific Ocean Division, Planning Branch, Environmental Resources Section, Ft. Shafter, Hawaii, 1985.

11. Yanaihara, T. "Pacific Islands under Japanese Mandate." Oxford University Press, London, 1940.

Strategic Planning In Water Supply
and Sanitation Sector Program Design

David Laredo*

1.0 Introduction

Private sector corporations have widely applied various forms of strategic planning (SP) as input to their decision making processes for the past 30 years, or so. Its application to public sector activities has been growing steadily over the past decade. This paper will briefly discuss the components of SP, and how they actually provide a process, the application of the process to the water and sanitation sector or program planning.

2.0 The Strategic Planning Process

Strategic planning is a long range comprehensive process which requires organizations to sharply define their purpose and direction related to the future generation of their products or services recognizing that market environments will be ever-changing. Policies, strategies and implementation requirements are thus designed to meet future conditions. A strategic plan can thus be viewed as an organization's process to position itself favorably to respond to a wide range of future market conditions. This favorable positioning should allow the capture of a desired market share, as the organization will be able to adapt to fluctuating external forces that significantly affect its market.

Successful SP thus allows shifts in the modes of operations against a wide range of "possible futures". As such SP is a systematic process used to generate key decisions relating to an organization's future vitality, and provides a unifying framework, over a given time period, for all other decisions to be made.

2.1 Elements in the Process

The SP process ordinarily requires continuous development of various combinations of the following activities:

- Develop Mission Statements -- establish guidelines for, and to direct strategy development
- Environmental Scan -- determine key trends and assess other external forces which will affect the organization's performance
- Key Issue Determination -- from the environmental scan, identify the critical issues affecting future performance

* Associate, Camp Dresser & McKee International, One Center Plaza, Boston, Massachusetts 02108

- o <u>External/Internal Analyses</u> -- identify the outside forces affecting the achievement of goals, and the organization's internal strengths, weaknesses and overall resources which will affect the goals

- o <u>Development of Goals and Objectives</u> -- determine what must be achieved, and how, when and where it can be accomplished

- o <u>Develop Action Plan for Implementation</u> -- formulate specific strategies, indicating a responsible group or individual, the resources to be employed and schedule for completion

- o <u>Implement Plan</u> -- perform the tasks required to implement plan

- o <u>Monitor Results and Update Plan</u> -- establish system to examine results and a mechanism to update and/or revise the plan based upon the results and lessons learned

The relationship of these activities in the strategic planning continuum is shown in Figure 1.

FIGURE 1 - STRATEGIC PLANNING PROCESS CONTINUUM

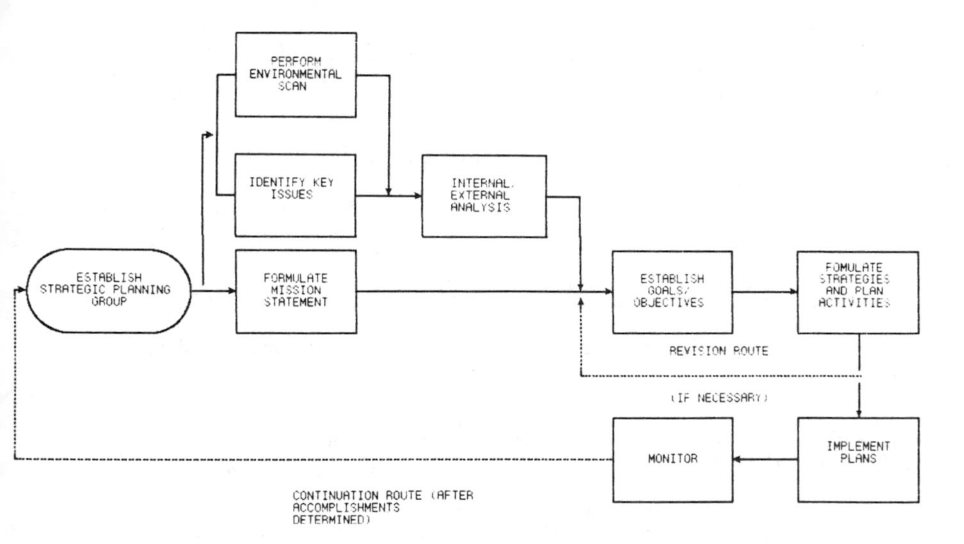

The key points to recognize about SP are:

a) it is a continuous process, with long term horizons, forcing the organization to continuously focus on how it will operate in the future

b) the process produces general long term strategies to reach a set of goals but recognizes the need to measure the long term by meeting intermediate objectives

c) the process inherently recognizes changing market conditions and forces strategies which will allow reactions across a wide range of alternative future conditions

3.0 Application to Water Supply and Sanitation Programs: Why is it Necessary?

The discussion in the previous section should not be misinterpreted. SP is not some high level, esoteric and difficult to apply process. Broken down into its barest components, a successful SP process provides an organization with a framework to analyze where it is in terms of its mission, where it should aim to be, and how to go about getting there.

The increased use of SP in water supply and sanitation sectors is understandable when SP is compared to the methods ordinarily utilized - that of "comprehensive master planning". Master plans are ordinarily directed towards determining, for a given point in the future, a particular level of service to be provided to a projected service population base. Sometimes intermediate goals/objectives are determined together with projections of human reneeds, and financial plans are usually prepared.

The basic problem with this approach is conceptual in that the planning outputs superimpose fixed targets generated from a relatively static plan onto a dynamic situation. There is little flexibility in the process which allows the plan to be updated - save preparing a new plan.

SP on the other hand is designed as a continuum, thus allowing updates as a matter of course. Further, by requiring a detailed examination of the key external factors which may constrain programs, the SP process almost forces the planners to consider the risks to program implementation including:

- the possibility of changing long range national planning priorities,
- changes in the basic desires of the service populations, leading to the political pressure which often leads to loss of control of programs,
- the effects on overall sector programming and implementation of the often competing demands of external donors and/or lending agencies,
- the linkage and coordination requirements vis-a-vis other sector or development programs, and/or ministries or organizations, and the reaction of these organizational hierarchies to the programs and their implementation requirements.

Too often, planning efforts have failed to include sufficient local input into their analyses (fortunately this is changing). Further, many plans have failed to include the possible effects on implementation caused by the enormous demands placed upon key managers by the almost daily operational problems and critical situations. These tend to sap enormous time and energy from the very managers who are responsible for implementing the programs.

SP is not a panacea, and adoption of the process will not insure problem free implementation. SP will provide a process, which has as its foundation a framework for program design which fosters a wide-ranging and integrated approach, and uses as a basis of plan formulation the inputs of the individuals and/or groups whose activities will directly affect the success of the plan. Further, the process methodology requires that programs be continuously evaluated and revised as required by changing conditions.

An example of how the process may be applied is discussed in the following sections.

4.0 Application of the Process

4.1 General

Basic to the application of SP to water supply and sanitation sector programming is the recognition of how the functional elements of the process relate to one another. These are graphically shown below in Figure 2.

FIGURE 2 - FUNCTIONAL PLANNING - IMPLEMENTATION HIERARCHY

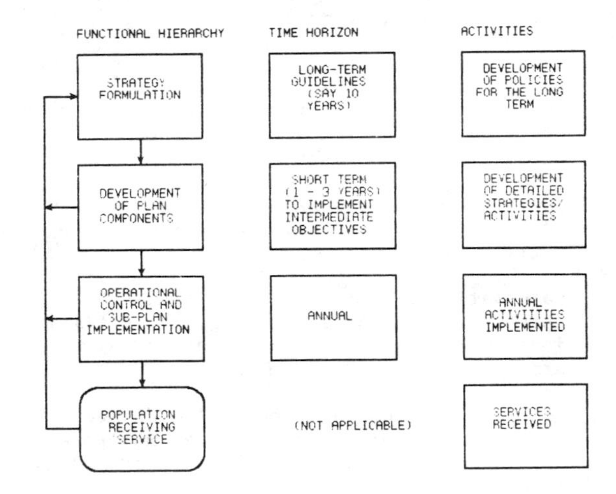

Generally speaking, the strategy formulation and development of plan components form the core of the SP process. However, the feedback mechanisms at all functional levels (see Figure 2) add to the process.

The process ordinarily starts with the establishment of a SP group or committee to oversee plan formulation and implementation. Its composition should encompass as wide a range of individuals as possible without becoming unwieldy. Members should include: high level managers even at the ministry level, and all or many of the managers who will be directly responsible for carrying out significant program components. In the water and sanitation sector, the group would probably be interministrial since urban and rural services rarely are the responsibility of one ministry. The process would be greatly enhanced if representatives of other ministries such as the Ministry of Finance, and the government's central planning organization are included.

As shown in Figure 1, mission statements, goals and objectives should be formulated for a specified planning horizon with inputs to the establishment of goals and objectives from the environmental scan, the key issue determination and the internal and external analysis. In order to monitor the plan, the objectives (defined herein as short term steps towards the longer range goals) should be formulated in a quantifiable manner.

Since any strategic planning process cannot be expected to produce a long term set of goals and objectives at "one sitting," the group's early sessions would be directed at preliminary formulations, assignment of specific research or analyses tasks to individuals to accomplish under a definite schedule. These tasks provide more definitive data for the formulation of specific goals and objectives.

Goals and objectives would then be further defined through strategy formulation and strategies converted to program components with estimates prepared for the resources required (human and financial).

This portion of the plan could also be handled on a preliminary manner, to be more specifically established in the development of the organization's "formal plan."

(Note that plan formulation is not a static exercise. Thus, if the data gathered as part of the process indicates the preliminary goals or individual objectives are too ambitious, they should be revised to fit the constraints.)

The plan's implementation (with close monitoring of results) is the next step. Results of strategic tasks should be monitored on a yearly or twice yearly basis, and the plan reviewed and revised as required, every two or three years.

4.2 Goals and Objectives - The Core of the Process

The most difficult part of the strategic process will be establishing the plan's objectives. The objectives will be the foundation of all program activities and will act to drive the management controls which measure progress. Objectives must be consistent with the policy framework established, and the resources available or expected. They should specify what is to be done, and when anticipated results should occur.

Thus, setting of objectives must be considered an iterative process. The following discussion illustrates a framework for this sub-process, and relates to Figure 3.

FIGURE 3 - OBJECTIVE SETTING FRAMEWORK

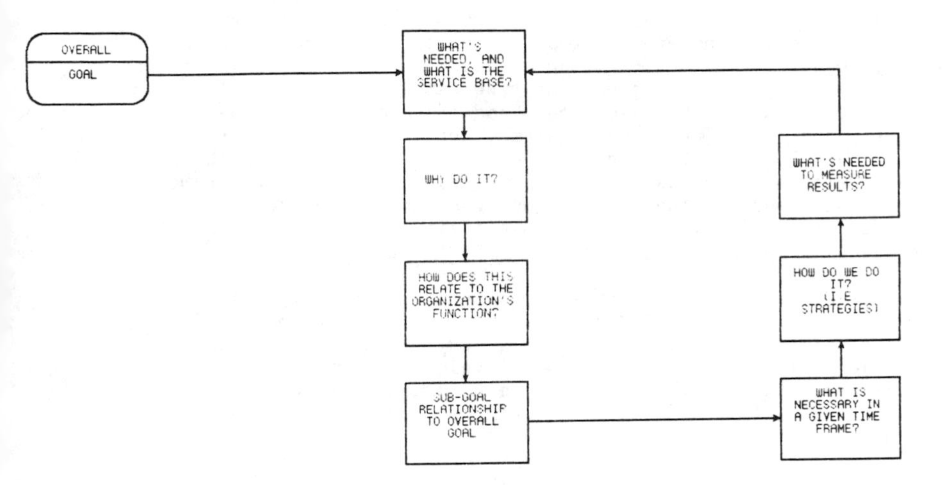

The initial phase of the objectives setting process - the first four items of Figure 3 delineates what's to be done (or what service is to be provided), the reason for doing it, the relationship of the task to the organization and the relationship of the objective to the overall goal.

The next step is to determine what has to be done to achieve the goal (i.e. what the actual objective is) and when the objective must be reached so that overall goal schedule remains valid.

How objectives will be reached pertains to the formulation of strategies - or activities that must be carried out to reach the objective. These include what must be done, its location and the individuals and organizational components responsible and the resources to be employed. The methods of measuring accomplishments are included as the final step in the sub-process.

Once the objective setting sub-process is complete it should be reviewed by going through each step to insure all points have been considered. An important point to keep in mind is that in formulating the objective the exercise may indicate inconsistencies, or may bring to light further points worthy of more detailed analysis not considered when the goal and or other steps in the objective process were formulated. Thus, further analysis may be required.

5.0 Example Application to Rural Water Supply and Sanitation Programming

The following example illustrates how the SP process can be applied to a rural water supply and sanitation (RWS/S) program.

Coincident or very soon after the SP group is formed and general mission statements and goals prepared, an environmental scan should be performed to provide the SP group detailed data. A typical scan for the RWS/S sector would include the type, size, number of villages and persons covered with WS/S services, and the number of villages and persons (and their location) requiring such services to meet the preliminary goals, a brief history of sector activity over the past 5 to 10 years including expenditures made by the national government, the villages and external donors and agencies, the sector's significant institutions, their capabilities and a description of their activities and procedures.

The environmental scan should lead to the identification of key issues. These may include the number of persons per year requiring service to meet the preliminary goals, estimated program costs and possible sources of funds, manpower levels and/or training requirements at both the central government and village level to implement the plan, and auxiliary programs required (i.e. health education, village level O&M training).

The key issues analysis should also indicate the risks involved in implementing various output levels of the plan. Specific risks include: poor funding history, an improper manpower mix in various sector institutions, a fragmented sector with many institutions involved, weak organizations in terms of structure and/or political clout, rivalries amongst organizations inhibiting action and poor absorbtive capacities in the sector.

Sector institutions should be examined in light of the data developed and their strengths and weaknesses, especially those affecting implementation analyzed. (Note that if the environmental scan has been properly prepared, an in depth picture of the sector will emerge, and the strengths and weaknesses of the sector institutions' should be clear and easily comparable to the risks and other key issues.) Preliminary objectives should be formulated and assigned to the institutions which appear to be logical choices as responsible groups to implement the objectives.

The SP group and the designated institution(s) should prepare a set of prelimianry strategies, and the institution(s) should convert the general strategies to operational plans. A prominent feature of the sub-processes for the formulation of strategies and their "conversion" to operating plans is the flexibility provided in these steps to anticipate, and thus mitigate risks through the design of the operational plans. For instance, if institutions responsible for implementation are weak, and/or badly organized, the first strategies should be directed towards improving the institutions. Another example would be the often encountered case where the most signficant risk is indefinate funding availability in the program's early phases.

Operational plans for various funding levels could be prepared for various phases of the program. The affects of curtailed funding could then be recognized and alternative strategy sets devised. Combinations of the alternative strategies could then be used, as required to keep the program on track if the provided funding was less than optimum.

The operational plans should include budget estimates (operating and capital) and human resource requirements necessary for implementation and a preliminary assessment of sources of funds. These should be reviewed by the SP group and the objectives and strategies revised if necessary.

Each revision of the objectives and/or strategies represent an alternative method of reaching the goals established. Thus, even if there appears to be no need for revisions, strategies should be tested by taking advantage of the process ability to accommodate testing a series of "what if" questions. These tests will provide a detailed study of alternatives for various combinations of strategies and sub-components of the plan. Once the plan is finalized, that is, the objectives are set for 2 or 3 years, approval should be obtained at the highest level of government necessary.

6.0 Imperfections and Shortfalls

The best laid plans often go awry. The countryside abounds with programs and plans that have been curtailed or deferred due to critical economic conditions, insurgency, improper coordination, delayed or low levels of external aid and/or lower than expected budget contributions from national treasuries. However, while not a cure-all, the SP process should have considered at least some of these possible constraints. Further, the data developed for the alternatives analysis discussed above should provide the planners with the basis for evaluating various combinations of strategies (or sub-alternatives) to meet future needs and be able to estimate the future requirements necessitated by delayed implementation.

The freebody in which water and sanitation sectors operate has always well illustrated the cliche - "the only certainty is change." The process discussed herein will not prevent changing conditions, however the approach may allow the practitioners in the sector to at least adapt to change more easily.

References:

The following listing provides general references pertinent to strategic planning applications in the public sector:

Olsen, J.B., Eadie, D.C. The Game Plan, Governance with Foresight, 1982, Council of State Planning Agencies, Washington, D.C.

Paul, S., Strategic Management of Development Programmes, 1982, (Management Development Series No. 19) ILO, Geneva, Switzerland.

Steis, A.W., Strategic Management and Organizational Decision Making, 1985, Lexington Books, D.C. Health and Co., Lexington, Massachusetts.

Steis, A.W., Management Control in Government, 1982, Lexington Books, D.C. Health and Co., Lexington, Massachusetts.

TRAINING IN WATER SUPPLY - PAHO'S ASSESSMENT

Horst Otterstetter*

1. INTRODUCTION

In Latin America and the Caribbean, those primarily responsible for water supply and sanitation have been the civil, the sanitary, and the environmental engineers; the professionals and technicians of several other disciplines have also collaborated. In the future, in order to take care of the changing set of problems in the countries, it is most probable that these professionals will continue to be responsible for water supply and sanitation, although there have been structural changes and redefinition of functions in the environmental health programs.

The demand and availability of sanitary, and environmental engineers with respect to present and future requirements is not documented nor well known in the Region of the Americas. Although some research has been carried out in the past, it is incomplete or has not been updated.

Recognizing the importance of promoting the preparation of trained professionals so that they perform the function of improving the health of the population, PAHO has just finished developing a regional directory of all the institutions devoted to the training of sanitary, public health, and environmental engineers. This directory should assist in improving the current situation and balancing the supply and demand of specialized human resources in the future. The need to mobilize new resources must be emphasized. Innovative mechanisms will have to be applied, such as the establishment of networks of institutions for professional training, and information and research to facilitate the optimal use of the available resources.

2. INSTITUTIONAL AND HUMAN RESOURCE DEVELOPMENT

Modern technology has given a new dimension to the responsibilities of the public health authorities because of the breakdown of the ecological balance and the creation of severe tensions resulting from industrial processes, the extensive use of chemical substances, the burning of fuel, the increase in human and animal wastes, and other factors. In order to find the solution to these specific problems, a series of specialized institutions with well-defined mandates have been created throughout the Region: institutions responsible for the drinking water supply, the collection and disposal of wastes, the control of water pollution, the control of

*Regional Adviser, Pan American Health Organization (PAHO), 525 23rd Street, N.W., Washington, D.C. 20037

air pollution, the management of natural resources, the control of endemic diseases, ecological development, etc.

Each one of these institutions has a well-defined institutional objective, corresponding to a mandate conferred by the community. The satisfaction of the very specific needs of the communities depends upon the achievement of these objectives; only so can the existence of these institutions be justified. Society possesses mechanisms that eliminate institutions which do not achieve objectives or guarantee individuals the minimum well-being necessary to fulfill productive commitments.

This viewpoint brings about a completely new situation with respect to the institutional development of the agencies that are devoted to supplying drinking water and the sanitary disposal of excreta.

Institutional development, defined as a planned process of change enabling companies to continuously fulfill the changing needs of the communities they serve, requires that the changing values and culture of the organization, and application of appropriate techniques and development of their human resources are incorporated and implemented.

For many years PAHO has been collaborating with the water and sanitation agencies of Latin America and the Caribbean in their institutional development. In the beginning the collaboration was limited mainly to organization and methods, and has now evolved into comprehensive institutional development with cooperation in the integrated development of the institutions and of their human resources as a key component.

The new developments require:

- an appropriate systemic approach to the programming of the different stages of diagnosis, analysis and formulation of models;

- the application, evaluation and adjustment of the programming;

- the financial resources for their execution and for the implementation of the recommendations and the models developed;

- legal resources;

- an agent of change;

- a methodology; and,

- human resources and mechanisms that guarantee the training of the personnel consonant with the changes in knowledge, experience and attitudes required for institutional development (Gónima A., 1982). Consequently, it is essential to implement manpower planning in the context of the overall planning of the institution. Only manpower planning that is included in the general planning of the institution can define the short and long term demand for personnel, from both the quantitative and qualitative points of view.

"For those effects, once the need to implement changes in the organization and in the employee's work environment is identified and the models of action are formulated through the institutional development mentioned above, it is necessary to prepare a job description for each position in the organization. This job description, compared with the personal profile of the different workers and employees of the organization, will make it possible to identify the needs for knowledge and abilities in relation to the new objectives and goals, technological advances, and changes in the organizational structure, in the functions and managerial activities and in the information systems and decision-making models, etc." (Gónima A., 1982).

The integrated concept of the development of institutions and their human resources can be observed through the analysis of Figures 1 and 2. They present, respectively, the matrix of organizational systems and managerial activities, and the occupational matrix of the same institution. Figure 1 indicates that the company, in order to reach its objectives and goals, requires the systematic execution of a series of organizational functions at four levels of managerial activities. The harmonious performance of the institution depends on the coordinated implementation of the managerial activities as part of the institutional functions and supported by a management information system.

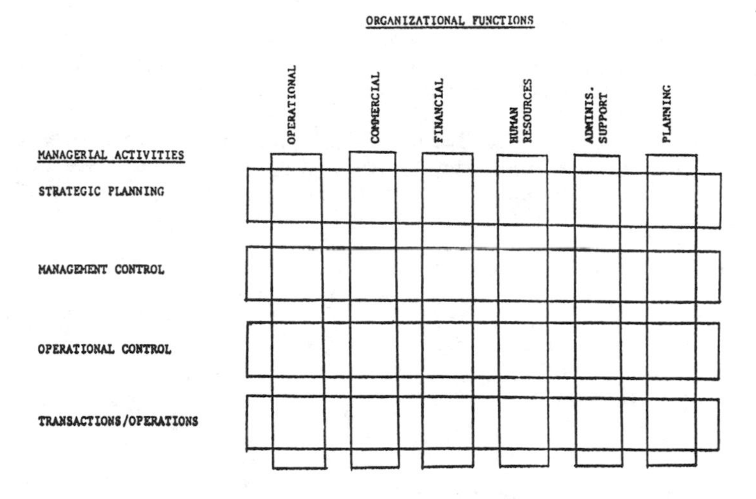

Figure 1. Matrix of Organizational Systems and Managerial Activities

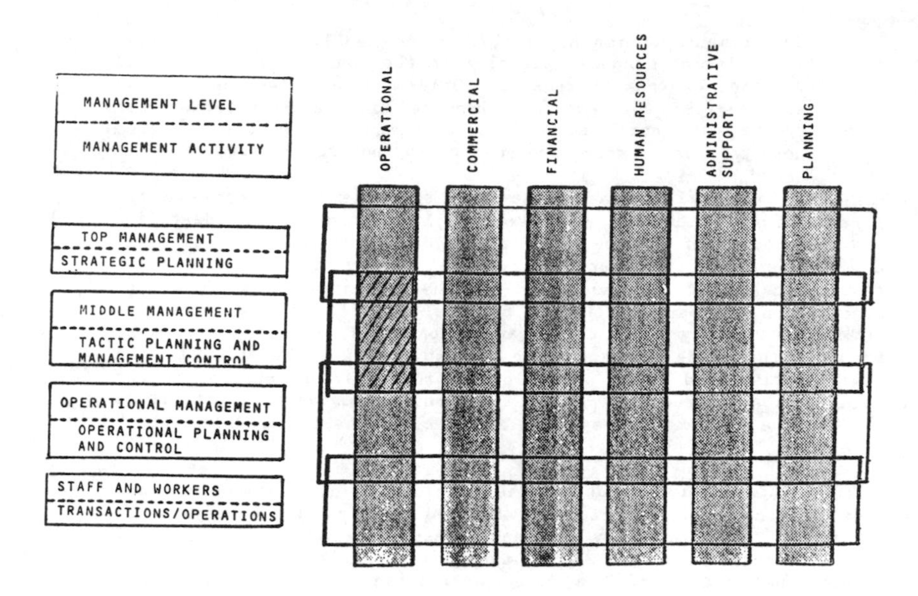

Figure 2. Occupational Matrix
Organizational Functions

Figure 2 shows that the intersection of the managerial activities with the organizational functions makes it possible to identify a series of tasks to be carried out, as well as the corresponding functional relationships, both horizontal and vertical.

This approach makes it possible to optimize the use of the training resources by directly relating the objectives of the training activities to the needs of the institutions.

This concept definitively incorporates human resources development into the institutional functions and rejects the idea that training can be developed as an ad hoc activity, determined by the supply with no attention paid to the institution's actual demand.

Projects for the development of institutions and their human resources are being carried out in a large number of water supply and sanitation organizations in Latin America and the Caribbean. PAHO, with long experience in these types of projects, participates as executing, collaborating, or advisory agency in several of these projects. Among them are:

BOLIVIA

Institutional Development of CORPAGUAS.

BRAZIL

Institutional Development of CAGECE.
National Institutional Development Program (BNH/CEF).
Managerial and Operational Strengthening of COSAN/BNH/CEF.
Studies of Technical Assistance and Institutional Reorganization of the Rural Sanitation Sector (IPEA).

DOMINICAN REPUBLIC

Institutional Development of INAPA.
Institutional Development of CAASD.

ECUADOR

Institutional Development of EMAG.

GUATEMALA

Strengthening of the Drinking Water and Sanitation Sector (COPECAS).

MEXICO

National Program for Control of Losses and Efficient Use of Water (SARH).
State Commission of Public Services of Tijuana-Tecate (CESPT-T).

PARAGUAY

Drinking Water and Rural Sanitation (SENASA).

MULTICOUNTRY PROJECTS

Project for training technical personnel in operation and maintenance - seven countries (Costa Rica, Dominican Republic, El Salvador, Guatemala, Honduras, Nicaragua, and Panama).

Water Management Project in the Caribbean Basin - 11 countries (Anguila, Antigua, Barbados, British Virgin Islands, Dominica, Grenada, Montserrat, St. Kitts and Nevis, Saint Lucia, St. Vincent, Turks and Caicos). Moreover, there are 12 projects in the process of approval and 12 were finished in the last 4 years.

An outstanding example is the institutional develop- ment of the water and sanitation companies in Brazil. A plan that was initiated in 1973 with two components, (institutional develop- ment -SATECIA and human resources development -SANAT-), with little coordination between them, was consolidated in 1979 into a single, integrated, institutional and human resource development project (PRODISAN).

This project provided an excellent opportunity for technical cooperation among Brazilian regions (TCDR). Pilot companies developed a given organizational function in depth and transfered it to other companies in the country. The strengthening of the water and sanitation sector in Brazil through this project is a model for the developing countries. As a consequence, a large number of professionals in the water and sanitation companies have been called to serve as consultants and instructors in several of the institutional and human resource development projects in which PAHO participates in other countries. This is a special type of technical cooperation among developing countries (TCDC).

3. SUPPLY AND DEMAND OF TECHNICAL PERSONNEL (PLANNING OF HUMAN RESOURCES)

Peter Drucker (1966) defined manpower planning as "the process through which maintenance of equilibrium between the demand for the personnel necessary for the different services in the institution and the real or probable supply of this personnel in the long, medium and short terms is sought."

As previously mentioned, manpower planning is done in the context of overall institutional planning, therefore it does not have an independent existence. Unfortunately, manpower planning has been, in general, an infrequent practice in the water and sanitation institutions in the Region. In addition, there are the political, social and economic instabilities through which most of the countries of the Region pass, and the result is that it is impossible to plan with any assurance for the human resources necessary for the water and sanitation sector in the medium and long terms.

Currently several national and regional efforts for the systematic acquisition of data on personnel working in the environmental health sector are being carried out. These efforts are recent and are limited to the field of water supply and sanitation and include the processes of monitoring and evaluating the International Decade of Water Supply and Sanitation were created. Table 1, presents an analysis of the data collected. For example in 1983 approximately 250,000 employees were responsible for the water supply for a population of approximately 240 million people (191 million urban and 49 million rural).

It is possible to determine and monitor basic indicators for the planning process only to the extent that historical data on the personnel in the institutions, and the corresponding numbers of persons receiving water and sanitary services, are obtained. Today it is the lack of data on the supply and demand for personnel that impedes the proper planning for human resources in the water supply and sanitation area. Several activities need to be done to change the present situation. Fortunately, some of these activities are already being initiated.

TABLE 1. PERSONNEL RESOURCES IN THE WATER AND SANITATION INSTITUTIONS

COUNTRY OR AREA	TOTAL NUMBER OF PERSONNEL
ANGUILA	11
ANTIGUA	124
ARGENTINA	31,775
BAHAMAS	273
BARBADOS	866
BOLIVIA	1,945
BRAZIL	78,316
CHILE	6,204
COLOMBIA	15,100
COSTA RICA	990
DOMINICA	130
DOMINICAN REPUBLIC	2,771
EL SALVADOR	3,885
GRENADA	152
GUATEMALA	3,915
GUYANA	670
HAITI	540
HONDURAS	1,587
MEXICO	62,000
MONTSERRAT	27
NICARAGUA	2,847
PANAMA	2,281
PARAGUAY	1,997
PERU	7,543
ST. KITTS AND NEVIS	92
SAINT LUCIA	282
ST. VINCENT	104
SURINAME	791
TRINIDAD AND TOBAGO	2,119
URUGUAY	3,928
VENEZUELA	15,418
VIRGIN ISLANDS	29
TOTAL:	256,707

SOURCES: Monitoring by IDWSSD, 1983;, CABES, 1983; Training Project GTZ/PAHO/CAPRE, 1984; CBNM Project - Yearly Manpower Inventory, 1981

The process of institutional development, which several institutions in different countries are undergoing, will cause manpower planning to be incorporated systematically into managerial activities. Moreover, the monitoring of the IDWSSD will systematize the collection of data at the regional level. The inventory of sanitary and environmental engineering education, to be incorporated in a regional directory, will make it possible to verify the supply of professionals on a permanent basis. The following schools, dedicated to training sanitary or environmental engineers at undergraduate and graduate levels, have been identified and inventoried:

ARGENTINA

Univ. Nacional de Buenos Aires

BRAZIL

Escola Engenharia de Sao Carlos, USP
Escola Politecnica - USP
Faculdade Saude Publica - USP
Universid. Fed. do Rio Grande do Sul
Universidade Federal da Paraiba
Universidade Federal de Minas Gerais
Fundacao Oswaldo Cruz
Universidade Federal do Ceara
Universidade Federal da Bahia
Universidade Federal do Mato Grosso
Universidade Federal do Para
Univ. Federal de Santa Catarina
Escola de Engenharia Maua
Inst. de Tecnologia da Amazonia
Pontificia Univ. Cat. de Campinas
Inst. Politéc. de Ribeirao Preto
Universidade Estadual de Campinas

CENTRAL AMERICA

Escuela Regional de Ingeniería Sanitaria, ERIS, Guatemala

COLOMBIA

Univ. del Valle, Cali
Univ. de Antioquia, Medellín
Univ. Nacional, Bogotá
Univ. P. Bolivariana de Medellín
Corporación Universitaria de Boyacá, Tunja
Univ. de la Salle, Bogotá
Univ. Javeriana, Bogotá
Univ. Distrital, Bogotá

ECUADOR

Esc. Politécnica Nacional, Quito

MEXICO

Univ. Nacional Autónoma de México
Inst. Politécnico Nacional
Univ. Autónoma Metropolitana
Univ. Autónoma de Nuevo León
Inst. Tecnológico y de Estudios Superiores, Monterrey
Univ. Autónoma de Yucatán

PERU

Univ. Nacional de Ancash, Huaraz
Universidad Nacional de Ingeniería, Lima

A directory, prepared by the American Association of Environmental Engineering Professors (AAEEP), indicates that there are 112 schools in the United States dedicated to teaching sanitary and environmental engineering at the graduate level.

Institutes providing continuing education, also make an important contribution to the training of professionals and technicians. In Latin America continuing education not only update knowledge but also makes up for training deficiencies that stem from the regular courses in sanitary and environmental engineering.

PAHO prepared a Directory of Training Programs for Latin America and the Caribbean in 1980. The Directory lists the following institutions offering short courses on varied subjects in the field of environmental health, including water supply and sanitation.

ARGENTINA

Comisión Nacional para Control de la Contaminación de Recursos Hídricos
Dirección de Saneamiento Ambiental, Secretaría de Estado S. Pública
Subsecretaría de Salud Pública Area de Recursos Sanitarios
Universidad Nacional de Buenos Aires
Universidad Nacional de Córdoba
Universidad Nacional del Litoral

BARBADOS

Barbados Community College

BRASIL

Companhia Tecnologia Saneamento Ambiental, CETESB
Fundacao Estadual Eng. do Meio Ambiente, FEEMA

CHILE

Universidad de Chile, Facultad de Medicina

COLOMBIA

Centro de Educación en Adminis. Salud, CEADS
Escuela Nacional de Salud Pública Universidad de Antioquía
Facultad de Ingeniería, Univ. de Antioquía

COSTA RICA

Escuela de Medicina Univ. de Costa Rica

GUATEMALA

Esc. Regional de Ingeniería Sanitaria, ERIS

JAMAICA

College of Arts, Science and Technology, CAST
West Indies School of Public Health

MEXICO

Univ. Autónoma de Nuevo León
Univ. Nacional Autónoma de México

PERU

Centro Panamericano de Ing. Sanitaria y Ciencias del Ambiente - CEPIS
Escuela S. Pública de Perú

VENEZUELA

Centro Interamericano de Desarrollo Integral de Aguas/Tierras - CIDIAT
Escuela de Malariología y Saneamiento Ambiental
Univ. Central de Venezuela

The civil engineering schools are another important segment of the educational system. In Latin America and the Caribbean approximately 200 of these schools teach sanitary engineering during the fourth and fifth years of undergraduate studies. Civil engineers who have specialized in sanitary engineering and traditionally worked in water supply and sanitation, far outnumber the sanitary and environmental engineers in the Region.

The associations of professionals in the field of sanitary and environmental engineering are continuously enlarging their role in updating their membership's knowledge. Two associations that have taken leading roles are ABES in Brazil and ACODAL in Colombia.

Another important effort that merits mention is the Training Network for Low Cost Sanitation Technologies under the coordination of the World Bank. In Latin America and the Caribbean specifically PAHO and the World Bank are joining their resources and efforts to help establish a strategically located network of centers that will disseminate the concepts and theories of low cost technologies to all the countries of the Region.

4. TECHNICAL COOPERATION AMONG INSTITUTIONS AND COUNTRIES

The three areas: institutional development, human resources development, and university education, are very appropriate for technical cooperation and technology transfer projects and activities. Since their beginning, the sanitary engineering schools of Latin America have benefited from basic collaboration agreements. ERIS of Guatemala and UNI of Peru have maintained ties with the School of Public Health of North Carolina. ERIS also had a relationship with the University of Lausanne, Switzerland. The University of Texas in Austin collaborates with the School of Public Health of Sao Paulo. Currently the Catholic University of Parana in Brazil has a cooperative agreement with UNIVALLE and the University of Boyaca in Colombia. The National Polytechnic School of Ecuador has signed an agreement with Clemson University in South Carolina.

The Inter-American Association of Sanitary and Environmental Engineering - AIDIS - is in a good position for promoting technical cooperation among countries in the Region. The special characteristics of AIDIS as an Inter-American Association with chapters in all the countries and the immense professional experience of its members enables it to transfer any type of knowledge, at any time, to any country or institution that desires so. This transfer of knowledge and experience also should promote the production and dissemination of a technical bibliography specifically oriented to the reality of Latin America and the Caribbean.

APPENDIX

1. DRUCKER, P. The Effective Executive. London, William Heinemann Ltd., 1966.

2. GONIMA, A. El adiestramiento como función institucional. In PAHO, "Educación médica y salud" 18 (3) 301-340, 1982.

Training of Workforce:
The Puerto Rico Experience

Ventura Bengoechea, P.E.*

Abstract

The Puerto Rico Water Pollution Control Association sponsored the creation of an operators' training group to benefit from the publications and technical information from the Water Pollution Control Federation and other organizations. The task was not easy due to language barriers and other difficulties which were overcome by a group of highly motivated operators and technicians who, voluntarily, shared their time and efforts. This experience could be directly utilized by other spanish speaking communities, or indirectly serve as a guide to other groups that want to create similar organizations.

Introduction

Training of operators is not always found at the top of the list of priorities of water and wastewater utilities, as it should. In order to emphasize the importance of training, I always tell both, operators and managers the following:

> **I HAVE SEEN PROBLEMATIC PLANTS WORKING AMAZINGLY WELL, THANKS TO GOOD OPERATORS, BUT I'VE NEVER SEEN A WELL DESIGNED PLANT WORKING WELL IF IT HAPPENED TO FALL UNDER A LOUSY OPERATOR.**

Being a firm believer of operators' training, I became interested in creating an operators' training organization under our local chapters of the American Water Works Association and the Puerto Rico Water Pollution Control Association. Our initial efforts were encouraged with the creation of the Professional Wastewater Operations Division (PWOD) of the WPCF. However, the initial steps were not easy and I would like to share with you some of the problems we faced.

---

* Senior Project Engineer, Consoer Townsend Harris Int'l, P.O. Box 7066, San Juan, Puerto Rico 00916

First of all, Puerto Rico has some unusual characteristics due to being at the crossroads of two worlds, both geographically and culturally. Geographically speaking, Puerto Rico has a priviledged position being at the focal point of the United States, Central and South America, and next to its sister islands of the Caribbean. Culturally speaking, it is at a crossroad too between the hispanic culture that runs through the veins of their people, and the american education that enhances their technical preparation.

But all these glamorous attributes have some setbacks too, especially when we refer to the subject that interests us today of human resourses at the operators' level.

Among the cultural problems of importing training programs from the States, is the language barrier and difficulties arising from the different idiosyncrasies and technical levels between U.S.A. and Puertorican operators.

But not all of the problems are imported, the main problem we faced when trying to implement the program for the operators came from local people. There was a group of dedicated engineers who, with their best intentions, vehemently oposed the creation of any organization for operators because they thought it was a waste of time and money.

In other words, the idea of having an organization for training operators on a voluntary basis received an initial thumbs down from the Board of Directors of the Water Pollution Control Association (WPCA). Since the PWOD allowed the creation of the local division through either the local WPCA or a parallel association, I threatened them with creating such an organization. At the end, to make the story short and concentrate on the other problems we faced, the organization was created against the odds, through the local WPCA, and has now about 100 voluntary members.

The initial main objectives of the organization were to hold technical seminars at different wastewater treatment facilities throughout the island and to publish a periodical bulletin with technical information. In order to pursue these goals, a group of people from both PRASA (the local water and wastewater utility) and the industrial sector volunteered their time and efforts. First of all we had to recruit a minimum of 20 operator members to start our program.

The task was fustrating at times because of the questions posed by various union leaders, and also because some discouraging responses from operators who claimed that: "they would only join the organization if the meetings would be held during working hours and special allowances paid for attendance". Obviously, this was not the spirit intended and to counteract this group, it was made clear that:

- The activities would be held on Saturdays.
- The members would have to pay some annual dues.
- The employer would not know whether they attended the seminars or not.

The response was very positive, although the dues appeared to be higher than we have expected (because of the annual fees of the WPCF), the operators overwhelmingly responded to a questionnaire saying that the dues were more than reasonable for the amount of information received. This unexpected response made us very pleased. But the main force behind the birth of this organization was the MOTIVATION of both the operators and the group of volunteers at the Committee level. I would like to emphasize the word motivation because when talking about technology transfer, we encountered barriers such as:

1. Lack of technical information for operators
2. Difficult-to-understand information.
3. Language barrier.

But all these barriers could be overcome with the motivation for producing and receiving the information.

How did we motivate the operators?

First of all, by making the organization meetings on Saturdays and having them pay some of their dues; by doing this, we automatically screened a great number of unmotivated people who otherwise would have joined the organization with no benefit for any one.

Second, the importance of being a good operator was always stressed, not for making them feel good, but because we believe on that. We emphasized too on the public responsibility of an operator: no matter how well designed a facility might be, if can produce a bad product if not properly operated.

Third, we encouraged the participation of the operators as speakers and to bring case histories with their experiences, so that they could be shared with their co-workers at the technical seminars.

Fourth, we encouraged the operators to use their imagination to solve the problems that may occur from deffficient design or maintenance. The problem with these ideas, however, is that they usually die in the same place where they were put to work, without the bene- fit of being shared among all the operators.

In order to motivate the operators to share these ideas, we created the "Genius Contest". If the ideas submitted by the operators are selected, they are given a "Genius" Diploma and a cash prize of $100.00. In addition, a description of their ideas is published in the local bulletin for the benefit of the rest of the operators. There is no established limit for the number of awards that may be given in a year period. In order to provide for reserves for the cash prizes, we requested contributions from local consulting firms, contractors and industries which responded with generosity.

As for the bulletin, we have to thank the WPCF for their cooperation and for allowing us to translate articles from the "Operations Forum" magazines into Spanish, at no cost.

The bulletins are divided in four sections:

1. Wastewater Treatment Processes
   1.1 Activated Sludge
   1.2 Biofiltration
   1.3 Sedimentation
   1.4 Sludge Handling and Disposal
2. Maintenance
3. Safety
4. Operators' Certification

I would like to call your attention to the fourth subject because it was requested to be introduced by the operators. Although certification of operators has not yet been approved in Puerto Rico, due to some opposition from a group of operators, we see that the highly motivated ones, who belong to this organization, are in favor of certification and want to be well prepared.

In summary, we have a large number of highly motivated operators who are thirsty for information, and another group of people, also highly motivated, who voluntarily translate or write technical materials for the operators. These materials are organized by subjects and published in a technical newsletter distributed among the members of the organization. In addition to these publications, seminars are being held at different facilities, both municipal and industrial, where the theories learned can be combined with the technical applications. Each seminar focuses on a particular process utilized at the plant being visited. Emphasis is given to the active participation of the operators as speakers or in the question and answer periods, and by means of their ideas as presented for the "Genius Contest".

Although the operators' organization is moving along very well, there are two thoughts that I want to share with you. First the group of people working at the committee level should be renewed periodically. The reason is not only for distributing the work load, but also to avoid that if anything happens to this group of people, the whole organization does not stop.

The second is that the benefits of the publications prepared in Puerto Rico are limited to the geography of the island when, with minimum additional work, they could be utilized by other spanish speaking communities. Therefore, I encourage you to contact myself or whoever is in charge of the Puerto Rico's PWOD to discuss ways of distributing technical materials.

Operators training is a most important subject and, as it was demonstrated through this case; it does not need to rely on big organizations. Small groups of people can start them anywhere if they have the four "M's" of the magic recipe of Dr. Abel Wolman: Motivation, Manpower, Management and Money, and the latter, believe me, is the least important.

ASSESSMENT OF THE OPERATIONS AND MAINTENANCE COMPONENT OF WATER SUPPLY PROJECTS

James K. Jordan*

INTRODUCTION

Many water supply projects in developing countries fail to function as designed, because provision for effective operations and maintenance (O&M) of the system is not made during the planning stage. One of the primary reasons for giving inadequate consideration to the O&M component is the absence of a methodology for assessing operations and maintenance.

The purpose of this paper is to describe a technique for project planners to use in analyzing the O&M needs of a water project during the design phase and to offer recommendations for correcting deficiencies that are likely to affect the continuing success of the project. The paper first discusses the concept of operations and maintenance, describes the central elements of O&M, and reviews the types of water supply projects that are most often constructed in less developed countries. It then recommends techniques for incorporating O&M assessment into project planning documents. Finally, the O&M assessment guide for one type of water system typically constructed in developing countries is presented.

Description of Terms

Certain terms relating to operations and maintenance and project planning appear regularly in this report Understanding their meaning as they are used in the assessment guides is important to using the guides properly. These terms are as follows:

External Support Agency- A bilateral or multilateral agency that provides funds in the form of either a loan or a grant to the government of an LDC to a PVO or NGO in order for a water supply project to be executed.

Implementing Agency - The agency with the responsibility for carrying out the objectives of the water supply project.

Project Documents - All binding and descriptive documents relating to the identification, design, financing, approval, implementation, and evaluation of the project.

Project Planning Team/Project Planner - The personnel who are responsible for designing the project and preparing the project documents.

Operations - A series of actions carried out by operators to make equipment and systems do the work it is intended to do.

Maintenance - A series of activities carried out to ensure that a piece of machinery or a system is able to do its intended work.

Breakdown Maintenance - Actions taken to either repair or restore equipment or systems to effective operating condition only after the equipment of system fails to operate. No preventive maintenance (PM)

*Camp, Dresser & McKee International, Inc., Arlington, VA

actions are performed on the equipment.

Corrective Maintenance - Actions taken to either repair or restore equipment or systems to effective operating condition. These actions may result from problems discovered during preventive maintenance or as a result of equipment or system failure during operation.

Preventive Maintenance - Actions performed on a regular and scheduled basis to keep equipment or systems operating effectively and to minimize unforeseen failures. These actions consist of inspections and or maintenance tasks.

Minor and Major Maintenance Tasks - For this workbook, minor maintenance tasks are those that are designated as the responsibility of the community. An example for handpump systems is the periodic lubrication of the above-ground parts of the handpump by the village caretaker. Major maintenance tasks are those that are the responsibility of regional maintenance crews.

Steps in Operating and Maintaining Water Supply Systems

In addition to lacking a methodology for assessing O&M, project planners may be inexperienced in developing an effective operations and maintenance program for water supply systems. This situation may lead to difficulty in promoting the need for local governments to support O&M programs. To assist the planner, the following four-step process has been identified that outlines the objectives of an effective O&M program:

1. The first step is to identify the operating standards of the system, that is, how the system is designed to operate. In many cases, the standards result from either a social or a political decision as well as from an engineering one.

2. The second step is to establish procedures that will enable those who have O&M responsibility to determine when maintenance work is needed. This step requires the development of a schedule for preventive maintenance (PM), and criteria for determining when corrective maintenance (CM) is needed, that is, to what extent can the operation of the system deviate from the operational standards before CM should be performed.

3. The third step is to provide the means to measure whether or not the system is functioning as designed.

4. The fourth step is to develop and implement a program that will enable the system to be restored to the standard in a timely manner and at an affordable cost.

To demonstrate how this process works, consider a community that is to be served by groundwater pumped from a well by a diesel-driven centrifugal pump to a storage tank and then distributed by way of a piped network to either standpipes or in-house connections.

For the first step, the water authority decides that water will be provided to the community continuously. This step requires that the pump be operated for a certain number of hours per day, depending on the size of the pump and storage capacity. A second pump is furnished for standby service. An alternate approach could have been to provide water only at selected times during the day.

The second step requires the preparation of a preventive maintenance task schedule and the criteria for performing corrective maintenance. One feasible criterion for scheduling CM is if the pump output drops 10 percent from its original production. This situation would trigger an investigation to determine the cause for the reduced output. The basis for planning and scheduling PM and CM activities is generally field experience and manufacturers' recommendations.

The means to measure the actual output of the pump against the corrective maintenance criteria, step three, could be a flowmeter or a gauge to measure the depth of water in the tank. Note that criteria for other parameters, such as excessive temperature, also need to be established.

The requirements for the first three steps should be identified by the planner of the water system and incorporated into the system design specifications. The fourth step -- the development and implementation of a maintenance program -- is the most difficult to complete successfully and is the reason for the preparation of the O&M assessment guides. For example, if the waterpump is no longer operating within acceptable bounds, it may require the coordination of the finance, tendering, supply, and maintenance sections of the water authority to obtain and install the replacement parts that will restore the pump to the operating standard.

The role of the project planner in the fourth step is to investigate the O&M capabilities of the local government and community. The planner should determine whether a program of operations and maintenance is in place that will enable the water system to operate effectively throughout its expected life. If the program is not in place, the planner should identify the steps that are necessary to establish an effective O&M program.

The objective of the O&M assessment guides is to assist the planner in completing the requirements of the fourth step by providing a tool to systematically analyze the central elements of O&M before the project is implemented.

KEY ELEMENTS OF OPERATIONS AND MAINTENANCE

Operating and maintaining a water supply system requires the effective interaction of a number of functions or departments within an organization as well as coordination between the agency responsible for water and the users. This situation is particularly true when the community is directly involved in the operations and maintenance of the water system.

The central elements that affect O&M are frequently the responsibility of several different entities. For example, the finance, tendering, supply, and operations sections of the government, as well as the community being served by the water system, may each provide support for O&M. Each must also be aware and carry out its required tasks regarding O&M.

An O&M program is composed of the key elements described below. Each element is addressed by the questions set forth in the assessment guides.

1. Institutional Capability

Both the governmental agency responsible for water and the community (or communities) receiving water service need to be actively involved in the water project if it is to be successful. The questions pertaining to this element focus on determining the commitment of government and community to operations and maintenance of the systems.

2. System Operations and Maintenance

The key to ensuring effective equipment maintenance is to make certain that responsibilities are clearly defined and that maintenance personnel have the tools and skills to do their job correctly. It is also essential to schedule preventive maintenance (PM).

3. Spare Parts and Supplies

Many water systems have failed because spare parts were not readily available to service equipment. Even the simplest water supply system requires a reliable source of supply for spare parts and other material needed to keep equipment in reliable operating condition. Numerous donors and the many types of equipment have compounded the problem of spare parts and created the need for large and diverse spare parts inventories. Because some parts may need to be imported, the necessity for a reliable inventory is potentially even more urgent in developing countries.

4. Logistics

The questions concerning this element consider the need for vehicles and workshops dedicated to the maintenance function. It is not unusual for the same group within a water authority to be responsible for both construction and O&M activities. In these cases, vehicles are not reserved solely for O&M and frequently are unavailable when needed. Such a situation may result in a poor response to equipment problems and lack of attention to preventive maintenance.

5. Finance

Before a water project is funded, the planner should address two issues relating to financing the recurring cost of the system as follows:

- How much will it cost to operate the system?
- Can the consumers and government afford this cost?

If the answer to the second question is negative, the project should be either redesigned (including the use of alternate financing) or abandoned. Project planners often assume that the host country is able to support O&M. If it is unable to do so, the result is a poorly maintained water system.

6. Records

Up-to-date and accurate records need to be maintained for all water supply (WS) systems. The type and number of records and reports needed is determined by the type of system. For piped systems with a large number of electrically powered units, an automated information system may be appropriate. For one involving either handpumps or protected springs with piped distribution, the requirements for records are quite different, yet equally necessary. Records and reports provide:

- System control enabling responsible officials to know the operational status of the system(s)
- O&M information for maintenance personnel
- Equipment operating history
- Information on parts and or supplies in inventory

7. Human Resources and Training

Training programs for equipment operations and maintenance are needed for all types of water systems. The technical content for training caretakers to maintain handpumps is, of course, less than for more sophisticated systems, but still must be planned. Training should be a continuing effort, particularly in LDCs where skilled technicians frequently learn a trade while employed by the water board and then seek higher paying work in the private sector. Ultimately, the success or failure of a water supply system will depend on the people who have the responsibility for operating and maintaining it.

These seven elements form the basis for a system of operations and maintenance. Each element must be investigated -- irrespective of the type of water system -- to ensure that O&M is adequately supported.

PROJECT PLANNING FOR OPERATIONS AND MAINTENANCE

Consideration of the long-term operations and maintenance of a water supply project should be initiated when the feasibility of the project is first being studied. For an AID-funded project, this means that the O&M capability should first be investigated during preparation of the Project Identification Document (PID). One effective approach for analyzing O&M for the project is for one of the technical members of the PID team to use the O&M assessment guides to: (1) obtain a general idea of the ability of the communities and central government to provide effective O&M for water supply systems and (2) prepare a list of items relating to O&M that need to be investigated by the Project Paper (PP) team.

The primary O&M investigation should take place before and during the PP preparation. Because part of the work of the PID team is to enumerate the strategy and responsibilities for preparing the PP, the PID should establish who should perform the required O&M investigations during the PP phase and how they will be carried out.

If there are numerous aspects of O&M that require investigation, an engineer with experience in the operations and maintenance of water supply systems should be included in the PP team. If there are few problems identified, the scope of work of the PP technical member (typically a water supply engineer) should include the appropriate investigation into the O&M capability. Other external support agencies may use the guides by incorporating O&M assessment technique in their project development procedures.

A second alternative is to have an outside organization, such as the Water and Sanitation for Health (WASH) Project, review the PP after completion and before approval is sought. The purpose would be to document potential problems in the O&M area that may not be addressed in the PP. Such a review would be based on use of the O&M assessment guide, the reviewer's previous O&M experience in LDCs, and interviews with the PP team. Because this approach is not based on field observations, it will not be as effective as the other alternative. It would, however, be less expensive, and the results would still help to highlight potential O&M problems in the project and might suggest modifications to the design of the project.

Using one of these approaches, in addition to the recommendations associated with the assessment methodology, will help to improve the O&M component of the project by identifying:

- Assumptions about the project and O&M
- Potential problems
- Possible project milestones and evaluation criteria for inclusion in the project agreement.

Methodology for Assessing the O&M Component

The technique for analyzing the O&M component is an iterative one requiring answers to questions concerning the O&M capability. Its purpose is to help the project planning team pinpoint problem areas regarding O&M and to develop strategies for incorporating solutions to these problems into the project design. To facilitate the use of the technique, the factors that influence

O&M, for example, long-term funding, spare parts, will be treated separately. Each step of the iteration will consist of the:

- o Question
- o Explanation of need for the question
- o Response (check YES, NO, or UNSURE)
- o Recommendation to satisfy needs identified by question if NO is the answer.

For example, question five in the Institutional Capability Section asks " Will a written agreement outlining the responsibilities of the village(s) in the project be made for each system among the government, the implementing agency, and the village(s)?"

The explanation step helps to clarify the intent of the question and its importance to successful O&M. For this question, the explanation is:

Explanation: While a written agreement detailing the responsibilities of the village is not essential to success, it is helpful because it:

- o Provides a clear statement of responsibilities
- o Increases the likelihood that assigned tasks will be carried out
- o Fosters the concept of village ownership of the project.

The next step in the iteration is to respond to the question by checking YES, NO, or UNSURE. The format is as follows:

YES_____ NO_____ UNSURE_____

Finally, after determining whether questions which are initially checked "Unsure" should be answered as "Yes or "No", the guide user may use the recommendation given as a guide for resolving No answers or proceed to the next question if the answer is Yes. For the sample question, the recommendation is as follows:

Recommendation: Investigate the feasibility of including a requirement for such a written agreement in the contract between the government and the external support agency.

Each of the four guides also contains an introduction that details the type of water supply system analyzed and lists assumptions that are made regarding the O&M component of the system.

Limitations of the O&M Assessment Guides

The Operations and Maintenance Assessment Guides are designed to assist project planners in pinpointing and resolving problems that are likely to limit the long-term success of a water supply system.

Successfully implementing a water supply project, however, requires that the project planner recognize that other components of the project that are not addressed in the guides also have an impact on effective operations and maintenance. Some of these components are as follows:

- o Design of the water system
- o Construction
- o Water quality
- o Health education
- o Social factors.

The final section of this paper presents a portion of the O&M assessment guide for one type of water supply system frequently found in LDCs - a number of villages using handpumps to draw water from groundwater sources. The introduction from the assessment guide plus one question* from each of the subsections representing the key elements are given. The complete guide for this type of water system and the O&M assessment guides for the other three types presented in the guide are available from the WASH information center.

* This guide contains a total of forty questions.

O&M ASSESSMENT GUIDE FOR SYSTEMS USING HANDPUMPS

Introduction

This Operations and Maintenance Assessment Guide has been developed for use in planning water supply projects in which a number of villages are to be provided with improved water systems by using handpumps to draw water from groundwater sources.

The following assumptions apply regarding system operations and maintenance:

- The water supply system will be owned by the communities it serves.
- All minor maintenance tasks are to be performed by one or more village caretakers.
- All major maintenance tasks (primarily those that require work on the part of the pump below ground) are to be performed by regional governmental maintenance crews.
- Treatment of the water from the source is not required.

The remainder of this guide is organized as follows:

Institutional Capability
System Operations and Maintenance
Spare Parts and Supplies
Logistics
Finance
Records
Human Resources and Training.

It is recommended that chapters one through four of this workbook be reviewed before one attempts to use the Operations and Maintenance Assessment Guide.

Institutional Capability

Both the community being served and the appropriate governmental agency need to assume part of the responsibility for operating and maintaining the water system in order for this type of water supply system to be successful.

Village Level

1. **Is active village participation included in the project?**

   Explanation: Previous experience with similar projects points to the absolute necessity of participation by the villages to give them a sense of ownership. This participation may involve money, labor, materials, or a combination of these. The villages should also participate in planning the water schemes for their community. Such planning may include site selection and choice of technology.

YES_____ NO_____ UNSURE_____

If UNSURE, study project plan. Status must be determined.

Recommendation: Community participation is essential to long-term success. The responsibility of the village for O&M must be clearly defined in the project documents.

ties for O&M and, for example, construction, to be in the same section. When this situation occurs, O&M generally does not receive its proper attention, because it does not have the same visibility as the construction of a new water system.

YES_____ NO_____ UNSURE_____

If UNSURE, determine whether a separate O&M section exists.

Recommendation: Effective O&M requires the meshing of several components of an organization. Bringing these components together requires the concentrated effort of a group within the water authority. This is true whether the water scheme is a piped system with treatment or one using handpumps. It is recommended that the contract between the ESA and government address this need, with specific requirements for identifying the group or individual who will have O&M as its only responsibility. This requirement should be used as a milestone for monitoring the commitment of the government to the water supply program.

System Operations and Maintenance

This section focuses of the need to ensure that the responsibilities for O&M are clearly defined and that the community and regional maintenance crews have the tools to carry out their designated maintenance tasks.

1. **Does the project include provisions for selecting villagers as pump caretakers?**

Assumption: Villagers are available to be employed as handpump caretakers.

YES_____ NO_____ UNSURE_____

If UNSURE, determine community involvement and act accordingly.

Explanation: The project has virtually no chance of long-term success if the village itself is not responsible for periodic maintenance.

Recommendation: One or more caretakers should be selected to perform regular pump maintenance. Further, the pump caretaker(s) should:

- o Participate in the installation of the well and handpump
- o Be selected after consultation with the village council
- o Be paid for the work.

The requirement that the caretaker be paid is probably unnecessary if the village has a practice of setting aside one day a week for the people to work on community projects. Maintaining the water system would be the caretaker's contribution to the community workday

Spare Parts and Supplies

The questions set forth in this section pertain to the ability of the community and regional maintenance crews to obtain the spare parts and supplies needed to maintain and repair the handpumps. Investigations in this area are particularly important if the pumps are imported.

1. **Is the hardware selected for use in the water supply project compatible with spare parts and supplies currently available in the country?**

   Explanation: The availability of spare parts and supplies is often one of the most critical factors in any O&M program for village water supply systems. The introduction of imported noncompatible hardware for the system necessitates the creation of a new import and distribution network, which may prove more difficult and less effective than adjusting technical specifications during the design phase to ensure hardware compatibility.

   YES_____ NO_____ UNSURE_____

   If UNSURE, determine compatibility.

   Recommendation: The project planner should weigh carefully the impact on long-term project success of project hardware compatibility with locally available parts and supplies.

   If the governmental water authority is not directly involved in project planning and implementation, it should be consulted regarding this matter.

   If noncompatible hardware is to be used, the supplier contract should include a clause specifying the selection of an import agency which is to guarantee spare part availability in the country for a specified period of time (ideally the life of the water supply system).

Logistics

Reliable transportation for pump caretakers, if they are responsible for more than one community and for regional maintenance crews, is an integral part of a successful maintenance program.

1. **Is reliable transportation available for locally based caretakers and regional maintenance workers?**

   Explanation: For the caretaker, a bicycle or motorbike is usually satisfactory if the caretaker is responsible for a number of pumps in several villages. Tools and supplies can be easily transported by these means.

   YES_____ NO_____ UNSURE_____

   If UNSURE, determine status.

For regional maintenance, appropriate and sufficient vehicles must be dedicated to maintenance and not used for other duties, such as monitoring construction activities.

Recommendation: The project preparation team needs to take two steps regarding vehicles:

a. Determine whether vehicles are available to be used for maintenance. If not,
b. Consider adding sufficient vehicles to support O&M to the commodities supplied as part of the project.

Finance

If properly maintained, the cost of servicing handpumps is not high. Some funds, however, will be needed to obtain replacement parts and supplies, to provide transport, and to pay salaries. The ability and willingness of the consumers and government to fund their part of the recurring cost of maintaining the water system must be determined before the project is approved.

1. **Has an estimate been made of the annual cost of operating and maintaining the WS system over the life of the system?**

   Explanation: Before addressing Question 2 of this section, an estimate of the recurring costs of operating the system needs to be made. If the maintenance program is planned (for example, preventive maintenance rather than breakdown maintenance), an accurate estimate of maintenance costs can be made.

   YES_____ NO_____ UNSURE_____

   If UNSURE, determine whether recurring O&M costs have been, or will be, estimated.

   Recommendation: The project preparation team needs to have the capability to estimate the annual cost of operating and maintaining the system. This estimate will require that the team include an engineer or financial analyst experienced in O&M.

Records

Accurate records of O&M activities enable both the community and government to estimate the funds needed to support the water system in the future. Because an estimate of recurring costs should have been made as the project was being planned, the purpose of maintaining records will be to document the operations and maintenance history of the handpumps and to prepare annual operating budgets more accurately.

1. **Will the caretaker be required to maintain records of PM done and materials used?**

   Explanation: The caretaker is more likely to provide proper maintenance if records are maintained on work completed and supplies and material used. If such records are maintained, the regional maintenance crew or village council must periodically inspect the records.

YES_____ NO_____ UNSURE_____

If UNSURE, determine status.

Recommendation: This requirement should be included in the maintenance plan for the handpumps. Appropriate written checklists and prepared O&M forms should be provided to the caretaker for this purpose. The most effective way to ensure that adequate records are maintained is to include a provision for them in the written agreement between the government and the village.

Human Resources and Training

Village and governmental personnel are the most important part of a successful operations and maintenance program. It is, therefore, imperative that they be given proper training to enable them to maintain the handpumps correctly and at a reasonable cost to the users and to the government.

1. **Has provision been made in the project for management training of village water committee members?**

Explanation: As noted above in Question 4 under Institutional Capability, the participation of the village in the project is essential to its success. It is equally important that the village committee responsible for operating and maintaining the community water system be able to:

- o Organize maintenance activities.
- o Arrange for and manage the collection of funds.
- o Purchase and store materials and supplies.
- o Maintain records.
- o Contract maintenance work to private enterprise, for example, village artisans.

This type of training is particularly important if the bulk of the maintenance work is performed by the village. As Village-Level Operation and Maintenance (VLOM) pumps are developed, the need for management training of village water committees will grow.

YES_____ NO_____ UNSURE_____

If UNSURE, determine whether training or extension work is planned for in the project.

Recommendation: One effective way to train the water committee is through community development or extension workers (CDW). The CDWs may be trained to impart the necessary skills to the village committees. The training of the committee should be completed before the water system is constructed. The use of project funds to train the CDWs is desirable, because this helps to emphasize the need for such training.

Reliability Engineering for System Design and Maintenance

P. M. Berthouex* and David K. Stevens**

Introduction

How often have we heard, "These investments in drinking water facilities in country A-to-Z were wasted. Half the systems installed in the last project are inoperable." Letting this happen not only harms the people who should have been served, it provides government budget makers with arguments for not providing badly needed services. Building systems that are reliable is vital to win adequate long-term financial support for water supply and to raise the level of health and happiness in the world.

All systems fail. The goal is not to prevent failure; attempting this would be technical and economic nonsense. A system that is weak and fails often, but which uses local materials, can be an excellent design if steps are taken in the design phase to increase reliability by using redundancy and maintenance and repair service. Reliability engineering is designing to balance the risk and cost of system failure against the cost of building a stronger, more dependable system.

The reasons for low reliability in developing countries are not hard to identify or understand. Local materials may be of low quality (e.g. strength or other measure of performance) and variation in quality may be high. Operating conditions are harsh. Spare parts and qualified maintenance personnel are insufficient. These factors are coupled with slow reporting of failures and an administrative system that reacts slowly to make repairs. It should be obvious that poor system service is guaranteed if these factors are not identified, quantified, and analyzed, so far as possible, when the system is designed. Reliability engineering provides techniques for estimating system behavior and, from this, maintenance requirements.

This paper suggests how reliability can be analyzed, primarily through examples. The method of analysis to be illustrated is Monte Carlo simulation, a simple but flexible and powerful technique.

We have implied that unreliable service in developing countries usually is not caused by lack of suitable technology but by failure to plan and design using the tools of reliability engineering. Having said this, we should also say that using these techniques requires information that is not generally available, such as failure rates, delivery times, repair times, etc. Thus, the reason that reliability engineering is not widely used is not the lack of technique, but lack of suitable information about system behavior. We are unable to provide this information, so this paper is written with the hope that

* Professor, Civil and Environmental Engineering, The University of Wisconsin-Madison, Madison, WI 53706
** Assistant Professor, Civil Engineering Department, Utah State University, Logan, UT

it might (1) show the designer/planner how to gain insight into the relative importance of factors that, although not precisely quantified, are known to affect system reliability and (2) encourage the collection and publication of information about system reliability.

The Cost of Unreliability

Harm caused by system failure may be reflected in some cases by monetary costs. More often intangible variables that indicate the level of well being of the service population are also important. The costs of not having good service are real and large, though largely incommensurate or intangible. Medical care costs more than clean water. The loss of happiness caused by sickness and infant mortality is great, even though its monetary value cannot be specified. Purchasing water from water dealers may consume ten percent of an income that cannot buy enough calories for a subsistence level diet. Time and energy are spent carrying water (which may be unsafe) long distances. Unreliable supply lead to buying water storage vessels and wastage of water. Few engineers, economists, or politicians believe these costs are lower than the cost of building reliable service systems. Still, we plan and build systems without making the necessary investment of resources to make them reliable.

Reliability

Reliability is operationally defined, for purposes of this paper, as the average length of time the system is out of service. The reliability of a routinely maintained system depends on (1) the inherent reliability of the individual system components, (2) environmental factors such as weather and rate of use, (3) routine inspection and maintenance, and (4) the capacity to perform repairs as required. The effect of each of these components must be quantified to accurately predict the reliability of a system.

Reliability theory differentiates between three kinds of failure: (1) early failures which result from poor manufacturing and can be eliminated by "burn-in" testing and quality control, (2) failure due to wearout which occurs because of old age, or when equipment is not properly maintained, and can be eliminated by planned timely replacement before parts fail, and (3) random (chance) failures which occur at random intervals and which neither burn-in testing or good maintenance or schedule early replacement can eliminate.

Random failures occur after burn-in and before wearout. These three kinds of failures follow different statistical distributions and therefore require different mathematical treatments. In this paper, only random failure is considered. This is not because random failures are necessarily the most important kind of failure. In developing countries, wear out, which is often premature, may be more important. There are no data on this.

Mean Time Between Failures

Where we are concerned with the overall failure rate of a large number of components with different lifetimes (some old and some new), it is usual to treat the failure rate, λ, as a constant. Failure rate is expressed as number of failures per unit of time (1000 hr, day, month, year, etc.). When a device fails at random intervals at a constant failure rate, λ, the probability that it will be functioning satisfactorily at time t is

$$R(t) = \exp(-\lambda\ t). \qquad (1)$$

Random means that the time of one failure of one device does not depend on, nor can it be predicted from, the failure of any other device. This formula is correct for all properly debugged (burned in) devices that have not yet suffered any degree of wearout damage or performance degradation due to their age. The life period of the device over which this formula is valid is the useful life of the device. Some useful references on the basic theory are Bazovsky (1961) and Dhillon and Singh (1981).

Environmental factors, such as weather and usage rates, affect reliability by increasing the failure rate, λ. For instance a pump that is sheltered should have a higher reliability than on that is exposed to the sun, rain, and dust.

The Reliability of Large Systems

Intuition almost always overestimates the reliability of large systems. For example, consider a system that comprises 1,000 simple pump/well installations in number of villages. If told that the reliability of each pump/well is such that the average time between failures is 200 days, many engineers will think the reliability is high. They will badly underestimate the number of wells that are out of service any given day and the average waiting time to repair (for a specified repair service program). Assuming all 1000 wells are operating at time zero, 50 wells out of 1000 will have failed within ten days and over 200 will have failed by day 50. The large system of 100 wells is inherently unreliable.

Once this is realized, the natural reaction may be "We have to get stronger pumps or in some other way greatly increase the MTBF (decrease λ)." This may or not be a wise design decision. Regular replacement of weak components, protection from environmental stress, or a good maintenance service may be better options. This needs investigation.

Time to Repair

The length of time the users are without service will depend on the MTBF and the available repair service. The time to repair (TTR) can be broken into four components: (1) the time until a failure is reported, (2) waiting time for service (including, delivery time for spare parts that must be ordered), (3) transportation time to the site of the failed system, and (4) repair time.

If repairs are to be accomplished quickly, failures must be reported promptly, properly equipped and qualified service personnel must be available, adequate spare parts must be on hand, and transportation of service personnel and equipment must be available. Only having all these elements in place will keep the out-of-service time short. Thus, the the overall quality of service and user satisfaction depends on inherent physical strength and a specified maintenance and repair program.

Using Simulation to Analyze System Reliability

The examples that will be presented in the next section will only deal with the waiting time for service, assuming that needed parts are on hand. In the first example it will be assumed that each well is inspected at a regular intervals of t_c days and that, if found inoperable, it is repaired. No other repair service are available. This obviously poor repair strategy (which is improved in the second example) will serve to explain the simulation approach to investigating system reliability

For this regular availability of repair, the time to repair (TTR) of a single well is

$$TTR = t_c - (t_f \ \mathbf{mod} \ t_c) \quad (2)$$

where t_f is the time of failure of the component measured with respect to an arbitrary time zero when this maintenance program is imagined to have begun. The quantity in the brackets is the time since the most recent inspection. The operator **mod** (for modulo arithmetic), calculates the remainder after dividing t_f by t_c. A sample calculation will make the assumptions and the arithmetic clear. For $t_f = 190$ days and $t_c = 60$ days, the time since inspection on day 180 is (t_f **mod** t_c) = 10 days. If repairs are available until the next inspection circuit, TTR = 60 - 10 = 50 days.

The mean time to repair (MTTR) for the entire system can be found using Monte Carlo simulation (Berthouex and Brown, 1969; Hahn and Shapiro, 1967). This is implemented by drawing uniformly distributed random numbers (every numerical value is equally likely to occur on any draw) with values between 0 and 1.0 to represent the reliability $R(t_f)$ of a particular pump/well. This reliability is then transformed in a failure time using the model $R(t_f) = \exp(-\lambda t_f)$ to calculate $t_f = -\ln(R(t_f))/\lambda$. The t_f values are used, as illustrated above, to calculate TTR. By repeating this a large number times, a large set of "data" on TTR is created. These values can be averaged to give the mean time to repair, MTTR. The errors associated with these averages can also be estimated. The name Monte Carlo suggests the probabilistic aspect of letting reliability of the wells vary randomly.

Analytical solutions can be derived for simple systems but the analytical solution has no advantage over simulation. Simulation has several advantages. It is intuitive and easy (it can be done by hand if necessary). Generating data on one aspect of system behavior makes it natural and easy to collect information on other aspects. It can be expanded to more complex situations. For example, adding spare parts inventory and random delivery of parts could easily be simulated.

Applications

Two example applications of this elementary reliability theory will be presented. They are loosely based on the maintenance program described by Bannerman (1980) and implemented in Ghana in 1974. The project installed 2400 water wells with hand pumps in several hundred towns and villages. Inspector, who did minor repairs, traveled by motorcycle to visit each pump/well every two months. A 1977 survey showed about 80% of the wells in working order, which corresponds to λ = 0.0037 days and MTBF = 270.

Example 1. Each well in system of 100 wells is to be inspected inspected every t_c days by circuit riders. The MTTR for the 100 wells is found by simulation as described above. The results of the simulations shown in Figure 1 was made by varying the average failure rate λ, from 0.0001 to 0.1/day (MTBF from10,000 days to 10 days). The inspection cycle time was varied from 15 to 60 days.

For all values of λ less than about 0.03/day (MTBF about one month) the MTTR changes very little with failure rate. This means that if the wells could be made substantially more durable, say by decreasing λ from 0.01 to 0.001/day, the MTTR would decrease by only about 7% even though there are one-tenth the number of well failures. By contrast, a change in the maintenance cycle time is accompanied by a proportionate change in the MTTR. This demonstrates that an improved maintenance system can give a large payoff in overall system reliability.

Of course, decreasing λ will decrease the number of failures and also the number of people put at risk by poor service. Still, those who were affected by a failure would scarcely see the difference. Of course, the number of wells failing, is generated by the simulation and the number of people affected could be tabulated. This example could be extended if space allowed.

Example 2. This example demonstrates that provision of repairs on call by emergency crews is highly beneficial. The basic maintenance program of example 1 is augmented by emergency crews. Now the simulation takes into account the possibility of several wells failing on the same day.

Three additional assumptions are made: (1) reporting and transportation times are zero, (2) a repair takes one day (a failure is reported and repaired on the same day and its TTR is 1 day), and (3) wells that cannot be repaired on the day they fail are placed in a service queue and are repaired as a repair crew becomes available or when the circuit rider makes his next visit.

Under these assumptions, and using a failure rate of λ= 0.01/day, the effect of cycle time and the number of emergency crews on the MTTR was assessed for a system of 100 wells over a period of 300 days (3/λ). Simulating a longer period would give results nearer the true system average; a longer simulation would be run to study a real system,

The results are shown in Figure 2. In example 1, halving the inspection cycle time roughly halved the MTTR, as shown in Figure 2 by the curve for zero emergency crews. Assuming there is one circuit rider, adding one emergency crew reduces the MTTR from 31 days to 15 days. Virtually identical results are seen for 2 and 3 crews, indicating no benefit to using more than one emergency crew. If there are 2 circuit riders and one emergency crew, the MTTR decreases from 15 to 7 days. Adding more circuit riders further decreases the MTTR, but there is little incentive to have more than four riders. Figure 2 also shows that if the number of circuit riders is large, say 8 to 10, the availability of emergency crews is not beneficial, when measured in terms of MTTR. Of course, this assumes

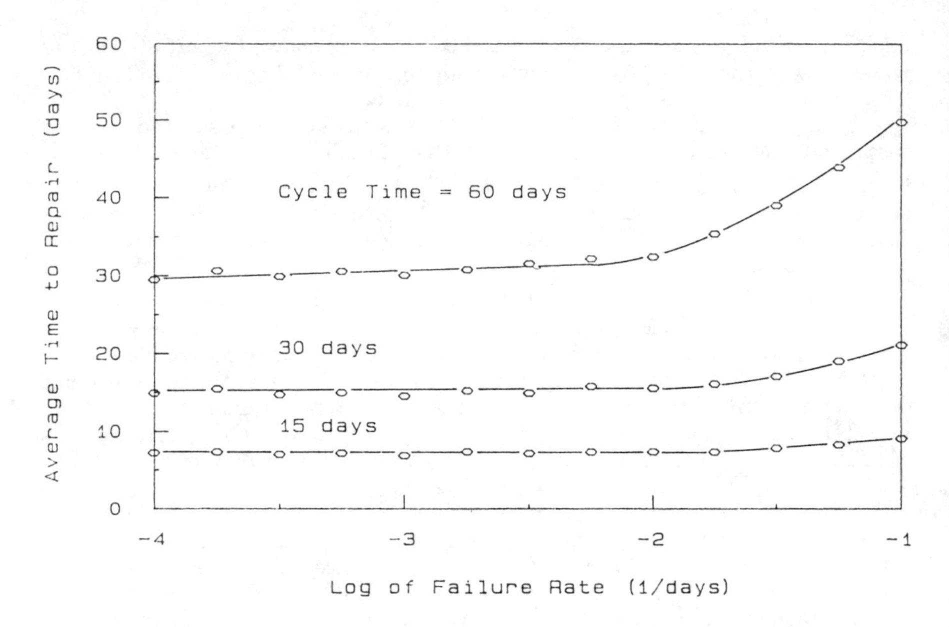

Figure 1 -- Average Time to Repair as a Function of Failure Rate -- Results of Example 1

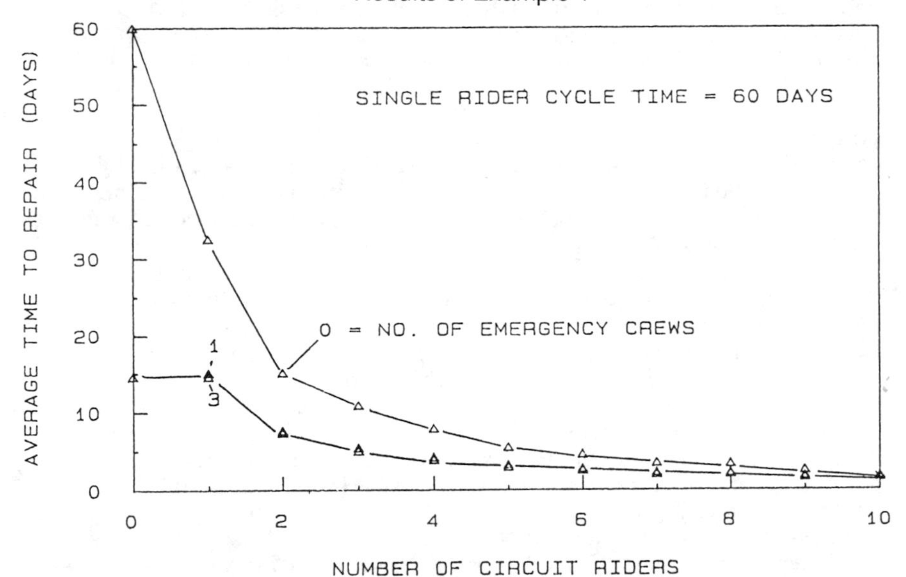

Figure 2 -- Average Time to Repair as a Function of the Number of Circuit Riders -- Results of Example 2

that the circuit riders can make all repairs that the emergency crew would make. This is unrealistic since these crews could carry heavier equipment.

Summary

Building a very strong and durable system may or may not be good design. Since even expensive systems fail, it is quite possible that the best way to get a reliable system is to plan for good maintenance and quick repair service. To make this decision, the balance between investment in physical equipment and in operations needs to be explored. Often, in the past, it has not been investigated- at least not formally.

Monte Carlo simulation is a flexible, powerful technique that can be used to gain insight into the performance of proposed systems. Different equipment specifications and different operating programs can be studied. Experience has shown that intuition is unreliable in estimating the reliability of large systems. Almost everyone, including experienced engineers, will generally be entirely too optimistic and underestimate the frequency of failures. Intuition and judgment need to be tempered with some analytical study. Simulation of even an overly simplified will make the designer aware of the magnitude of the problems that may exist with regard to reliability and move him in the right direction.

Admittedly the examples presented are too simple to be of direct practical use. They were intended to demonstrate (1) how reliability engineering can be used to evaluate a proposed maintenance program and (2) that is it important and possible to evaluate maintenance and repair programs as part of design. If they have served this purpose, perhaps in the near future engineers will start to collect and publish information about the reliability of existing systems so these methods can be put to better use.

References

Bannerman, R., R. (1980) "Maintenance of Regional Rural Water Supplies - The Ghanian Approach," in Proc. 6th Conference of the WEDC, Nigeria.

Bazovsky, I. (1961) Reliability Theory and practice, Prentice Hall, Inc., Englewood Cliffs, New Jersey.

Berthouex, P. M. and Brown, L. C. (1969). Monte Carlo Simulation of Industrial Waste Discharges," Jour. Envir. Engr. Div., ASCE, Vol. 95, pp 345-356.

Berthouex, P. M. and Stevens, David K (19882). Reliability Engineering for Appropriate Environmental Technology. Proc. 1st Int'l Symp. of Appropriate Tech. for Water and Sanitation in Developing Countries, Istanbul.

Dhillon, B. S. and Singh, C (1981). Engineering Reliability - New Techniques and Applications. John Wiley and Sons, New York.

UNDP/World Bank Rural Water Supply
Handpumps Project (INT/81/026)

Saul Arlosoroff*

Abstract

The importance of providing safe water to the millions of people in rural areas of developing countries who lack this basic service has been repeatedly stressed by national governments and international agencies. Among the activities of the International Drinking Water Supply and Sanitation Decade designed to address this problem is the UNDP/World Bank Rural Water Supply Handpumps Project. The Project is focusing on options which provide the least-cost solutions to water supply problems now existing throughout the developing world, while maintaining reasonable levels of service until national or local governments can afford to support higher levels. Such efforts result in a policy of "some for all instead of all for some" and provide coverage for the majority of populations in need.

Introduction

An estimated 1,800 million rural people will have to be provided with improved water supplies in the fifteen years to the end of the century if developing countries are to approach full coverage targets of the International Drinking Water Supply and Sanitation Decade (IDWSSD, 1981 to 1990). The first half of the Decade has seen increases in the percentages of the rural population with access to safe water supplies in Africa, Asia and Latin America, but only in Asia has the pace been quick enough to envisage a target of essentially full coverage by the end of the century (ten years later than the original Decade goals). In Africa, present progress rates would leave half of the rural population still without safe water in the year 2000, while in Latin America, it may be 30 to 40 years before widespread coverage is achieved unless progress improves dramatically.

Accelerated progress is hampered by financial and technical resource constraints faced by many developing countries, and the problem is aggravated by the growing number of completed projects which are broken down and abandoned, or functioning well below their

*Chief, Applied Research and Technology, Water Supply and Urban Development Department (UNDP Projects Manager), World Bank, 1818 H Street, N.W., Washington, DC 20433. The views and interpretations in this paper are those of the author and should not be attributed to the World Bank, the UNDP, or their affiliated organizations.

intended capacity. Attempts to increase the pace of providing improved rural water supplies have often been frustrated because the technology used has proved impossible to sustain in village conditions. To make a lasting impact on the urgent needs, rural water supply (RWS) strategies must be based on sustainable and replicable technologies and systems, and must take account of the pace at which resource constraints can be overcome.

The "Package Approach" -- Recommended Strategy

Successful RWS programs involve a combination of hardware and software -- technology and institutional/organizational support elements -- matched in such a way that each community recognizes the benefits of the improved supply, can afford at least the costs of operating and maintaining it, and has the skills, spare parts, materials and equipment available to sustain it. To maximize health benefits, parallel investments in health education and sanitation programs should accompany implementation of the RWS improvements.

This approach involves consideration of a number of key issues, each individually important, and with the final project put at risk if they are neglected or inadequately addressed:

- Direct involvement of the community in the design, implementation, maintenance and financing of planned improvements, with promoting agencies providing technical assistance and support services as needed. Villagers' needs and wishes have to be reconciled with their capacity and willingness to pay for the level of service planned.

- Provision for operation and maintenance cost recovery, with support of construction costs for poorer communities offset by complete cost recovery, gradually implemented, where higher service levels are provided.

- Maximum involvement of in-country industry in the supply of services and materials for project construction and maintenance (e.g., supply of pumps and spare parts, servicing and repairs), with the important proviso that quality control and reliability should be assured.

- Technology chosen to match the resources available.

- Institutional and manpower development programs matching the needs of the planned water supply system.

- Parallel programs in health education and sanitation improvements.

Service Level and Technology Choice

The decision about the level of service to be provided for a particular community or district involves consideration of many of the issues listed above. Choices may have to be made between surface water and groundwater as the source, and then from yardtaps, public standpipes or handpumps as the method of distributing the water to the beneficiaries.

Groundwater has distinct advantages over surface water as a source for RWS improvements, the main two being that no treatment should generally be needed to produce safe water and an abundance exists throughout the regions. The resource demands of water treatment plants needed to safeguard supplies from surface water sources are beyond the reach of most rural communities, and use of untreated surface water represents an unacceptable health risk. In cases where an upland catchment can be protected against contamination, a gravity-fed system can be reliable and safe, but only a small percentage of the rural population in need of improved supplies live close to such sources. It will therefore be rare for RWS programs to be based on surface water as the source, and the technology choices analyzed in this report are focused largely on groundwater-based RWS systems, which will be the right choice for the great majority of rural areas.

The three main technology options (handpumps, standpipes, and yardtaps) represent progressively increasing service levels, and call for increasing financial and technical resources for their implementation and maintenance. The choice of appropriate service level for a particular project or program can only be made when resource constraints have been taken into account, including the capability of the users to operate and maintain the proposed system.

Capital costs of the three technologies generally range from US$10-30 per capita for wells equipped with handpumps to US$40-60 per capita for motorized pumping and standpipes and US$80 per capita or more for yardtap services. In global terms, that means that cost estimates for meeting rural water supply needs to the year 2000 range from US$50,000 million to US$150,000 million, depending on the choice of technology. With the obvious difficulties of mobilizing financial resources for this scale of investment, rapid progress in meeting basic needs can be achieved only if a large proportion of the rural population in need receives services at the lower end of the cost range. Upgrading to a higher service level may then be financed by the community later, as benefits from the initial investment and from other sources increase available resources.

Analysis of other resource demands of the different technologies also points to a substantial role for handpump-based systems in meeting present needs. The most significant difference between handpump projects and those based on standpipes or yardtaps is the switch to mechanized pumping, and the consequent need for dependable power/fuel supplies and skilled pump mechanics.

In cases where reliable low-cost electric power is available from a central grid, an electric pump can be a relatively inexpensive and operationally simple means of lifting water. Communities which have the financial and technical means available to implement and sustain projects based on electric pumping should be given every encouragement to do so, as this frees scarce public sector funds and external aid for projects serving poorer communities. However, the number of rural villages with dependable electricity supplies is presently small -- well below 10 percent of the total rural population in Africa and only a little higher in most countries in Asia.

In the absence of reliable electric power, the alternative power source for mechanized pumps is diesel engines. The logistical problems of ensuring dependable diesel supplies for dispersed communities have rarely been successfully overcome, and there are few examples of diesel-powered RWS systems operating successfully in the long term. The cost of trucking diesel fuel along hundreds of kilometers of rural roads will usually prove prohibitive. Future developments in solar technology may eventually make solar pumping economical for drinking water supplies, but at the moment such schemes have very high initial costs.

Adding the institutional constraints and the severe shortage of skilled mechanics in developing countries, it is clear that systems involving mechanized pumping are appropriate for only a minority of those in need of new supplies in the coming years. For the rest, it seems clear that drilled or dug wells equipped with handpumps will be the right choice, which makes it vitally important that handpump-based projects are planned and implemented in ways which will ensure that they perform reliably and can be sustained in the long term and widely replicated.

The Handpumps Project

In 1981, as one of the activities in support of the Decade, the United Nations Development Programme (UNDP) and the World Bank initiated a Global/Interregional Project for the Testing and Technology Development of Handpumps for Rural and Urban-Fringe Water Supply (the Handpumps Project). Other projects being carried out under the World Bank/UNDP Decade Program include the Development and Implementation of Low-Cost Sanitation Investment Projects (INT/81/047), the Research and Development in Integrated Resource Recovery Project (GLO/84/007), the Information and Training Program in Low-Cost Water Supply (INT/82/002), and the Preparation of Water Supply and Sanitation Investment Projects (RAS/81/001, RAF/82/004).

The main objective of the Handpumps Project is to promote the development of designs and implementation strategies which will improve the reliability of schemes based on groundwater and handpumps, and which will enable schemes to be managed by the communities and replicated on a large scale. Technology was thought to be at the root of past problems experienced with handpump-based RWS systems, and the Handpumps Project has carried out laboratory tests in the U.K. and field trials in 17 countries to measure the performance of a total of 2,800 handpumps. Field trials lasted at least two years on each pump, with some 70 different pump models represented in the lab and field trials.

From the beginning, the Handpumps Project has promoted the concept of VLOM (Village-Level Operation and Management/Maintenance) as a means of overcoming some of the major obstacles to sustainable water supply systems. Now recognized as one of the fundamental principles of handpump design and RWS project planning, the VLOM concept seeks to avoid the high cost, long response time, unreliable service and other operational difficulties in the repair of handpumps through central maintenance systems. Many past failures of RWS systems can be blamed on the inadequacies of central maintenance, in which a water authority dispatches teams of skilled mechanics with motor vehicles from a base camp, often serving a large district, to respond to requests for repairs or to carry out routine maintenance. Instead, maintenance should be a community responsibility, and this in turn means that the pump design has to be suitable for repair by a trained caretaker or area mechanic with basic tools, and that spare parts should be affordable and readily available to the community.

The Handpumps Project strongly advocates that pump maintenance should be delegated partly or wholly to village committees and that pumps used should be suitable for village-level maintenance. Developing country governments and donor agencies are increasingly changing their policies to include these principles in projects or programs. This is a significant departure from previous practice, particularly in Africa, where unsuitable pumps have often been brought in through donor assistance, and recipient agencies have taken on unmanageable maintenance commitments depending on public-sector mobile maintenance teams.

Planning and Implementation

Few handpump system failures can be blamed solely on the pump. Other major causes are inadequate or unrealistic provisions for maintenance; poor well design or construction, allowing sand to enter and damage pumping elements; and the corrosive effects of groundwater, which are much more extensive than had previously been suspected. Experiences in the field trials and data from many other RWS projects have enabled the Handpumps Project to formulate guidelines for the planning and implementation of RWS projects using wells equipped with handpumps. The five critical elements dealt with in the guidelines are as follows:

Community management of maintenance. Under the recommended system, the community organizes and finances all repair and routine maintenance of the handpump. Work is carried out either by a designated community member with minimal training and basic tools, or by an area mechanic (usually with a bicycle or moped) covering several pumps. The public authority has an important role to play in the training of caretakers and mechanics and in the organization of an adequate spare parts distribution system, but should then hand over maintenance of the scheme to the beneficiaries.

Community involvement. The highest potential for sustainability is achieved when the community is involved in all phases of the project, starting from the planning stage. If the scheme is to continue to operate satisfactorily, villagers have to recognize the need for the improved service, be able and willing to pay for the maintenance cost (and eventually the construction cost), and be willing to manage its maintenance.

Aquifer analysis. Competing demands for other water uses, such as irrigation pumping, have to be taken into account when evaluating aquifer potential for handpump projects. The well needs to be deep enough to allow for seasonal and long-term lowering of the water table, but no deeper, because of the additional cost and complexity. Legislation and administrative enforcement are needed in some areas to prevent overpumping for irrigation leading to future loss of groundwater as a source of domestic supply.

Well design and construction. Wherever the rock is not fully consolidated, screens and filter packs are essential to prevent sand and silt intrusion. Otherwise rapid damage will occur to commonly-used types of seals and valves. The right choice of drilling equipment, backed by appropriate organization of drilling, can significantly reduce drilling costs and result in more dependable wells.

Handpump selection. Quite a number of factors influence handpump selection, in addition to the cost of the pump itself. Among the most important are the suitability for the intended maintenance systems (e.g., can it be repaired by a trained pump caretaker?), durability, and discharge rate. Pump choice will depend on the required lift and the planned number of users per pump. Standardization of one or a few pump types for any one country can have a significant impact on maintenance and is an important selection criterion; and corrosion resistance has to be taken into account when groundwater is aggressive.

Today's Handpumps

The standard test procedures used in the laboratory and field trials revealed many shortcomings in existing handpump designs. Manufacturers responded well by modifying their products and introducing new models, and there are now many more pumps on the market which are durable and which allow for substantial involvement of villagers in pump maintenance. Manufacturers from industrialized countries are also being encouraged to combine with enterprises in developing countries to manufacture pumps under licensing or joint-venture agreements. Local manufacture strongly improves the likelihood that spare parts will be available when needed, and facilitates standardization on pump types in a country to simplify caretaker training and stocking of spare parts.

Encouraging as these developments are, there remains a scarcity of handpump models which can be described as VLOM and are suitable for lifting from depths of more than about 25 meters (though the majority of the rural population lives in regions where the water table is not so deep). The depth of installation and heavy pump construction make removal of downhole components difficult. An added problem is that, due to the high cost of the well, deep pumps tend to serve more people per well and so suffer rapid wear.

It is clear that some pumps are much more suited than others to conditions in developing countries, and that as pump lift increases, the number of pumps suitable for village-level maintenance declines rapidly. Nevertheless, the Handpumps Project has shown that, even from the pumps presently on the market, it is possible to design handpump-based water supply systems for the vast majority of conditions prevailing in developing countries, which can be sustained in reliable operation without dependency on a significant level of support from a central authority.

The Abidjan Statement

One of the most important outputs of the Decade, which supports the Project's promotion of community-based water supply systems, is "The Abidjan Statement." The Statement resulted from an important international seminar on low-cost community water supply, organized by the Project, UNDP, and the Government of the Cote d'Ivoire, and held in Abidjan in October 1986. The Statement, endorsed by the approximately 100 participants representing developing and developed countries and international agencies who attended the Seminar, is already having a major impact throughout the world. It outlines the recommended strategy for planning and implementing low-cost rural and urban-fringe water supply programs and supports the VLOM concept being promoted by the Project. Although focusing on Africa, the Statement is applicable to all countries in the world. The preamble summarizing the Statement is as follows:

> Lasting health and economic benefits for the rural and urban-fringe populations of Africa can be achieved through increased community management of water supply and sanitation systems based on proven low-cost technologies. African governments and donors are urged to identify and commit adequate resources and provide all necessary support for the direct involvement of communities in choosing, managing and paying for their water and sanitation systems.
>
> The main elements of the five-point strategy are as follows:
>
> (1) the role of governments and donors, policies to standardize technology and socio-economic approaches, sustainability and replicability, inter-agency coordination;
>
> (2) the involvement of communities -- especially women -- in decision-making and management; affordability of water supply systems;
>
> (3) community water supply as an integral part of primary health care;
>
> (4) choice of appropriate technology, in-country manufacture and distribution of handpumps and spare parts; and

(5) community-based maintenance, supported by a national strategy of standardization of spare parts.

The Future for Point Source-Based Community Water Supply

The need to accelerate large-scale implementation of RWS schemes to meet urgent needs calls for a more systematic evaluation of strategies and for the preparation of detailed guidelines for implementation at the regional and possibly country level. Lessons and conclusions about the implementation, operation and maintenance of point source-based community water supplies have to be implemented initially through demonstration projects in specific regional conditions. The demonstration projects will also include evaluation of measures to enhance the benefits from rural water supplies, and to develop recommendations on synchronizing related health and other interventions with water supply improvements. The proposed comprehensive RWS package therefore includes:

1. Well design, construction and development.

2. Implementation of projects with VLOM handpumps.

3. Community participation in planning, construction and management of maintenance.

4. Selection and training of caretakers, establishment of incentive schemes, and an increase in the role of women.

5. Spare parts supply and distribution.

6. Implementation of sanitation components.

7. Health education.

8. Cost recovery by the community to cover at least recurrent costs.

9. Measures to reduce capital and recurrent costs.

10. Non-domestic water use, such as micro-irrigation and cattle watering, wherever applicable.

A joint effort is needed by donors and developing country governments to initiate demonstration projects on a large enough scale to permit development and analysis of country- or region-specific ways of implementing relevant items of the package. There may be, for example, several different ways of organizing spare parts supply and distribution which make best use of private and public sector activities in particular countries.

These demonstration projects and other parallel activities should clear the way for large-scale implementation of community-managed rural and urban-fringe water supply systems.

TRAINING NEEDS IN EGYPT

Henry R. Derr* P.E., M. ASCE

Training programs in the operation and maintenance of Third World public works utilities have, at best, produced mixed results. People at all levels have received training, but whether this training translates into improved facility operation depends on the continuous motivation and skill of the operating and supervising staffs. This paper discusses some reasons why present programs may not meet long-term training needs and suggests ways to expand traditional training programs to make them more effective.

Introduction

Many public works projects undertaken by A/E firms include some form of operation and maintenance training for completed facilities. Training programs are vital in projects in which technologies and practices that have been developed in and for Western cultures are applied to a Third World setting. Western A/E firms have become more sophisticated in the development and conduct of training programs in these nations.

Despite the increasing sophistication of these programs, there are doubts about the long-term effectiveness of any training program in the Third World. These doubts spring from problems that are inherent in many water supply and sanitation utilities in developing nations and are not improving through present assistance programs. The low prestige given to water supply and sanitation careers, plus low salaries, discourage qualified people from entering these fields, damage morale, dampen motivation, and encourage the best and most capable people to leave for greener pastures.

The future effectiveness of a training program will depend on whether or not trained personnel at all levels retain their motivation to stay in the field and apply their newly acquired skills. Present training programs do not adequately address the long-term issues of pay and prestige, but they should if assistance programs are to help the host country. While these

* Project Manager, Metcalf & Eddy International, 10 Harvard Mill Square, Wakefield, Massachusetts, 01880.

issues may seem hard to manage, or at least beyond the reach of most training efforts, assistance programs can be redirected to get closer to the root of the problem.

The Problem in Egypt

Careers in the water supply and sanitation fields are not highly regarded by engineers and scientists in developing nations. This attitude prevails in Egypt because of cultural and societal factors.

Egypt's rich heritage is based on rural life, as are many other Middle East countries. Traditional attitudes towards manual labor, personal hygiene, and the role of the individual in the rural community are opposite those of the present rapidly growing and shifting population. The rural traditions have not adapted easily to the country's growth into a more urban society.

The explosive growth of urban populations placed a premium on new housing construction and, with it, the abilities of structural and construction engineers. Because these abilities fit into the cultural traditions of this area, jobs were easily filled. The emphasis on industrial development, typical of most Third World societies, resulted in the need for expertise in a variety of technical and managerial fields. These, too, could largely be accommodated by traditional attitudes.

The highly technical positions in the chemical and petroleum industries were also particularly desirable. As the society's upper levels were attracted to the emerging technologies, these new fields of expertise acquired prestige within the society. This resulted in higher pay which also helped to attract the most qualified people. The status attached to these fields by the educated and ruling classes eventually filtered down through the rest of the society The pay scales were raised at all levels and the better qualified people from every social level began working in these fields.

The water supply and sanitation fields have not received a similar level of attention and esteem in Egypt. A career in the sanitation field is considered undesirable. Without the support of governing bodies and the patronage of the upper classes, the fields of water supply and sanitation remain unprestigious, low paying, and unattractive. The effects are noticeable at every level, but the impact is most keenly felt in the education system.

The reputation associated with an occupation is directly tied to the quality and strength of the university programs in that field. Programs in structural, chemical, petroleum, mechanical, and electrical engineering are usually very strong in most Middle East universities. These programs lure the top professors and the brightest students, and are well funded. On the other hand, the typical sanitation engineering program has to scramble for qualified professors and funding, and cannot attract enough of the best and brightest students. For example, an engineering student's progress through the educational system is increased by the results of year-end comprehensive exams. The students with the highest scores can select the programs they wish to pursue. Inevitably, the top achievers select the more prominent and better funded engineering programs. Less prestigious and poorly financed programs continue to struggle.

The negative impact on the educational system has further intensified the shortage of qualified managerial and technical personnel. There are many skilled and dedicated people in the field, but the number at all levels is still inadequate. The situation becomes worse as water supply and sanitation systems are pressured to expand because of increasing population. These pressures, coupled with the problems of low prestige and poor pay, make it very difficult for qualified people to remain in the field.

Present Practice

The typical water works or sanitation training program covers both technical and managerial aspects. Technical training is provided in a classroom setting and in hands-on, or on-the-job training. Managerial and administrative training is usually conducted in offshore visits, which include the course, site visits, and professional conference attendance. A simplified breakdown of the staff classifications and the type of training provided is shown in Figure 1.

Classroom and hands-on technical training is generally designed for plant staff, including semi-skilled and skilled labor, foremen, and some plant supervisors. Staff engineers are sometimes involved as well. Offshore training is generally directed at selected administrators, staff managers, staff engineers, and plant supervisors. As Figure 1 illustrates, the staff engineer and plant supervisor levels generally receive the most comprehensive training.

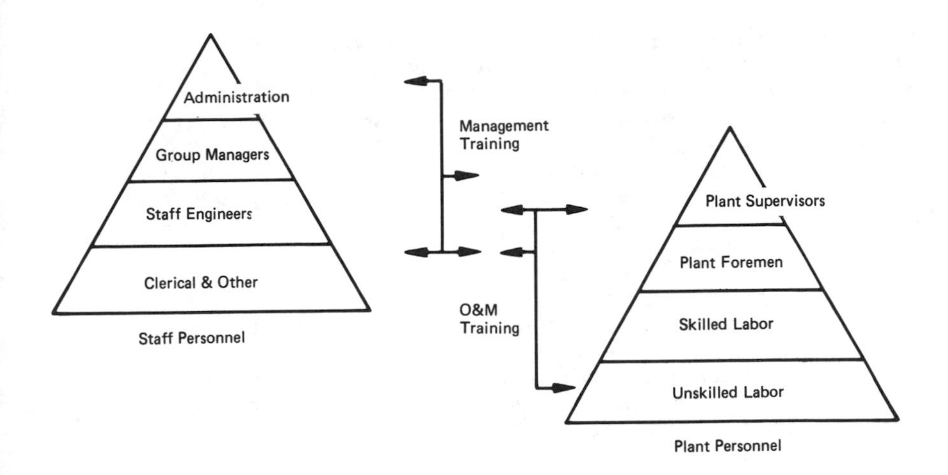

FIGURE 1. TYPICAL TRAINING BY CLASSIFICATION

The difficulty is not in the quality or the scope of present training practices, but in the underlying working conditions that await the people. They have changed (for the better, hopefully) but their working conditions have not. Salary is the most immediate problem. It is difficult, if not impossible, for a public works agency to raise these pay scales, even for these valuable, newly trained employees. It is not surprising that many of the newly trained staff seek jobs in other countries where they can earn 2 to 3 times their present salary. If they have to remain in their same jobs at the same salary, they will gradually grow discontented and lose the motivation to apply their new skills. Either way, the utility loses.

Attacking the Problem

The heart of the problem facing most public works agencies in the Middle East and other areas is the cultural attitudes towards the sanitation fields. It is impractical to alter these attitudes but it is possible to attack their societal manifestations. The best opportunity for improvement is changing the perception of the field at the social structure's higher levels. At the lower social levels, the cultural attitudes are more pervasive and more strongly held. The power to change these attitudes rests with the upper levels of society. The higher educational systems offer the best opportunities for improvement and may even be relatively easy to approach.

University programs in environmental engineering, encompassing fields of study in water supply and sanitation, are poorly funded and understaffed in Egypt. But there are two ways to improve this situation: through governmental assistance programs, and through joint associations with successful environmental programs in the west.

Governmental Assistance

Typically, aid programs have concentrated on the planning, design, and construction of needed water supply and sanitation facilities. More recently, operations and maintenance training programs have been included as well. These programs only provide short-term relief, because the underlying cultural and social issues have not been addressed. A fairly modest extension of these aid programs could go a long way towards providing long-term improvement. Aid should be extended to cover improvements in existing academic programs in water supply and sanitation and, more importantly, in establishing new programs. Governmental aid funds should go to the rehabilitation of existing facilities where appropriate, and to the design, construction, and equipping of new classrooms, laboratories and research facilities. Modern and well-equipped facilities will help attract the better professors and students. Yet, better facilities alone will not be enough.

Aid should also be used to set up and finance permanent funds for endowments, scholarships, fellowships, and research grants. These funds could be set up to be self-perpetuating and independant of future aid programs. These funds could also help defray program operating costs, such as classroom materials, laboratory supplies, clerical staff salaries, and facility maintenance. Exchange programs for students, research assistants and professors could be established with these funds.

The combination of modern and fully equipped facilities, and permanent funds for program staffing, maintenance, and operation will improve the status of environmental engineering as a career choice. Starting a new program with new facilities will also help provide a break from past attitudes.

Private Assistance

Another method for improving environmental programs in Egypt is through cooperative ventures with university programs in the west. Collaborations have been tried in the past, usually on a project basis. More permanent relationships would tremendously benefit these fledgeling programs and the students and staff of the western university. Governmental aid programs could fund efforts to attract people who were born in the host country, were educated in the West, and have established professional careers in the West. Luring these successful professionals back to their native country either as visiting lecturers or in tenured posts would do much to improve the quality and competiveness of the new environmental programs.

The participating western university would have the opportunity to observe the engineering problems and solutions of another culture, and to participate in these efforts. A more attractive benefit is that needed funds could be obtained. The governmental aid programs could be set up to encourage or even require the participation of qualified western environmental engineering programs in the development, operation, and staffing of the new facilities in the host country.

Summary

The construction of new water supply and sanitation facilities in Egypt providing much needed assistance to aging and decaying infrastructures overwhelmed by exploding population pressures. The recent attention of aid programs to the need for training in the operation, maintenance, and administration of the new facilities has further improved the effectiveness of these projects. However, the effects of these aid programs may be short-lived if the underlying problems of pay and prestige in the fields of water supply and sanitation are not addressed. A likely approach to these difficult problems may be through improvements in the country 's academic programs. By improving the quality and motivation of the mid- to upper level professionals in the water supply and sanitation agencies, the perception of the entire field will be improved. Improved prestige will lead to improved pay scales. Greater respect and better pay will then attract the more qualified people of all job classifications, resulting in the long-term improvement and stability of the water supply and sanitation field.

Preparing Engineers To Help World Development Through A Curriculum In Appropriate Technology

G. William Zuspan*

Abstract

Drexel has introduced a new curriculum in Appropriate Technology in its Engineering College. Students enrolled in this curriculum will receive a basic Engineering education with an emphasis on developing a global perspective. The students will be concerned with working on real problems within developing countries and working on those problems in the field during their co-op period.

The basis of the engineering profession has always been to apply technology to the real and basic needs of the world. In our high technology society this aspect of engineering seems diminished since many of our developments are so complicated and removed from the heartbeat of life that they are not crystallized in the awareness of the people.

The needs of the world as represented by developing countries is enormous and complex. The developed world places great pressure upon the less developed country to make a quantum jump in development from their present state to the world of high technology. Such a change is not only difficult to accomplish but may in fact destroy the potential for a solid based development program which will result in self-sufficiency for a country.

Many efforts have been made since the close of World War II to help nations which suffered economic and social devastation. The Marshall Plan was successful in helping nations which were developed countries prior to World War II to regain and even surpass their previous level of advancement.

When our nation shifted its help to less developed countries little effort was made to change the style of help which was largely aimed at making a country into a facsimile of the USA. Because of complex social, cultural and political problems these efforts have been less successful than first anticipated. Development work has received a bad name in many places and especially that portion called Appropriate Technology.

Development work in the past which has flown under the banner of Appropriate Technology may in fact have been less appropriate than other measures. However, the basic concept of Appropriate Technology is valid for when it is applied to local needs as expressed by local people through local infrastructures, then it is successful and can be the real answer to development that will maintain a local base which will sustain it over the years.

Drexel University became interested in Appropriate Technology in 1983 and began the process of developing a curriculum within the Engineering College. At its root is the concept that highly trained

*Professor of Engineering, Drexel University, 32nd & Chestnut Sts., Philadelphia, PA 19104

technical people whose vision has been expanded globally and who have been indoctrinated into the theory and practice of Appropriate Technology can in fact help nations create the devices and systems that are necessary to help them move from their present stage of development to a more advanced stage.

In order to accomplish this, Drexel modified the General Engineering Curriculum to include not only the basic fundementals of engineering but the opportunity for students to carefully select a program which would develop their expertise in a particular field such as water supply and at the same time give to them the understanding needed in order to work effectively in a devloping country.

The only special courses in the curriculum are the four courses in Appropriate Technology taught by an Engineering Professor and a Professor of History and Politics. The first course is offered at the freshman level with three other courses, one each in the second, third and fourth years. The thrust of the Appropriate Technology courses is to help the students develop an understanding of the principles involved and to apply them to real problems from a particular country. The capstone is a Senior Design course which is directed towards the solution of a problem from a devloping nation.

Drexel University is a Cooperative Education School and the students in Appropriate Technology spend eighteen to twenty-one months in a work assignment in a developing country. To date students have worked in Haiti, Zaire and Appalachia. The co-op program is expanding and groups such as CARE, Catholic Relief Services, The Peace Corps, Save The Children, Habitat For Humanity, Lutheran World Relief, The Heifer Project International and others are accepting Drexel co-op students and are providing projects which need solutions from a wide variety of countries.

Currently, Drexel students are working on a housing design to withstand typhoon force winds for the Solomon Islands, a water harvesting project for the Heifer Project, a human powered planer for Habitat For Humanity, a mini-irrigation system for a community garden in Haiti, a rat and thief proof grain storage unit for Save The Children in Gambia.

Drexel students in the past two years have designed and implemented a pig raising project incorporating a biogas digester and fish raising pond, a swivel bridge to eliminate wash-out during the rainy season, a church building, a large vehicular bridge, a bicycle powered table saw and grinder, a solar food dryer, a bicycle powered dry cell battery charger. They have also worked on many other designs producing a preliminary engineering report incorporating creative approaches to the solution of many different problems.

The intent of Drexel's Appropriate Technology program is to begin to channel engineering expertise and creativity into the real problems facing developing nations. This requires engineers with a broad global perspective and a desire to help people who are suffering the agonies of hunger, thrist and poverty. The key is to begin to provide developing nations with a pool of engineers especialliy educated to help, not hinder the gentle but positive development process

Considerations for Engineering Mobilization in Colombia

Stewart M. Oakley*

ABSTRACT

Engineering mobilization for the development of appropriate technologies is an important element in solving water supply and sanitation problems. While the sociocultural aspects of appropriate technology have been widely discussed in relation to project beneficiaries, less attention has been paid to their effect on engineering mobilization. Using Colombia as an example, this paper examines how historical and sociocultural processes influence the ability of the engineering community to adequately address specific problems. It is concluded that unless these sociocultural influences are taken into consideration, the engineering community will play a limited role in developing appropriate technologies.

INTRODUCTION

Since the United Nations Water Conference at Mar del Plata in 1977, there has been a worldwide rise in interest in appropriate technologies for water supply and sanitation in developing nations. Field experience has clearly shown, however, that technology alone cannot succeed without addressing critical sociocultural factors. Many failures of projects that appear to be technically feasible, for example, have been attributed to failures to consider the role of cultural values and behavioral aspects, especially in relation to community participation (6).

An important, and perhaps decisive, element in solving water supply and sanitation problems is the mobilization of engineering capabilities for the design, construction, and operation of appropriate technologies. While the sociocultural aspects of appropriate technologies have been widely researched and discussed in relation to the intended beneficiaries of projects, less attention has been paid to the ways sociocultural factors influence the ability of the engineering infrastructure to adequately address specific problems. Engineering capability in any society must be considered in its technical, cultural, and historical context before it can be successfully mobilized; using Colombia as an example, this paper examines the influence of these factors on engineering mobilization.

Colombia faces many of the severe water supply and sanitation problems typical of countries throughout Latin America. Excreta-related and water-related infections are a major cause of morbidity and mortality for all age groups. Many urban areas severely lack capacity for

*Associate Professor, Department of Civil Engineering, California State University, Chico, California 95929-0930, U.S.A.

water distribution and sewerage; while water treatment is practiced in cities to varying degrees, municipal wastewater treatment is essentially nonexistent. Rural areas suffer greatly from lack of potable water and sanitation services.

In an attempt to improve this situation, engineers from Colombia and the international community have been retained by development agencies to initiate numerous projects in urban and rural areas that emphasize appropriate technology; many of these have failed. The failures, however, seem to have as much to do with the nature of engineering as it exists in Colombia as with the cultural and behavioral aspects of project beneficiaries. The complex interplay of cultural values, social structure, technical expertise, and historical processes within the Colombian engineering community--areas which have seldom been considered because they represent a "cultural background" among professionals--plays a key role in the probable success or failure of any appropriate technology project.

In this paper I wish to examine how historical and sociocultural processes within the Colombian engineering community influence its ability to be mobilized to develop appropriate technologies. Because of the pervasive influence of U.S. engineering on Colombian engineering, the context in which environmental engineering in the United States has been developed will be contrasted with what is needed in Colombia. It is concluded that until these sociocultural influences are considered and accommodated, Colombian engineering, which should be counted on as a valuable resource, will be severely limited in its ability to be mobilized effectively.

ENGINEERING AND THE CONCEPT OF APPROPRIATE TECHNOLOGY

Much of the discussion on appropriate technology has focused on hardware and has ignored the decisive role engineering capabilities play in shaping the direction of technological development. This is unfortunate since technology, in its broadest sense, consists not only of hardware but also engineering capability; this comprises the entire body of knowledge, experience, methods, and organizational forms of a technological system. The hardware, for the most part, is developed as a result of existing engineering capability.

The engineering component of technology is the result of historical processes and the nature of the sociocultural system in which it was developed; it thus carries with it implicit values that orient the focus of engineering work in specific directions. These subtle influences must be fully appreciated before engineering knowledge from one society can be adopted elsewhere, or mobilized to develop a different form of technology. As Jequier has aptly noted, it is not necessarily the hardware component of technology which causes problems for developing countries, but a lack of "appropriate engineering" in the form of knowledge, experience, organizational forms, and institutional structures (2). While transferring hardware from one country to another may be an easy task, transferring engineering capability is much more difficult because it is culture-specific. The development of engineering capability in the form of knowledge, experience, and organizational structures commensurate with a given culture's values, needs, and resources is the essence of appropriate technology.

Unfortunately, the culture-specific orientations of engineering as it has been developed in the U.S. are frequently not appreciated by either Colombian or U.S. engineers. Indeed, it is commonly assumed that the technology and engineering methodologies developed in the United States are also appropriate elsewhere. Many Colombian professors have studied engineering in the United States, and all universities which have engineering programs model curricula after their U.S. counterparts. Since indigenous authors are rare, these engineering schools for the most part use translated textbooks originally published in the U.S.; the one environmental engineering textbook published by a Colombian engineer relies exclusively on material published in the U.S. (1).

Unless fundamental changes occur, it is doubtful that the Colombian engineering community can be effectively mobilized if it continues to use the U.S. model of engineering, which is still not sympathetic to the concept of appropriate technology. It is therefore paramount that the origins and context of engineering capability developed in the U.S. be fully understood and compared with Colombian circumstances before an appropriate technology that is likely to be successful can be developed.

HISTORICAL AND SOCIOCULTURAL FACTORS

United States

At the beginning of the nineteenth century the U.S. was essentially a developing nation that had to import many essential items from England (2). During the Napoleonic Wars it became isolated from its British suppliers for almost two decades. Gradually, as a result of the interruption in trade, the U.S. began to develop its own small scale industries which paved the way for its industrial revolution. These small scale industries, which originated both in the cities and in rural communities, developed their own skilled labor. Engineers were trained in the "shop culture", which involved an apprenticeship placing emphasis on practicality and acquisition of skills in the "school of experience."

In the latter part of the nineteenth century, engineering education became formalized into higher education, resulting in the formation of the "school culture". When engineering first entered the domain of the university, however, it was viewed with positive disdain from other disciplines who believed technical training had no place in academia (3); this view is certainly prevalent even today. To gain an air of academic respectability, the school culture gradually adopted the rigor of the sciences as the best way to train engineers; the increasing complexity of engineering problems at the turn of the century, coupled with the needs of the newly developed science-based industries, gave further credence to the scientific and analytic orientation of the school culture (3).

For some time there was a debate between the shop culture, which argued for a practical, hands-on approach to engineering, and the school culture, with its emphasis on science and mathematics. The school culture obviously triumphed in the U.S., and the shop culture of engineers has been relegated to roles as skilled technicians and technologists, who receive their training at vocational schools, technical institutes, and junior colleges. Many of the values of the shop culture, however, continued to play a role in engineering education up to recent times. Textbooks published before World War II are replete with descriptive material, visual aids, practical applications, and design criteria based on experience and rule-of-thumb calculations. Shop and field work with emphasis on technical skills, although now eliminated from most curricula, were fifty years ago an integral aspect of engineering training. Unfortunately, the modern day school culture now exhibits the same disdain towards practical, hands-on training that was once directed at it when it entered academia.

Colombia

While the lower- and middle-class Europeans who colonized the United States came there to work and develop a society based on the fruits of their labor, the Spanish who colonized Colombia had something quite different in mind. The Spanish colonies were regarded as places rich in primary materials to be exported to Spain, which, in turn, exported industrial products to Colombia. An indigenous industrial revolution thus never took place. To this day Colombian industrialization is based primarily on imported technology, in the sense of both hardware and engineering.

The class society resulting from the Spanish colonization has also had a significant effect on technological development. The concentration of economic resources in the upper classes has perpetuated a social system that discourages small scale technical innovation (5). Manual labor has been identified with a subservient class and has always been given a low status; as a result, there has never been a shop culture in Colombia approaching anything like that in the U.S. The school culture, however, has a high status, and it has been relatively easy to adopt the U.S. system of engineering education throughout Colombia (5).

Although Colombian and U.S. engineers may speak the "same language" in the sense that their educational backgrounds are very similar, it is important to realize the differences in the ways engineering functions in their societies. In the U.S., engineering has a vast infrastructure of skilled technicians and workers who have a relatively high social status, are paid well, and who frequently have technical responsibilities comparable to those of engineers. Although the school culture frequently looks down on this shop culture in educational settings, in practice and historical context there is sufficient precedent to acknowledge its status and importance.

In contrast, Colombia has a minimal infrastructure of skilled technicians and workers. There are very few vocational schools, technical institutes, or junior colleges to train persons to operate the types of technologies developed in the U.S. There is also no historical precedent of a shop culture, which lacks social status. Colombian engineers thus have had little practical experience with important technical considerations such as operation and maintenance. The lack of a shop culture has important ramifications for the development of appropriate technologies using manual or semi-skilled labor; this type of work is not likely to be readily accepted by anyone, and engineers who come from higher classes often have a cultural barrier between themselves and the lower classes they may be trying to help.

THE ROLE OF ENGINEERING EDUCATION

Since the end of World War II, in a continued effort to improve its academic stature by emulating the sciences, the school culture in the U.S. has undergone changes that now give almost exclusive emphasis to methods of analysis based on scientific theory. Traditional engineering based on the shop culture has either been eliminated by modern technology or relegated to a lower status (in the view of engineers) where it can be dealt with by technicians and skilled or semi-skilled workers. Engineers trained in the U.S. thus have a bias against the shop culture approach to engineering, even though the historical roots of modern technology were founded on it. At the same time, these engineers implicitly assume there will be an organizational infrastructure, including a cadre of skilled technicians, that will fulfill the operation and maintenance requirements of the sophisticated technologies they have been taught to design.

Today environmental engineering in Colombia, because it has adopted the U.S. model of engineering education, focuses almost exclusively on the scientific and mathematical analysis of complex systems emphasizing urban areas. Proven technologies that are simple, and may even be in widespread use, often are not considered to fall under the purview of engineering if they do not fit into the present scientifically oriented value framework.

The types of technologies which are more appropriately related to many of the problems of water supply and sanitation in Colombia are thus not studied in engineering. What is worse is that they have been relegated, in the view of many engineers, to a lower status for sanitarians, public health workers, and technicians; this is due, of course, to the shift in emphasis in engineering education away from the practical, shop culture approach to the scientific, school culture approach. In Colombia, however, there exists neither a shop culture which could aid in the development of indigenous appropriate technologies, nor an organizational infrastructure with skilled labor to operate and maintain the technologies developed in the U.S. These deficiencies are compounded further since manual and skilled labor have a low social status, and because engineers, due to class divisions, are limited in their ability to adapt designs to local conditions. It is urgent for engineers to be aware these influences. Until this is done, it seems unlikely that the mainstream engineering community will take an interest in appropriate technology, or, if it does, it will be effectively mobilized.

APPROPRIATE TECHNOLOGY IN WATER SUPPLY AND SANITATION

Environmental engineering in Colombia currently focuses on the scientific analysis and design of large-scale water and wastewater systems which have been successful in meeting the water quality and treatment objectives of urban areas throughout the U.S. Although there have been numerous articles discussing the success of water supply and sanitation projects in Colombia, little attention has been paid to the role of the engineering community in fostering the design, construction, operation and maintenance of appropriate technologies. For example, although over 90 percent of all urban areas have house connections or easy access to treated water supplies, these services chronically lack sufficient capacity for demand and suffer sporadic service; in addition, there are thousands of urban dwellers in shantytowns who have no easy access to treated water. While about three-fourths of the rural population has easy access to a drinking water supply, as many as 40 percent of these systems may not be working at any given time (4). Clearly, there is still much room for innovative engineering for systems which are easy to design, install, operate, and maintain. In this area, environmental engineering as taught and practiced in Colombia has little to offer. Historical practices, however, such as the use of the slow sand filter in the U.S., may have special relevance for the development of appropriate technologies.

While in principle the technology suitable for urban areas in Colombia could be similar to that in the U.S., in practice there is a serious problem with operation and maintenance. U.S. technology has been developed using sophisticated equipment requiring a cadre of specialized workers to operate and maintain it. Almost all water and wastewater treatment plant operators and technicians have received specialized training at universities, junior colleges, technical institutes, and formal seminars; these personnel also have somewhat of a professional status through their affiliation with such organizations as the American Water Works Association. This type of organizational infrastructure does not exist in Colombia, and there is a severe shortage of qualified personnel; those who do operate treatment plants often are not paid well, are ill-trained, and lack a professional-type status which would motivate others to enter the field.

Rural areas pose the greatest problem for the development of appropriate technology because of cultural influences. Many rural systems that have worked well have been abandoned after a breakdown (Figure 1); this has usually been attributed to the failure of community participation in the project, and the lack of an organizational infrastructure for operation and maintenance. A more subtle influence is the class difference between the designers of a project and the intended beneficiaries. College educated engineers from urban areas will often have a difficult time communicating with -- and being accepted by -- campesinos. Because of the low status of manual and semi-skilled work, rural residents will not usually be enthusiastic about operating and maintaining a technology being given them by someone who has a high status because they have a university degree and have avoided manual labor! The role of gender is also significant since in many rural areas the responsibility to collect water is delegated to women; if discussions on water supply systems only include men, as is typically the case in Latin American culture, most projects will be doomed to failure from the start (Figure 1).

Figure 1. A broken down water pump in rural Colombia that was not repaired; instead the well cover was removed so water could be collected with a bucket. This well belonged to a man who had just received three years of training in water supply and sanitation, and was supposed to install handpumps in his community! In this particular area, it was the responsibility of women to collect water; because of cultural factors, however, only men received training. The pump was not repaired by the man because he would never think of collecting water; his wife, of course, could not repair it and thought, quite logically, that it was easier to collect water with a bucket anyway. The university engineers who designed the pump and did the training never visited the rural community they were attempting to help.

A shop culture approach to water supply and sanitation is badly needed in rural and urban Colombia. For this to happen it would be desirable for engineers to examine the history of water supply and sanitation in the U.S., and to adapt these older systems, or new ones developed indigenously, to the existing sociocultural circumstances. This will be no easy task unless fundamental changes occur.

CONCLUSION

For the engineering community to be mobilized effectively to develop appropriate technologies in water supply and sanitation will call for a different form of engineering -- an ingeniería latina -- more appropriate to existing conditions. To do so it will be necessary to develop, indigenously, a knowledge and experience base, and the ability to experiment and innovate with technologies on a trial-and-error basis. A shop culture approach adapted to Colombian circumstances is likely to be the most appropriate way to accomplish much of this.

The greatest problem is overcoming the pervasive influence and biases of engineering as practiced and taught in the U.S. Fundamental changes are likely to occur only when engineers become more sensitive to value assumptions, cultural differences, and the urgent need for appropriate technologies. Until this is done, the engineering community will likely play a limited role in the development of appropriate technology.

APPENDIX -- REFERENCES

1. Bedoya, J., El Hombre Y El Ambiente, Universidad Nacional de Colombia, Medellin, 1981.

2. Jequier, N. "The Major Policy Issues," in N. Jequier, Ed., Appropriate Technology: Problems and Promises (Organization for Economic Co-operation and Development, Paris, 1976) pp. 16-112.

3. Noble, D., America By Design (Oxford University Press, Oxford, England, 1977).

4. Pan American Health Organization, Environmental Health: Country and Regional Activities in the Americas, Environmental Series No. 2, Washington, D.C., 1982.

5. Safford, F., The Ideal of the Practical: Colombia's Struggle to Form a Technical Elite (University of Texas Press, Austin, Texas, 1976).

6. van Wijk-Sijbesma, C., Participation and Education in Community Water Supply and Sanitation Programmes: A Literature Review, Technical Paper No. 12, WHO International Reference Center for Community Water Supply, The Netherlands, 1979.

The International Training Network for Water and Waste Management

Michael Potashnik *

Introduction

The International Drinking Water Supply and Sanitation Decade has passed the halfway point. General assessments which have been conducted indicate that although substantial improvements have been achieved towards the provision of potable water and sanitation facilities, the gap between the demand and service continues to grow. With this observation, many countries have begun to restudy their current policies and strategies to redirect development efforts. To assist in accelerating this process, the World Bank and the United Nations Development Programme initiated the International Training Network for Water and Waste Management (ITN). ITN has been established to help improve the effectiveness of investments in the sector and to promote and encourage more extensive use of alternative approaches and appropriate, affordable technologies through research, training, and information dissemination.

Current Situation

The Decade, which traces its origins to this region, has undoubtedly raised the level of awareness and enthusiasm among sector agencies, donors, and beneficiaries. Increased investments and activity have characterized the sector world-wide. Many countries have placed a high priority on investments to make water supply and sanitation facilities available to the most remote rural areas. The Decade Report of the U.N. Secretary General of April 1985, indicated that Latin America and the Caribbean registered the most impressive performance in urban water supply and sanitation.

About 85% of the population in urban areas have adequate water supply and 80% have sanitation facilities. In rural areas, however, only 49% have water supply and 20% have sanitation facilities.

Institutional arrangements for rural development are not as clearly defined as those for urban development. Existing institutions manifest a strong orientation towards conventional technologies applicable to urban areas. This is expected to aggravate the imbalanced development between urban and rural areas. Many national and local institutions have recognized this situation and are considering alternative affordable technologies and innovative approaches.

---

* Chief, WUDTR, Water Supply and Urban Development Department, World Bank, 1818 H Street, N.W., Washington, D.C. 20433.

	Population x 1.0 M.		Population Covered							
			Water Supply				Sanitation			
			1980		1983		1980		1983	
	1980	1983	Pop	%	Pop	%	Pop	%	Pop	%
URBAN	234	254	183	78	215	85	131	56	203	80
RURAL	124	126	52	42	62	49	25	20	25	20
TOTAL	358	380	235	66	277	73	156	44	228	60

Sources : 1980 Report of the Secretary General concerning the Decade; 1983 WHO surveys

In Tegucigalpa, Honduras, the Pan American Health Organization (PAHO) has conducted a review of the urban water supply sector and is now looking at sanitation in urban poor areas. UNICEF has expressed interest in assisting low-cost sanitation efforts in Honduras. In Colombia, the city of Cartagena plans to extend the city sewerage system to two low-income communities using affordable low-volume pour flush toilets (the "taza campesina") and small bore sewers. In Guatemala, UNICEF has planned a comprehensive sector review in rural sanitation. It has previously supported the construction of latrines which has gained wide public acceptance. Guatemala plans water supply to both urban and rural areas through a mix of service levels, and sanitation through sewerage and latrines. Bolivia, with a large unserved rural population, plans to provide minimum but safe water supplies to rural areas through handpumps and protection of water sources and sanitation through latrines. UNICEF is similarly preparing to assist Barbados, Surinam, and Guyana in rural water supply and sanitation. These examples reflect the emerging response to needs in rural and marginal urban areas in the region.

Analysis

Specific policy and approach issues must be addressed to adequately meet the needs of rural and urban depressed areas.

1. Financing and Cost Recovery: Investments, though already large, are still inadequate due to the highly capital-intensive nature of the conventional technologies currently applied. More institutions now realize the need for efficient and equitable cost recovery policies and mechanisms to develop and construct systems. Many systems become affordable when appropriate technology alternatives are considered.

2. Approach: If project implementors agree that improved health is one of the prime objectives, then a serious study into socio-cultural factors is indispensable. Research shows that improvements in health correlates better with increased literacy, level of female education and income rather than with level and availability of water supply and sanitation

services. In sanitation project particularly, the impact of socio-cultural norms and values cannot be overlooked. The need for a multidisciplinary approach becomes very clear.

3. Management, Organization, Development, and Training: Institutions and people are key elements for success in any undertaking. Effective planning, decision-making and control systems, and the exercise of leadership are crucial processes. Staff must have the appropriate skills, knowledge, and attitude to do their tasks competently. Many of the constructed systems are either underutilized or are breaking down and are being abandoned due to poor operation and maintenance, inadequate training, and inefficient management of resources.

4. Appropriate Technology: Appropriate technologies do exist. Incentives for its adaptation and application are, however, inadequate. Conventional technologies, which have become expensive in the backdrop of depressed rural and poor urban economies, have tended to require either high levels of government subsidy or infeasible cost recovery schemes. The upgrading of engineering education and professional training programs can make a significant impact on policy makers, practicing engineers, and students in recognizing the value of appropriate technologies.

5. Community Involvement: Many implementation and operational problems have arisen due to the lack of community participation in project development and implementation. Current practices have tended to make passive observers out of the beneficiaries.

Concerned policy makers, donor institutions, and planners have begun to challenge their current policies, objectives, and strategies with a view towards improving the performance of the sector.

The Network

To assist in analyzing these issues, the World Bank recently initiated the International Training Network for Water and Waste Management (ITN). This is a joint project of bilateral and multilateral development agencies coordinated by the Bank. It was established in 1984 to help improve the effectiveness of investments in the sector and to promote and encourage more extensive use of and further research into low-cost technologies and the multidisciplinary approach in the development of water supply and sanitation.

In collaboration with existing institutions, to be called Network Centers, in developing countries, the Network seeks to:

1. inform decision-makers, educate and train professional and

student engineers, and other field staff on the use of low-cost appropriate water supply and sanitation technologies;

2. promote the use of a multidisciplinary approach, emphasizing socio-cultural and health aspects in the planning, implementation, and management of water supply and sanitation systems;

3. support the active collection and dissemination of information on low-cost technologies and their successful applications; and

4. undertake research to improve the cost effectiveness and replication of sound water supply and sanitation programs.

Network Centers are established in educational centers, sector implementing agencies or professional organizations who have a widely acknowledged reputation for excellence. They are selected based on their previous experience in administering formal or nonformal education and training programs. The Center must have the necessary facilities and equipment to conduct workshops for instructors, staff and those of other institutions. A senior instructor is assigned from among the institution's multidisciplinary staff to direct the activities on a full-time basis. Other Center personnel are recruited as needed. Senior management's interest and support is ascertained through its commitment to provide partial funding in cash or kind towards the cost of the Center's activities and full funding by the end of the fifth year when the Center is expected to be self supporting and all external support normally terminates. Universities who host a Center, not only strengthen their existing curriculum, but also create new programs based on the Network materials and current research. Assistance for faculty and student research work may be available through the Network.

The Network Coordination Unit promotes the establishment and development of these Network Centers by providing technical and management guidance. It arranges with potential bilateral and multilateral donor agencies for assistance in the start-up activities of the Centers. It coordinates the participation of development oriented Associated Institutions in the conduct of initial instructors' workshops. It monitors and reviews the plans and activities of the Centers.

The Network Centers extensively use the audio-visual modules and manuals developed based on studies done by the World Bank. These materials are available in various media (16mm films or video tapes and slide/cassettes) supplemented by Participants' Notes, Instructors' Guides and Reading Materials.

The information and training materials are designed for three different audiences:

1. Decision-makers in government or financing institutions, to make them more aware of the alternatives which are available for meeting the water supply and sanitation development objectives.

2. Professional engineers, staff, and students who will be responsible for technical and operational decisions, to familiarize them with a broader range of technologies.

3. Project staff who work with the communities to be served. The materials are designed to give project staff an insight into how to mobilize community participation, carry out hygiene education, and improve communication techniques.

Network Centers for Latin America and the Caribbean

Some institutions in the region have expressed the need for the creation of Network Centers in the region. A proposal has been drafted for consideration of potential donor agencies.

The proposal envisages the formulation of a regional training strategy with clearly identified priorities for human resources development and local institutions who can operate as Network Centers in order to meet the needs. At least three Network Centers will be set up initially; more Centers will be established as the demand increases. Training and education activities will be planned and implemented by the Network Centers for decision-makers, engineers, students, instructors, community based workers, and environmental health personnel on affordable water supply and sanitation. The Network also envisages the adoption of new concepts and approaches to low-cost water supply and sanitation, community participation, and health education into the existing education and training programs of universities, polytechnics, and sector agencies. Additionally, audio-visual information and training materials on affordable water supply and sanitation technologies, including actual case studies on their application will be prepared. These activities all aim to achieve an increased understanding of affordable technologies and the financial and institutional arrangements that are appropriate to Latin America and the Caribbean.

Several institutions have expressed their interest in hosting Network Centers. In Guatemala, the Escuela Regional de Ingenieria Sanitaria y Recurcos Hidraulicos (ERIS), Universidad de San Carlos, which has been recognized as a leading center for post-graduate education and training in sanitary engineering and water resources for the past 21 years in Central America, appears to have the essential requirements for hosting a Network Center. The Centro Panamericano de Ingenieria Sanitaria y Ciencias de Ambiente (CEPIS) of the Pan American Health Organization (PAHO) has been invited to assist in planning and conducting research, training, and information dissemination activities on appropriate technologies. In Brazil, the Network will cooperate closely with ABES, the Companhia de Tecnologia de Saneamento Ambiental (CETESB) and similar institutions.

PAHO, with funding from the German Agency for Technical Cooperation, GTZ, will soon finish the Spanish translations of the training and information materials. In the future, PAHO will provide technical and financial support through CEPIS.

In the coming months, the Network Coordination Unit will jointly prepare a regional strategy for training in water supply and sanitation in Latin America with the Latin America and Caribbean (LAC) Projects Division and the Economic Development Institute (EDI) of the World Bank. Reconnaissance, institutional assessment and project preparation activities for setting up Network Centers in Central America and in Brazil will be done. Arrangements will be made with CEPIS for adapting materials, convening workshops on curriculum development, facilitating technical information exchange and similar activities. A regional workshop on low-cost water supply and sanitation for instructors from prospective Network Centers is also planned.

Conclusion

The development challenges in Latin America and the Caribbean lie in the expansion into rural and depressed urban areas through affordable technologies and innovative approaches and in the development of effective institutions who can plan, implement, manage, and maintain numerous, but small scale, water supply and sanitation projects. Decision-makers and professionals are urged to look closely and seriously into the innovations and opportunities for low-cost technologies and approaches in order to meet the needs in the rural and urban poor areas. The establishment of Network Centers will hopefully accelerate this re-orientation process.

Mr. Potashnik is a staff member of the World Bank.

India's Experiences and Plans in Water and Sanitations Systems Development

K.S. Murty*

Because of its sheer size, India is probably the most important single country for the International Water and Sanitation Decade, 1981-90. One in every three people lacking clean water and sanitation in the world is an Indian. India's commitment to the UN Drinking Water Supply and Sanitation Decade is definite and unhesitating. The plan allocation rose from Rs.490 million in 1951-56 for water and sanitation to Rs.41177 million in the Sixth Plan(1980-85). The targets recommended for the decade are total coverage of the rural and urban population in respect of water supply, 100 per cent coverage of the population of Class I cities in respect of urban sewerage and sanitation, and 50 per cent of the population in Class II cities and other towns. The overall coverage in each State would be 80 per cent of the urban population by means of sewerage or simple sanitary methods of disposal. In respect of rural sanitation 25 per cent of the rural population would be covered with sanitary toilets. The financial resources needed to achieve these targets has been estimated at Rs.150,000 million based on 1980 prices. The seventh plan(1985-90) provides Rs.22530 million in the States/Union Territories sectors and Rs.12010 million in the Central sector to cover about 39,000 of the 'problem villages' as a priority item. A provision of Rs.300 million has also been made under each of the Rural Landless Employment Guarantee Programme and National Rural Employment Programme for construction of 0.5 million sanitary latrines in rural areas during the seventh plan.

Introduction

India's sheer size makes it the most important single country for the International Water and Sanitation Decade, 1981-90. The 1981 Census recorded a population of over 685 million and if the present annual exponential population growth rate of 2.25 per cent continues for the remainder of the century, by 2001 India will have a population of about 1074 million. In about 45 years from now this might be double its present size(Mathew, 1986). At the time of Independence India's population was about 341 million and in less than four decades it has more than doubled. It now constitutes almost one sixth of mankind. The trends and characteristics of this growth, density per sq.km and others show interesting variations(Table 1).

*Univ. Dept. of Geology, Nagpur 440 001. India

Table 1

India's Population : Trends and Characteristics

Year	Population (millions)	Decadal variation (%)	Average exponential growth rate (%)	Density per sq. km.	Urban Population (%)
1901	238.3	..	..	77	10.8
1911	252.0	5.75	0.56	82	10.3
1921	251.2	-0.31	-0.03	81	11.2
1931	278.9	11.00	1.04	90	12.0
1941	318.5	14.22	1.33	103	13.9
1951	361.0	13.31	1.25	117	17.3
1961	439.1	21.51	1.96	142	18.0
1971	548.2	24.80	2.20	178	19.9
1981	685.2	25.00	2.25	216	23.3

(Govt. of India, 1972, 1983, 1984).

It can be observed that India's population grew at a very low rate till 1921, but during 1921-51 the rate picked up and yet remained below 1.50 per cent annum. However, in the following decades it kept steadily acclerating from 1.96 in 1951-61 to 2.20 in 1961-71 and 2.25 in 1971-81. There are enormous variations among the States regarding size, density and several other crucial demographic parameters(Table 2).

Table 2.

Decadal Growth Rates : Statewise

	Growth Rate 1961-71	Growth Rate 1971-81	% of India's population, 1981
INDIA	24.80	25.00	..
Andhra Pradesh	20.90	23.10	7.82
Bihar	21.23	24.06	10.20
Gujarat	29.39	27.67	4.97
Haryana	32.23	28.75	1.89
Karnataka	24.22	26.75	5.42
Kerala	26.29	19.24	3.71
Madhya Pradesh	28.67	25.27	7.62
Maharashtra	27.45	24.54	9.16
Orissa	25.05	20.17	3.85
Punjab	21.70	23.89	2.45
Rajasthan	27.83	32.97	5.00
Tamil Nadu	22.30	17.50	7.06
Uttar Pradesh	19.78	25.49	16.18
West Bengal	26.87	23.17	7.97

(Govt. of India, 1982)

Among these fourteen large States, the 1971-81 decadal growth rate varied from 17.50 for Tamil Nadu to 32.97 for Rajasthan. While the intercensal growth rates for 1971-81 were lower than that for 1961-71 in many States, and particularly in the States of Kerala, Tamil Nadu and Orissa where it was significantly lower, the growth rates went up in quite a few States including U.P. and Bihar both of which put together constitute 26.38 per cent of the total population of India.

Water Supply and Sanitation

Every civilised country has placed a provision of a protected water supply as the single most important public health measure. Unfortunately in India it was till the fifties a neglected problem. According to the Bhore Committee report, the population in urban and rural areas which could bank on safe water supplies was 6.6 per cent in Madras, 7.3 per cent in Bengal and 4.1 per cent in U.P. In the rural areas of the Punjab only 0.8 per cent of the population could boast of a protected water supply. A few large cities in India like Bombay, Calcutta and Madras had safe water supply arrangements which ensured laboratory tested clean water supply by pipe. After India gained Independence and planned development was initiated, the problem of water supply received attention. Towards the end of the First Five Year Plan(1951-61), the National Water Supply and Sanitation Programme was formulated to provide a protected water supply and adequate drainage for the entire population of the country in the course of the next two or three decades. A sum of Rs.127.2 million was made available for allotment in the first plan period then remaining, 1954-56, for the urban phase while a sum of Rs.60.0 million was made available for giving assistance to the States for the rural phase of the water supply and drainage programme in the First Five Year Plan. Twenty three States participated in the urban phase of the programme, submitting 287 water supply schemes and 79 drainage schemes for technical scrutiny and of these 196 urban water supply and 58 drainage schemes were approved by the Central Government for sanction of loans. Twenty one of the States participated in the rural phase of the National Water Supply and Sanitation Programme and 134 out of the 170 schemes submitted were approved. The achievements were significant; out of a targeted populations of 2,02,67,451 and 44,41,684 respectively in the urban and rural phases, 1,60,29,661 and 25,38,698 received the benefit. The progress would have been better but for lack of resources like equipment and trained personnel. External assistance became a necessity and under an Indo-U.S. Operational Agreement No.25 and its supplement signed in 1954 and 1955, assistance to the tune of $3,100,000 in the U.S. fiscal year 1954 and $1,800,000 during the fiscal year 1955 was provided for the purchase of equipment and materials and transport to India for the National Water Supply and Sanitation Programme through the T.C.M. The U.S.A. was also to make available under this agreement and supplement additional funds necessary to pay the salaries and other expenses of public health engineers and other sanitary personnel employed by the U.S.A. and assigned to the Government of India and the States of India for the purpose of providing technical assistance for the National Water Supply and Sanitation Programme. This was perhaps the first instance of International assistance to India in this programme and in subsequent years other countries too extended similar assistance(Borkar, 1957).

In the second five year plan, a provision of Rs.480 million was made for carrying out urban water supply and sanitation schemes. A sum of Rs.319.3 million was spent in the first 3 years of the 2nd plan. 275 water supply schemes and 61 sewerage schemes were executed under the National Programme. The States provided Rs.28 crores for rural water supply and sanitation schemes for which the Central Government made matching contribution equivalent to 50 per cent of the cost of the schemes under the programme.(Govt. of India, 1960).

Under the Third Five Year Plan, a sum of Rs.670 million was set apart for the supply of drinking water. Of this, Rs.160 million went to urban areas for provision of piped water supply under health programme; Rs.120 million was given to Community Development for rural water supply; Rs.40 million was earmarked to meet the drinking water needs of the backward classes; and Rs.350 million was set apart for water supply schemes under local development works. The 1966 surveys showed that it would require about Rs.6000 million to make good water available to one-third of the total rural population of India. About 660 city water supply projects with an estimated cost of Rs.1120 million were started in the 2nd plan to provide safe water to a city population of 15 million and this programme spilled over to the third plan(Govt. of India, 1961). The 1961 census counted 5,59,000 villages with 390 million people living in them. Two-third of them live in the areas which are far away from rivers or canals or well-maintained tanks(Govt. of India, 1967). 150 new urban water supply and sanitation schemes were taken up and 1764 schemes in rural areas were completed. The work done in 1961-69 ensured piped water supply to 6,000 more villages.

Before the fourth five year plan, rural water supply schemes were undertaken as part of the programmes of Community development works and welfare of backward classes. They were supplemented by the National Water Supply and Sanitation Programme of the Ministry of Health. As a re-result of the mid-term appraisal of the fourth plan, a

Central scheme to accelerate the programme of rural water supply in the States sector was introduced in 1972-73 as a part of the overall programme of special social welfare scheme. Under this scheme, 100 per cent assistance was given to the States and Union Territories for extending the water supply to villages or areas where the problem was most acute. Investigations were carried out by the Geological Survey of India and Central Ground Water Board in Rajasthan and Gujarat for drinking water. UNICEF gave 118 high speed diesel rigs. As a result of the effort made during the fourth plan and the earlier plans, nearly 36000 difficult villages benefitting a population of nearly 21.6 million have been provided with water supply.

The allocation for water supply programmes rose from Rs.1057 million in the third plan to Rs.4058 million in the fourth plan(1969-74). On the eve of the fifth plan(1974-79) it was expected that 85 per cent of the urban population would have piped water supply and nearly 0.116 million villages with a population of 61 million would still not have even the most elementary water supply system. As regards sewerage, only 38 per cent of the urban population, chiefly in the metropolitan cities, would have sewerage, while most medium and smaller towns would have no sewerage systems! The fifth plan outlays naturally had to be raised for water supply and sanitation(Table 3).

Table 3.

Fifth Plan Outlays for Water Supply & Sanitation

Sector	Outlay(millions) Rs.
1. State & Union Territories Plan	
(a) Rural Water Supply Programme	5730
(b) Urban Water Supply & Sanitation	4310
Total	10040
2. Central Sector	
(a) Centrally sponsored projects	166
Grand total	10206
3. Major Projects of Metropolitan Cities.	
(i) Bombay Water Supply & Sanitation	925
(ii) Delhi Water Supply & Sanitation	600
(iii) Madras Water Supply & Sanitation	275
(iv) Bangalore Water Supply & Sanitation	150
(v) Hyderabad Water Supply & Sanitation	156

At the beginning of the fifth plan, the water needs of the rural population were estimated against the groundwater potential in the States(Table 4).

Table 4.

Water needs of rural population and groundwater potential in States

State	Net groundwater recharge availa- ble(after irriga- tion) M.A. ft.	Area in sq.km per 200 rural popula- tion	No. of wells needed to be dispersed
Andhra Pradesh	13.6	1.58	17,500
Assam	16.7	1.46	68,490
Bihar	19.5	0.69	253,400
Gujarat	6.1	2.04	95,900
Haryana	2.7	1.07	40,990
Himachal Pradesh	N.A.	3.52	15,910
Jammu & Kashmir	4.0	11.77	18,860
Kerala	5.4	0.44	89,070
Madhya Pradesh	22.5	2.56	174,400
Maharashtra	9.2	1.78	173,160
Karnataka	9.0	1.73	110,740
Nagaland	N.A.	7.33	2,320
Orissa	15.8	1.55	100,600
Punjab	3.6	0.97	51,310
Rajasthan	1.4	3.23	105,970
Tamil Nadu	8.0	0.91	143,280
Uttar Pradesh	17.6	0.77	379,980
West Bengal	15.7	0.53	167,560
Union Territories	N.A.	2.70	25,940
All India	170.8 + N.A.	1.11	2,035,460

Note: It is presumed that a well serves 200 people. The number of wells needed to be dispersed includes existing wells.
(Govt. of India, 1972).

In the fifth five year plan a new programme was introduced, the Minimum Needs Programme. The concept emerged and crystallised out of the experience of the previous plans that neither growth nor social consumption can be sustained, much less accelerated, without being mutually supportive. This programme lays down the urgency for providing social services according to nationally accepted norms within a time bound programme. Its allocations are earmarked and it seeks to ensure the nessary provision of resources. It has eight components, including rural water supply. In the fifth plan, this component had an expenditure of Rs.5830 million. In the sixth plan(1980-85), under

Minimum Needs Programme the outlay was raised in the States and union Territories Plan to Rs.14070 million with an additional out lay of Rs.6000 million in the Central Plan and the target was to cover all the remaining problem villages by 1985 excepting in some difficult areas like hilly and desert regions. The Urban Water Supply & Sanitation was allocated Rs.17535.60 million, compared to Rs.5391.70 million in the fifth plan. While towns with nearly 84 per cent of the urban population were provided with drinking water facilities, the population coverage is partial and uneven. Even in thelarger cities many of the newer settlements and areas inhabited by the economically weaker sections continue to be without adequate water supply. Further, out of the 1027 towns still lacking drinking water supply facilities, as many as 902 belong to the group of towns which have a population of less than 20,000 and 50 per cent only of this population is served by drinking water facilities. The position in regard to urban sewerage and sanitation is even less satisfactory. Out of the 3,119 towns, only 198 have been provided with sewerage facilities. Even in respect of class I cities(population of over 100,000), only 46 per cent have arrangements for sewerage and sewage treatment. The overall coverage in the urban areas is about 20 per cent of the population. Some effort has recently been made to evolve low cost techniques for urban sanitation. The UNDP Global Project in India is intended to assist and promote the installation of water-seal latrines in 110 towns in 7 States, viz,, Assam, Bihar, Gujarat, Maharashtra, Rajasthan, Tamil Nadu and Uttar Pradesh. Pilot projects have been taken up in these States to provide low cost water-seal latrines with on-site disposal of human waste. During the sixth plan, it was envisaged to complete 930 urban water supply schemes and 120 urban sewerage and drainage schemes and to take up new schemes of water supply in about 550 towns and sewerage schemes in 110 towns. The total outlay in the sixth plan was more than Rs.39220 million(Government of India, 1980), raised later to Rs.411775 million. The seventh plan(1985-90) is tentatively at Rs.65224.70 million for water supply and sanitation. Under the rural sanitation programme, it is proposed to construct sanitary latrines in all village level institutions like health sub-centres, schools, 'anganwadis' etc., and, to the possible extent, provide sanitary latrines in all rural housing projects sponsored by States. A provision of Rs.300 million has been made under the programmes each of Rural Landless Employment Guarantee and National Rural Employment for construction of five lakh(half a million) sanitary latrines in rural areas. The Seventh Plan aims to provide potable drinking water to the entire rural population by 1990. It is proposed to involve voluntary agencies and mass media for health education and promotion of use of sanitary latrines and arrange a training programme for masons in low-cost sanitary latrine construction.

The International Drinking Water Supply and Sanitation Decade is everybody's concern.

References

Borkar, G. 1957. Health in Independent India. Ministry of Health, Government of India, New Delhi. Ix+224.
Government of India, 1960. Second Five Year Plan- Progress Report, 1958-59. p.164. Planning Commission.
Government of India, 1961. Towards self-reliance Economy. India's Third Plan(1961-66), Planning Commission.p.286.
Government of India, 1967. Towards Better Life - Drinking Water for Millions. Govt. of India Publications.p.201.
Government of India, 1972. Census Centenary 1972 India. Pocket Book of Population Statistics.
Government of India, 1972. Report of the Irrigation Commission, Vol.1, Ministry of Irrigation & Power.
Government of India, 1980. Sixth Five Year Plan, 1980-85. Planning Commission, p.222-227 & 397-401.
Government of India, 1982. Census of India 1981, Series 1 India, Paper 1 of 1982. Final Population Totals.
Government of India, 1983. Key Population Statistics Based on 5 Per Cent Sample Data.
Government of India, 1984. Census of India 1981, Series-1 India, Paper-1 of 1984. Population Projections for India 1981-2001.
Mathew, Thomas. 1986. India's Population : the worsening scenario. Yojana, Vol.30, No.19, p.4-7 & 14.
Yojana. 1986. Major rural sanitation programme to be launched. Vol.30, No.5, p.2.

WATER SUPPLY AND
PUBLIC HEALTH FOR SAIDPUR, BANGLADESH

Donald M. Schroeder, P.E.*

Abstract

In 1976, the urban community of Saidpur, Bangladesh, recognized that the lack of safe drinking water and inadequate sanitation were the most pressing public health problems facing the entire community. The local government officials approached the Mennonite Central Committee and requested them to provide technical and economic assistance to the community toward alleviating these problems. The Mennonite Central Committee agreed to assist the community and a cooperative program was begun that has resulted in the construction of both public and private sanitary latrines, a sanitary nightsoil disposal area, as well as household drinking water supplies. This paper will detail the history and development of the assistance program in Saidpur and will explain the current activities that are being directed toward the improvement of the health of the entire community.

Definitions

It will be helpful at the start to briefly define a few of the terms that will be used in this paper.

Aqua-privy--A public latrine utilizing a large, compartmentalized, water filled, storage tank for the digestion of human excreta.

Methars--Nightsoil collectors belonging to the Hindu "untouchable" caste.

Nightsoil--Human fecal waste in its most concentrated form, without the dilution from water that would normally be associated with an underground sewage system.

Pourashava--A unit of government most closely related to a municipality.

Purdah--The seclusion of women from public observation.

Service latrine--The predominant form of latrine in Saidpur. Defecation takes place on a raised platform and the feces pass through a hole in the platform to a bucket kept in the space below.

Union--A governmental sub-division of a pourashava. Saidpur is divided into six unions.

*Engineer, Environmental Division, Barge, Waggoner, Sumner and Cannon, 162 Third Avenue North, Nashville, Tennessee 37201.

Introduction

Mennonite Central Committee

A word of introduction is in order concerning the Mennonite Central Committee (MCC). MCC is the cooperative relief and service agency of the Mennonite Church in North America. MCC is a Christian organization that carries out community development, peace making and material aid projects in some 45 countries of the world. The over 800 voluntary personnel have expertise in agriculture, education, engineering and health-related fields. MCC began their work in Bangladesh in 1971 in response to the tidal wave that occurred in late 1970, which is estimated to have claimed 400,000 lives.

City of Saidpur

MCC began their work in Saidpur in 1972. Their initial involvement came at the request of the International Red Cross, who needed assistance in staffing the maternal-child health and nutrition centers that had been established in Saidpur. The City of Saidpur is located in Nilphamari District which is in the northwestern corner of Bangladesh. The current population of Saidpur is 150,000, with a total Pouroshava land area of 8 mi^2 (21 km^2). The local population is primarily composed of two distinct ethnic groups, the Bengalis and the Biharis. The Bengalis are Muslim and Hindu people who have occupied that area for many hundreds of years. Their language is Bengali. The Biharis are Urdu speaking, Muslim people who came to the area now called Bangladesh at the time of partition in 1947. Many of these people settled in the Saidpur area due to the presence of a large railway workshop in that city. At the time of the civil war, in 1971, the Bihari people remained loyal to the government in West Pakistan. After the defeat of this government, and the formation of Bangladesh, the Biharis were looked upon as traitors and much communal violence erupted between the two groups. Seeking safety in large numbers, there was a sudden influx into Saidpur of some 50,000 Bihari people. In response to this influx of Biharis, the International Red Cross set up 60 refugee camps. The maternal-child health and nutrition centers that originally brought MCC to Saidpur were established to provide assistance to the most destitute of those located in the camps. The details of life, and death, in Saidpur in the first few years after 1971 must be left to another paper, but suffice it to say, the overall city could easily have been called a disaster area.

Water Supply and Sanitation in Saidpur

The majority of the population gets their drinking water from hand-dug open wells that are lined with burnt clay rings. These wells are at an average depth of 20 feet and are typically in the center of the household courtyard. Most are constructed without platforms or covers and are considered heavily polluted based on microbiological examination that indicated average fecal coliform densities greater than 50/100 ml (Feachem, 1977). A piped public water supply does exist, but due to the age and state of disrepair of the system, it is only able to provide water at low pressure, three times a day; for two hours in the morning, an hour at noon, and one hour in the evening. The condition of the underground pipes and the long periods of no flow allows the ingress of pollutants into the system and tests carried out indicate unacceptable fecal contamination of the water supply, which receives no treatment.

The majority of the houses in Saidpur have service or bucket latrines to meet their sanitation needs. These latrines are generally in various stages of dilapidation and the system of nightsoil collection has very nearly broken down. Many of the latrines are without buckets and the excreta is swept into the open drains, which become, in effect, long open cesspools. These excreta-filled drains are a significant source of direct and indirect disease transmission. Seasonal flooding provides for increased contamination of the open wells, and the above-ground presence of human wastes provides a vector route for animal and airborne disease transmission. The service latrines that do contain buckets are emptied by the methars on an irregular basis. The methars bring the buckets to different fixed points throughout the town. The buckets are dumped into carts and trailers that are then taken by either buffalo or tractor for final disposal to a large low-lying, swine-infested field outside of town.

Initial Proposals

In March of 1976, the Divisional Medical Officer for the Bangladesh Railway in the Saidpur area, and the Saidpur Pourashava Chairman approached MCC for assistance in improving the water supply and sanitation facilities of the community. The proposal of the local officials consisted of rough drawings and a cost estimate for a five million dollar underground sewage collection system. Due to the level of technology and maintenance required for this type of collection system, it was judged not appropriate for implementation in Saidpur.

To provide an alternative to the proposal, MCC requested OXFAM, a relief and development agency based in England, to provide some assistance in developing a program that was appropriate considering the local conditions in Saidpur. In January of 1977, two consultants from OXFAM came to Saidpur and studied the local problems and conditions and made recommendations to MCC. The proposal from OXFAM called for immediate implementation of the following:

a) Construction of good quality public lavatories with washing facilities, if thought necessary, at selected points throughout the town. Excreta would be contained in a tank and emptied daily via a rail car.
b) Reorganizing the nightsoil and refuse collection systems and basing them on rail transport, with specially designed rail cars and loading points.

c) Improvement of the open drains for the conveyance of sullage and stormwater.
d) Construction of a properly managed disposal area where nightsoil and refuse could be composted either separately or together.

There were several aspects of this proposal that received significant criticism from the local authorities. The main objection centered around the railway being used to carry nightsoil. Muslims would not be willing to maintain or repair equipment used for the transportation of nightsoil, and the rail authorities would not be responsible for cleaning up the spillage from any accidents. In addition, it was felt that the general population would be strongly opposed to the railway nightsoil collection points and the presence of human wastes in rail cars that would have to pass through heavily populated areas of the city.

MCC Pilot Project

In response to the various proposals, and in consideration of the objections raised, MCC decided to implement a pilot scale project in one of the unions to test the feasibility of some of the aspects of the OXFAM proposal. OXFAM agreed to provide funding assistance for the pilot program, and in fact, is still providing funding to the currently on-going program.

The centerpiece of the pilot project was a new toilet/shower/ laundry facility which was to service approximately 2,000 people. This facility was of the aqua-privy design and was located off of a main walkway between a residential area and the main bazaar. The toilet block was constructed in 1980 at a total cost of U.S. $52,000. In addition, to shore up the nightsoil collection system, four trailer carts and a tractor were purchased for U.S. $9,800 and a nightsoil stabilization pond was constructed at a cost of U.S. $6,700. The stabilization pond was to improve on the practice of indiscriminate dumping of nightsoil carts and trailers in the fields outside of town. The ponds were sized to serve an area of 20,000 people, consisting of two anaerobic ponds which are used alternately; when one is full, the second is used, while the first pond is desludged after digestion is complete. The anaerobic ponds drain into a facultative pond and then a final maturation pond.

In conjunction with the physical improvements of the pilot project, MCC implemented a sanitation education program in the same union. It was felt that any improvements made in a vacuum of ignorance concerning habits of cleanliness in defecating, food preparation, and washing, would have little lasting effect. The details of this education program will be discussed later on in the paper.

A year after the completion of the pilot project, an evaluation was made as to its effectiveness and its potential for application to the entire community (Chowdhury, 1981). The large toilet block was found to be successful, in that it was used daily by around 1,500 people. However, the people using the facility were not those living in dwellings situated near the facility, but were rather members of a transient population such as truck drivers and village folk coming to town to sell their produce. Thus, while meeting a real need, the public toilet block was not fulfilling its intended purpose of providing a sanitary place for the discharge and storage of excreta for the residents of the community. It was hoped that a nearby sanitary facility would encourage people to abandon their dilapidated service latrines and provide for a safer living environment; this proved not to be the case. This was especially true among the women and children of the area. While this group account for around 75% of the population, they only accounted for 5% of the users of the facility. Due to the observance of purdah, women rarely venture out in public for any reason.

The trailers and tractor that were purchased to shore up the nightsoil collection system were well used, but were not being maintained by the Pouroshava. It was estimated that an additional 15 trailers would be required to transport the nightsoil being produced in the community but this would require an additional tractor as well, and it was felt that the Pouroshava did not have the personnel or finances to maintain all of this additional equipment in the long run. The evaluation found that the stabilization pond was also being used and was a significant improvement over the prior practice of indescriminate nightsoil dumping.

So, while the major components of the sanitation program were beneficial, the objective of providing an effective sanitation system for the residents of a union of Saidpur was not achieved. As sort of an aside to the original pilot project, three service latrines were converted to pour-flush, water seal latrines. These three conversions were included as part of the evaluation and were found to be quite successful. A decision was made to implement this conversion program on a larger scale and offer it to the whole Saidpur community.

Latrine Conversion Program

The idea of converting service latrines to pour-flush latrines did not originate with MCC (Pathak, 1981). The converted latrine consists of a concrete pan installed in the space where the bucket used to be. This pan is connected by a concrete pipe to an underground soak pit made of brick construction. The pits are typically 5 ft. 6 in. (1.7 m) deep with the bottom 3 ft. (0.9 m) of open-joint brick construction to allow for the seepage of water into the soil. The outside of the pits with open-joint construction is backfilled with 3 in. (8 cm) of brick chips to prevent dirt from entering the pit. All of the pipes, bricks, and pans are made locally in Saidpur. The cement must be imported but it is readily available in the bazaar. This type of conversion is suitable for those houses where the existing service latrine superstructure is of adequate construction. Through the first five years of the conversion program (1982-1986), a total of 1,732 private, household latrines were converted. MCC provided a 50% subsidy to motivate individuals to participate in the program. In 1986, the subsidized price for an average sized conversion was U.S. $40. MCC inspected and supplied all of the materials used in construction and provided construction supervision of the conversion work itself. All of the craftsmen, supervisors and inspectors were local people, employed by MCC on a contract basis. The size of the pits used depended on the number of users and MCC developed a reputation for delivering a quality product for the price agreed to on the front end.

During the first year of the conversion program (1982), it was noted that a significant percentage of the population either had no latrine or had such a dilapidated superstructure that installing a conversion would not be possible. In most cases, these households were also the financially poorer members of the community. It was at this time that a low-cost, complete latrine was developed. This low cost latrine was constructed with a concrete squatting slab, concrete rings to support the slab, and a concrete ring soak pit. The bottom rings in the pit were constructed with holes in them to allow for the seepage of water. The superstructure above the slab was constructed of bamboo poles and matting, with an asbestos cement sheet for the roof. As with the converted latrines, all local labor and material were utilized and MCC assured a consistent product at a fair price. MCC provided a 50% subsidy for these latrines also. In 1986, the subsidized price for an average low-cost latrine was U.S. $30. Since many people were not able to come up with 50% of the cost on the front end, an installment plan was implemented to allow individuals to pay back their 50% over a six month period. Through the first four years of offering the low-cost latrine (1983-1986), a total 360 of these units have been installed. It should be noted that MCC is currently in the process of developing both a manual and electric pump that will be suitable for emptying the soak pits when they become full with solids.

From the beginning of the assistance program in Saidpur, it was recognized that in addition to having a sanitation problem, the town also had a water supply problem that was having a negative impact on the overall public health. The majority of the residents received their drinking water from open wells, which were heavily contaminated with fecal organisms. It is for this reason that beginning in 1983, MCC also offered hand-pump, household tubewells at a 50% subsidy. In the first four years of their offering, a total of 497 household tubewells have been installed. Once again, a choice of two different models of tubewells were offered to try and meet the ability to pay of the local people. The more expensive model is the Standard No. 6 tubewell that has found broad application in developing countries. The subsidized price for this model, installed with a concrete platform, was U.S. $32 in 1986. MCC staff members developed a lower cost head, No. 3 size, that was installed with plastic pipe and filter. This model was offered at U.S. $22, which included a 4 ft. x 4 ft. (1.3 m x 1.3 m), concrete platform.

Since the converted latrines and new tubewells were to be installed close together, there was some concern that contamination of the groundwater could be occurring from the water leaching out of the soak pits. Saidpur was fortunate to have a relatively high groundwater level through all of the seasons and a 20 ft. (6.1 m) pipe with a 6 ft. (1.8 m) well screen was adequate to provide water year round. The sandy soils also allowed the tubewells to be sunk by hand at a relatively low cost. To determine if any contamination was occurring, an extensive microbiological testing program was performed on the existing open wells and the new tubewells. Water was sampled and tested for fecal contamination both before and after new latrine pits were constructed in an area. In no case were the soak pits, regardless of their proximity to the water supplies, responsible for any increased levels of fecal contamination of the drinking water. The sandy soils of the area were apparently quite effective in both physically and biologically purifying the water that would leach out from the pits.

Public Education Program

From the beginning of the pilot program, until the present, there has been a public education aspect to the work in Saidpur. During the pilot program, the education work concentrated in the same union that the toilet block was built in. Once the conversion program began in 1982, the education effort was expanded to the whole community. The current program is conducted by three male and three female field workers. The field workers' message emphasizes self-reliance to the community; that despite the prevailing unsanitary environment in the community, changes in personal practice can improve the health of the family. This basic message has been packaged into a number of topics such as dangers of bad sanitation, communicable diseases, safe water, refuse, diarrhea, ORS preparation, flies, scabies, worms, etc. The workers use many types of teaching aids, including a slide series, filmstrips, serial posters and several motion picture films.

The biggest part of the field worker's time is spent on household visits and group meetings. Each household is visited several times and a series of lessons are presented. After these household visits have been completed in an area, a group meeting is arranged for the showing of a motion picture film. Time for questions and discussion are allowed. The underlying message in all of the educational effort is to show the individual what he can do to improve his own health and that of his own family. In addition to household visits, the field workers:

- Conduct meetings in the classes of the primary schools,
- Hold street corner meetings directed at those children not in school,
- Conduct weekly rickshaw lectures using loudspeakers and signboards,
- Conduct regular sessions at the government-sponsored nutrition and health clinics,
- Contact local religious and community leaders to encourage their support in increasing the public health awareness of the general population, and
- Place educational advertisements in movie theaters.

In addition to their education emphasis, the field workers are the front-line publicists for the subsidized sanitary latrines and tubewells being offered by MCC. The workers spend a considerable amount of time in motivating people to actually purchase one of the converted latrines and tubewells. They all carry copies of promotional brochures, price lists, and application forms to all of their meetings with the public. It is felt that by combining education concerning the public health effects of poor water supplies and bad sanitation, with the offering of a means to eliminate these problems, that a real improvement can be made to the quality of life of the citizens of Saidpur.

Conclusions

The pilot program in Saidpur demonstrated the ineffectiveness of providing public facilities to meet the water supply and sanitation needs of this community. The emphasis must be on the household level. People should be educated as to the health benefits of pure drinking water and the sanitary disposal of excreta. At the same time, they should be offered affordable and reliable means to implement these health benefits for their own family. It has been the experience of MCC in Saidpur that when these two factors are brought together, the community will respond. This work is currently on-going in Saidpur and current plans call for implementing decreased levels of subsidies over the coming years.

Appendix

Chowdhury, S. and Schroeder, D. M. and Williams, D. F., "Private Cleanliness and Public Squalor," 1981. (An evaluation of the Saidpur Pilot Sanitation Project for OXFAM and the Mennonite Central Committee.)

Feachem, R. and McGarry, M. and Mara D., Water, Wastes and Health in Hot Climates, John Wiley & Sons, 1977.

Pathak, B., Sulabh Shauchalaya, 1981. (Hand-flush water seal latrine, a simple idea that worked.)

PRESENTATION OF THE PROWWESS PROGRAMME

Siri Melchior*

(Editor's note: The following is an outline of an unplanned, impromptu presentation given by the author. The conference organizers are grateful to the author for making this presentation on short notice.)

"PROWWESS", Promotion of the Role of Women in Water and Environmental Sanitation Services, is a programme of field demonstration projects, reporting to the Steering Committee of the IDWSSD, and located in the UNDP (United Nations Development Programme). At present, it has a dozen ongoing field projects, geared toward showing examples of:

1. How can one promote participation by communities, and in particular women, in water and sanitation projects?

    a. What are the different functions that communities (women) can perform (e.g., needs assessment, planning, construction, production of hardware, operations, maintenance, health education, financing, fund raising, evaluation)?

    b. What methods can be used to get this participation?

    c. What indicators can be found to measure this?

2. How does participation influence the effectiveness of water/sanitation projects?

    a. Rate of implementation (e.g., wells dug, latrines built)

    b. Rate of usage

    c. Sustainability (maintenance, "cost-recovery")

    d. What are the implications for the management of the "hardware project" (budget, timeframe, decision-making)?

    e. How does one measure this, particularly (b) and (c)?

3. How do the resultant water/sanitation projects affect people's lives?

    a. Health (attitudes, practices, morbidity)

*Programme Director, PROWWESS, U.N. Development Programme, 304 E. 45th Street, New York, New York 10017

b. Burden (reduced time, effort, and how this liberated energy is used - leisure, sleep, income-generation for self, work for others, child care, etc.)

c. Attitudes and community organization (do individuals and communities display a change in how they meet development challenges generally? Is two-way communication authorities/communities strengthened?

d. How do you measure this?

4. Is the process replicable, and has it been replicated?

a. Does it continue, after outside assistance is withdrawn, <u>within</u> the project area?

b. Have any lessons learned been applied beyond the project area - by government or international organization - any inputs in national plans and policies?

The emphasis of the programme is on supporting demonstration projects and then disseminating results, rather than creating guidelines.

A few examples of ongoing projects which address a number of the above questions were given:

KENYA In collaboration with the UNDP/IBRD hand pumps testing programme and with UNIFEM, PROWWESS has helped KWAHO, a national NGO, undertake community mobilization efforts. Water committees (including women) have been created (only in those communities which choose to participate) and pump caretakers (women) trained. This is facilitated by the fact that IBRD is developing easily maintained pumps. Each committee had collected $100-800 for maintenance, a third have deposited in banks. Although the pumps are only now to be handed over, and long-term experience needs to be documented, as testified in a report by SIDA (which funds the hardware) of December 1986, the KWAHO involvement is "probably the most positive factor in implementation of the programme". It is planned to extend the experience to a larger area during this year.

LESOTHO In collaboration with an IBRD executed rural sanitation programme with the Government of Lesotho, PROWWESS has provided a health educator and training in participatory techniques for extension workers. There has been a measurable increase in participation at traditional meetings, special working groups set up on sanitation which include women, and an acceleration in acceptance of latrines (which are constructed and paid for by the beneficiaries) during this last year of PROWWESS input. There is no proof (yet) that this is due to the "software" component, but it is a good sign that it will now be taken over as part of the core programme, and the approach will be accepted nationally.

INDIA PROWWESS was asked by the Government of India to work with local research institutions to develop participatory research with villagers on the type of sanitation requirements they had - the plan in the one completed project was accepted and funded by Government/-IBRD, the remainder only recently started. We plan to go back to see whether this actually made a difference in the functioning of the project.

A set of documentation, albeit preliminary and partial, is available from the author.

TRAINING WITHIN A WATER SUPPLY & SANITATION AGENCY

Amy A. Titus*

This presentation discusses the process and methodology for designing the training component of an institutional rehabilitation project for a Central American urban water supply and sanitation agency. It outlines the development of a unique, cost-effective approach which specifically addresses the critical need for human resource development within the agency. It also highlights the issues encountered and the rationale behind the various design decisions made in the project.

Introduction

Increasingly we recognize the value of competent human resources within a water supply and sanitation agency. In fact, human resources and institutional development have been established as priority concerns for this decade. The Basic Strategy Document on Human Resources Development (World Health Organization, 1984) underscores the importance of developing people to strengthen the sector. Too often effective management and employee development are neglected in an agency; human resources are not used to capacity. The results are inefficiency and low morale. One solution is agency rehabilitation, paying close attention to the development of the institution supported and facilitated by training.

Background

The city where the agency is located is presently experiencing very rapid growth. Water service, deficient in quantity and quality, is only intermittently available in most parts of the city, and the pressures are unevenly distributed. The rapid increase in the demand for the agency's services has urgently brought to light the long-overdue need to reorganize policies and procedures in order to satisfy this newly-enlarged scale of operation.

Over the past five years, however, the agency has faced serious management, production and financial problems and these have hampered its ability to rapidly to rapidly expand its services. Intervention was required

* President, Titus Austin, Inc., 1511 K St., NW, Suite 417 Washington, D.C. 20005

to help the agency become financially self-sufficient and competent. The project that was subsequently designed included technical assistance, training, and associated operational improvements. This presentation will describe the training interventions that took place and the benefits expected from them.

Description of the Training Component

Prior to project implementation, agency training was ad hoc, not linked to the organization's performance needs. A needs analysis revealed that there were performance problems at all levels of the organization, primarily due to deficiencies in skills and knowledge. These deficiencies were exacerbated by a weak personnel system. The agency required integrated institutional development supported by training. Thus the training component's central purpose will be to provide agency staff with the skills and knowledge necessary to support institutional rehabilitation and staff productivity. Furthermore, the training system will not be limited to only meeting project-related training needs, but will also provide the capability for continued identification and satisfaction of manpower development requirements beyond the project period, thus supporting future institutional development efforts.

Because approximately 50 percent of the agency's staff vitally need training to perform their jobs over the five years of the project, a small training unit will be created immediately, with the appropriate facilities and technical assistance to plan, coordinate, monitor, and evaluate training activities. At the same time, monitoring groups will be established to identify training needs and monitor training activities conducted in the unit. A central committee composed of mid- and top-level managers will also be formed to develop an overall training policy and then guide and monitor the training program's progress. These three groups will work together annually, to develop a training program based on an institution wide training needs analysis. This program, coordinated by the training unit, would describe who is to receive training, when, and where. All three groups would perform the essential task of monitoring training implementation.

The training unit, to be located within an institutional development unit, will report to the agency manager. Drawing on organizational support, the training unit will annually identify training needs, draw up a training plan, and arrange training activities. The skills of management and high-level, administrative, technical, and operational staff will be developed through three means: 1) On-the-job training by trained trainers from within the agency, mainly in operational areas, 2) Training in local training institutions, or by local training institutions at the agency, mainly in specialized areas, and 3) Training at a foreign agency (i.e., twinning agency), mainly in financial, commercial, technical, and managerial issues.

(1) On-the-Job Training (OJT)

Training needs at the agency's operational and lower levels will be addressed by on the job training. OJT will be conducted at the worksite using real work activities. To implement OJT, the training unit would first develop a training of trainers course; over five years, this course would train approximately one hundred staff from all areas of the agency. Participants would be drawn from operational levels and would either be supervisors, or more outstanding employees, capable of training their colleagues. OJT in areas receiving technical assistance would be conducted jointly with the consultants on-site. During the TOT course, participating staff would design both the training manual and the on-the-job performance checklists that they would later use to train approximately twelve to fifteen persons in their own work area.

2) Local Institutions

Local vocational, training, and educational institutions will be utilized in two ways. First, in areas such as accounting and budgeting where identified training needs cannot be met by agency personnel and the expertise exists locally, staff at all levels would attend selected courses at local training and educational institutions. Second, when agency staff have special needs that require customized instruction, local training and educational institutions would be contracted to deliver a course on site. On site courses would be designed to fit the schedule of agency staff and facilitate reinforced learning through practice.

3) Visits Overseas

Managerial needs often extend beyond training requirements. For example, introductions to new systems and approaches would best be met through visits to a water supply agency similar to but more developed than the project agency. For short periods of time employees will work with counterparts at the twinning agency to develop key identified skills. Technical assistance for on the job training will also be drawn from the twinning agency. Municipal and governmental staff will also briefly visit the foreign water agency to observe its operation.

Through the combination of these three feasible, cost-effective approaches, the agency personnel vital to institutional improvements will be trained. The three approaches will also be integrated with training activities sponsored by bilateral and multilateral agencies working with the project. Through this cohesive approach, a solid and consistent development plan can be conceived, implemented, and maintained.

Development of the Training Component

The training component was designed based on information gathered in comprehensive interviews of managers and staff from all levels of the organization. On each interview, the consultant was accompanied by the agency counterpart who would direct the training unit. Interviews were held with the heads of every division: finance, commerce, operations and maintenance, construction, administration (personnel, purchasing, warehousing and supplies, vehicles), planning, and management information systems. Each division head was asked about the current work situation and the successes, problems, and issues of the division. In addition, future goals were also explores. Meetings with their staff on training needs provided further discussion of problems and issues.

A performance analysis of the organization was conducted based on these conversations. The performance analysis compares current performance to desired performance and determines various ways to bring the two together, such as the provision of training, equipment, management support, information, or resources. In the analysis, gaps that could be filled through training were identified and were central in designing the overall organizational training plan.

In various meetings with top level managers of the organization, data about their views of current and future training needs was collected. Their views were also compared with the data collected from personnel interviewed throughout the agency. This activity not only provided data but also familiarized staff with the process of collecting a needs analysis and increased their awareness of the importance of constant internal agency communication.

Many discussions were also held ensuring integration of this component with the other rehabilitation efforts of the overall project. Training timelines had to be in sync with the provision of technical assistance and equipment.

Issues

The training component faces two major risks in its implementation. First is the training unit's potential inability to coordinate the activities specified in the training program. Unless systems are in place and continual top level support is maintained, the unit could encounter problems. However this risk is minimized by the provision of technical assistance in training for 24 man months and the specification that other technical assistance will support training activities. The second risk is the potential lack of follow-through on the part of the units and departments sending personnel to training or conducting on-the-job training according to schedule. To minimize this risk, the central training committee, the unit monitoring groups, and management must visibly support the execution of training programs.

Benefits

The benefit of a training component is the establishment of an ongoing institutionalized process for systematically identifying and meeting training needs. Training can also contribute to increased productivity and morale. In order to be effective and self-sustaining, water supply and sanitation agencies must explore ways to improve productivity and performance. Training, if institutionalized and given top level support, can be a useful vehicle for helping to achieve these vital goals.

References

World Health Organization. Human Resources Development Handbook, Geneva. September 1984.

CARIBBEAN BASIN WATER MANAGEMENT TRAINING PROJECT

Ronald A Williams[1]
O K Yhap[2]

1. INTRODUCTION

Over the years, as part of its hemispheric responsibility, the Pan American Health Organization (PAHO) has delivered a technical cooperation programme in health to Caribbean countries, while the Canadian International Development Agency (CIDA) has developed its own broader external assistance bilateral programme to Caribbean Commonwealth countries.

In 1975 CIDA and PAHO began a joint collaboration with seven Eastern Caribbean countries in the institutional development of their water utilities through the medium of human resource development. This training effort was called the Caribbean Basin Water Management Project (CBWMP). Since that time the project has evolved through three phases; and influenced by regular evaluation exercises, has changed directions from time to time to accommodate or match developments in the water sector and indeed in the participating countries. This also applies at this time (early 1987) to its further extension into a Fourth Phase (1987-89).

This review of the project shows up the flexibility in the evolution of some of its policies despite the determined rigidity in employing the principle of Technical Cooperation among Developing Countries (TCDC) and of achieving increased training activity at the local water utility level.

2. FIRST PHASE (1975-77)

2.1 Description

The project commenced in mid-1975 under the sponsorship of the Canadian International Development Agency (CIDA) and the Pan American Health Organization (PAHO/WHO), and with the following 7 Eastern Caribbean countries participating: Antigua, Dominica, Grenada, Montserrat, St Christopher/Nevis, Saint Lucia and St Vincent.

The methodology used was fellowship training at locations outside of the Caribbean where waterworks problems and solutions were inappropriate to Eastern Caribbean agencies. However, in two years, as many as 50 academic and 80 short-term fellowships were awarded, the emphasis being placed on individuals at the senior level in

1 - Area Engineer Advisor, Pan American Health Organization
2 - Project Manager, Caribbean Development Bank

the agencies (4% of total labour force). It was clear that training was knowledge-oriented and benefits were assumed to be positive without the use of evaluation procedures.

2.2 Assessment (1977)

In December 1976 a preliminary evaluation of the project was carried out and a number of changes were suggested. Then in April 1977 an in-depth assessment was carried out and findings included:

- The water agencies had no training policy or programme.

- Training funds were in the main spent on senior employees (3.8%), while little or no training was provided for the labourer employees (81.3%).

- Returning fellows were not utilized in training their fellow workers, and appropriate instructional materials did not exist.

Some of the key recommendations from the assessment were:

(a) Develop a self-sustaining training delivery system for the water agencies, making optimum use of existing expertise and institutional experience.

(b) Train trainers, by providing technicians with the communication skills and instructional techniques, to maximize home-country training.

(c) Develop appropriate training/job manuals; and stress manager/supervisor training.

These recommendations were accepted by the participating countries (and sponsoring agencies), and the project took a new direction in September 1977.

3. SECOND PHASE (1978-81)

In this phase the original 7 Eastern Caribbean countries were joined by: Anguilla, Barbados and the British Virgin Islands.

3.1 Description

New institutional arrangements for redirecting the project in this Second Phase included the following:

- A project management office in the PAHO subregional office in Barbados, staffed with a Project Manager and Secretary.

- Training Coordinators appointed on a full-time or part-time basis in each of the water agencies.

- Personnel in each participating water agency, equipped with the communications and instructional skills to train staff.

A proposal to establish an Office of Training Coordination in the then Barbados Waterworks Department was never realized.

In order to maximize training activities and their effects, the new programme incorporated the following:

- Emphasis on achieving a broad multiplier effect at all levels in each water agency by using a training-of-trainers approach, making training performance-oriented and evaluating events for effectiveness on employee performance and agency operation.

- Utilization of a broad range of procedures: on-the-job training, local workshops, short courses in the Eastern Caribbean, and limited external fellowships.

- Adherence to the principles of TCDC and appropriate technology to develop self-reliance: by West Indians training West Indians, increased training in local venues, personnel from participating water agencies developing training materials and serving as trainers, etc.

In addition to CIDA and PAHO, external assistance was received from the Government of the Netherlands and the British Development Division.

3.2 Activities and Achievements

Despite the fact that the project was inactive throughout 1979 due to lack of funds, a considerable amount of activities and achievements have been reported for this Phase (1978, 1980 and 1981). In addition to the training activities/achievements reported in Table 3-1, there were the following:

- Supply of reference materials to libraries in the water agencies;

- 38 Caribbean nationals developed (with project assistance) various waterworks training documents: 15 manuals, 7 job-aids, 3 guides and 2 handbooks, and these were shared with each of the agencies;

- Provision of leak detection equipment to 8 of the countries and the commencement of on-the-job training in leak detection;

- Conduct of 3 evaluation missions;

- Production of audio-visual materials (e.g. tapes, field charts, etc.) and publication of a monthly Project Newsletter.

TABLE 3-1 SECOND PHASE TRAINING ACTIVITIES/ACHIEVEMENTS

TRAINEES		TRAINING EVENT	
		CONTENT	DURATION
147	Supervisory personnel	Training of trainers	1-week
50	T of T graduates (1978)	T of T refresher	1-week
35	Top-level managerial/ supervisory staff	Management	1-week
12	Trainer/Consultants	Middle-management	2-months
115	Middle-management	Man/supervision	Six 4-hour
20	Accounting Dept. personnel	In-service	9-months
11	Miscellaneous fellowships	Attachments	4-6 weeks
8	Supervisors	Leak detection	1-week
10	Training Coordinators	3 Annual Workshops	1-week

3.3 Evaluation (1980-81)

In April 1980 a mission was mounted "to evaluate the effectiveness of the Training-of-Trainers workshops in all the participating countries as well as to obtain data to recommend follow-up activities". Visits were made to all the countries, interviews were held and questionnaires were used. The exercise was very thoroughly reported, and it was concluded that the Training-of-Trainers effort of the project "could be termed a moderate success". However, the report summarized its many recommendations with the following: "The future and greater success of the project depends essentially on the effort to be made the utilities. Trainers must be supported by a set of appropriate policies and standard procedures besides the financial/economic, material and human resources".

In May 1981 the main evaluation of the Second Phase was carried out. Its objectives were, in brief, to assess project relevance, effectiveness and the need for follow-up, and to identify constraints. The evaluation team concluded that while training activities were effective, a "self-sustaining training delivery system was not yet achieved". Also that "project management arrangements were inappropriate and that insufficient allocation of funds had been made".

As a result, the team made several major recommendations, the most important of which were: a Third Phase of the project with the participating countries sharing costs on an agreed formula basis and with project management transferred from PAHO to an indigenous subregional or national institution.

Early in 1982, the Caribbean Development Bank (CDB) agreed to manage the project and later that year appointed Eng. O K Yhap as Project Manager.

4. THIRD PHASE (1984-86)

4.1 Transition Period (1982-83)

The years 1982 and 1983 were a transition period for the project, and activity was confined to negotiations and arrangements for the transfer of project management from PAHO to CDB, including new Memoranda of Understanding between CDB and the original sponsoring agencies - CIDA and PAHO. It was also necessary for all 10 participating countries to agree on policy, managerial and financial details for the new phase.

The PAHO Project Manager (Eng. N Carefoot) departed in February 1982, the CDB Project Manager (Eng. O K Yhap) was appointed in October 1982 (after which essential project documents, equipment and funds were transferred), and new Memoranda of Understanding were signed in November 1983. In 1983, in addition to project re-mobilization activity (e.g. collection of country contributions, policy meetings, etc.), some training activities for Training Coordinators, and in Water Quality Control, were carried out.

4.2 Description

In this phase the project was managed by a senior CDB engineer in consultation with an Advisory Committee of participating countries and contributing agencies (CIDA and PAHO), which met annually to review the previous year's work and plan the activities of the next year.

In addition to this management change, the project's activities concentrated on local training workshops/courses, attachments to local agencies, and a few fellowships: and the development of appropriate training material (e.g. manuals) was no longer pursued. Work on leak detection did continue in such areas as further situation analysis and project development. In January 1985 the Turks & Caicos Islands joined the project, and the TCDC policy continued to guide all training activities, including the use of summer course-work at the College of Arts, Science & Technology (CAST) in Jamaica.

The 1986 financial contributions by participating countries were budgeted as follows:

Anguilla	US$ 1,316
Antigua & Barbuda	20,740
Barbados	57,984
British Virgin Islands	4,364
Dominica	9,832
Grenada	17,260
Montserrat	4,252
St Christopher/Nevis	7,088
Saint Lucia	23,144
St Vincent & the Grenadines	14,420
Turks & Caicos Islands	1,316
	US$161,716

In addition, this 3-year phase called for a cash input of CAN$200,000 from CIDA and a technical cooperation input of US$95,000 from PAHO, which means that the countries are making a greater contribution than the donor agencies.

4.3 Activities and Achievements

The training activities (and achievements) of this Third Phase are listed for the three years (1984-86) in Table 4-1, which shows a variety of subjects, locations and course lengths (1 day to 6 weeks). The number of trainees dropped in 1986, but this may be of little significance.

What is worth noting is that a "definite lack of organized training at many Utilities" was reported in 1984 by a 2-member consultant team on the use of Training Manuals, Job Aids, etc. produced in the previous phase; but in 1986 most countries reported holding in-house training activities - special events and on-the-job training.

In terms of project activities, the administrative and coordinating effort required to successfully plan, conduct and report training activities, collect country contributions, etc. should not be under-rated.

4.4 Evaluation (1985)

As required in the Memoranda of Understanding between CDB and CIDA, and between CDB and PAHO, a 4-man evaluation team was appointed in July 1985 to represent CDB, CIDA, PAHO and participating countries:

- to assess current progress of the project in terms of effectiveness and efficiency towards achieving its objectives;
- to make recommendations for the remaining period of the project life; and
- to assess the need for any continuation of the project beyond the Third Phase (November 1986).

The evaluation followed the same methodology as before (e.g. country visits and interviews using standard guidelines). As a result of their findings and conclusions, the evaluation team made a number of positive recommendations which can be summarized as follows:

- Strengthen local in-house training programmes and subregional work attachments (Third Phase).
- Extend the project into a Fourth Phase (1987-89) with existing planning and management structure.
- Expand project scope to include waste-water collection and disposal.
- Increase country contributions by 10% (1985-86).

TABLE 4-1 THIRD PHASE TRAINING ACTIVITIES/ACHIEVEMENTS

ACTIVITIES (Workshops/Courses)	1984			1985			1986		
	LOCATION	DURATION	PARTICIPANTS	LOCATION	DURATION	PARTICIPANTS	LOCATION	DURATION	PARTICIPANTS
WATER METERS	Saint Lucia	2 weeks	20				Puerto Rico	1 week	19
TRAINING OF TRAINERS	Barbados	3 days	30						
PR AND CUSTOMER SERVICE	Barbados	1 week	30	Antigua	1 week	42			
HUMAN RESOURCES DEVELOPMENT	Dominica	3 days	11	St. Kitts	3 days (two)	32	St. Kitts	1 week	18
CAST SUMMER COURSES	Jamaica	6 weeks	18	Jamaica	6 weeks 1 week 2 weeks	16 9 19	Jamaica	6 weeks 1 week	12 11
WATER QUALITY CONTROL	5 islands (on-the-job)	1 week each	52 total	Grenada	1 week	25			
PREVENTIVE MAINTENANCE	Antigua	1 week	37	St. Lucia	1 week	36			
MANAGEMENT INFO. SYSTEMS				Anguilla Barbados	1 day 4 days	10 24			
GROUNDWATER DEVELOPMENT							Antigua	4 days	20
DISTRIBUTION SYSTEMS: O + M							St. Lucia	1 week	32
W & S PROJECT PLANNING							Barbados	2 weeks	22
ATTACHMENTS/FELLOWSHIPS	Local External	1 week 2 months	2 2	BVI	1 week		Grenada St. Lucia	1 week 2 weeks	1 1
		TOTAL	202		TOTAL	216		TOTAL	136

Note: CAST Summer courses are: General water/sew - 6 weeks, Certification - 1 week, Pumps - 2 weeks
W & S means Water and Sewerage

5. FOURTH PHASE (1987-89)

In November/December 1986 a consultant to CIDA carried out a mission to the Eastern Caribbean and submitted a proposal for a Fourth Phase for the project.

Preliminary discussions in January 1987 suggest that the following recommendations are likely to be agreed upon by the three project sponsors (CDB, CIDA and PAHO) and the participating countries (including new member country, the Cayman Islands):

(a) The basic training programme of workshops and attachments should continue on an annual "most-need" basis, as decided by the Advisory Committee. This should include courses at CAST, Jamaica, giving priority to training at management and mid-management levels, using a modified DACUM approach, etc.

(b) The accent should be on completing the development of self-reliant and sustained training programmes in the water agencies of member states, including: programme development and operation by training committees; increased local delivery of all levels of training; and more effective monitoring and evaluation of programmes on a standardized basis.

(c) Financial contributions from participating countries as well as contributions by sponsoring agencies should continue in cash (CDB, CIDA) or in kind (PAHO) at existing levels.

Indications are that this may be the final phase of the project. If so, the project should achieve self-reliance in Phase Four to satisfy future training needs without the benefit of a major cash contribution from an external source.

6. CONCLUSIONS

During the years of the project (1975-present) a large cadre of waterworks personnel has been trained to meet the needs of increased and expanded water supply systems in Eastern Caribbean countries. Water utilities and governments have appreciated the project and have repeatedly expressed their satisfaction.

Throughout its three phases the project has experienced progressive changes in its effort to meet the felt-needs of participating water agencies; and these changes have been influenced by formal project evaluations from time to time. But while methodology and management have changed, there has been no change of the early concept of utilizing appropriate technology based on the principle of TCDC in the Caribbean subregion.

In Phase Four, annual work programmes will continue to be appropriately selected, efficiently implemented, and managed to the full satisfaction of all participating countries and agencies. In addition, national training programmes in the water sector are expected to mature and maximize the benefits from this very successful water training project.

Implementation Effectiveness of a Water Supply Scheme : An Empirical Study on Human Dimensions

T V Jacob*, M ASCE

Abstract

This paper tries to probe certain areas of human dimensions relating to implementation effectiveness. The human dimensions explored are: (i) leadership style of the project leader, (ii) leadership style of immediate superiors, (iii) communication, (iv) motivation, (v) environment (constraints/supports) and (vi) systems and methods. The primary data for the study was collected by a questionnaire seeking responses/reactions of field technical personnel working for a water supply scheme under two successive project leaders. The data thus collected were statistically analysed to find out significant differences, if any, under the two psychological environments created by the project leaders. The study findings indicate that with proper and appropriate leadership styles and effective systems and methods, it is possible to create a favourable environment for productive performance even in a bureaucratic government set-up.

1.0 Introduction

1.1 The water supply scheme under discussion was a project designed with two objectives, viz : (i) immediate supply of water to Cochin Refineries Ltd. (abbreviated as CRL) and other industrial units in Ambalamugal area, Kerala, (India) in the first stage and (ii) augmenting supply of drinking water with necessary additional treatments to rural population benefiting five hundred thousand people in the area in the second stage. The total works of first stage of the project were divided into: Part I costing Rs.70 million** for producing and conveying water upto CRL and Part II costing Rs.6.1 million for works beyond CRL. A special division with two sub divisions was formed in October 1981 for implementation and the actual works commenced in April 1982 with the target date of 31st December 1983 for providing water to CRL.

2.0 Implementation : regular phase

2.1 Implementation of the project from October 1981 onwards progressed in the normal way as was usual in a governmental set-up. By June 1983, formalities of finalising the award of contracts for most of the critical works of part I of the project were completed, and

* Superintending Engineer, World Bank Project Circle, Kerala Water Authority, Alwaye, Kerala, India.

** Rs. = Indian Rupees; 100 Indian Rupees = 9.94 US$ (1983)

the work on intake well was commenced. Considering the importance and urgency of supplying water to CRL, which had taken up a massive expansion programme, an empowered implementation committee with all administrative and financial powers was constituted by government on 1st July 1983. The empowered committee created additionally one office of the superintending engineer with two sub divisions for implementation within the target date originally fixed. The office, under the superintending engineer (who had control earlier continued exclusively for the project) which came into effect from September 1983, consisted of one division and four sub divisions. The executive engineer in charge of the division was also replaced by another one who took charge in September 1983. The expenditure on the project upto August 1983 (termed as regular phase in this study) was Rs.15.1 million.

3.0 Implementation : crash phase

3.1 All the critical works - intake well, pump house, electrical house, laying of 6 km. of 600 mm asbestos pipe and 13.2 km. of 700 mm. concrete pipe etc. - to pump water from the river directly to CRL without preliminary treatment were completed during October 1983 to February 1984, at a total cost of Rs.29.1 million, under the leadership of the new executive engineer. CRL had made it known that the water would be required only by the end of March 1984. Pumping commenced during the first week of March 1984 well ahead of the revised target date. The work in this period (termed as the crash phase in the study) which enabled supply of water to CRL within the revised target date is the focus of the present study. It is pertinent to point out that quantitative techniques like PERT were employed neither during the regular phase nor the crash phase. The study aims at probing certain areas of human dimensions on implementation effectiveness.

4.0 Scope and methodology of the study

4.1 The human dimensions explored in this study are: (i) leadership style of the project leader (executive engineer), (ii) leadership style of the immediate superiors (officers below the rank of executive engineer), (iii) communication among the technical operative personnel, (iv) motivation, (v) environment (constraints/supports) and (vi) systems and methods. The methodology adopted was a questionnaire eliciting the responses of engineers and technical personnel who worked both in the regular phase and in the crash phase. The questionnaire consisted of 81 statements regarding the six human dimensions mentioned above. The response to each statement was in the form of 'yes' or 'no'. Score of (+)1, (-)1 and 'zero' were assigned to positive, negative and neutral responses respectively. The engineers and technical personnel were classified into two groups, viz: (i) common group (those numbering twelve who worked in the regular phase right from the beginning and continued with the project in the crash phase till March 1984) and (ii) crash phase group (those numbering nine who were working elsewhere in the department and who joined around the time when the new project leader took over charge and worked through the crash phase). Personal interviews with some of the technical personnel and the contractors were conducted to collect additional data on their personal views and feelings on project implementation effectiveness.

5.0 Results of the responses to the questionnaire

5.1 The differences in scores between the common group and the crash phase group in the crash phase and the difference between the regular phase and the crash phase of the common group for each statement in the questionnaire for each human dimension were calculated. The values were statistically analysed using 'student's t' to determine whether the observed differences between the two groups/phases were statistically significant or not. The data are furnished in table 1.

5.2 The findings revealed that the two groups, viz: the common group and the crash phase group, showed no significant difference in responses to different dimensions in the crash phase (table 1 line t1). This indicates that everyone - members of both the common group and the crash phase group - was completely involved in the work programme under the new leader. It further reveals that the climate and the leadership style in the crash phase was perceived as similar by both the groups irrespective of their background work experience in the project. The responses of the common group in the six human dimensions in implementation both at the regular phase and at the crash phase reveal significant differences (table 1 line t2). This implies that climate (psychological atmosphere that affects the perception, feelings and attitude of members towards the work situation) created by the project leader in the crash phase was conducive to better effectiveness of performance.

6.0 Perceptions and feelings of the team members as revealed in the personal interview

6.1 The findings reported above were supported by the perceptions of the team members as revealed in the personal interview. The work team was not aware of the time-bound nature of the project before the crash phase and hence there was no urgency in the implementation of the project. The members in the regular phase did not feel that the project leader had the conviction of completing the project on time. They perceived the leader as procedure-oriented than performance-oriented resulting in lack of facilities and funds. However, he was perceived as a person of integrity. The project leader was also perceived as one who did not encourage his subordinates to make independent decisions. On the other hand, during the crash phase after the initial readjustments, the work team became cohesive with high team spirit and all the members felt a sense of urgency in the implementation, as the new leader convinced them that achievement of target was within reach. They appreciated the performance-orientation, quick problem solving style, willingness to listen to subordinates without abdicating responsibility, clarity on objectives, awareness of all details, quick decision-making after discussion and with-

Table 1 : Differences in the perception on human dimensions in implementation

Group (s) and phase (s)	Sample constants	Human dimensions: Leader ship style of the project leader (29 items)	Leader ship of immediate superiors (15 items)	Communication and interaction (14 items)	Motivation (13 items)	Environment (7 items)	Systems & methods (3 items)
(1)	(2)	(3)	(4)	(5)	(6)	(7)	(8)
Common and crash phase groups in the crash phase	n_1	12	12	12	12	12	12
	n_2	9	9	9	9	9	9
	$\bar{x}_1$	21.08	11.92	8.08	9.00	2.75	1.33
	$\bar{x}_2$	20.55	11.33	6.67	8.00	2.56	2.00
	S_1	4.79	1.89	3.17	2.55	2.42	1.25
	S_2	6.27	1.16	2.98	2.58	2.40	0.94
	S^2	33.11	2.89	10.58	7.26	6.44	1.40
	t_1	0.21@	0.797@	0.99@	0.85@	0.171@	-1.28@
Common group during the regular and crash phase	n	12	12	12	12	12	12
	$\bar{d}$	-10.000	-3.667	-2.000	-2.750	-2.000	-1.833
	Sd	8.256	5.691	3.605	5.290	2.580	2.477
	t_2	-4.010*	-2.137 f	-1.839 f	-2.350**	-2.570**	-2.454**

$\bar{x}_1$, $\bar{x}_2$, $\bar{d}$ denote mean, S_1, S_2, Sd denote standard deviation

$$S^2 = \frac{n_1 S_1^2 + n_2 S_2^2}{n_1 + n_2 - 2}, \quad t_1 = \frac{\bar{x}_1 - \bar{x}_2}{\sqrt{S^2(1/n_1 + 1/n_2)}}, \quad t_2 = \frac{\bar{d} - 0}{\frac{Sd}{\sqrt{n-1}}}$$

@ not significant (table value to be significant at 0.20 level ± 1.528)

* significant at 0.01 level (table value to be significant at 0.01 level ± 3.106)

** significant at 0.05 level (table value to be significant at 0.05 level ± 2.201)

f significant at 0.10 level (table value to be significant at 0.10 level ± 1.796)

out ignoring subordinates and the goal-directed approach of the new project leader who encouraged subordinates to make independent decisions by giving adequate support. The new leader created a climate of participation, maintained hierarchy and facilitated communication by the introduction of the new system of everyday review meetings for an hour at 8 a.m. with all technical field officers at the project site office. The team members also felt that the new project leader enjoyed complete support from the higher ups and his goal-directed approach enabled him to get all the facilities, resources and funds required for completing the task within the stipulated period. In addition, the new project leader was perceived as one with high integrity and high commitment to the project. He was also seen as one who could persuade local leaders and other departmental officers to attend conference and solve local problems that hindered the progress of work.

6.2 In short, the personal style of the new leader, who was accepted as a true leader, was perceived as instrumental for achievement of the target on time. The personal style was perceived as one that created enthusiasm rather than fear at immediate subordinate level. Lower level subordinates felt both enthusiasm and fear and at the lowest level there was a strong feeling of fear. Team members felt that the leader would take strong action against indiscipline and the action of the leader would be supported by officials at higher levels. The members further perceived that the new leader acted as a facilitator, a trainer of the subordinates and a representative of the group to the higher ups. Team members felt that the contractors' feeling of power and influence at higher levels dwindled, and the contractors feared that they might be put into disadvantages if they did not obey the decisions of the officers and had to withdraw their resistance to high speed of work. The enthusiasm and feeling of confidence of the members were revealed by their expression of willingness to continue with the team under the project leader with the psychological atmosphere created for the completion of the project.

7.0 Perception of the project by the contractors as revealed by the personal interview

7.1 Personal interview with the contractors revealed that they were not aware of the time-bound nature of the project during the earlier period of project implementation as they were given usual periods for completion of the works entrusted to them. But in the crash phase, they became aware of the time-bound nature of the project and the need to complete the works within the programmed target date. They also experienced certain differences in the project in comparison with other works they had undertaken. Availability of all required facilities, physical presence of the officers at the work spot, prompt payments, supply of specialised materials on the spot, prompt decisions, settlement of labour disputes by involvement of higher officials, technical support for problem-solving by officers for reducing the work time, technical support of officers by developing new methodology for problem-solving, non harassment from any source, officers' enthusiasm to complete the work on time, etc. were perceived as the positive nature of the project. Contractors, however, felt that there was no change in the attitude of the contract workers. The project leader was perceived as competent and committed to work with a high degree of integrity. Hence they did not feel the need for approaching higher ups - ministers, political leaders and higher level officers - to settle disputes or to get favours, though they confessed that they were capable of involving and indulging in such practices.

8.0 Discussion

8.1 In a governmental set-up, it is generally assumed that the culture, which depends, among other factors, on the philosophy and policy of the top management is unfavourable as the government approach usually is more conservative than innovative. External factors that affect culture include inter-dependence on other organisations and departments, technology, availability of equipments and raw materials, labour attitude etc. Among internal factors important ones are characteristics of employees, expertise, number of people and the organisational climate created by the management style.

8.2 The formation of the empowered committee which took quick decisions was a change from the normal set-up. The committee gave necessary support for actions taken in genuine interest of the work. These initiated the development of a favourable culture. This was further strengthened by the positive actions of the project leader in getting support from other departments, making available equipments and necessary materials, giving expertise in solving technical problems etc. But this alone was not sufficient. For effectiveness, a favourable climate should be developed by creating favourable attitude amongst the team members towards the project. This was achieved by the project leader by the daily meetings and his managerial style of providing personal support and encouragement. The daily meetings provided a venue for interaction among the team members, and this created better understanding among themselves. The gradual development of a favourable climate is evident from the remarks of the team members that even though there was resistance for daily meetings in the early stages, it became well accepted later.

8.3 The results in table 1 (col 3 line t2) revealed that the difference in leadership style of the project leader in the two phases was significant at 0.01 level and therefore highly correlated with performance effectiveness. The acceptance of the leader in the crash phase was a major factor for the good performance. The project leader believed in self direction and self control of the immediate subordinates at higher levels and external control and external direction at the lower levels of personnel. This was revealed by the fact that the immediate subordinates felt enthusiasm rather than fear, while at the lowest level there was a strong feeling of fear. (This is appropriate as the psychological maturity of individuals with higher need levels at the immediate subordinate level of the project leader is higher than at lower levels. Under immaturity conditions external direction and control are appropriate). He provided ample opportunities for the growth and development of the subordinates which induced inspiration and created a feeling of satisfaction in the minds of the team members. By providing adequate facilities and making timely payments, the contractors also worked with earnestness. The overall team work and mutual understanding were responsible for the effective and efficient implementation of the project.

8.4 The immediate subordinates during the regular phase were not trained adequately and were waiting for instructions at every stage from the project leader. In the crash phase, during the daily meetings which enabled communication and participation, important decisions were made by the group and the necessity of immediate superior giving instructions were very much reduced. However, responses indicated that the differences in the areas of 'leadership of immediate superiors' and 'communication' in the two phase were significant at 0.10 level (table 1 cols 4,5, line t2).

8.5 Inducing motivation to perform well depends on the managerial style of the leader. The managerial style of the leader in planning, problem-solving and decision-making through discussions in the daily meetings created feelings of involvement and commitment which helped in developing a motivating climate. Development of a cohesive team for group motivation helped to overcome deficiencies by providing necessary interpersonal support. The difference in the area of motivation was found to be significant at 0.05 level (table 1 col 6 line t2). The areas 'environment' and 'systems and methods' were also found to be significant at 0.05 level (table 1 cols 7,8 line t2) indicating relevance of 'environment' and 'systems and methods' to effective performance.

9.0 Conclusions

9.1 It is the human effort that determines the effectiveness and efficiency of any result-oriented programme. Project leaders usually assume that human efforts will automatically follow. When the human efforts are not forthcoming to the desired level the quality of human material is blamed without exploring why the people are not giving out their best. The present study clearly reveals that in a governmental set-up, where rules predominate performance and employees are generally not sufficiently motivated, effectiveness with the available people in the organisation is possible even without the application of scientific quantitative methods, provided the right type of work climate is created by the project leader by his style, attitude and decision-making skills.

Acknowledgement

The author wishes to express his deep sense of gratitude to Prof. P R Poduval, School of Management Studies, Cochin University of Science and Technology, Cochin-22, Kerala, (India) for his advice and support at various stages in making this study a reality. The assistance rendered by the technical personnel of the project in responding to the questionnaire and expressing their personal views in the interview is also acknowledged.

TRAINING SYSTEM FOR OPERATORS OF WATER SUPPLY SYSTEMS IN BRAZIL

Maria Lúcia de Souza Lobo*

This paper aims to present a Training System for operators for Water Supply Systems developped in Brazil and to divulge its methodology which may be used for attainments in other sanitation areas.

1. Historical Development

The implantation in Brazil of the National Plan for Basic Sanitation (PLANASA), since 1971, leds the sanitation sector to undertake an accelerated process of expansion aimed to provide brazilian cities with Water Supply Systems in order to eliminate the existing deficit.

On 1981 a survey was carried out and showed that the building of a great number of Water Supply Systems associated with the lack of an agile training for operators generated several problems in the operational area. It was then identified the necessity of training operators so that the sector needings were fulfilled.

Considering water as the most important factor to the quality of life it becomes indispensable to turn out able manpower to operate Water Supply Systems.

2. Training System Development

After the identification of needs it was decided to elaborate a Training System for the operators of the Water Supply System.

Aiming the final perfection of the training system at the last stage, each stage was tested in field after its conclusion.

2.1 Stages of the Training System Elaboration

2.1.1 Problem Formulation

The realization of each stage is a preliminary task of the Training System development and includes the following steps:

2.1.1.1 Delimitation of the problem

Definition and preliminar specification of the population chosen to be trained in order to delimitate the problem emphasized in the Training System.

2.1.1.2 Agents responsible for the Project Development

The guidance and development of the project

* Engineer, Banco Nacional da Habitação, Av. República do Chile, 230
17º andar - Rio de Janeiro - CEP 20039 - Brazil

was under the responsability of a team composed by professionals from: Banco Nacional da Habitação - BNH (National Housing Bank; Associação Brasileira de Engenharia Sanitária e Ambiental - ABES (Brazilian Society of Environmental and Sanitary Engineers) and of the Núcleo de Tecnologia Educacional para a Saúde - NUTES (Educational Technology Center for the Health).

Seve state sanitation companies representing different Brazilian regions participated in the project elaboration: AGESPISA, Piaui/ CAERN, Rio Grande do Norte/ COPASA, Minas Gerais/CORSAN Rio Grande do Sul/ SABESP, São Paulo, SANEMAT, Mato Grosso and SANEPAR, Paraná. Technicians from these Companies followed the project development and turned out feasible specialized technical assistance necessary to its elaboration.

2.1.1.3 Characterization of the target population

This characterization consists on the socio-cultural survey and determination of the population profile chosen to be trained.

On the course of the survey it was detected that the operators had not much diversified culture and were not used for read and write. Consequently they presented bad concept elaboration, difficulty to translate abstract signs (written words, for instance) into thoughts and speech.

2.1.1.4 Methodology determination

The technology adopted that permits the elaboration of a Training System in large scale is based on the methodology of behaviour analysis of the tasks. Its usage is justified in the following situations:

a. Existence of a great number of persons to be trained at a reasonable cost;

b. minimum standard of knowledge, capability, skilfulness and attitudes to be trained so that could be attained a satisfactory and homogeneous result, so assuring a minimum acceptable performance of the functions, tasks and behaviours;

c. descentralization of the training, runned on the work site, without the presence of the instructor, becoming feasible the immediate application of knowledge learned and valuated by the supervisor, which avoids cost for trainees displacement.

2.1.2 Definition of the Project Objectives

Project objectives were defined as:

a. to develop modules of training for the jobs of Water Supply Systems operators; b) to provide instruction material for the training; c) to implement the Training System application; d) to valuate the Training System results.

2.1.3 Identification of Functions and Sub-Functions

When the former stages were concluded the phase corresponding to the Project Execution started. The first step relied on the definition of the functions and subfunctions to be trained.

The objective of this stage is to identify and to select the knowledges, capability, skilfulness and attitudes that the trainees must show in order to train and actualize them with the work they area carrying out.

2.1.4 Specification of the Terminal Bahaviour

The specification of the terminal behaviour corresponds to an item of testes that evaluates the trainee performance when exercising his function. The items of test, for the methodology adopted, represent the expected bahaviour specifications and also act as items of judgment for the posterior instruction material elaboration.

2.1.5 Items of the test validity

The next step after the elaboration of the items of test is a proceeding related to their validity, whose finality regards the finding of existing problems in comprehension of the items by the operators through the answers and judgments.

2.1.6 Behaviour analysis of the tasks

It consists on the specification of each expected behaviour regarding knowledge, capabilities and skills in the sequence through which the trainee executes them. This sequence of connections of stimulus and answers is named behaviour chain. In this stage are also specified all concepts which must be teached as well as the learning of the pre-requirements.

2.1.7 Ways of Instruction Selection

The objective of this stage is to determine the way of instruction which will be used during the training.

Taking into consideration the operators elementary school education, undoubtly the best way to diffuse the informations should be the moving picture (movie or video- tape). Nevertheless a great number of Institutuions has no condition to own adequate equipments (that is video-cassetes) in sufficient quantity allowing to run the training on service).

Facing this fact, it seemed to the project team that the most useful format should be an illustrated book including short texts with colour photographs prints which represents better and concretely a static reality.

Viewing to support this decision it was decided to elaborate three prototypes on the format chosen. The prototypes were tested in field. The result of the test showed that the material were easily consulted by the operators and motivated them to read it and to solve the exercises proposed. It was also emphasized the high level of the context comprehension.

Then was decided that the instruction material should be elaborated in the prototypes format. The high cost of the initial production should be recovered by the great quantity of the issue.

2.1.8 Specification of the Logical Sequence of Teaching

After the selection of the most adequate way of instruction it is necessary to order the contents in a logical sequence according to the pedagogical concepts.

In this stage the operators functions were grouped by correlated subjects, taking into consideration the volume of informations that might be transmitted which should originate the auto-instruction modules.

2.1.9 First Version Production

The first version of the Training System was published in books with colour photographs prints and short texts.

The context written in a personal and colloquial manner considered the daily concrete activities of the operator. In possession of the material published, the operator has also to solve the problems proposed to him and this creates a more dynamic relation between them and assures that he will learn the concepts necessary for taking decisions.

2.1.10 First Version Test

The first version was tested in field under a controlled experimental situation. The data of this experimental control were computed, analysed and originated modifications and improvements of the instruction books.

2.1.11 Final Version Production

At that time, the supposition that the high cost of the material should be recovered by the big issuance of it were no more real. By the experience obtained along the project execution, it was clear that the material had to be modified and adapted to different realities.

Then it was decided to substitute the colour photographs printes by black and white pictures and to prepare colour slide sound programs with the same book content.

The final version of the Training System consists of auto-instruction material encompassing thirty two functions related to the tasks to be executed by the operators, grouped in the following sets:

a) SET 1 - PUMP; b) SET 2 - WATER MAIN; c) SET 3 - HOME CONNECTIONS; d) SET 4 - MEASURINGS -DEVILES; e) SET 5 - TREATMENT. (Annex-1)

The material is presented in:

a) Twenty six books illustrated with pictures; b) Twenty six colour slide-sound; c) five books containing items of test for the diagnostical and somatical valuation.

2.1.12 Final Version Test

The final version test was applied to 19 operators.

It was observed that:

a) It is easy to understand the material and the quality of answers given to the items of test demonstrated and good learning of informations;

b) there is a great interest of the operators to read the books.

3. Conclusions

Within the specific objectives of this project the following goals were attained:

- the development of training through modules for the functions of the Water Supply System operator.

- the production of the instruction material for the training.

The training system elaborated was apllied and evaluated during several stages of its development but under experimental conditions.

Its effective application and valuation is being done by institutions which operate Water Supply Systems and consequently they need to turn out able manpower to this objective.

We know that the application of an specific training system to be used in differente realities needs suitable adaptation and the institutions need to believe that the basic technical concepts are homogeneous, independently of equipments, material and proceedings used in each case.

Some proceedings described by the material elaborated may be different under different realities. Our opinion is that this fact does not render invalid the material because through it can be transmitted the technical concept allied to the why of it and to the importance of the function being executed, even in the extreme case of different execution.

It is essential for the usage of the material elaborated that the operational and human resource groups of the institution shoud be well acquainted with the material and really want to utilize it as a work tool and to establish training programs, plans and strategies.

It is expected that the instruction material elaborated reformulates the practice of the operators in service, conducting them to adopt new attitudes and capabilities contributing for their better functions performance, for a better Water Supply System operation and for rendering services of better quality to the communities.

It is also expected that this work arouse the interest for a training methodology which can be utilized for the production of other instruction materials similar to this one presented and related to other usages.

ANNEX-1

- INTRODUCTORY MODULE
- SET 1 - PUMPS
  - 1.1 Energizing the system, switching on and switching off the horizontally shafted motor-pump unit
  - 1.2 Energizing the system, switching on and switching off the vertically shafted motor-pump unit
  - 1.3 Maintenance and priming of pump
  - 1.4 Switching on and switching off motor-pump unit according to water level in reservoir.
- SET 2 - WATER MAIN AND DISTRIBUTION SYSTEM
  - 2.1 Periodical flushing of water main and of distribution system, to clean pipes
  - 2.2 Periodical inspection of water main
  - 2.3 Operating metering unit on distribution system
  - 2.4 Lowering sections of the distribution system
- SET 3 - HOUSE CONNECTIONS
  - 3.1 Measuring water pressure
  - 3.2 Connecting house pipe to distribution system
  - 3.3 Applying extension to distribution system
  - 3.4 Placing and replacing water units
  - 3.5 Stopping supply to unconventional connections
  - 3.6 Checking house connections to detect leak
- SET 4 - METERING
  - 4.1 Metering and registering water flow
  - 4.2 Metering and registering pressure flow with macro-meter
  - 4.3 Metering static level of well
  - 4.4 Metering dynamic level of well
- SET 5 - WATER TREATMENT
  - 5.1 Preparing desinfectant solution
  - 5.2 Switching in, operating and switching off chlorine misture pump
  - 5.3 Maintenance of chlorine mixture pump
  - 5.4 Operating, stopping and cleaning chlorine hydro-ejector
  - 5.5 Operating, stopping and cleaning mixture unit of constant level
  - 5.6 Collecting samples for bacteriological analysis
  - 5.7 Washing reservoir

Introduction to Finance and Economics

Harold R. Shipman and John Kalbematten

Sessions on the Economic policies followed by international agencies in the evaluation of water and sanitation projects include presentations by representatives of major financing institutions as well as by consultants and specialists. Prospects for future financing of projects in this sector as well as likely policies to be applied are covered. The means for predicting ability to pay for water at charges reasonably reflecting costs of supply are the subject of several papers covering studies either under way or completed.

The availability of financial resources for water and sanitation projects to be undertaken in the next five to ten years are discussed along with actions needed to improve financial performance and operation. The means for implementing a successful cost recovery scheme applied to one rural project in a developing area is also discussed.

It is the intent of the Financial/Economic sessions to convey the current best information on policies to be applied by the major financial sources in their evaluation of water and sanitation projects for the immediate future. This will prove of great importance to consultants and developing country planners alike, in any actions they may take to prepare projects for presentation to external sources for financing.

Information on those actions which are most likely to achieve results in the improvement of financial and economic performance of water system operation in developing areas should also prove rewarding. It is also the intent of these sessions to give a better understanding of the problems of management in dealing with the urban and rural poor in developing countries on questions of payment for water used and sanitary facilities to be built.

Finance and Economics

Harold Shipman
John Kalbermatten

Introduction

In planning this part of the program, an effort was made to devote a portion of the time alloted to a review of past practices and policies applied to the financing of projects in the water and sanitation sector by international funding sources over the past ten to fifteen years. The program which finally evolved, reflected this effort only partially since a number of papers were included which, although relevant to the topic, did not address specifically the question of past and future policies in the international arena. Papers presented, therefore, ranged from the highly tedchnical, such as how to optimally design a water distriution system, to the ecomomic justification of water supply investments., Discussions were equally wide ranging, from how to do suggestions to philosophy; they were also very spirited, leading at least one partifipant to suggest that interdisciplinary debates should stress learning from experience rather than parceliong out of blame for past mistakes.

Although the presentations and discussions were as wide ranging as the background of the participants was varied, a motif, if not a consensus developed quickly: users are able and willing to contribute to, and participate in, the development of water supply and sanitation to a far greater extent than planners and functionaries realize. This then became one of the major points of discussions, with two other topics, economic and financial justification and institutional issues, also receiving a great deal of attention.

Context

There are about three and a half years left in the International Drinking Water Supply and Sanitation Decade. Participants in the ASCE conference on Resource Mobilization, therefore, had an opportunity both to assess progress made so far and to look beyond the Decade. Clearly, resource generation for the Decade has been insufficient. However, progress has been made in the use of low cost technical alternatives. Equally important, officals and users alike are now willing to use these alternatives. As a consequence, the resource gap is not the major impediment to progress it once was thought to be and the stage is set for significant progress during the remaining years of the Decade and the years beyond. Conclusions drawn from participants presentations and discussions indicate clearly that substantial increases in service coverage can be achieved if low cost alternatives are consistently used and if user participation plays a major role in project development, implementation and operation.

Willingness and Ability to Pay

The willingness of the users to pay for water supply and sanitation services and their ability to do so was an issue raised and remained a priciple discussion topic thereafter. The main conclusions of this interchange can be summarized as follows:

Both willingness and ability to pay are usually greater than assumed by project developers;

Project proponents-financiers, governments and their water agencies- must make greater efforts to determine willingness and ability to pay; they must then reflect these findings in the design, financial and institutional arrangements covering the project.

Dialogue with users must be established to prevent the design and implimentation of projects which often result in decreasing, even destroying, the users willingness to pay and thus leading to project failure;

Tariff structures should be designed to satisfy economic, financial and social considerations including internal subsidies from the richer to the poorer users;

There is need for broad dissemination of reliable information on matters of willingness and ability to pay . All project developers and decision makers should be exposed to the facts surrounding these issues. More attention should be given to conveying this information to the NGO's.

Economic and Financial Justification

International agencies involved in the financing of water supply projects in the past have usually required that a sound economic justification for investment be established. They have found that because of the difficulty in measuring the benefits of water in the reduction of disease, in reducing fire losses, and in its contribution to a other community amenities, the only precise measurement was the financial benefit as measured by the revenue generated from water sales. It has been common therefore to note that in the justification of water projects financed by, among others, the World Bank, a statement is carried which says that the economic return which will include any benefits to health will be over and above the financial return. In the discussion of this issue, it was recognized that clear distinctions do exist between the economic and financial justification of projects. It was also recognized that there is much that needs to be done to bring more substance to the means for measuring the economic benefits which water can bring to a community. Among the areas needing exploration are those of water's contribution to production of raw materials, its benefits to commerce and industry, particularly small industry; and its ability to attract greater resources of other types needed for national development.

The discussions also brought out the need for improving technical and administrative arrangements through reductions of costs of operations and the improvement of efficiency. Better use of manpower, better retention of capable people, establishing causes for, and acting to reduce the high amounts of non-revenue (unaccounted-for) water, better billing and collection practices, and better metering policies were just a few of the areas needing attention.

There was agreement that grants and subsidies for water supply should not be used for payment of operation and maintenance costs. Where government policy provides financial assistance to poor areas and to special areas confronted with unusual problems, it is possible to apply such contributions to capital costs without seriously disrupting efforts to stimulate self support by the water agency.

Background and Experience related to WHO publications on rural water and excreta disposal, and to recent work on latrine and handpump design being carried out under UNDP funding by the World Bank.

Some of the broad observations made covering the overall presentations on this topic can be summarized as follows:

The basic public health concepts incorporated in the latrine designs shown and described in the original WHO publication have not changed. Recent work on improving designs have been directed primarily at giving these facilities a greater appeal and attraction to encourage better acceptance by the people. The major problem confronting everyone working in this subsector is that of changing habit patterns and stimulating use and acceptance to build and use latrines. This remains as it has been over the years the number one problem. In this regard, it appears that the support which the World Bank has given to latrine programs, added to WHO's continuing promotion of safe excreta disposal practices, are showing some encouraging signs. It seems likely, however, that the urban fringe and unsewered central areasmof cities will feel the need and be the places where greatest activity is likely to be focused for the foreseeable future.

The WHO publication on rural water supply, like its counterpart on excreta disposal, has not been revised since publication in 1959. It remains as a basically sound treatise on the public health justification for safe water, and on the various types of well and spring improvement suited to protecting them against contamination. Recent work on hand pump improvements have been directed primarily at the operation and maintenance aspects of well water supplies and have not altered to any extent the basics of well protection presented in the WHO publication.

One area of progress brought out in the discussions was that of the improvements realized in recent years in the work of water exploration and in well drilling. Current methods have literally revolutionized the rate at which wells can now be put down.

Hopefully the results of the very extensive work done on hand pump design will begin to show up in longer lasting pumps and fewer breakdowns. However, there is no evidence that the well or the handpump have yet been designed which do not require maintenance and repair. The means by which these are to be done is being concentrated at the village level and the hope is that emphasis on the village capability for maintenance, will bring a major improvement to the task of keeping these facilities in operation .

Funding the Decade Water/Sanitation Efforts

Annexes 1 to 3 present an overview of the sources and amounts of funds estimated to have come from the various organizations, agencies and institutions in support of projects, technical assistance, and coordination work to date. There is also presented some estimates of needs for the future.

Recommendations

Given the pervasive governmental control of the sector, it is tempting to address recommendations to governments. It is equally tempting to recommend actions on all of the problems identified. Although the following recommendationsd can be implemented by governments, it is more important that they be acted upon with out delay by conference participants and by those active in the sector without governmental actions and policy changes. The number of recommendations is kept to a minimum in the hope that this will lead to a concentaration of efforts and thus lead to visible results within a reasonable time. The recommendations are:

1. Project design should be based on willingness and ability to pay. This requires better studies in urban areas and greater dialogue with rural user communities than has been customary up to now. Results of this approach should be disseminated widely permitting everyone active in the sector to learn from the experience.

2. A greater effort should be made to establish the economic justification

of water and sanitation projects. Among other approaches, attempts should be made to quantify the value of services provided to commerce and industry, and to ensure that proposed investments represent the most efficient solution through a mix of service standards and technologies suited to the situation.

3. There should be strict adherence to the doctrine that water systems

charge for water the full cost. The means by which sewer service charges should be charged and collected remains a major problem where more experience and more information is needed and must be collected.

A final conclusion and recommendation

The water and Sanitation Decade when it closes in 1990 will find few countries with targets and goals fully reached. ASCE should should join with all the other national and international agencies in advocating the continuation of a follow-up program which will support and encourage all countries to renew their work in this sector by establishing new targets, new financial needs, and new manpower training plans in the fields of water supply, sewerage, drainage, and waste disposal. Whether the target date is set at 1995 or the year 2000 is not important. It is important to support the developing nations of the world in their continuation of an effort which must be on-going and essential to the health and welfare of their peoples.

Annex I

Funding the Decade

Martin Beyer, UNICEF

There is a need to provide an idea of the world-wide dimensions of:

a. the total funding needs in order to reach the goals of water supply and sanitation for all;

b. the present level of funding from all sources: domestic (national government and communities) plus external (loans and grants); and

c. the sources of funding.

The following figures may give an idea at least of the order of magnitude.

They represent only a very personal quesstimate, but based on WHO statistics with adjustments following field experience.

Projections for the next 5-10 years are based on the recent major policy shifts of the majority of governments in the developing countries jointly with the international donor community.

These projections, however, are for new installations only, not for rehabilitation or replacement of old systems, which would add considerably to the below figures.

An informal UNICEF working paper is under preparation with more details on global funding needs. In spite of the source, any figures still remain rough guesswork. The realistic possibilities hinge on many other concerns.

A. TOTAL FUNDING NEEDS

For all developing nations:
(US dollars in 1985 values)

$150 billion	
of which for	
conventioanl urban systems	$ 80 billion
poor urban and rural	70 billion
Total	$150 billion

For a ten year period, this would mean $15 billion per year. This is not much above present funding levels but implies wholesale use of lowest-cost technologies and massive community inputs in cash, kind and labor (and decision-making/planning).

• 2 •

Annex II

B. PRESENT LEVEL OF FUNDING

of water and sanitation installations
in all developing countries:

Total $10-12 billion per year

Assuming $10 billion per year, this investment distributed as follows for first half of water/sanitation Decade; the years 1981-1985:

Number of beneficiaries (new users) added (million persons):

	Water	Sanitation
URBAN	200M	150M
RURAL	200M	100M
TOTAL	400M	250M
Investment $billion	25	25
per capita $	62.5	100

C. SOURCES OF FUNDING PER YEAR
present level (1985)

1. Total	$ billion	Percentage
Domestic	7.5	75%
External	2.5	25%
Total	10	100%

2. External inputs per year: by sources:
(US$ millions)

Input	Source	Trad. HC	Low-cost	Total
LOANS	World Bank	500	100	600
	Regional Development banks	400	-	400
	Intergovt. (Arabs, etc.)	130	20	150
	Bilateral [3]	200	50	250
	Commercial	300?	-	300?
	Subtotal loans	1530	170	1700
GRANTS [1]	UN system[2]	20	130	150
	Bilateral [3]	50	450	500
	NGO/PVO's [4]	-	150	150
	Subtotal grants	70	730	800
LOANS + GRANTS TOTAL		1600	900	2500

Notes: 1 - Equipment, materials, "non-supply" (cast, etc.)
2 - See table 3. for breakdown
3 - E.g., USAID and other government donor agencies.
4 - Non-governmental organizations/Private voluntary organizations.

• 3 •

Annex III

3. United Nations system inputs per year
(grants in US$ million)

Source	$million	Work Scope
WHO	30	Methods, health research, standards,
ILO, UNESCO,FAO	3	human resources development, training,
Reg. Econ. Comm.	2	education, policy promotion, situation monitoring and evaluation.
UNDP	5	As above, global development and coordination
UNCDF	5	Implementation (financing)
UNTCD (Wat. Res. Sec.	10	Implementation/W. Africa & water res.
UNICEF	65	Implementation/global
Subtotal	120	
Emergency inputs	30	Draughts, earthquakes, conflicts, etc.
Total	150	

4. United Nations system 1987:

Number of water/sanitation/environmental specialists employed (guesstimate):

Organization	Headquarters		Field	Total	Remarks
UN Secretariat (DIESA)	New York	1	-	1	Coordination
Regional Economic	Santiago	1	-	1	)
Commissions	Geneva	1	-	1	) also work
	Addis Ababa	1	-	1	(with water
	Baghdad	1	-	1	) resources
	Bangkok	1	-	1	)
UNEP (Env. Proj.	Nairobi	1	-	1	
WHO (Health)	Geneva	6	70	76	
UNDP/World Bank (Dev.)	Washington/ New York	20	60	80	
UNTCD (Water Resources)	New York	6	30	36	
UNICEF (Children)	New York	3	150	153	
ILO (Labour)	Geneva	1	-	1	HRD/Training
FAO (Agriculture)	Rome	1	-	1	Micro-img.
UNESCO (Educ. & Sci.)	Paris	1	-	1	Education
INSTRAW (Women)	Santo Domingo	1	-	1	Social Research
TOTAL (approx.):		46	310	356	

LOD:pj
jmk-lb

LESSON LEARNED ON THE SOCIO-ECONOMIC AND FINANCIAL POLICIES OF PAST WATER/SANITATION PROJECTS

Klas Ringskog *

Abstract

It is likely that many developing countries will have lower service levels in 1990 at the end of the International Water Decade than they had at the beginning. Among the reasons are the misdirected socio-economic and financial policies applied in the sector. In the course of discussion of these policies the author provides recent data on past and projected per capita sector investment levels in Latin America and the Caribbean.

Introduction

The International Water Decade is coming to a close. Given the lead time necessary to prepare and implement water supply and sanitation projects it is possible to have a fairly good idea of the service levels that the developing world will have by 1990 - the last year of the decade. Although many countries will be able to look back at the Decade as one of solid progress, in others service levels will likely have dropped. For instance, WHO[1]/ estimates that in 1970 66% of the urban population in Africa had water coverage but that the level had dropped to 61% in 1983. Likewise, in the West Pacific in 1970,75% of the urban population had water coverage but the level had fallen to 70% in 1983. Similarly, a comparison between World Bank Social Indicator Tables 1985 and PAHO sector studies conducted in the years 1978-79 indicates that at least Argentina, the Dominican Republic, Honduras, Mexico and Nicaragua had lower shares of their urban population connected to piped water systems in 1984 than in 1977.

What went wrong? It is true that the International Water Decade had the bad fortune to coincide with a decade of debt crisis and depression in many developing countries. However, the economic crisis showed up the very inadequacies of some of the socio-economic and financial policies applied in the water supply and sanitation sector in many countries. It is therefore fitting to analyze the myths behind the policies in the light of the truths that have been learned from the implementation of past projects. Many of these myths and misconceptions have hampered the progress of the sector and the sooner they are dispelled the better.

---

* Principal Economist and Human Resource Specialist, Latin America Water Supply Division, World Bank, 1818 H Street, N.W., Washington, D.C. 20433. This views and opinions expressed in this article are those of the author and should not be attributed the World Bank.

The First Myth

The first myth has it that it has been mainly lack of funding that has held back the sector and that it is impossible to raise the money needed to reach the targets of the Decade.

The truth is slightly different at least for Latin America. An analysis of the investment requirements for Latin America reveals that in order to reach targets of 90% of their urban and rural concentrated dwellers with house connections and 70% of the urban population with sewerage by the year 2000, the required investment rate amounts to only US$8 per capita a year-0.3% to 1.6% of GNP per capita. (see Annex 1). During the 1970's the same countries invested at rates of 0.2% to 0.9% of GNP per capita which is not very far from the required levels. As a matter of fact many of the countries in the Latin America region invested at for higher rates in the 1970's. For instance, Venezuela invested at an annual rate of US$26 per capita in 1978 prices of its total population in the 1971-78 period.2/ This is a level that is clearly above that needed to reach the targets for the year 2000. The fact is clear: rather than looking only at the absolute level of funding it is more revealing to study the relative efficiency of the investment funds in connecting additional population to piped water and sewerage systems.

A series of sector studies conducted by the Pan American Health Organization for the Latin America region analyzed the investment efficiency over the 1971-78 period by comparing investments to the absolute increase in numbers served. The following table provides these data for the major countries in the region:

TABLE - Historical Incremental Investments Costs, 1971-1978
(US$ in 1978 prices)

	Investments 1971-1978 US$ million	Additional Served* Millions			Per Capita US$
		Water	Sewerage	Total	
Total	11,060	22.2 +	15.4 =	37.6	290
of which:					
Argentina	1,470	0.7 +	1.0 =	1.7	860
Brazil	3,080	10.9 +	7.3 =	18.2	170
Colombia	640	2.9 +	2.0 =	4.9	130
Mexico	2,450	3.2 +	3.3 =	6.5	380
Venezuela	2,540	1.2 +	0.4 =	1.6	1,600
Remainder	1,190	3.5 +	1.4 =	4.9	240

* With either water supply or sewerage.

The differences in investment efficiency are remarkable. Colombia and Brazil managed to connect relatively many per investment dollar, whereas, Argentina had to spend about 5 times as much, and Venezuela 10 times as much to connect each additional person. Mexico and remaining countries occupy intermediate positions. The explanation for Agentina's and Venezuela's high figures is probably their higher cost level, price distortions, and overly conservative design parameters which translate into very high per capita costs. The use of more appropriate low-cost technology and shorter design periods could alleviate much of the presumed shortage of investment funds in the sector.

The Second Myth

The second myth tells us that it has been the insufficient level of external assistance that can explain the relatively disappointing results of the International Water Decade.

The truth is that external funding has played only a complimentary role in funding the sector's requirements. As a matter of fact, the external funding in the Latin America region declined from 20% in 1971 to about 8% of total sector investments by the end of the 1970s and stood at about 11% during the 1981-85 period. External assistance, although modest, has played a catalytic role in mobilizing resources and focusing on sector development. The main financial burden, however, has fallen and will continue to fall, upon the countries themselves. This is not to say that external aid could not finance a greater share of total funding needs. For this to happen, however, project preparation in the countries would need to improve and the financial and institutional policies would need to be modernized. External lending has as a rule been held back by a lack of well-prepared projects and by the relative absence of well-managed sector institutions.

The Third Myth

The third myth is that the sector has suffered unfairly in the allocation of investment funds since it is a social infrastructure sector rather than a sector of directly productive investments or productive infrastructure.

If the sector is perceived as "soft" then the sector itself must assume a large share of the blame. It has been content to look for Government grants for its investment needs rather than emphasize internal cash generation and loan funding that has been the norm common in the power and telecommunications sectors. Given the water supply and sanitation sector's dependence on grant financing it is little wonder that it became a prime candidate for budget cuts when the harsh macro-economic climate of the 1980's forced the developing countries to put greater emphasis on raising public sector savings. The sector has made efforts in recent years to ease the pressure on central government budgets by preparing sector institutions capable of gradually covering their own operating costs and financing a major share of their own investments. These efforts to develop sound financial policies must be sustained, especially in the urban subsector. Subsidization of capital construction in urban areas should be discontinued to free funds for works in villages and rural areas, where subsidies are likely to remain necessary.

Where grant funding is available in urban areas it should be used to create revolving funds where the initial investment and the operations and maintenance costs are recovered out of operating surpluses. The belief that the water and sanitation sector is different and can be run inefficiently because it is a social sector is misplaced. Clearly, the greatest social good is achieved through efficient provision of services to the poor. If a water agency can serve its customers with only a third as may employees as it presently employs it should do so. Expenditures on superfluous staff can then be channelled into connecting the unserved within the sector - or reducing their water bills. And the released staff can be redeployed in productive occupations elsewhere. This is precisely how a sector adjusts and grows strong and competitive.

The Fourth Myth

The persistent myth of overwhelming reliance on budget resources is rooted in the belief that consumers cannot afford to pay the higher tariffs which it would take to make the sector financially autonomous.

This hypothesis does not stand up to empirical tests. Metered consumers have been known to pay up to 6% of their income to receive water and wastes disposal services - many times what they are asked to pay and certainly enough to cover the full cost of safe water connections. The argument that tariffs must be kept low because of the poverty of consumers is especially cruel. In one country after another we find that the lowest income population is systematically deprived of house connections and that they often pay more for scarce and unsafe water from water vendors. In other words, the myth that consumers cannot pay is equivalent to blaming the victim. The myth is sometimes reinforced by the belief that water consumption need not be metered. Failure to meter and charge the full cost of water usage simply means that those who consume the most receive the largest subsidies. Usually, they are also the wealthiest consumers since water consumption is income elastic. Much too often consumers are often charged less than the cost to the economy of providing services. Consumption and investments necessary to meet excess demand are significantly higher than those that would prevail under a rational tariff policy. There is no such thing as a free lunch, so why would there be a free glass of water in the economic sense?

Conclusion

In order to explain its shortcomings the water/ sanitation sector has often recurred to four myths: those of insufficient total financing, of insufficient external financial assistance, of insufficient domestic budget allocations, and of insufficient payments capacity of its customers to pay tariffs that reflect the full cost of service. More often than not these myths do not stand up to critical empirical analysis.

The conclusion is that the sector, in order to progress in a sustained fashion, will have to apply realistic policies: charge tariffs that reflect the full cost of service to its consumers, estimate demand on the basis of such tariffs, design the systems to meet demand at the least possible cost, and finally strive for full financial autonomy which in turn will make possible institutional autonomy and higher efficiency.

REFERENCES

1/ "IOWSSD Review of Regional and Global Data (as of December 31, 1983)", WHO Offset Publication No. 92, 1987 - Tables A.3.2.a - 3.

2/ Venezuela Water Supply and Sanitation Sector Study, 1979
Pan American Health Organization, Washington, D.C.

3/ Summarized in "Latin America and the Caribbean and the International Drinking Water Supply and Sanitation Decade" Course Note-863 of the Economic Development Institute of the World Bank, September 1980.

Water Supply Investments in Developing Countries:
Some Technical, Economic, and Financial
Implications of Experience

Harvey Garn[1/]

In 1977, at the United Nations Water Conference at Mar del Plata, the decade 1981 to 1990 was designated as the International Drinking Water Supply and Sanitation Decade. The Decade was to be the time during which governments of developing countries, working with bilateral and multilateral assistance would extend the benefits of safe drinking water and adequate sanitary facilities to all their people. The Decade is now more than half over, but its work is not half done. The purpose of this paper is to comment on some lessons that can be learned from the experience of water supply support agencies that have significant implications for the future development of the sector. This paper will focus on the links between economic and social factors which determine the ability and willingness to pay for water, on the one hand, and the scale, timing and types of technical solutions and the financial implications of such solutions, on the other hand.

Some of the Lessons

A progress report on the Decade, presented to the U.N. General Assembly in 1985, discussed the need for addressing major constraints that were impeding achievement of the goals of the decade. Among the constraints cited, a common thread was the problem of ensuring that cost recovery from the population to be served is suficient to support the particular technical options chosen to satisfy expected levels of per capita or per household consumption. In many cases, relieving these constraints requires designing systems and pricing services in ways that ensure that the population is able and willing to pay for the level of services being provided. Unless the common failure to do so in developing countries is overcome, the ability to achieve the most fundamental and widely shared goals of the International Drinking Water Supply and Sanitation will be seriously impaired.

1/ Harvey Garn, Consultant, in the Water Supply and Urban Development Department of the World Bank. The author is indebted to Alfonso Zavala, Senior Sanitary Engineer, also in the Water Supply and Urban Development Department, for helpful comments and suggestions. The views presented herein are those of the author, and they should not be interpreted as reflecting those of the Bank.

Evaluations of the past experience of bilateral and multilateral agencies which fund water supply programs increasingly emphasize the central importance of the ability and willingness to pay in selecting technical options and charging systems. In reviewing it experience with water supply projects, the Federal Republic of Germany concluded, for example:

> An analysis of all evaluations in the water supply sector which was undertaken in 1981 came to the conclusion that in spite of all feasibility studies and planning efforts a large number of projects showed serious defects...while the technical execution of the projects had normally been satisfactory, the economic aspects had normally been neglected.

The World Health Organization in its 1983 report on the Decade cited similar findings by the Australian Development Assistance Bureau, the European Economic Community, and the Canadian International Development Agency. More recently (April, 1986), the WHO Study Group on Technology for Water Supply and Sanitation concluded that:

> To ensure the feasibility, acceptability and success of the planned water supply service, the adopted (design) criteria must be responsive to the needs and constraints of the community concerned. Thus, design criteria comprise technical, health, social, economic, financial, institutional and environmental factors which determine the characteristics, magnitude and cost of the planned system.
>
> In view of the wide differences in the conditions, needs and constraints of the various communities in developing countries, design criteria, other than those related to the minimum public health requirements, cannot be standardized[1/] and applied indiscriminately.

A retrospective review of World Bank-supported water supply projects has shown that there have often been substantial shortfalls between connection and sales volumes expected at appraisal and actual volumes over equivalent future time periods. A significant implication is that there may have been mismatches between the capacity provided and the amount of access and water usage for which people were prepared to pay. On the financial side, the review showed that the rapid rise in operation and maintenance expenditures (in local currency terms) coupled with the lower than expected sales volumes has often resulted in an inability to reach cost recovery targets even in cases where larger real tariff increases have been obtained than were expected.

1/ This should not be taken to imply that basic technical aspects of design cannot or should not be standardized.

Thus experience clearly has shown the need to have a well documented approach to the identification of technological packages most likely to provide appropriate levels of service and be maintained on a sustainable basis at costs which can be met by the served community. The great diversity of communities being served (e.g., in their population levels and growth, density, income levels, and current water use) makes it necessary to anticipate substantial diversification of technologies employed in different settings.

In recognition of these concerns, the World Bank is undertaking a major study of water consumption and the factors which influence it in several countries and sites within each country coupled with an assessment of how improved information on willingness to pay for water supply can be effectively incorporated into technical design (e.g., level of service, capacity, and timing) and financial management (e.g., charging systems, revenue forecasting, and unaccounted for water reduction) options. The research will be conducted by the Operations and Policy Research Division of the Water Supply and Urban Development Department in collaboration with the Bank's Regional Project Divisions and with the support of the UNDP program and several bilateral donors.

Capital Costs and Quantities Produced

During the retrospective review of World Bank-supported water supply projects referred to earlier, data from 54 projects, which were started between 1966 and 1981, were examined extensively. As part of this review, data were gathered on project costs, production volumes, sales volumes, volume and percent of unaccounted for water, numbers of connections, sales revenue, operating costs, depreciation, and interest payments. The data on project costs and volumetric outcomes of the projects permits an indicative estimate to be made of the relationship between capital cost and production volumes in the design of water supply projects. This estimate is from projects with a variety of components. Some projects emphasize production and transmission while others include substantial distribution components. The estimate, therefore, shall be treated as average values.

One purpose of the estimates was to provide a sense of the economies of scale expected to be achievable in the projects. Consequently, the function estimated was an exponential function with the form:

$$\text{Cost} = \text{Constant} \times (\text{Quantity Produced})^{b}$$

where:

Cost = Project Cost in millions of 1980 U.S. dollars

Quantity Produced = Expected Production at capacity in millions of m^3 per year

b = the economy of scale exponent

The estimated equation, assuming an average design horizon of 10 years to reach capacity and using appraisal estimates of costs and production, is:

$$\text{Cost} = 4.76 \times (\text{Quantity Produced})^{0.6722}$$

The estimated equation, if an average design horizon of 15 years is assumed, is:

$$\text{Cost} = 3.48 \times (\text{Quantity Produced})^{0.6747}$$

The economy of scale parameter, b, of about 0.67 is plausible when compared with other engineering estimates of economies of scale and implies that a 10 percent increase in output could be obtained with about a 7 percent increase in costs for these projects. The function, when evaluated at the mean value of production as estimated at appraisal, assuming a 10 year design horizon, shows that the average design production of 83.6 million m^3 per year (or 2.65 m^3/SEC) would cost about $93 million.

The function can be interpreted also as a means of estimating the capital cost consequences over a given design horizon of either over- or under-estimating the production which will be required to satisfy demand. To illustrate this, additional data from the review will provide some examples. As suggested earlier, the review showed that actual production levels did not generally reach the levels anticipated at appraisal. Moreover, unaccounted for water (production - sales) tended to increase in volume as production expanded and, in percentage terms (unaccounted for water/production), remained relatively constant.

The actual average level of production for the projects in the illustration above, assuming once again a 10 year design horizon, was about 60 million m^3 of which about 35 percent was unaccounted for by sales. Had the actual levels of production, including unaccounted for water, been anticipated at appraisal; the cost function indicates that this level of output could have been attained at a cost of about $75 million. This would represent a cost saving of about 20 percent relative to the average cost of about $93 million estimated at appraisal.

The reduction of unaccounted for water, could play a substantial role in holding down new capacity requirements or in postponing the time when it would be necessary to undertake capacity expansions. To illustrate the cost saving potential of accurate sales forecasts and aggressive efforts to reduce unaccounted for water, a further calculation from the review data is instructive. If production capacity had been designed to meet actual sales plus the amount of unaccounted for water anticipated at the time of appraisal (about 24%) rather than the larger amount which was actually unaccounted for; the capital cost requirement (again estimated from the cost function) would have been about $61 million -- which would represent about a 35 percent cost saving relative to earlier estimates.

The absolute magnitudes of the potential cost savings cited above should be treated with some caution, being both retrospective and based upon a relatively limited data base. Moreover, they do not take account of the fact that some of the unaccounted for water is, in fact, being used through unauthorized illegal connections. They are sufficiently striking, however, to suggest that the possibility exists in many circumstances to save scarce capital resources by tailoring capacity additions to more accurate forecasts of production requirements than is often the case. Since production requirements are based upon anticipated sales and unaccounted for water, a relatively high value should be placed on accurate sales forecasts and on identifying cost-effective means of reducing unaccounted for water.

Related Financial Issues

The short-term financial solvency of water supply institutions and their longer term prospects of financing capacity expansions, when needed, depend critically on sales revenues and their relationship to on-going operation and maintenance costs, depreciation and costs related to amortization of debt. Sales revenues, in turn, depend upon the rates of water consumption and the effective unit charges on water consumed.

The issues discussed in the previous section clearly have an important bearing on the financial situation of a water supply authority. The relationships, however, are multi-faceted and complex. The supply institution's costs are associated with its production capacity, the extent of use of that capacity, and the financing arrangements used when it adds capacity. The revenues it can achieve depend, however, primarily on the number of consumers it has and the volumes of water they use at the rate at which they are charged. They depend, in short, on the actual and potential users willingness to pay to connect to the system and use it at a certain level. To put the point another way, the revenue requirements are intimately connected to the expectations about consumer demand which are reference points for the technical design and the revenue possibilities depend on the degree to which these expectations are realized.

The experience of many water supply institutions in developing countries has been that the sales expectations at the time of system expansion have not been fully realized and high levels of produced water which is not accounted for by sales have persisted. Many supply institutions have responded to this by raising tariffs on the water which is sold only to discover that one of the results of this is that the rate of consumption per connection does not rise at the rate anticipated at the time system expansion was undertaken due to the response to price.

A plausible picture of the financial effects of all these factors can be obtained from the review data discussed earlier. These data, aggregated on an annual basis from the date of project implementation for the subset of the 54 projects for which all the relevant data are available, are shown for an average project in Table 1.

Table 1

Annual Averages from Year of Implementation
(1980 US$ per m^3 sold)

		Costs		
Year	Revenue	O and M	Depreciation	Interest
1	0.23	0.17	0.05	0.04
2	0.22	0.18	0.06	0.04
3	0.23	0.19	0.06	0.04
4	0.24	0.19	0.06	0.04
5	0.21	0.18	0.06	0.03
6	0.20	0.18	0.06	0.02

The figures in Table 1 illustrate several important points. First, in spite of substantial inflation in many of the countries from which the sample is drawn, real revenue per m^3 sold declined by less than might have been expected. Second, real revenues per m^3 sold exceeded operation and maintenance costs per m^3 in all years but the margin between the two declined steadily making it progressively more difficult for the supply authority to cover its other costs from revenue. Third, with the exception of the first year, real revenue was insufficient to cover operation and maintenance costs plus depreciation. Fourth, the relationship between revenue and costs per m^3 indicates that supply institutions would be unable, on average, to make a contribution to future capacity expansion.

The figures in Table 1 are calculated on the basis of revenue and costs per m^3 sold and, therefore, do not directly reflect two further concerns. In many cases, sales volumes are substantially below the levels anticipated by the supply institution when a supply expansion is undertaken. For the projects included in Table 1, for example, in year 6 the sales volumes were 20% below the anticipated sales volumes. Such shortfalls have obvious implications for total sales revenue. The second concern is the persistently high share of production which is unaccounted for in sales. As indicated in another paper prepared for this conference by the author and Mr. Saravanapavan, "Reduction and Control of Unaccounted for Water in Developing Countries", unaccounted for water in World Bank supported projects has been, on average, about 34% of production. The relevance of high levels of unaccounted for water to this paper is that both the operations costs and asset costs are related to production capacity and production levels. If unaccounted for water could be reduced in a cost-effective manner and the additional water sold total revenue would increase and/or the costs per m^3 sold would decline, improving cost recovery by the supply institution. It is estimated, for example, that a reduction from 34% to 30% in unaccounted for water could increase revenue (if there is sufficient demand) by about $700,000 a year or reduce costs (if there is excess supply) by about $600,000 a year in an average project.

Conclusions

The experience of both multilateral and bilateral agencies providing support for water supply investments in developing countries have shown that there is often a significant mismatch between the production requirements and sales volumes anticipated at the time investments in new or additional capacity are made. where such mismatches occur there may be substantial opportunities for savings in capital costs and opportunities for improved cost recovery. Similarly, the persistence of high levels of unaccounted for water, suggests that improved management of the assets and aggressive reduction programs in unaccounted for water may have substantial payoffs in producing net revenue gains or cost savings.

A central aspect of reducing the probability of a mismatch between production capacity and the production required to meet demand is the development of cost-effective approaches to the estimation of future willingness to pay for water and, therefore, future demand. The research effort at the World Bank, mentioned earlier, is beginning to develop the primary data and analysis to improve the ability of sectoral institutions to accomplish this. The full range of possibilities will not be realized, however, so long as costs are incurred to produce water which remains unaccounted for in sales or when the levels of unaccounted for water require new production capacity earlier than is necessary.

ASSESSMENT OF INTERNATIONAL LENDING POLICIES

David B. Bird and Eugene S. Churchill*

As a result of policies imposed by international lenders regarding the existence and operation of utilities, some operational improvements are being noted. Well-run water and wastewater utilities exist; however, they are a small minority, and the others need assistance in order to develop satisfactorily for the benefit of the recipient country its customers, the public-at-large, as well as the lending institutions. The recommended policy improvements are:

- o Provide true autonomy to the operating agency;
- o Assure that all potential income sources are used and controlled to maximize agency revenues; and
- o Involve the operating agency in project evaluation and approval during negotiation.

INTRODUCTION

This discussion consists of the authors' personal observations and conclusions developed while serving as consultants to water and sewerage agencies in developing countries. The paper deals with the effects of current international lending institutions' policies on economic and financial improvement of the water/wastewater agency and within the developing country as a whole. Lender policies are viewed in light of evolving policy changes and their actual impact on sector improvement.

The views presented in this paper highlight the strengths and weaknesses of current policies especially in relation to the real constraints that exist in the recipient countries. It also must be acknowledged that many political issues and considerations face the international lender that cannot be readily overcome or resolved in a short period of time or if at all.

LENDING AGENCIES AND THEIR POLICIES UNDER DISCUSSION

International lending agencies have been playing an increasingly important role in promoting improvements within the water/wastewater sector organizations of developing nations. The lending policies of the World Bank (IBRD - International Bank for Reconstruction and Development), the International Development Association (IDA), Inter-American Development Bank (IDB), Caribbean Development Bank (CDB), the United States Agency for International Development (USAID), and the Canadian International Development Agency (CIDA) are included in this discussion.

*Vice President and Project Manager, respectively, James M. Montgomery, Consulting Engineers, Inc., 250 N. Madison Ave., Pasadena, CA 91101.

To varying degrees, these agencies influence the economic and financial climates in the water sector by their lending policies as follows:

- o Pre-loan project requirements and other conditions
  - Satisfactory economic benefit analyses of project.
  - Acceptable cost/benefit ratios.
  - Satisfactory operating institution existence or formation.
  - On-lending arrangements and repayment responsibility.
- o Loan conditions and loan covenants
  - Revenue coverage of O&M, debt service and/or depreciation.
  - Revenue reserves for future projects.
  - Universal metering and collection enforcement.
  - Reduction in unaccounted-for-water.
  - Tariff adjustments (increases).
  - Improved enabling laws and regulations.
  - Operating agency staff and efficiency improvements.
  - Overall institutional improvements.

APPARENT POLICY GOALS OF INTERNATIONAL LENDERS

The goal of requiring ever-increasing conditions on water/wastewater sector financing seems obvious. The policies are directed at:

1. Developing the potable water supply/wastewater sector into a financially viable, self-sustaining utility. This permits the recipient country to allocate its limited resources and foreign exchange towards social betterments and important programs that cannot be revenue supported.

2. Institution building to achieve financial viability and to provide adequate, safe water supply and wastewater disposal service to the greatest possible number of citizens. Proper institutionalization will also assist in protecting the large investments in plant assets and will defer future capital obligations caused by premature deterioration.

To achieve these basic goals, the lenders attempt to initiate the following changes through loan conditions.

1. Establishment of an operating agency that is semiautonomous and that can operate efficiently without direct involvement by central government in its day-to-day activities.

2. Collection of all revenues owed to the operating agency that are based on realistic, fair, and equitable assumptions.

3. Establishment of tariffs for all goods and services that will generate sufficient revenues to meet all revenue requirements.

4. Capability within the operating agency to attract, recruit, and retain qualified and motivated technicians, managers, and other staff.

SUCCESS OF CURRENT LENDING POLICIES

The increase in specific loan conditions is showing positive results through noticeable improvements in the sector operating agencies. The following trends have evolved over the past decade:

1. More operating agencies are becoming semiautonomous and operate outside the direct control and supervision of a government ministry or department.

2. Tariffs are becoming more realistic and equitable yet still within the ability to pay of the various income groups.

3. Collection enforcement has tightened and cash revenues are on the increase.

4. Institutional improvements are apparent in many areas of agency activity.

5. Unaccounted-for-water is under tighter control and is being reduced.

6. The operating agencies are gradually evolving into a businesslike utility, and management is becoming aware of the need for financial viability.

The lending policy changes over the last 10 years have promoted sector improvements. This has benefited the recipient countries by forcing change through specific loan conditions. However, all loan covenants are not satisfied due to real constraints within each recipient country.

CONSTRAINTS TO LENDING POLICY GOALS

The fact that clear and concise conditions and loan covenants are imposed and agreed upon does not insure compliance by the borrower. The lender cannot, and does not, stop project loans midstream because of unsatisfied loan covenants and will rarely withhold future project loans for this reason. Generally, the borrower is aware of this. Reasons behind the performance shortfalls are understandable but difficult to overcome. The economic and financial policies directly impact the institutional capabilities, and the two are inseparable. The most frequently unrealized loan covenants deal with the following important improvements.

Establishment of a Truly Semiautonomous Agency

For a water sector operating agency to become efficient and businesslike, it must have certain autonomy from direct government control and interference and must be able to organize and staff its agency effectively. This requires the ability to employ capable and qualified personnel and to manage this personnel in an effective manner. Government civil service practices usually prohibit this freedom and restricts the quality of staff available to the operating agency. This denies the agency the ability to compete with the private sector for qualified and efficient employees in the country's labor market. Without this capability, the operating agency is shackled with a costly and overinflated staff that is inefficient and largely unproductive. Efficiency and financial viability are almost impossible to achieve.

Constraints. As long as an operating agency relies on substantial government subsidy for its operations and/or improvement programs, the central government will be extremely reluctant to permit "real" autonomy that allows employment benefits in excess of those provided to direct service government employees. Allowing a highly subsidized agency the autonomy to establish superior salaries and benefits could cause unrest and dissatisfaction throughout government service.

Establishing Tariffs to Meet Revenue Requirements

With existing levels of efficiency in the operating agency, this frequently means rates and charges that are at, or beyond, the "ability to pay" or "willingness to pay" of the various income groups. This could create unrest among the customers and users and may cause abandonment of piped water supply for less safe and reliable alternate sources.

Constraints. The central government, even in a totalitarian situation, is very political in nature and sensitive to the feelings of the masses. Any unpopular increase in rates and charges for water supply or wastewater disposal services would probably not gain government approval even if it means continued subsidy to the operating agency from limited, or non-existent, financial resources. More than one country has experienced a turnover in government solely because of support for unpopular water tariffs.

Loan Negotiations and Responsibilities

The lender negotiates directly with the central government for water sector loans that are usually on-lent to the operating agency at a predetermined interest rate. The negotiations may or may not include the direct participation of the operating agency.

Constraints. The operating agency should have a direct voice in the loan program under consideration including the project size, priority, and timing and the ability of the agency to service the additional on-lending debt obligation. Without this participation, programs could be financed that are not needed now, are overbuilt, or use money needed elsewhere by the agency. Also, if the agency is not directly and intimately involved in approval of loan programs, it will feel that repayment obligations belong to central government and are not its responsibility.

Institutional Improvements

This wide array of organizational, managerial, and procedural betterments is essential to develop an efficient businesslike agency. These improvements can only be realized with direct support and participation by qualified managers and supervisors on the agency staff.

Constraints. Without the ability to achieve real autonomy for the operating agency as discussed earlier, the chances for real and lasting institutional improvement are slim, at most. Lack of a workable organization plan supported with qualified and motivated middle management and supervisors will negate the ability to improve the institutional activities.

Financial Viability

These other constraints directly impact the ability of the operating agency to become financially viable. Therefore, the important lending institution goals are at the mercy of these real constraints that must be overcome before these goals can be realized.

POTENTIAL LENDING POLICY IMPROVEMENTS

The need to overcome constraints to goal achievement is obvious to all concerned. As consultants on the inside looking out, we feel that certain approaches are possible to overcome the constraints. These are set forth here in order of their importance.

Achieving Real Autonomy for the Operating Agency

Since this is the key to most goal attainment, it should receive the greatest thought and emphasis. Programs should be developed that are palatable to central governments, and which provide real autonomy under specific agreements setting forth benchmark achievements within distinct time frames. As an example, autonomy would be granted under the following conditions and legal authority.

a. The agency could establish employment benefits for specific positions equal to the practices of the private sector within the country.

b. The agency would be exempted from certain civil service regulations concerning employment, promotions, dismissals, and disciplinary action. Earned benefits of long-standing employees would have to be resolved in some satisfactory manner.

c. The agency will meet a distinct and realistic timetable in reaching financial viability, i.e. -

- Sufficient revenues to cover all operating costs within 2 years.
- Sufficient revenues to cover debt servicing costs within 4 years.
- Sufficient revenues to establish future capital improvement reserves at a determined percentage of total revenues within 6 years.
- Complete financial independence within 8 years without need for government subsidy.

The agreements would condition the continuing autonomy privileges on the performance of the agency in meeting the revenue-producing goals and would tend to spur the management and staff to improved efficiency and good practices. However, conditions must also be set in the event that central government blocks agency attempts to establish realistic and acceptable tariff schedules. Figure 1 graphically depicts the proposed methods of achieving sector goals.

Maximizing Operating Agency Income

The simple approach of increasing tariff schedules may not be equitable or politically acceptable. The lending institution should insist that all avenues be investigated for potential increases in income from existing tariffs. Additional income sources could include the following:

Uniform Collection Enforcement. Assuring that all beneficiaries of agency goods and services pay their debts in a timely manner.

Illegal or Unrecorded Connections. Assuring that all water users and sewage connectors are properly recorded and billed for service.

No Exemption From Tariffs. Insisting that all beneficiaries of service pay their fair share of costs--whether government, important personages, social groups, schools, religious institutions, or agency employees and relatives.

Accurate Consumption Metering. If meters are used, assuring that all meters measure consumption accurately and cannot be tampered with for financial gain. This may involve locating, specifying, and acquiring a distinct type of water meter that meets all accuracy requirements and other agency needs.

Proper Control Systems. Adequate internal controls to eliminate errors and eliminate or minimize any anomalous activity by agency employees.

Other Charges and Fees. Assuring that other charges are established at reasonable levels to offset all involved costs, whether current or potential future liabilities. This would include connection charges and the costs for extension of service facilities.

Project Identification and Loan Negotiation

An established operating agency should be directly involved with project evaluation and selection and financing negotiations. The agency will be responsible for operating and maintaining the completed project and, at least theoretically, will be responsible for servicing the debt. Project economic benefit analyses and cost/benefit ratios do not consider practical priorities, all alternatives, the agency's capacity to run the project, or the financial ability of the agency considering all future debt obligations. The operating agency should be directly involved in appraising and approving:

a. The project in relation to other priorities.

b. Project alternatives.

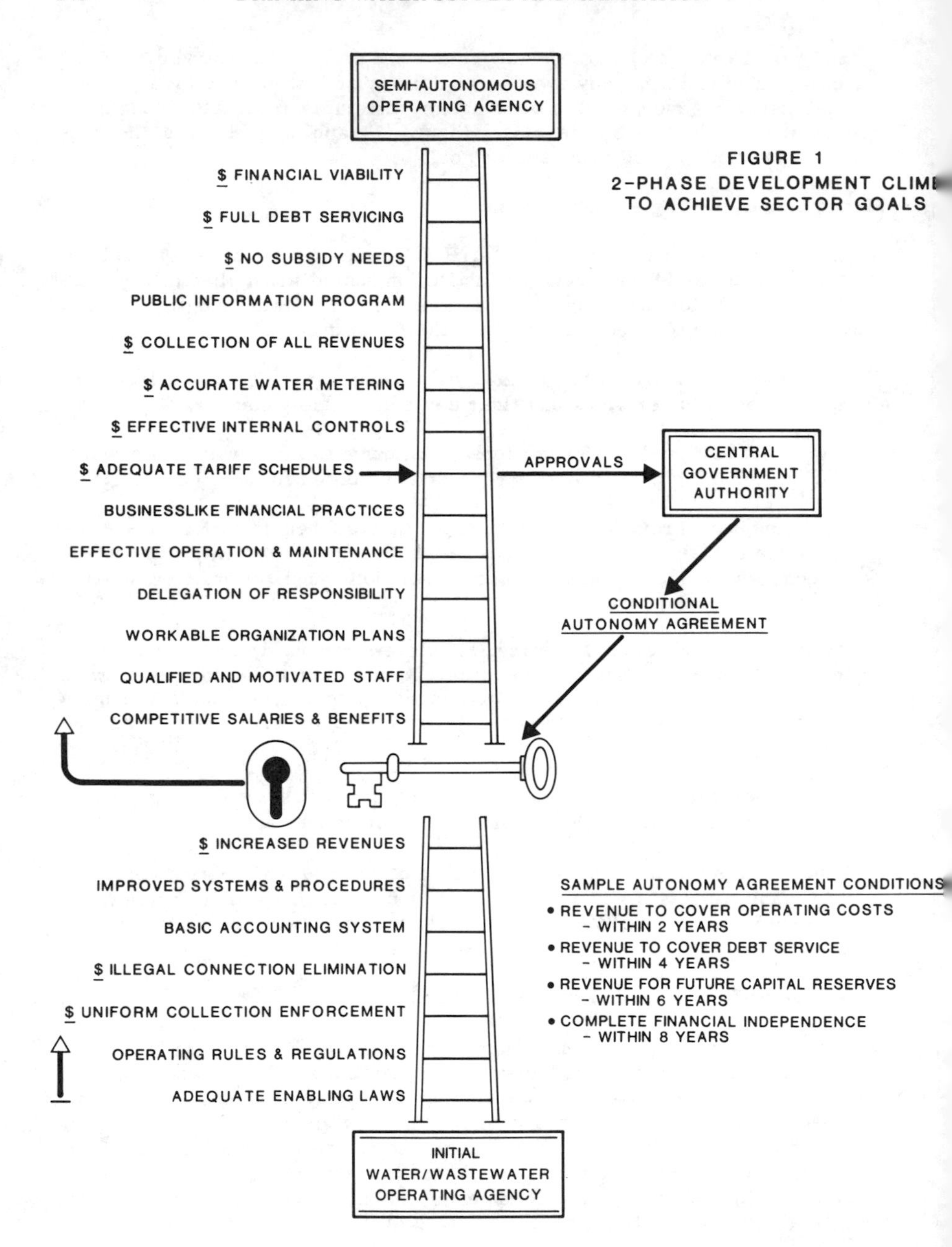
SEMI-AUTONOMOUS OPERATING AGENCY
FIGURE 1
2-PHASE DEVELOPMENT CLIMB TO ACHIEVE SECTOR GOALS
$ FINANCIAL VIABILITY
$ FULL DEBT SERVICING
$ NO SUBSIDY NEEDS
PUBLIC INFORMATION PROGRAM
$ COLLECTION OF ALL REVENUES
$ ACCURATE WATER METERING
$ EFFECTIVE INTERNAL CONTROLS
$ ADEQUATE TARIFF SCHEDULES
APPROVALS
CENTRAL GOVERNMENT AUTHORITY
BUSINESSLIKE FINANCIAL PRACTICES
EFFECTIVE OPERATION & MAINTENANCE
DELEGATION OF RESPONSIBILITY
CONDITIONAL AUTONOMY AGREEMENT
WORKABLE ORGANIZATION PLANS
QUALIFIED AND MOTIVATED STAFF
COMPETITIVE SALARIES & BENEFITS
$ INCREASED REVENUES
IMPROVED SYSTEMS & PROCEDURES
BASIC ACCOUNTING SYSTEM
$ ILLEGAL CONNECTION ELIMINATION
$ UNIFORM COLLECTION ENFORCEMENT
OPERATING RULES & REGULATIONS
ADEQUATE ENABLING LAWS
SAMPLE AUTONOMY AGREEMENT CONDITIONS
• REVENUE TO COVER OPERATING COSTS - WITHIN 2 YEARS
• REVENUE TO COVER DEBT SERVICE - WITHIN 4 YEARS
• REVENUE FOR FUTURE CAPITAL RESERVES - WITHIN 6 YEARS
• COMPLETE FINANCIAL INDEPENDENCE - WITHIN 8 YEARS
INITIAL WATER/WASTEWATER OPERATING AGENCY

c. The size and timing of the project.

d. Operating and maintenance capabilities.

e. Capability to service the on-lent debt burden over time.

f. Other effects on the existing organization and its development plans.

This direct involvement will help assure that the project is needed and is properly developed for operation by the agency. Also, the agency management will feel more responsible for the project's operating results and the need for repaying its on-lending obligation with government. A well-managed agency may wish to completely abandon a project for good reasons.

SUMMARY

The authors would like to acknowledge that international lenders have improved their lending policies over recent years and that the results from this action have been positive. It is now possible to find a few water sector operating agencies in developing countries that could compete with many United States utilities for efficiency and good management. However, these are a small minority and the others need to be assisted in order to develop satisfactorily for the benefit of the recipient country, its customers and users, the public-at-large, as well as the lending institutions. Again, in order of priority, our recommended policy improvements deal with the following objectives:

1. Provide true autonomy to the operating agency so that it can organize, staff, and operate effectively. This could be achieved through specific goal targets in a revocable agreement with central government.

2. Assure that all potential income sources are thoroughly exploited and controlled to maximize agency revenues. Only then, promote tariff increases.

3. Directly involve an ongoing operating agency in project evaluation and approval and in the negotiations with international lenders.

Many new water sector operating organizations, agencies, corporations, and authorities have been established in developing countries at the recommendation and insistence of planners and international lenders. Yet most of these bodies are unable to exercise all of the legal provisions in their enabling laws due to the local constraints discussed earlier. In many cases, the legal instrument is available but the authority to operate under its provisions is not given by central government. This is the major obstacle that must be overcome to achieve the desired water sector goals.

Experiences in Predicting Willingness to Pay on Water Projects in Latin America

Christian Gómez 1/

This paper describes the use of demand functions to predict willingness to pay, and the application of cost-benefit methods in the planning process of expanding or improving water supply systems. The results are compared across countries, between the rural an urban settings and between different income level beneficiaries. Finally, issues that still need to be resolved are discussed both methodologically and with regards to empirical research in this field.

Introduction

On February the 3rd, 1961, the Inter American Development Bank (IADB) approved its first loan of US$ 3.9 million, to be used in the expansion of the water supply and sewerage systems in Arequipa, Perú. Since then the Bank has continued financing and giving priority to projects in social sectors, such as Health, Sanitation and Education, as well as in other sectors which have been traditionally associated with multilateral lending such as Transportation, Energy, Industry and Agriculture.

The By-laws of the IADB state that all projects financed by the bank should contribute to the economic development of member countries, i.e. add to their aggregate welfare. This means, among other things, that economic benefits derived from Bank financed investment should compensate and override all costs implied. Similarly to the way commercial banks check for the borrowers ability to repay a loan, economists at the IADB make sure that potential projects are chosen by the borrowing country using appropriate prioritizing criteria, and that both timing and scale of projects are kept in step with the relative scarcity of resources as well as keeping pace with existing and future needs. In order to accomplish these tasks it has become essential to develop ways in which to estimate, in a quantitative fashion, the benefits provided by all the projects being financed.

It is relatively easy to assess the impact of projects carried out in productive sectors such as agriculture or industry, by valuing

---

1/ Economist, Infrastructure Division, Inter American Development Bank, 1300 New York Avenue, N. W., Washington D. C. 20577. The views expressed in this paper are exclusively those of the author and not necessarily those of the Bank.

output at market or border (international) prices. When trying to estimate the value of benefits provided by investments in the social sectors, however, one is confronted with difficulties arising from the fact that prices charged for services are usually subsidized and therefore underestimate the value of the service being provided. The IADB has moved toward its objective of financing economically viable projects by developing methods for measuring benefits in social projects. An example of the methods currently in use is the SIMOP model(Powers, 1977, 1978 and 1980) which was developed in 1978 to estimate the social profitability of water supply projects in urban areas. The model uses demand curves for water to estimate the benefits of increasing the capacity of water systems. A demand curve describes the maximum prices a consumer is willing to pay for each unit of water consumed, or conversely, the maximum quantity he is willing to consume at each price charged. The area below a demand curve is therefore the value of the benefits derived from increasing the availability of water. See figures 1, 2 and 3 for examples of demand curves.

The exercise of demand function estimation with information gath ered through surveys has been carried out successfully since 1980 at the IADB in 11 Latin American countries. It is now common practice in the IADB to carry out surveys, in order to correctly estimate the benefits to be derived from investments in the sanitation sector. In what follows, we present an intuitive explanation of the significance of demand functions in the process of planning the expansion of water supply systems, followed by a description of the estimation process. Finally we present demand functions estimated for eight countries in Latin America.

Demand Function and its Uses

A demand function for water describes the influence that variables such as prices, family size and income have in determining the amounts of water consumed on a monthly basis. An econometric model describing this relationship looks as follows:

$$Q = f(P, Y, N, e) \qquad (1)$$

where:

- Q = Cubic meters of water consumed by a family in a month
- P = price charged for each additional unit of water consumed
- Y = Monthly family income
- N = Family size
- e = Statistical error term

The model purports that, among other factors, there is a negative relationship between quantities consumed and prices charged, and a positive correlation with income levels, and family size. Also, for an average sized family with a particular level of income, one can use de demand function to plot how quantities of water consumed would decrease if prices charged were to be increased. This relationship is of extreme importance in the analysis of water projects since every point in the curve, expresses the willingness to pay (WTP)of users for the water being demanded. The WTP is also the value in terms of monetary income that the user assigns to the service being provided.

Demand functions should be central in the efficient management of water supply systems, both from a day to day financial and operational point of view as well as in planning the expansion of a water system. In Annex figure 1 we present a planning scheme for a water system.

The process starts when the planner sets initial prices for water. Population and income level projections are exogenous inputs which together with the price level provide the necessary information to be plugged into the demand function and arrive at consumption levels for each year of the planning period. Industrial, commercial, public water use and losses are then added to complete the projection of demand to the system.

Different technical alternatives that satisfy this demand can then be set up in order to compare costs and select at the minimum cost solution. Traditionally minimum cost analysis has been limited to comparing investment and maintenance costs of different water sources, technologies used in designing treatment facilities and location of the different components of the system. The basis for establishing the different alternatives has also been limited to those projects which satisfy the demands in the planning period.

Demand functions provide the tool to estimate the cost of not entirely satisfying demand in one particular point in time. Usage of demand functions in the minimum cost analysis lets the planner choose the least costly alternative not only with respect to investment costs but also with regards to the cost to society of only partially satisfying demand at a particular point in time. A much wider array of alternative investment possibilities thus becomes available in the analysis, since an alternative that by chance does not cover the demand of particular period is not automatically discarded, and is taken into account with an additional cost assigned as a penalty for each cubic meter it is not able to supply. 1/

This system allows other alternatives, such as loss reduction programs and staging of large projects to distribute investment costs through time, to be considered, even if demand has to be rationed during at a particular point in time. What becomes crucial in the decision of whether to sacrifice water consumption or not is the comparison of failure costs vis a vis incremental investment and operational cost of satisfying demand at all times.

Once a minimum cost expansion path is chosen, it should be evaluated to ascertain whether the benefits provided by its implementation

---

1/ The demand function properly reflects rationing costs to the user in a medium or long term context in which the consumer is allowed to adjust his consumption patters to the expected scarcity level.

are larger or equal to the costs involved. Once again in this evaluation exercise, demand curves are critical in determining benefits, since each point in the curve represents the WTP and therefore the "use value" assigned to that unit of water.

If benefits are shown to be greater than costs, the planner should recommend that particular expansion path at the original tariff level. If benefits of the minimum cost alternative do not override costs, pricing levels should be checked against forward looking marginal costs 1/. If tariffs are lower than marginal costs they should be increased, in order to contract future demand levels and in that manner decrease the cost of supplying it. These adjustments are carried out until the expansion path designed is both a minimum cost solution and shows to have a positive impact on welfare of society. 2/

Estimation Process

The estimation of demand curves for water requires a wide array of observations of water use behaviour from different types of consumers, at different price levels. In developed countries this has been done successfully using mainly billing data from utilities (Agthe, 1980, CV Jones, 1984, Danielson, 1979, Hanke, 1984, How 1967 & 1982, Primeaux, 1978). In Latin America, however, this source of information by itself is not very useful since very little variance exists in water prices charged by water utilities. In fact, marginal water prices are generally subsidized at low levels and are zero in many cases.

In order to find more variance in prices charged, the market has to be searched specifically for family units who are not being served by the water utility and therefore have to pay higher prices, to other suppliers. In the upper price range we can find water users who have to resort to water vendors, or spend many hours per day hauling water from public taps or sources that often lie distant from their homes. The cost of hauling is estimated on the basis of the hourly per capita family income of some other measure which could be shown to represent the "leisure cost" of having to spend time in this endeavor instead of using it either productively or in leisure. These users are generally low income and live in areas where the water supply system has not yet expanded to.

In the mid price range we obtain information from water users who have irregular service from the public utility and even though

---

1/ For a very illustrative summary of different types of marginal costs that can be computed and their meaning, see Saunders 1977.

2/ A positive net benefit does not necessarily mean that the chosen expansion is optional from a purely economic standpoint, which requires that prices be set to marginal cost. There are, however, valid practical administrative and financial constraints which generally preclude the use of long run marginal costs pricing.

enjoying normal indoor plumbing and appliances, have to purchase water from private sources who deliver by truck. Generally, such users own cisterns with enough capacity to fulfill family needs for periods of up to two weeks, and prices paid are above those charged by the public utility. Also in this price range we find households with private wells who have to pay high variable pumping costs for their water in addition to other fixed costs (wells, pumps, storage tanks, etc.).

In the lower price range we use information obtained from the water utilities. From billing data we select only those customers that have 24 hour service, and are charged for each additional unit of water consumed. A random sample of these accounts is selected, upon which field work is performed in order to enquire about income and family size.

Information on income, family size, quantities and cost (monetary or otherwise) for users paying mid and upper prices for water is gathered with surveys. The first part of a typical survey questionnaire is dedicated to ascertain family income. It is always very difficult to obtain accurate information for this variable from surveys. Apart from it being a very private matter, problems arise from the fact that income can be perceived as affecting taxes or utility pricing. Also, inaccurate information results from incomplete answers, as the respondent excludes or forgets income derived from all of the household members. In order to avoid these issues it becomes necessary to list all members of the family and enquire separately about each of their incomes. In the rural areas it is also important to account for any in-house agricultural production consumed by the family.

The second section of the questionnaire is generally deviced to obtain information about water use habits and prices being paid. Since one of the main objectives of the survey is to get data on high prices paid for water we refer the respondent to water sources used at times of scarcity. Each possible answer to the source question leads the interviewer to a new set of questions which ascertain the price levels and quantities consumed. At this level, careful separation of answering options for questions related to water prices and quantities is of vital importance in order to produce usable data once the field work is done. For example, in the case of water hauling it is important to ask detailed questions such as number of trips per day, duration of the round trip and waiting time, types of containers used, their capacity, and how much was charged for filling them.

Similar questions are levied in surveys to be carried out in urban settings, where we target population located in areas suffering from water scarcity, and therefore alternative water sources are used. In the urban areas there generally is a larger array of possible options regarding water sources, as for example, bottled water, public standpipes, neighbors, private wells and water vendors, among others.

Results of the Estimation Process

Annex Table 1 presents twelve demand equations which have been used

Table 1
Elasticity Values derived from Demand Functions

Year	Country	Functional Form	Elasticities 1/ Price	Income	Family Size	Average Values Income 2/ (US$)	Family Size	Exchange rate (per US$)	Quantity Consumed 3/ (M3/month)
1980	Chile	Logarithmic	-0.547	0.101	n.s.	349	5.0	39.00	11.0
1981	Guatemala	Logarithmic	-0.626	0.541	n.s.	695	5.7	1.00	21.0
1983	Bolivia	Linear	-0.060	0.208	0.586	528	6.1	162.80	21.0
1983	Brazil	Logarithmic	-0.597	0.784	n.s.	626	6.3	420.00	13.0
1984	Costa Rica	Logarithmic	-0.444	n.s.	0.581	559	5.0	45.00	13.0
1985	Chile	Logarithmic	-0.310	0.430	0.169	457	5.0	150.00	16.0
1985	Haiti	Linear	-0.021	0.207	0.657	157	5.6	5.00	10.1
1985	Honduras	Linear	-0.033	0.422	0.083	611	6.2	2.00	15.3
1985	Mexico	Logarithmic	-0.377	0.323	0.325	825	5.2	180.00	10.0

n.s.: Statistically non significant.
1/ Elasticity values for Bolivia, Haiti, and Honduras correspond to families of average size and income at a price of US$ 0.30.
2/ "Per capita" per year. For Haiti it represents food expenditures.
3/ Quantities consumed are estimated at a price of US$ 0.30 per cubic meter.

to estimate WTP in project analysis in nine countries in Latin America. All of them are expressed in local currency, and the table notes point out the units in which quantity consumed is expressed in each case. To aid the reader, the exchange rate which prevailed at the time the function was estimated is also presented in the table.

Both linear and logarithmic functional forms were experimented with in the estimation process. The form finally chosen to estimate WTP is the one which captures the greatest variance, and provides the best fit, taking into account the location of the observations in the the PQ space. In general, linear forms seem to provide better fit when many observations contain near zero marginal prices.

All demand functions chosen with the exception of those estimated for Bolivia, Haiti, and Honduras are of the logarithmic form. Thus the regression coefficients which appear beside Log P, Log Y, and Log N, correspond respectively to price, income and family size elasticities of demand. Table 1 presents a summary of information derived from the estimation process. Price elasticity values show a large variance, ranging from -0.31 in rural Chile to -0.63 in urban Guatemala. As is the case with all linear demand equations, elasticities vary depending on the value of independent variables. For a price of US$ 0.30 per cubic meter, price elasticity ranges from -0.02 in rural Haiti, to -0.06 in urban Bolivia. Price elasticities for Bolivia, Haiti and Honduras rise with prices charged.

Income elasticities have an even larger range of variance, ranging from 0.101 in rural Chile to 0.784 in urban Brazil. Income elasticities for Bolivia and Haiti were 0.207 while Honduras had a much higher value of 0.422. In Costa Rica income was shown not to be statistically related to consumption of water.

With the exception of the 1980 Chile and the Guatemala functions, family size entered as a significant determinant in water demand. Values for this coefficient range from 0.169 in rural Chile, to 0.581 in Costa Rica. In the case of linearly estimated functions, family size elasticities ranged from 0.083 in Honduras to 0.657 in Haiti.

The expansion of rural water supply in generally financed through programs that operate on a Country wide basis, for large numbers of localities which have different characteristics with regards to water use. Instead of surveying all of these towns, it is more cost efficient to gather information on a sample of representative households, which account for all the differing characteristics. Variables that seem to influence overall behavior in this sense are weather and geographic conditions,traditional water scarcity, distance to the source, and cost, size and economic development .

Econometric techniques can be used to group localities according to these or other characteristics which might appear pertinent, to test the hypothesis that some of the regression coefficients which describe the relationship between water demand, price, family income and size, are in fact related to them. These techniques are used in the 1980 Chile study, as well as in Costa Rica and Honduras. As can be observed

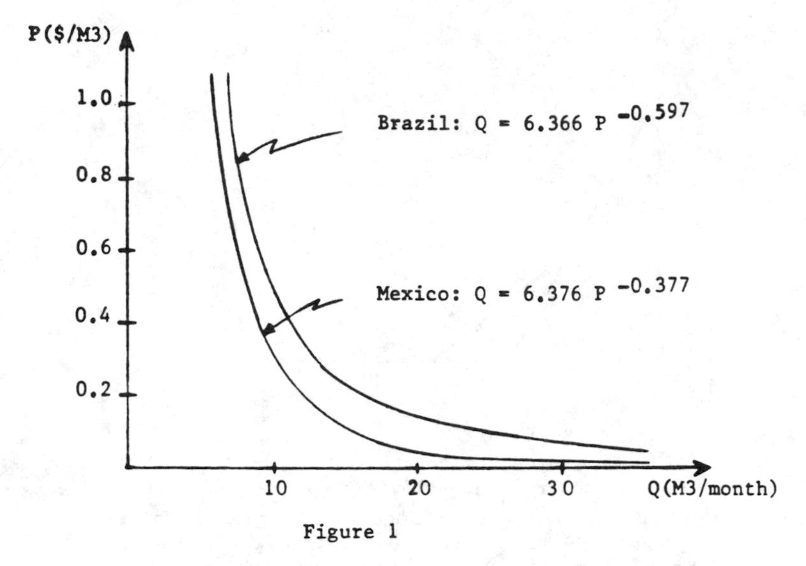

Figure 1

Demand Curves for Brazil and Mexico

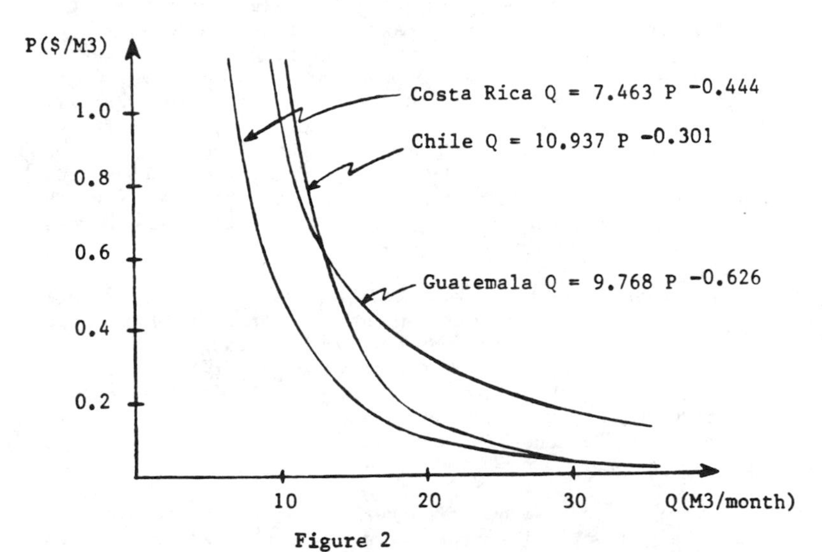

Figure 2

Demand Curves for Costa Rica Chile and Guatemala

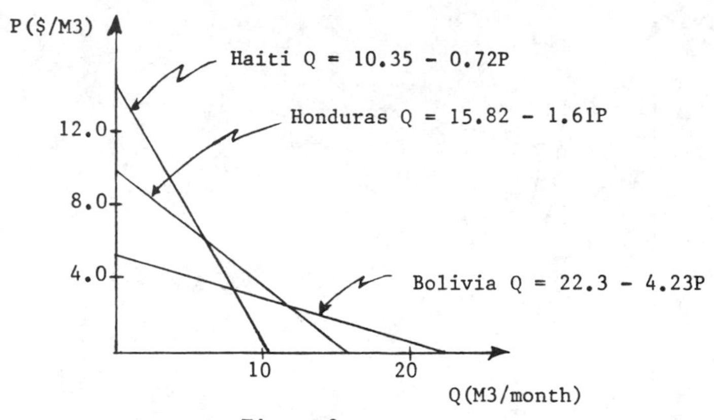

Figure 3
Demand Curves for Haiti
Honduras and Bolivia

in Table 1, the Chilean study concludes that regional differences account for marked variations in all coefficients. In the case of Costa Rica, altitude of localities, influence the equation's intercept and price elasticity of demand. These same parameters are affected in Honduras by size of localities served.

Another way of comparing these results is to view a graph of how water consumption varies with price for each of the cases analyzed. We do so in figures 1, 2 and 3, separating logarithmic from linearly estimated functions, and the former in two categories, according to Country size. Figure 1 presents the curves for Mexico and Brazil, figure 2 the curves for Chile, Costa Rica and Guatemala, and in figure 3 the cases of Bolivia, Haiti and Honduras. Units used are US$ for prices and cubic meters per month per family for quantity consumed. These curves represent the behavior of an average family, in which income and size are set at a level which is comparable across countries.

Figure 1 presents demand curves for cities with approximate populations of 1.5 million. The city analyzed in Mexico however is located in a very dry and hot area, where water has traditionally been considered a scarce commodity, which is not the case for the Brazilian case, where quantities demanded are comparatively higher at all price levels. A hypothesis could be made that habits and traditional use of water in the household, account for the differences in willingness to pay in these two cities.

Figure 2 presents a comparison of curves for Chile, Costa Rica and Guatemala. In the case of Chile, the curve was estimated using rural household data, while mid size urban localities served as a sample for Guatemala's function. Costa Rica's curve describes behavior for both the urban and the rural areas. The highest WTP levels are seen in Guatemala, the lowest levels in Costa Rica and Chile lies in the mid range, showing the highest level for extreme scarcity situations. This comparison serves in part to confirm a belief that urban areas tend to have larger consumption levels at all prices, given the widespread existence of indoor plumbing, and ownership of water using appliances.

Lastly in figure 3 we present the linear functions. In case of Bolivia, the curve shows demand behavior for a relatively small urban area, and therefore we observe high consumption levels at low prices and low willingness to pay at scarcity levels, relative to the other two countries depicted in the graph. In the case of both Honduras and Haiti, where the curves depict rural settings, the function is able to capture higher levels of willingness to pay at scarcity levels, since these situations are extremely prevalent. In fact, in many rural areas of these countries, many hours every day are spent in the process of fetching water from distant sources. Water becomes so "expensive" that relatively small quantities are consumed per person on a daily basis. At the same time when piped water does become available at low prices, consumption stays at relatively low levels, given the traditionally few essential uses it has such as cooking and washing.

Some of the general conclusions that one can draw from putting together these studies carried out in diverse countries at different periods of time are the following: It is possible to estimate demand functions in less developed countries complementing data from utilities with surveys to capture prices paid in scarcity situations. Secondly, one need not carry out extremely large surveys, so long as the objective population is chosen properly from the stand point of obtaining sufficient variance in the independent variables such as prices income and family size. Thirdly, even though one could make preliminary hypotheses about the influence that weather, geographic, and cultural characteristics, have on willingness to pay, there is no apparent reason that explains the large variance observed in the elasticities estimated.

Data used for the estimation of these functions is still available, and the next step in the statistical process should be to use other more sophisticated techniques to enquire about the reasons for the divergencies that have been described above. This would involve the manipulation of large data sets, and should be subject of another more extensive research endeavor, whose main purpose would not be the analysis of projects in particular situations, but to arrive at generalized conclusions that the could be used as basis for studies where surveys were not feasible.

This paper has studied the behaviour of residential water users, taking into account the effect variables such as prices, income and size have on levels of consumption. An area that merits continued research is the methods of estimating demand curves for non residential users, such as industry, commerce and institutional facilities.

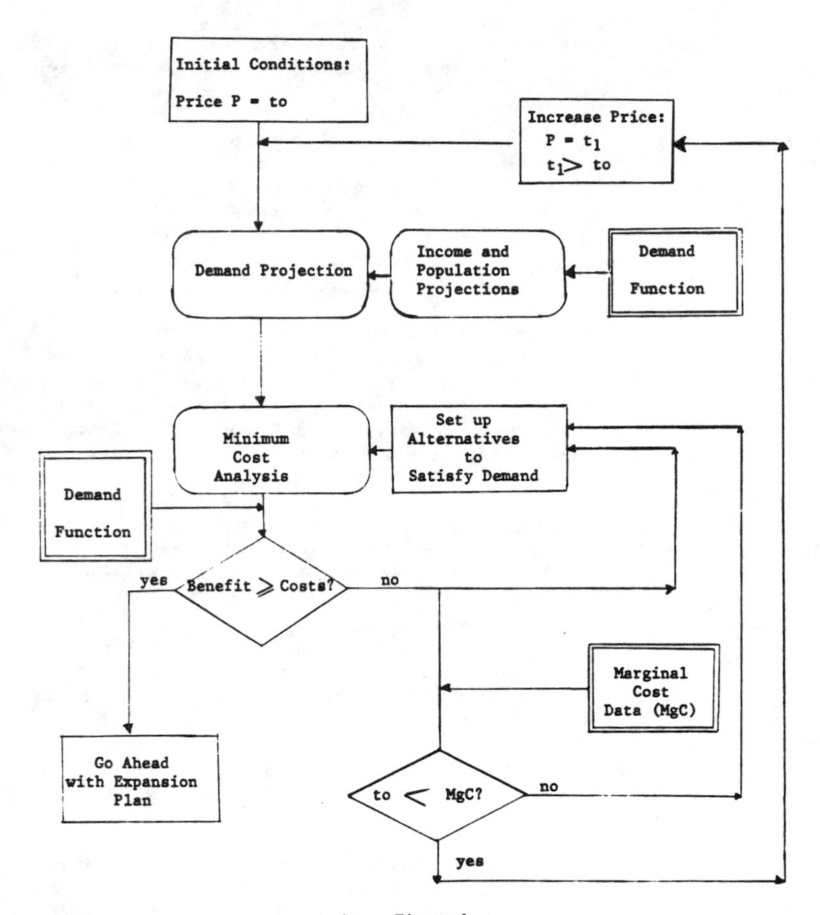

Annex Figure 1

A Planning Scheme for the Expansion of a Water System

Annex Table 1
IADB Estimation of Demand Functions for Project Analysis

Year	Country	Type of Project	Estimated Demand Functions	Survey Size	Exchange Rate (per US$)
1980	Chile	Rural	$\text{Log } Q = 4.776 - 0.547 \text{Log } P + 0.101 \text{Log } Y$ (1) $\text{Log } Q = 4.338 - 0.424 \text{Log } P + 0.134 \text{Log } Y$ (2) $\text{Log } Q = 5.198 - 0.489 \text{Log } P + 0.134 \text{Log } Y$ (3)	1042	39.00
1981	Guatemala	Urban	$\text{Log } Q = 1.836 - 0.626 \text{Log } P + 0.541 \text{Log } Y$	678	1.00
1983	Bolivia	Urban	$Q = 5.600 - 0.026\ P + 2.018\ N + 0.0001\ Y$	213	162.80
1983	Brazil	Urban	$\text{Log } Q = -0.329 - 0.597 \text{Log } P + 0.784 \text{Log } Y$	408	420.00
1984	Costa Rica	Rural & Urban	$\text{Log } Q = 2.636 - 0.366 \text{Log } P + 0.581 \text{Log } N$ (4) $\text{Log } Q = 2.765 - 0.444 \text{Log } P + 0.581 \text{Log } N$ (5)	4452	45.00
1985	Chile	Rural	$\text{Log } Q = -0.738 - 0.310 \text{Log } P + 0.430 \text{Log } Y + 0.169 \text{Log } N$	837	150.00
1985	Haiti	Rural	$Q = 1.594 - 0.143\ P + 0.0250\ Y + 1.199\ N$	630	5.00
1985	Honduras	Rural	$Q = 265.430 - 26.460\ P + 0.338\ Y + 6.750\ N$ (6) $Q = 343.460 - 26.460\ P + 0.338\ Y + 6.750\ N$ (7)	1653	2.00
1985	Mexico	Urban	$\text{Log } Q = -0.302 - 0.377 \text{Log } P + 0.323 \text{Log } Y + 0.325 \text{Log } N$	480	180.00

Units: See (1) in page 2. Prices and income are in local currency. 1980 Chile and Guatemala, Q = liters/person/day. Haiti, Y = weekly food expenditures. Brazil and Honduras Q = liters/family /day.
(1) For Regions I, II, III, VI, VII, and XI. (2) For Regions V, VIII, IX, and X. (3) For Region IV.
(4) For Localities situated below 500 meters above sea level.
(5) For Localities situated above 500 meters above sea level.
(6) For localities with less than 2000 inhabitants.
(7) For localities with 2000 inhabitants or more.

REFERENCES

1. Agthe, Donald E. and Billings, R. Bruce, "Dynamic Models of Residential Water Demand". Water Resources Research, Volume 16, No. 3, 1980.

2. CV Jones, et al. :Residential, Industrial and Commercial Water Demand:. in: CV Jones, JJ. Boland, JE Crews. CF DeKay and JR Morris: Municipal Water Demand: Statistical and Management Issues. Studies in Water Policy and Management No. 4. Westview Press. Boulder. 1984.

3. Danielson, L.E.. :An Analysis of Residential Demand for Water Using Micro Time-Series Data:. Water Resources Research, 15(4), August 1979.

4. Hanke, S. and de Mare, L.. "Municipal Water Demands" in: Modelling Water Demands. ed. J. Kindler & C.S. Russel. Academic Press, 1984.

5. Howe, C.W. and Linaweaver, F.P.. "The Impact on Residential Water Demand and Its Reaction to System Design and Price Structure". Water Resources Research, 1967.

6. Howe, Charles W.. The Impact of Price on Residential Water Demand". Water Resources Research, Volume 18, No. 4, pp. 713, 1982.

7. Powers, Terry A.. "Guide for Appraising Urban Potable Water Projects". Inter-American Development Bank, Analysis Paper No. 4, Washington D. C., February 1977.

8. Powers, Terry A.. "Benefit - Cost Analysis of Urban Water Projects." Water Supply and Management, Vol. 1, pp. 371, Pergamon Press, Great Britain, 1978.

9. Powers, Terry A.. "SIMOP Urban Water Model". Inter-American Development Bank, Papers on Project Analysis No. 5, Washington, D. C., July 1980.

10. Primeaux, Walter J., Jr. and Hollman, Kenneth W.. "Price and Other Selected Economic and Socioeconomic Factors as Determinants of Household Water Consumption". Proceedings of the First World Congress on Water Resources, Volume 3, pp. 189-198, 1978.

11. Saunders, Robert J., Warford, Jeremy J and Maun, Patrick C. "Alternative Concepts of Marginal Cost for Public Utility Pricing: Problems of Application in the Water Supply Sector. World Bank Staff Working Paper No. 259, 1977.

COST RECOVERY AND USER CHARGES FOR COMMUNITY WATER SUPPLY

by
Ved Prakash[1]

The Context

Urbanization is basically a twentieth century phenomenon. In 1920, of the world's total population of 1.9 billion, only 360 million or 19 per cent lived in urban areas. By 1950, urban population increased to 712 million or 28 per cent of the total world population. Currently the urban population is around 2 billion. At the end of this century, it is anticipated that it would be around 3.2 billion and constitute 51 per cent of the total world population. Although the level of urbanization has been and continues to be higher in more developed countries than in the less developed countries, the average annual rates of growth of the total as well as the urban population have been significantly higher in the developing than the developed countries. In 1980, for the first time, there were more people living in urban areas of the less developed countries (972 million) than in the more developed countries (834 million). By the year 2000, it is expected that two out of every three urban dwellers would be in the developing countries. Between 1980 and 2000, the developing countries would have to absorb more than 1.2 billion additional persons in the urban communities. A salient feature of the urbanization process in the developing countries has been the invariable trend towards concentration of population in the capital cities and large urban centres. By the year 2000, 44 per cent of the total urban population in the developing countries will be in one million plus cities. Of the world's fifteen largest cities in 1950, eleven were in the developed countries. By 2000, twelve of the fifteen largest cities will be in the developing countries (Hauser and Gardner 1982).

Rapid urbanization, low levels of income and net savings, and high costs associated with infrastructure, shelter and urban services are the critical factors contributing to the deteriorating urban environment in most of the developing countries. The seriousness of the urban situation is recognized, but public policy, planning, programming, resource allocation, and financing (resource mobilization) aspects have generally been disjointed and unintegrated. In many of the developing countries, even if shelter and urban services are accorded the highest priorities, fiscal resources may be deficient. Some of the problems may be caused by the adoption of inappropriate standards. Unconventional and far-reaching changes are essential towards efficient and equitable provision and financing of urban development and services (Prakash 1985, Linn 1983, and Stokes 1981).

Standards for urban services have been adopted on the basis of "desirable quality targets" rather than relative to costs, economic capacity (income levels and distribution), and existing and potential financial measures. Most developing countries are finding it difficult to defend decent standards of urban development in the face of rapidly growing needs and extremely limited means. If standards adopted are extravagant relative to economic capacity, opportunity costs associated with any level of investment are bound to be high, and it would be impossible to develop appropriate fiscal instruments for financing urban services. There is ample evidence to suggest that

[1]Professor of Planning, Department of Urban and Regional Planning, University of Wisconsin, Madison, WI 53706 USA

although in humanitarian terms the standards may not appear to be high, in fact they strain both the economic capacity of the public sector and the ability or willingness to pay on the part of the private sector. Low-income households have increasingly limited access to urban services, large-scale subsidies have become inevitable, and they have accrued largely to the middle- and upper-income households (Prakash 1985, Saygal 1981, and Clark 1984).

Assignment of responsibilities for providing different urban facilities and services and methods of financing them vary from one country to another. The initial capital outlays for urban facilities are generally financed through public borrowing (including external and international resources) or on pay-as-you-go basis or through a combination of the two. The recurring expenditures (including repayment of principal and interest on loan financing) are financed primarily through public revenues raised by one or more levels of government through taxation and receipts from utilities and other revenue producing public sector activities.

The size of the public sector is small relative to the national income in many of the developing countries. In about half, total public expenditures (at all levels) constitute less than 25 per cent of the Gross Domestic Product. The total public expenditure relative to GDP has changed only slightly during the last two decades in most of the developing countries. Government revenues constitute less than 20 per cent of the GDP in a majority of these countries.

It is important also to note that these countries rely heavily on indirect taxes. As pointed out earlier. urban infrastructure, shelter and services are heavily subsidized and primarily benefit the middle- and upper-income groups. The end result is inequitable redistribution of net benefits. Generally the tax structures exhibit relatively low elasticities/buoyancies. A consensus seems to be emerging that greater reliance on nontax revenues may lead to greater mobilization of resources relative to service requirements. It is also believed that greater utilization of user charges is likely to improve efficiency and equity in the provision and financing of urban facilities and services (Linn 1983 and Schroeder 1986).

A related problem is poor cost recovery in respect of urban/projects and services through tax and non-tax measures. Generally, at the time of project planning, emphasis is on capital costs and their financing. Very little attention is paid to the impact of capital projects on the long-run recurring maintenance and operating expenditures (including debt service), and their financing. Consequently, cost recovery measures are inappropriate. There is ample evidence also that in many of the cities of the developing countries, administration and collection of even the real estate taxes and water and other utility rates has worsened over time in spite of the skyrocketing of property values and substantial increases in the provision (consumption) of services usually financed through user charges.

It must be underscored that affordability, cost recovery, and replicability are interrelated and interdependent concepts in both theory and in practice. Affordability is the key to cost recovery, and cost recovery is the key to replicability. Examination of recent experience with respect to urban projects suggests that affordability and cost recovery concerns were largely ignored, thereby severely restricting the replicability of these projects and programmes.

Finally, it should be pointed out that the concern for efficiency is not necessarily at odds with equity considerations in providing the poor with greater access to urban services and thus substantially increasing the chances for replicability.

Role of User Charges

User charges belong to the family of non-tax revenues. The non-tax revenues consist of two major catagories: (1) fees and fines, and (2) rents and prices and special service charges. Fees and fines include fee for licenses and permits and fines and forfeitures (Bird 1976, Miller 1953, Netzer 1966 and 1970, Stockfish 1960 and Bahl and Linn forthcoming). The rents and prices include rent of publicly owned property and service charges for utilities and other urban services provided by the public authorities.

Rents and prices or user charges accrue to public bodies, responsible for provision of urban facilities and services, either from their ownership of real estate (land, markets, houses, etc.) or from their operation of public utility and semi-commercial undertakings (water supply, sewerage, city transport, etc.). The user charges are theoretically based on the benefit principle and affect only those who really avail of the service or benefits provided. They embody a fair element of quid pro quo and to this extent are different from taxation. However, in some cases taxes may be used as proxies for benefits received (e.g. real estate taxation to recover costs in respect of water supply, sewerage, garbage collection etc.).

As regards user charges for public utility services, some economic theorists consider these as inappropriate for financing what are defined as "pure public goods"--services whose dominant characteristics are such that no one is denied their benefits regardless of whether or not one pays for them (Vieg et. al. 1960). It is further argued that in some cases even when a particular public service is not a "pure public good" user charges may be feasible but not desirable. The user charges are also of limited applicability where the service produces substantial public benefits, in addition to those enjoyed by the user as such, e.g., libraries (Netzer 1970).

Notwithstanding the merits of these arguments, the theory of "pure public good" would need to be viewed as a relative concept. It is doubtful if it would be advisable to apply this approach to developing countries in the initial stages of their development and in the face of rapid urbanization. The emphasis has to be on resource mobilization, capital formation, and cost recovery. Few developing countries can at present afford to provide, for example, potable water free of charge or on a highly subsidized basis. If user charges are not related to the costs of public services, the community will have to finance them through taxation which in most developing countries may well be regressive and unstable (Schroeder 1986). While in taxation the citizen has no options, in user charges one has some options in the sense that he/she may, if possible, not consume the service, or one can at least economize in its consumption. Moreover, most user charges, where desirable, can certainly be made differential in favour of the underprivileged sections of the community (cross-subsidization, life-line tariff and/or minimal or no charges for water supply provided through standpipes etc.).

Economic theory dictates that most efficient use of the economy's resources is achieved when the price of any product equals the marginal cost of producing the last unit sold. The desired result is achieved through the forces of competitive market for a given distribution of income. Such an outcome may or may not coincide with other public policy objectives--equity and social justice. Government interventions through taxation and subsidies are aimed at the above policy objectives. In the case of most goods and services produced in the public sector, the market does not set prices. The governments do. Pricing policy criteria to achieve economic efficiency for publicly provided goods and services, would be the same, i.e., setting price equal to marginal cost. However, besides efficiency, pricing policy for public sector revenue-earning enterprises also incorporates equity and financial (resource mobilization) objectives.

In real world, economic theory, at best provides a set of suggestive principles which are rarely adhered to in the public policy arena. Baum and Tolbert correctly point out that:

> Such intervention is not easy; the fact is that it is not always possible to know what the "right" price is in a particular sector or activity. Calculating marginal costs is often difficult in practice--although a rough approximation may suffice for many purposes--while determining the most desirable trade-off among efficiency, equity, and revenue considerations may entail difficult matters of judgement. But even when the "right" prices are known, they are often not adopted for political reasons. Economic planners have to recognize the political pressures operating on decision-makers, and to understand that policies the economist perceives as irrational are often perfectly rational from a political point of view; such necessary wisdom brings greater realism to the process of devising policies and programs. However, political considerations are all too often used as an excuse for failure to analyze fully the costs of departing from sound pricing policies, to see to it that the costs are adequately examined at the proper level of authority, or to ascertain whether the efficiency benefits forgone are clearly compensated by the political benefits

obtained. Even when these steps are taken, implementing the "right" decisions may be painful and call for political will of a high order. But, as we shall see below, the rewards can be substantial (Baum and Tolbert 1985).

Cost Recovery Through User Charges for Water Supply & Sanitation

Potable water supply and sanitary disposal of residential, commercial, and industrial waste water are recognized as one of the most critical basic needs of society. On per capita basis daily consumption of twenty to fifty litres of safe and convenient water (for drinking, food preparation and personal hygiene, and providing ways to dispose of excreta) is considered essential to safeguard human health and environment. The lack of adequate water supply and sewerage disposal systems is a far more serious problem in urban than in rural areas. According to the World Health Organization estimates, in 1980 about 73 and 32 per cent respectively of the urban and rural population had access to safe water. Comparable figures for safe sanitary disposal of wastes during the same year were 48 per cent in the urban and 12 per cent in the rural areas. It should be recognized that regional as well as cross-country variations are substantial. During the last decade or two, important advances have been made in providing access to potable water supply and sewerage disposal. However, much remains to be done. It may be noted that the current decade has been declared by the United Nations as the International Drinking Water Supply and Sanitation Decade. The overall stated objective of this declaration has been to eliminate the vast service backlog by 1990. According to the World Bank estimates this would require an annual investment of $30 billion (in 1983 prices). UNDP estimates suggest that the 1983 level of investment was actually about $10 billion. Even if the level of investment during the rest of the century can be substantially increased, the existing gap is likely to continue or even become greater amongst the low-income countries. Some observers also point out that in spite of the additional investment, these services may have actually deteriorated in some urban areas due to deferred and insufficient maintenance.

As pointed out earlier, there are important linkages between water supply and sewerage, which in turn necessitate joint investment planning, cost recovery, and institutional responsibility. Bulk of the public costs for sewerage are related directly to the amount of water use. However, variations between sewerage costs relative to water across different cities are considerable (McLure 1977). Thus charges for sewerage are usually directly related to water charges and/or consumption. Both these services are suitable for financing through user charges--directly or indirectly.

The gap between identified needs and available resources in water supply and sewerage services is more critical than most other sectors of the economy in most of the developing countries. In large part, this is due to inadequate cost recovery and tariff structures, and inappropriate standards and technologies. For example, in the

Republic of Gambia, provision of water supply, sewerage, electric, and gas services are entrusted to the Gambia Utility Corporation. Since its establishment in 1972, it has incurred substantial losses each year except in 1977-78. Currently, its receipts do not even cover the cash operating expenditures and their tariff structure does not make provision for interest, amortization of debt, and depreciation. This results in across-the-board subsidies benefitting the middle and upper income groups disproportionately. The situation in the Gambia is typical of many other countries. However, there are a few notable exception, e.g. Tunisia, Nicosia, Managua, and Singapore (Baum and Tolbert 1985). The gap between service requirements and resources can by narrowed only by working simultaneously on mobilization of more resources reducing investment costs, and minimization of across-the-board subsidies.

Appropriate system of urban water supply and sewerage charges should be so designed as to best meet efficiency, financial viability, and equity goals. The studies dealing with price elasticity of water demand, in the developing countries, are far and few in between. Findings of the few available studies suggest that price elasticity of water demand is quite low (Bahl and Linn forthcoming). The tariff system for water supply and sewerage should include: (1) capitalization and full cost recovery of development costs (off-site and on-site infrastructure costs) through land development charges to the largest extent possible, irrespective of whether the lot is actually connected to the water and sewer mains (Doeble 1982, Buch 1982, Prakash and Sah 1974, Grimes 1982, and World Bank 1980 and 1983); (2) connection fee either as a lump sum charge upon initial connection or a fixed periodic fee; and (3) consumption charge related to the quantity consumed. Economic efficiency can be best achieved, when charges for the above components are set to equal their marginal costs. In recent years, some of the developing countries have adopted the Average Incremental Cost as a basis of pricing water. This is a further refinement of the marginal cost pricing principle and is aimed at overcoming the problem of lumpy investments and at providing smoothing of price fluctuations.

Given the externalities associated with water supply and sewerage service as well as the fact that many poor households lack ability to pay for these services, cross-subsidization is justified. This has been done in many developing countries by the introduction of "life-line tariff" (Warford and Julius 1977). For example, "a life-line tariff on water would consist of very low tariff for an initial consumption block equivalent to, say 20 to 40 litres per capita daily, with consumption beyond this amount charged at full marginal cost" (Linn 1983). In many cities water charges are based on rising block rates. The predominant practice also is to charge the commercial and industrial users at higher rates.

In the United States, charges for water supply are usually based on consumption blocks. Unit rate is highest for the smallest blocks (usually single family residences). Water and sewerage charges constitute a small fraction of household income--less than 2 per cent--in most cases. The unit rates decline with increased consumption. This by and large correspond to the marginal cost pricing principle. Given the low level of household incomes, income distribution and inadequate affordability for a large number of households the rate structure would have to be significantly different in developing countries.

A related question with regard to water supply is whether metering is appropriate on economic and administrative efficiency grounds. There is no clear cut answer to this question when different cities are examined. The advantages and disadvantages of whether to meter or not (or meter house connections--including commercial and industrial users--and not meter standpipes) must be carefully weighed in each case. Consensus seems to be that there should be full cost recovery from house connections and that no charges be levied with respect to standpipes. However, in extremely poor countries, the vast majority of the population can be provided access to water supply only through standpipes. In these cases, a strong case can be made for cost recovery in respect of water provided through standpipes. Standpipes can be metered and leased to individuals (preferably to women) in the neighborhoods. Such an approach would,in all likelihood,significantly reduce wastage.

Conclusion

The role and structure of user charges in financing urban development and services must be examined in the broader context of the rapid urbanization process in the developing countries, low levels of income and net savings, unprecedented investment requirements for urban infrastructure and shelter, and the anticipated recurring maintenance and operating expenditure for urban services. These considerations have to be dovetailed with affordability or ability/willingness to pay of the different income groups, which in turn provide an appropriate framework for planning urban investments and appropriate cost recovery (resource mobilization) measures to insure replicability of urban projects.

Developing countries recognize the seriousness of the deteriorating urban environment. However, in most of them, approaches to public policy, planning, programming, resource allocation, and financing have lacked an integrated framework. Standards adopted for urban development projects and programs have invariably been excessive. Consequently, opportunity costs and associated efficiency and equity losses have been extremely high. Concern for efficiency is not at odds with equity considerations in providing the poor with greater access to urban services. Potential exists for significantly improving both of them through appropriate public policy interventions.

Currently user charges do not play a significant role in financing urban facilities and services. If efficient pricing policies are adopted, user charges can provide a dynamic resource for financing capital outlays as well as recurring expenditures. An integrated approach for urban development projects as opposed to sectoral approach is preferable and it may facilitate financing sizeable proportions of on-site and off-site infrastructural development costs through land development charges. Water supply, sanitation, public mass transit and parking are key services which should be financed through service charges. The practice of across-the-board subsidies for urban services should be replaced by cross-subsidization amongst consumers and also in the case of related services.

Political, administrative and management problems for adopting new strategic approaches (including significantly enhanced cost recovery through user charges and other revenue measures) may be insurmountable, at least in the short-run. Current practices, tend to subsidize heavily the middle- and upper-income groups. Improvements in the efficiency and equity of urban development would drastically reduce the benefits currently enjoyed by overlapping economic and political elites. More time spent on policy analysis could provide invaluable information to decision-makers on the consequences and implications of alternative policies and programs. This in turn would engender greater support from the people concerned, as wellas from other affected groups for cost recovery measures; it would also increase chances of success in reorienting urban policies and programs.

Appendix--References

Bahl, Roy and Johannes F. Linn, *Urban Prospects and Processes in Developing Countries* (Washington, DC: The World Bank, forthcoming).

Baum, Warren C. and Stokes M. Tolbert, *Investing in Development: Lessons of World Bank Experience* (New York, NY: Oxford University Press, 1985), p. 33.

Bird, Richard M., *Charging for Public Services: A New Look at an Old Idea* (Toronto: Canadian Tax Foundation, 1976).

Buch, M. N., "Improvement of Human Settlements with Special Reference to Urban Land Problems--As Asian View," International Seminar on Urban Development Policies: Focus on Land Management, Nagoya, Japan, October 13-18, 1982 (sponsored by the United Nations Centre for Regional Development, Nagoya.

Clarke, Lawrence, "The Financing of Shelter and Infrastructure, Issues, Options and Prospects for Nigeria,"--Paper presented at National Seminar on the Formulation of Shelter Strategy for Nigeria, Lagos, December 1984.

Doebele, William A., "Providing Land for the Urban Poor: An Overview." Keynote Address: International Seminar on Land for Housing the Poor in Asian Cities, Bangkok, January 18-31, 1982, sponsored by the Asian Institute of Technology, Bangkok.

Grimes, Orville F. Jr., "Financing Urban Infrastructure in Developing Countries," in William A. Doebele (ed.), *Land Readjustment* (Lexington, MA: Lexington Books, 1982), pp. 207-212.

Hauser, Phillip M. and Robert Gardner, "Urban Future: Trends and Prospects" in Hauser et. al. *Population and the Urban Future* (Albany, NY: State University of New York Press, 1982), pp. 1-58.

Linn, Johannes F., *Cities in the Developing World: Policies for Their Equitable and Efficient Growth* (London: Oxford University Press, 1983).

McLure, Charles E. Jr., "Average Incremental Costs of Water Supply and Sewage Service: Nairobi, Kenya," (Washington, DC: World Bank, Urban and Regional Report No. 77-13, 1977).

Miller, J. Maurice, Service Charges as an Important Revenue Source," *Municipal Finance*, Vol. 26 (August 1953), pp. 49-53.

Netzer, Dick, *Economics of the Property Tax* (Washington, DC: The Brookings Institute, 1966).

Netzer, Dick, Economics and Urban Problems (New York, NY: Basic Books, Inc., 1970).

Prakash, Ved, "Affordability and Cost Recovery of Services for the Poor," Regional Development Dialogue, Vol. 6, No. 2 (Autumn 1985), pp. 1-39.

Sanyal, Biswapriya, "Who Gets What, Where, Why and How: A Critical Look at the Housing Subsidies in Zambia," Development and Change, Vol. 12 (1981), pp. 409-440.

Schroeder, Larry, "Local Government Tax Revenue Buoyancy and Stability in Developing Countries," A Paper Presented at Expert Group Meeting on Financing Local and Regional Development, UNCRD, Nagoya, Japan, March 17-20, 1986.

Stockfish, J. A., "Fees and Services as a Source of City Revenue: A Case Study of Los Angeles," National Tax Journal, Vol. 13, No. 2 (June 1960), pp. 97-121.

Stokes, B., Worldwatch Paper No. 46 -- Global Housing Prospects: The Resource Constraints (September 1981).

Vieg, John A. et. al. California Local Finance (Stanford, CA: Stanford University Press, 1960), p. 210.

Wayford Jeremy J. and DeAnne Julius, "The Multiple Objectives of Water Rate Policy in Less Developed Countries," Water Supply Management, Vol. 1, 1977, pp. 335-342.

Don't Blame The Poor: Cost Recovery for Rural Water

Jerri K. Romm, Aff. ASCE*

Abstract

Findings from Nepal and Bolivia dispprove the common assumption that poor rural communities are unable to recover the costs of projects. Engineering, economic and institutional errors and decisions contribute to problems with cost recovery. Poorly designed and constructed systems increase capital and O&M costs and limit benefits. Methods to evaluate financial feasibility and strategies to improve project cost recovery are presented.

Introduction

Nobody likes to talk about cost recovery except the World Bank. In an analysis of feasibility studies of 43 water projects financed by 13 donors in Sri Lanka in 1982, only the Bank looked at cost recovery. It is commonly assumed that rural communities are too poor to pay the O&M, let alone the capital cost of projects. Since water is a basic need, it is implied that it is bad taste to bring money into the discussion because communities which need service most will be left out. In addition, engineers and the occasional social scientist involved in project planning, are not trained to evaluate cost recovery.

This combination of lack of knowledge and fear of what might be discovered if studies were actually carried out, results in the best planners training O&M personnel, setting up a water committee and leaving it with vague words about its responsibility to collect some money. In the absence of regular appropriations from an already overcommitted central government operating budget, the engineer reappears on the scene when the village system comes up for rehabilitation and expansion, actually a euphemism for rebuilding a system which has had little attention since the day it was constructed. If so many villages were not still in line for initial systems, this day of rehabilitation could easily arrive anywhere from one day to six months after the system was completed for 50% of the rural projects constructed under the Decade. Few professionals revisit completed systems, however, and constructed project statistics grow fatter.

Background: Cost Recovery and Project Benefits

The purpose of cost recovery is to sustain constructed water systems on a long term basis and to provide funds to support the future construction and expansion of additional systems. Limited government financial resources make rural water supply a low priority, and recurrent costs are increasingly difficult to allocate from the government's operating budget as the number of systems increases, unless there is some revenue from those using the service.

---

*Consulting Economist and former World Bank Financial Analyst for Water and Urban Department. San Anselmo, California.

Project benefits only accrue from properly functioning systems. If the costs to maintain them are too high, the benefits will not be equitably distributed or the system will fall into disrepair and the value of the project investment will be lost. When the argument is given that "something is better than nothing" an answer comes from the People's Republic of China about the ill-conceived Baoshan Steel Works: "Does it matter that the intentions were good, if no one benefitted?" When resources are scare, money inefficiently allocated has also been lost for positive use in other communities.

Contrary to the prevailing assumption about the financial limitations or collection skills of rural communities, evaluation of three hundred systems in Nepal and Bolivia indicate that a real constraint to implementing successful cost recovery programs is the failure of professional staff and executing agencies to consider critical technical, economic and institutional factors rather than the financial limitations of the villages. In communities ranging in size from 200 to 10,000, 50% of newly constructed projects were found to have poor or non-functioning systems, underutilized output and no health benefits attributable to the project to compensate for significant stress placed on community financial and human resources.

These conditions were common to two different countries with completely different implementing agencies, policies and engineering approach. What both projects had in common was their location in the poorest sections of remote areas of mountainuous countries. The Bolivian systems were funded by a bilateral donor, implemented by an international NGO using a decentralized approach with staff Bolivian engineers and American supervision. Some of the Nepalese projects had a similar orientation using UNICEF funds and supervision; others were funded by the central government and designed and implemented by the national water agency.

The status of the Bolivian projects was especially interesting since they already incorporated aspects of project design which typically are meant to respond to problems in operation and maintenance and long-term system viability: strong community mobilization, and participation in the construction and supply of financial resources, the training of village operators, the organization of a village water committee responsible for the future system and collection of water fees.

An Evaluation of Non-Financial Factors Affecting Cost Recovery

An evaluation of the systems showed that about 50% had operating problems within days and 6 months of completion. Only 30% of the communities were found to be collecting any household fees, and these were insignificant, although they had been impressed with the importance of doing so. None of the communities had purchased a basic set of tools to use for maintenance. The NGO concluded from these results that the comunities were too poor to collect adequate funds, but since the systems had been handed over to the communities, the staff turned to the construction of other similar systems.

It was found that there was not a simple correlation between communities responsible for their systems, failure to maintain them and lack of financial resources to pay the price of doing so. In fact the planning, design and construction process which had been the responsibility of engineers had overlooked a number of factors which directly affected the community's ability to pay for the continued operation and capital expansion of the system. Failure to consider them left communities with inefficient systems, wasted their capital investment of 30% and threatened to be a drain on future resources. All affected the potential cost recovery on the system which would give it long term value.

Planning and Design Factors

In the planning process certain decisions were made:

1) The project only supplied one service level of patio connections. No public standpipes were provided. People were expected to contribute 30% of the capital cost of the project in labor and funds, and also purchase a patio connection in order to participate in the system. Even in the poorest communities the NGO assumed that at least 30% of the capital cost of the project of "more than basic" standard was affordable. Those who did not join the system were simply assumed to be malcontents and were permanently left out, although all projects were designed to effectively capture the full flow of the source. The results were that between 30 and 60% of the population was excluded from systems and were worse off than before, left to using inferior sources. Had the policy permitted a combination of standpipes and patio connections, an analysis of the alternatives would have demonstrated that the incremental cost of a few standpipes was very small, whereas the revenue base from which to collect household fees using a two-tier system to reflect level of service, would be greatly broadened.

2) The project was to supply communities with populations of 200-2000. However the NGO based its funding proposal on popualations of 400 and then committed itself to serving a certain number of communities for the funds received. This forced staff to allocate materials to small communities with small systems but few economies of scale. Thus where more people could have been served at lower cost per person in a community of 600 than two communities of 300, the NGO selected the small communities to ensure it met its quota. Distribution systems were also arbitrarily cut off in the middle of villages when they found they had used up a certain amount of pipe. The effect was to leave the communities with systems which immediately required extension using their own capital. The use of incremental analysis on the cost of service area expansion would have shown that more people could be served for less, than moving on to the next village. As in the example above the revenue base for cost recovery would have been expanded beyond the increased incremental cost of extending the distribution system. Where the community was left to extend the system itself, it clearly lacked the technical skill to do this, and was likely to lower the quality of existing service in its attempts.

3) No studies were carried out as to the ability or willingness of the villagers to pay. It was assumed they were able to. However the number of households which failed to join indicated that the cost was too high for many, and the failure to collect fees in communities strongly motivated and organized to do so indicated that financial resources had been exhausted by the high cost of the initial contribution.

4) In the design phase the community was not involved in the selection of alternative or service level. Had a study of ability or willingness to pay been combined with capital and O&M cost estimates for various alternatives, including more basic levels of service, the NGO and the community may have confronted the fact that they couldn't afford the system selected. Engineers could have tried to make the alternative more cost effective. Communities may have selected a more affordable option. As it was no one

had the facts until it was too late. No one was seriously looking at cost recovery for O&M and system expansion.

5) The NGO selected diesel pumps over electric pumps because the capital cost was lower. However the communities were unable to pay the higher operating cost of fuel and began to limit operations, rendering the system inefficient, in order to limit its costs.

6) Although the NGO told communities they had to collect fees, they never estimated the materials, equipment and labor needs nor their combined costs. The communities had no idea what was required or whether it could be afforded. The engineers never knew either. Knowledge at the planning stage that there is a gap between system requirements and the ability to meet them does not have to mean the project can't be built. Instead a re-evaluation must take place. Once the system is designed and constructed it is difficult to make adjustments.

Other problems with design reflected more standard engineering problems but still led to systems that didn't function: Sources were overestimated, the maximum day water demand was based on the wrong demand period, and tanks were constructed where they captured only a small portion of the source while the villages had to ration water and systems operated intermittently. In these cases the communities had committed their resources to systems which were poorly designed, and then were left to pay the higher price of trying to operate and maintain them.

Construction and Operation and Maintenance

In construction the problem was similar. Construction supervision was inadequate and systems suffered failures constantly which the communities had to pay to fix.

In training for operations and maintenance, communities were not trained in accounting and collection procedures, how to know when to purchase materials to maintain an inventory. They had no way of keeping the household fees current with the needs of the sytem over the years.

As a result of the problems found in the systems, the donor financed a "rehabilitation program" for about 50 systems which were less than a year old. This became part of a follow-up rural water supply project for the NGO.

Evaluating Strategies for Cost Recovery

The elements of a cost recovery program include assessment of annual systems costs of operation, maintenance and replacement, development of a water charge structure, an evaluation of the balance between projected annual costs and revenues, the ability of households to pay the projected monthly charges, and the potential for a working surplus. Issues involve how the costs will affect the poorest households' access to project benefits and strategies to lower the financial impact.

The following are some practical examples of how to assess the financial feasibility of rural systems. These are the result of preparation of projects in Bolivia and Nepal following the evaluations noted. The explanation and details are not intended to be complete. The full reports can be obtained from the author, CARE ("Bolivia Child Survival Project" August 1986) and The World Bank Water and Urban Department ("Nepal Western

Development Region Rural Water and Sanitation Project." April 1984).

Operation, Maintenance and Replacement Costs

Coming up with reasonable annual costs requires no specialized knowledge of the engineer. It is only necessary that he think out, for a given system, a) routine costs such as wages, materials, travel to purchase materials, fuel and b) contributions to the replacement for major or extraordinary repairs or replacement of worn out parts of the system, calculated on the basis of their economic life. The most important aspect of these costs is that they be based on the type of system constructed, the real physical conditions and village treatment of the system. Costs should represent the "worst case". It is better to assume that the village operator will be paid. If he will do the work voluntarily, all the better. These estimates should be made during the investigation and design phase and discussed with the water committee. The table "Bolivia Rural Water Supply: Annual Costs and Revenues" shows the cost of a Bolivian system to serve 50 families.

In Bolivia the instability of the Bolivian peso, with a 10,000% rate of inflation, convinced engineers it was meaningless to try to estimate future costs. Two actions could have been taken. First, the engineers still could have estimated the materials, tools, pipe which would be needed at given intervals and given this to the committee, which would make periodic visits to the market and assess the need to raise service fees in accordance with rising prices, and second, a district level storehouse could be set up with materials and equipment purchased in bulk to serve all the systems within one day's walk. This would minimize the impact of price changes although full replacement costs would still have to be charged to keep the storehouse stocked.

Water Service Fees

It is the responsibility of the project to offer guidance to the water committee in setting fees which reflect the funds necessary to keep systems operating properly, the villagers' ability to pay, and the level of service each would receive. In the first 200 projects in Bolivia the engineers had given up this effort, eventually convinced the villages could not generate the necessary funds. (This did not, however, lead them to change the policy of patio connections and high capital investment.) The lack of goals for the community, unaware of what they should buy, when and how much to collect to be able to do so, permitted small repair problems to build up and become large ones. The very vagueness of their instructions to "collect money" discouraged otherwise motivated people.

The table shows several categories of monthly household water charges to recover costs and equitably reflect through charges the different levels of service received. The most important aspect of the water charges is that no charge should be so high as to discourage use of sufficient quantities of water to achieve health objectives of the project. On the other hand, charges should not be so low as to encourage waste by users, prematurely signalling the need for additional investment. (Ideally the charge should reflect the true economic cost of development for each incremental cubic metre. This is best accomplished with average incremental cost(AIC). AIC is equal to the present value of the investment divided by the present value of the incremental water production during the useful life of the project). The committee should also be warned to raise the charges at appropriate intervals to keep up with price increases.

BOLIVIA RURAL WATER SUPPLY
Annual Costs And Revenues
(in Millions of Pesos- 1986)

	Monthly Charge	Number of Connections	Annual Revenue
Connections			
Semi-public tap (4 households)	1.5	10	180
Semi-public dispersed	3.0	4	144
Patio connections	3.5	30	1260
Patio conn. w/ water seal toilet	6.0	5	360
Industrial (chicha, etc)	8.0	1	96
Additional taps	6.0	0	0
Water Fee Income			2,040
Other (connection fees)			20
TOTAL REVENUES			2,060

	Annual Costs
Wages	600
Chemicals	15
Materials	196
Replacement of civil work, plant & equipment	954
Other (administration)	94
Other (major repairs, reserves for bad accounts)	40
TOTAL COSTS	1,899
NET SURPLUS (+)	161
Consumption (m^3/connection/month)	9.1
Volume (thousand m^3)	5.5
Average Tariff (millions P/m^3)	0.372
Average Cost / Volume Sold	0.346
Operating Ratio (%)	92

The proposed rates show a logical relationship between the charge and the service received through increased quality and quantity and capital contribution. The fee set for the lower income group receiving semi-public service (one tap per four households) should be a nominal one. When a household in this group demonstrates increased ability to pay by applying for a patio connection, it would be charged accordingly, as would any other household which improves its service to the next level.

System Revenues and Costs

The table shows that even a system with relatively high annual costs serving a small village population is financially feasible. Water use was assumed to vary from 30 lcd for semi-public use to 75 lcd for patio connections with water seal latrines. It is also assumed that even during the life of the project, some households will begin to upgrade their service as they save money and see the advantages of higher levels.

The results of the income statement show that with a range of monthly charges to households of $0.75 to $3.00 depending on the level of service, the amount of water used and the ability to pay, the community can adequately cover its annual costs and have a surplus of at least $100.

Affordability of Project Services

Ability to pay and willingness to pay should be based on estimates of the lowest household income in the poorest departments served by the project. The water fees in the Bolivia case were found to be affordable. Three sources were used and cross-checked to arrive at a realistic basis for determining the low income household's ability to pay: an unpublished report by IFAD of Gross National Product Per Capita for the relevant departments was used to establish the average monthly household income in the poorest department. To establish the lower end of this income distribution scale, the daily rate and minimum monthly rate for farm and unskilled labor were used. Under the worst possible case the lower end of the income scale was established at $17-$26 per household per month. The highest level service, with water-seal toilet, could be achieved for less than four percent of monthly income. The lowest level could be attained for the value of one day's unskilled labor. These were found to be reasonable charges which communities should be both willing and able to pay and project designers could proceed with confidence.

Nepal Cost Recovery Findings and Strategy

The findings for potential cost recovery in Bolivia are supported by similar findings in Nepal. In Nepal the philosophy was completely different. Only a basic level of service was to be provided. Improved service levels in the form of patio connections would be justified only if they recovered the full incremental cost and generated sufficient revenues to effectively lower the monthly charges required of the lower income households.

NEPAL RURAL WATER SUPPLY
Monthly Cost Per Household
(NRs. 1984 prices)

Project	All Public Taps	Public Taps/ Yard Taps	Charge as % of HH Income	Yard Taps as % of Total
ArlangKot	6.0	3.6/13.0		
Daugha	2.7	0/10.5		
Dhurkot Bastu	7.7	0/0		
Musikot	3.2	0/12.3		
District Average				
Year 1990	4.8	2.9/13.0	<0.5 - 1.5	18.1
Year 2000	3.5	1.1/13.0	<0.5 - 1.5	18.9

The results showed that where only public taps were provided the costs in different villages ranged from $0.10 to $1.10 per month. In communities of over 2500, providing improved service was justified since it resulted in reducing the charge to low income households to $0.00 to $0.70 (patio connections were $0.88 per household per month plus a connection fee and the cost of labor and materials.) The Nepal project concluded that the most equitable and beneficial method of cost recovery was the use of a geographically based approach. Within each district, a district-wide flat rate would be applied per household for each service level. The rate for patio connections was $0.88 per household with that of public tap users at $0.18 to $0.60 per household. Small communities with high costs per capita would be subsidized by larger communities benefitting from economies of scale and a broader revenue base.

The justification for the cross subsidy between different user groups was that it constituted a benefit tariff proportionate to convenience and increased water use. The justification for district-wide rates essentially providing a cross-subsidy from larger communities to smaller ones was that by centralizing certain technical services for all systems in a district, all would be better served at a lower cost. An evaluation of ability to pay found that the charges represented 2 to 3 percent of monthly income for the poorest households. In five of the eight Nepalese districts it was found that a net operating income would exist to provide a source of investment capital for new rural works or the sanitation program.

Conclusions

An evaluation of rural water systems in Bolivia and Nepal found that engineering, economic, and policy decisions adversely affected the ability of villages to recover costs. Financial plans incorporating economically efficient design, realistic assessment of O&M costs, a broad revenue base, equitable, multi-tiered tariff structure, and an analysis of ability to pay by the lowest income households, demonstrate that even poor rural villages can recover all recurrent and significant capital costs.

COST MODELS FOR SMALL SYSTEMS TECHNOLOGIES: U.S. EXPERIENCE

Robert M. Clark*, Jeffrey Q. Adams** and Richard G. Eilers***

Introduction

The objective of water treatment is to provide safe and aesthetically acceptable water to customers in sufficient quantities at reasonable costs. Communities with an abundance of safe water generally have little trouble in meeting the above objectives. Communities that have a limited supply or a source of water that must be treated may be faced with many problems in meeting those objectives. These problems are amplified in small communities that have insufficient money or qualified personnel to construct and operate a water treatment facility.

In the United States, over 37,000 systems serve fewer than 500 people. A significant number of these systems have difficulty in providing water that meets the Maximum Contaminant Levels established under the U.S. Safe Drinking Water Act. In addition to quality problems many of these systems have financial difficulties as well. The cost of technologies required to meet the requirements of the act have raised many concerns among water utilities in general.

In response to concerns about impacts of cost on drinking water utilities, the Drinking Water Research Division of EPA initiated a study to develop standardized cost data for 99 water supply unit proceeses.[3] The approach was to assume a standardized flow pattern for the treatment train and then to estimate the cost of the unit processes. This approach requires assumptions about such details as common wall construction and amounts of interface and yard piping required. After the flow pattern was established the costs associated with specific unit processes were calculated. As built designs and standard cost reference documents were used to calculate the amount of excavation, framework, and materials such as concrete and steel. Information from existing plants and manufacturers was used to calculate the costs of equipment associated with a unit process.[3] Once basic information had been calculated, capital cost curves were developed. In 1984, three years after the first set of reports was issued, another report was issued containing cost curves for "small systems technology", using the same methodologies.[2]

*Director, Drinking Water Research Division, Water Engineering Research Division, 26 West St. Clair Street, Cincinnati, OH 45268
**Research Engineer, Systems and Cost Evaluation Staff, Drinking Water Research Division, Water Engineering Research Divison, 26 W. St. Clair Street, Cincinnati, Ohio 45268
***Operations Research Analyst, Systems and Cost Evaluation Staff, Drinking Water Research Division, Water Engineering Research Division, 26 W. St. Clair Street, Cincinnati, OH 45268

Derivation of Cost Curves

The construction cost for each unit process considered in this study was presented as a function of the process design parameter that was determined to be the most useful and flexible under varying conditions. Such variables as loading rate, detention time, or other conditions that can vary because of designer's preference or regulatory agency requirements were used. For example, GAC contactor construction costs are presented in terms of cubic feet of contactor volume, an approach that allows various empty bed contact times to be used. Contactor operation and maintenance curves are presented in terms of square feet of surface area because operation and maintenance requirements are more appropriately related to surface area than to contactor volume. Such an approach provides information more than if the costs were related to water flow through the treatment plant.[3]

Development of Cost Equations

To make this data more useful and transportable, the cost curves for the small technologies were converted to a set of equations.[1]

The functional form of the estimating equations used in this development is as follows:

$$Y = a + b\,(USERATE)^c(d^z) \qquad (1)$$

where

Y = operating and maintenance or capital cost
USERATE = design or operating variable
a,b,c,d = coefficients determined from nonlinear regression
z = 0 or 1 used to adjust cost function for a range of userate values.

The dummy variable effectively changes the slope of the function shown in equation (1) for a given value of the independent variable.

If the data when plotted on a log-log scale fits a straight line or has a mild bend or curve it does not need a "d^z" factor. Deciding if the d^z factor is needed is done on a case by case analysis.

For example the following equation gives the construction cost of a small system steel pressure contactor:

$$CC = 16125 + 7632.0\,(CUFT)^{0.5229}\,(1.10)^z \qquad (2)$$

where

CC = construction cost for steel contactor
CUFT = reactor volume in cubic feet
$z = 0$ for CUFT ≤ 400
$z = 1$ for CUFT ≥ 400

The various technologies being studied are given in Table 1.

TABLE 1. TECHNOLOGIES STUDIED

Item	Item
Package Complete Treatment Systems	Contact Basins - Direct Filtration
Cation Exchange Softening	Ultraviolet Light Disinfection
Anion Exchange Nitrate Removal	Clearwell Storage
Activated Alumina Fluoride Removal	Slow Sand Filters
Granular Activated Carbon Adsorption	Packed Tower Aeration
	Chemical Feed Systems
Chlorine Dioxide Generation and Feed Systems	Disinfection Systems
	Sludge Handling Systems
Oxone Generation, Feed and Contacting Systems	Water Wells
	Pumping Systems

Cost Equations

Equations have been derived for each of the technologies listed in Table 1. Table 2 contains a summary of the parameters in equation (1). Most of the exponents are given with four significant digits to maintain precision with conceptual design cost estimates.

TABLE 2. CONSTRUCTION AND O&M COST FUNCTION PARAMETERS

Process/Function	Variable	a	b	c	d	Range for z = 0
Steel Pressure GAC Contactors						
CC	ft^3	16125	7632.0	0.5229	1.102	<400
PE	"	50	0.1855	1.075	-	-
BE	"	1300	553.1	0.4563	1.319	<140
MM	"	100	34.21	0.6005	-	-
OL	"	186	11.90	0.5176	-	-
Steel Gravity GAC Contactors						
CC	ft^3	40000	664	0.8662	-	-
PE	"	0	0.380	0.9748	-	-
BE	"	3230	50.13	0.8815	-	-
MM	"	625	3.161	0.9307	-	-
OL	"	0	158.1	0.1906	-	-
Package Lime Softening						
CC Single Stage	gal/day	90240	0.1251	1.038	-	-
CC Two State	"	154700	0.1017	1.089	-	-
PE Single Stage	"	7400	0.0010	1.255	-	-
PE Two Stage	"	12600	0.0027	1.210	-	-
BE Single Stage	"	2130	0.0049	1.048	-	-
BE Two Stage	"	4210	0.0093	1.053	-	-
MM Single Stage	"	1840	0.0601	0.8059	-	-
MM Two Stage	"	3785	0.000003	1.568	-	-
OL Single Stage	"	1075	0.000093	1.146	-	-
OL Two Stages	"	1460	0.00015	1.110	-	-

TABLE 2. (Cont'd)

Process/Function	Variable	a	b	c	d	Range for z = 0
Cation Exchange						
Softening						
CC	ft^3	8400	3512.2	0.6243	1.003	≥ 30
PE @ >185ft^3	"	100	0	1	-	-
BE	"	1030	90.1	0.7261	-	-
MM, Daily Regen	"	75	18.65	0.8647	-	-
MM, Regen every 2 days	"	100	15.20	0.8496	-	-
MM, Regen every 4 days	"	0	199.4	0.0165	-	-
OL, Rengen daily	"	0	101.1	0.0146	-	-
OL, Regen every 2 days	"	0	50.6	0.050	-	-
OL, Regen every 4 days	"					
Slow Sand Filters						
CC - covered	ft^2	147920	73.58	1.043	-	-
CC - uncovered	"	165000	13.63	1.083	-	-
BE	"	0	4713.9	0.2428	-	-
DF - uncovered	"	0	2.831	0.3901	-	-
MM - covered	"	421	1.331	0.8954	-	-
MM - uncovered	"	371	0.0149	1.160	-	-
OL - covered	"	175	0.2375	0.8580	-	-
OL - uncovered	"	195	0.4748	0.7979	-	-
Anion Exchange						
Nitrate Removal						
CC	ft^3	10100	3567.2	0.6859	-	-
BE	"	920	90.0	0.7257	-	-
PE, 2 Regen/day	"	60	1.835	1.086	-	-
PE, 1 Regen/day	"	0	2.114	0.9700	-	-
PE, every 2 days	"	0	8.606	0.6163	-	-
MM MTL, 2 Regen/day	"	100	97.38	0.9884	-	-
MM MTL, 1 Regen/day	"	135	74.59	0.9888	-	-
MM, every 2 days	"	155	62.42	0.9908	-	-
OL, 2 Regen/day	"	390	0.3846	1.0	-	-
OL, 1 Regen/day	"	200	0.1923	1.0	-	-
OL, every 2 days	"	100	0.0961	1.0	-	-
Activated Alumina						
CC	ft^3	5000	6980.0	0.5300	-	-
BE	"	1020	50.96	0.8681	-	-
PE	"	0	11802.0	0.0710	0.5	≥125
MM, Regen @ 4.5 days	"	340	20.48	0.9667	-	-
MM, Regen @ 8 days	"	430	10.08	0.9850	-	-
MM, Regen @ 12 days	"	415	8.037	0.9653	-	-
OL, Regen @ 4.5 days	"	1310	1.44E-9	4.020	-	-
OL, Regen @ 8 days	"	950	2.86E-6	2.812	-	-
OL, Regen @ 12 days	"	850	1.10E-4	2.229	-	-
Gas Feed Chlorination						
CC	lb/day	20200	13.75	1.0	-	-
BE	"	1450	-	-	-	-
PE	"	1250	6.897	1.458	-	-
MM	"	0	303.4	0.0671	-	-

TABLE 2. (Cont'd)

Process/Function	Variable	a	b	c	d	Range for z = 0
Complete Package Plant						
CC	ft^2	61275	8067.5	0.7746	-	-
BE	"	3610	308.9	0.9315	-	-
PE	"	0	342.8	0.6221	-	-
MM, 1bw/day	"	785	57.49	0.9337	-	-
MM, 2bw/day	"	920	154.7	0.7642	-	-
OL, $2gpm/ft^2$,1bw/d	"	350	47.75	0.6204	-	-
OL, $2gpm/ft^2$,2bw/d	"	365	51.90	0.6078	-	-
OL, $5gpm/ft^2$,1bw/d	"	240	267.5	0.3256	-	-
OL, $5gpm/ft^2$,2bw/d	"	550	50.25	0.6141	-	-
Filter Media						
CC, rapid sand	ft^2	0	160.0	0.8289	-	-
CC, dual media	"	0	275.0	0.8028	-	-
CC, mixed media	"	0	575.0	0.7142	-	-
Packed Tower Aeration						
CC	ft^3	18000	1609.1	0.5336	-	-
PE,A:W 10:1	gal/day	3345	0.0257	1.105	-	-
PE,A:W 20:1	"	4100	0.0135	1.161	-	-
MM	"	265	0.00033	1.031	-	-
OL	"	180	1.0 E-9	1.878	-	-
Ultraviolet Light Disinfection						
CC	gal/day	6900	32.31	0.5400	-	-
BE	"	800	3.506	0.4976	-	-
PE	"	615	0.0923	0.9481	-	-
MM	"	160	0.0177	0.9277	-	-
OL	"	66	0.0160	0.5082	-	-
On-site Hypochlorite Generation System						
CC	lb/day	17250	1033.9	0.6361	-	-
BE	"	1800	0.3224	1.640	-	-
PE	"	0	912.0	1.0	-	-
MM	"	200	96.93	0.6505	-	-
OL	"	17	13.18	0.7555	-	-
Chlorine Dioxide Generation System ($\leq$ 77 lb/day)						
CC	lb/day	36400	-	-	-	-
BE	"	1800	-	-	-	-
PE	"	0	5716.6	0.3473	-	-
MM	"	25	484.2	0.1158	-	-
OL	"	175	0.1201	2.354	-	-
Basic Chemical Feed Systems						
CC	lb/day	1000	229.6	0.5175	-	-
PE hyrated lime	"	200	900.5	0.3095	-	-
PE (other)	"	100	28.8	0.4990	-	-
MM	"	0	17.61	0.3246	-	-
OL	"	200	0.7070	0.9693	-	-

Table 2 (Cont'd)

Process/Function	Variable	a	b	c	d	Range for z = 0
Ozone Generation System						
CC, contactors	gal	14170	39.96	0.7009	-	-
CC, oxygen feed	lb/day	103065	2275.6	0.8076	-	-
CC, air feed	"	112460	4192.3	0.8140	-	-
BE	"	6870	20.43	1.043	-	-
PE, oxygen feed, 3.5 kwh/lb		55	1274.1	1.0	-	-
PE, air, 5 kwh/lb	"	0	1823.4	1.0	-	-
PE, air, 8 kwh/lb	"	0	2920.0	1.0	-	-
MM, oxygen	"	2190	381.1	0.9816	-	-
MM, air	"	2150	164.6	0.8444	-	-
OL	"	300	159.1	0.2439	-	-
Clearwater Storage						
CC, above ground	gal	6775	88.39	0.5770	-	-
CC, below ground	"	6520	39.15	0.6797	-	-
Water Wells						
CC, 250 ft. deep	gal/day	48400	0.0394	1.021	-	-
CC, 250 ft. deep	"	68100	5.50	0.681	1.5	<750,000
PE, 250	"	270	0.2965	1.0	-	-
PE, 500	"	0	0.7177	0.9969	-	-
MM, 250	"	970	0.0174	0.8336	-	-
MM, 500	"	1100	1.0	0.5560	-	-
OL, 250	"	425	0.00017	1.0	-	-
OL, 500	"	475	.00017	1.0	-	-
Package Raw Water Pumping						
CC	gal/day	29000	0.0035	1.139	-	-
PE	"	0	0.0810	1.0	-	-
MM	"	90	0.030	0.6326	-	-
OL	"	50	0.000016	1.107	-	-
Sand Drying Beds						
CC	ft^2	225	28.73	0.8358	-	-
DF*	"	0	0.2705	0.7907	-	-
MM*	"	0	0.3652	0.9124	-	-
OL*	"	0	0.1776	0.7923	-	-
Sludge Dewatering Lagoons						
CC	ft^3	1695	0.3193	0.9739	-	-
DF	ft^3/yr	0	0.0144	0.9551	-	-
MM	"	0	0.0950	1.0	-	-
OL	"	5	0.0039	0.9950	-	-
OL, with liner	"	20	0.0036	0.6621	-	-

*multiply function by [(No. of cleanings/year)/5.0)]

In table 2 the following terms are defined:

CC = Construction Cost, $
BE = Building Energy, kwh/year
PE = Process Energy, kwh/year
DF = Diesel Fuel, gal/year
MM = Maintenance Material, $/year
OL = Operational Labor, hours/year

In equation (1) the Engineering News Record Index Construction Cost Index (CCI/383) and the Producers Price Index (PPI/287) are used to index Capital and Operating Costs respectively.

Application of Equations

In order to illustrate the use of the cost equations an example has been constructed based on a comparison between package lime softening and cation exchange. It is assumed that both systems will be used to remove hardness from 250 ug/L as to 80 ug/L as $CaCO_3$. The cost assumptions for the two technologies are listed in Table 3. It is assumed that the package lime system is a single stage treatment system. Design assumptions are shown in Table 4.

Table 3. COST ASSUMPTIONS FOR EXAMPLE

Item	Cost
CCI	4351 (1985)
PPI	290 (1985)
Interest Rate	10%
Amortization Period	20 years
Engineering Fees	15% of Construction Cost
Power	8¢/kwhr
Labor	14 $/manhour
Chemicals (lime)	740 lb/thou gal @ $145/ton
CO_2	42 lb/thou gal @ $220/ton
Salt Price	
<20 tons/yr	$100/ton
20-2000 tons/yr	$90/ton
>2000 tons/yr	$80/ton

TABLE 4. DESIGN ASSUMPTIONS

Item	Value
Cation Exchange	
Loading Rate	7.5 gpm/ft^2
Depth	8 ft
Salt Required	3.0818 (mg/L Hardness Removed) X mgd
Regeneration	every two days

Using these appropriate equations yields the cost curves in Figure 1. As can be seen the lime softening has very well developed economies of scale. Cation exchange is less expensive then lime softening at the 1.0 gallon/day systems size. These equations can be very useful for preliminary planning purposes.

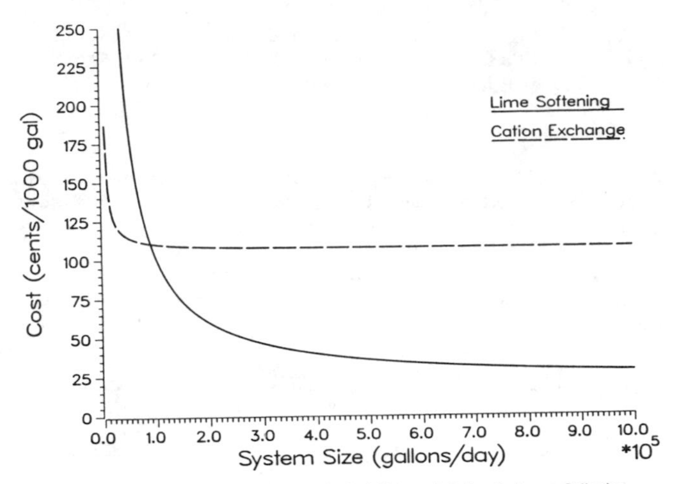

Figure 1. Comparison Between Cost of Lime and Cation Exchange Softening Versus System Size

Summary and Conclusions

Several systems in the U.S tend to have more difficulty in meeting drinking water standards than do larger systems. In addition to having quality problems many small systems are unable to acquire an adequate resource to install technology to meet Safe Drinking Water Act MCLs. Because cost is a major factor in meeting MCLs DWRD has invested considerable effort in developing cost data for the technologies most commonly used by small utilities. This report has developed cost equations that could be extremely useful in making estimates for these small systems technologies.

References

1. Clark, Robert M. and Dorsey Paul, A Model of Costs for Treating Drinking Water", Journal of the American Water Works Association, Vol. 74, No. 12, pp. 618-627, December 1982.
2. Gumerman, R. C., Burris, Bruce E., Hansen, S. P., Estimation of Small System Water Treatment Costs, U.S. Environmental Protection Agency, EPA-600/S2-84-184, Mar. 1985.
3. Gumerman, R. C., Culp, R. L. and Hansen, S. P., Estimating Water Treatment Costs, Volume 1, - Summary, Environmental Protection Agency Research Series, EPA-600/2-79-1629, U.S. Environmental Protection Agency, Cincinnati, Ohio 45268.

Acknowledgments

The authors wish to acknowledge Mr. James Westrick and Mr. Douglas Endicott for their helpful suggestions. Ms. Patricia Pierson and Ms. Diane Routledge assisted in preparation of this manuscript.

Simplified Optimum Design of Branched Pipe Network Systems

Arun K. Deb F.ASCE*

Introduction

The decade of 1980 has been declared as the United Nations Drinking Water Supply Decade. The goal is to supply safe drinking water at minimum cost in cities and villages of developing nations. Most of the cost of a water system is in the distribution system. Therefore, reducing the cost of the distribution system would be one criterion used when designing a water system for a developing nation.

Closed-loop water distribution pipe networks are now widely used for public water supply systems. Though closed-loop pipe networks are always more expensive than corresponding branched pipe network systems, a closed system is generally used for its reliability and flexibility of operations. In developing countries, particularly for small village communities' water supply, a branched network consisting of a pump and over head reservoir is used. In addition, a branched pipe network system is also used for irrigational water supply. The usual practice of design is based on the assumption of suitable sizes of pipes for various sections of a distribution system, and then required hydraulic conditions are checked. The design of a branched water distribution system by this method may not be economical. A linear programming technique was used to find an optimum design of branched pipe network systems. Deb also developed a method of obtaining the combination of head loss and pipe sizes for least cost for a branched pipe network system incorporating the pipe, pump, and energy cost functions. This method also becomes complicated, depending on the complexity of the branching structure. In this paper a simplified methodology for optimum design of a branched pipe network system has been outlined. The efficacy of the design method has been illustrated with an example, and the simplified solution is compared with rigorous optimum solution.

Model Formulation for a Single Pipeline System

In the design of a pipeline system, the costs of the system consist of the cost of pumps, pipe operation and maintenance (O&M), and energy. The size of the pipeline is an important factor in the whole pumping system. Smaller diameter pipelines will result in greater friction heads and more energy consumption. Again, with the increase in pipe size, energy costs will decrease but the cost of the pipelines will increase. The cost benefit from saving energy costs will be gradually reduce with the increase in pipe size and will reach an optimum size beyond which energy costs are not sensitive to the total cost of the

*Vice President, Roy F. Weston, Inc., Weston Way, West Chester, Pennsylvania 19380, U.S.A.

system. The optimum size of a pipeline is a function of the flow, pipe, and energy cost functions, and the fluid friction head loss equation. For a single pipeline and pumping system, an optimum velocity of flow can be obtained which will result in an optimum total cost.

In developing a mathematical model of a pumping system, the total cost is divided into two parts: capital costs and O&M costs (i.e., maintenance and energy costs). The total capital cost of an installed pipeline (including laying, jointing, etc.) can be expressed as a function of diameter:[1]

$$y_1 = k_1 D^{m_1} \tag{1}$$

where

y_1 = capital cost of the pipeline
D = diameter of the pipeline
k_1 = a coefficient
m_1 = an exponent of the cost function

On the basis of average 1986 cost data, when y is expressed in dollars/meter length and D is in millimeters, the value of k_1 = 0.0766 and m_1 = 1.27.

For water supply, the Hazen-Williams equation is widely used and can be given as:

$$H_f = \frac{k\ LQ^p}{C^p D^q} \tag{2}$$

where

H_f = friction head loss in meters
L = length of pipeline in meters
Q = flow in 1,000 cubic meters per day
C = Hazen-Williams coefficient
D = diameter of the pipeline in millimeters
p = 1.85
q = 4.86
k = 1.116×10^{12}

Average annual energy cost can be calculated using an expression of power requirement of the pump as:

$$y_2 = \frac{0.114\ Q\ 24\ c\ 365\ (H_R + H_f)}{E_m E_p} \tag{3}$$

where

H_R = Residual head in meters
c = cost of energy per kwh
E_m = mechanical efficiency of the pump
E_p = electrical efficiency

Combining the Hazen-Williams equation for head loss, equation (2) with equation (3), the annual energy cost Y_2' can be expressed as:

$$Y_2' = \frac{0.114\ Q\ 24\ c\ 365}{E_m E_p} \times \left[H_R + \frac{k\ LQ^p}{C^p D^q} \right] \tag{4}$$

The O&M, other than the energy cost, can be assumed to be f times the energy cost. Thus the total annual O&M cost, including energy cost, can be written as:

$$Y_2 = \frac{(1 + f)(0.114)\ Q\ 24\ c\ 365}{E_m E_p} \times \left[H_R + \frac{k\ LQ^p}{C^p D^q} \right] \tag{5}$$

For water pumping systems, the value of f has been suggested by Linaweaver[2] as equal to 0.08. To obtain the total annual cost of the system, the capital pipe cost is converted into annual capital recovery costs, Y_1, considering an interest rate of i and repayment of period of n years (useful life of pipeline) and added to the annual O&M costs. The capital cost of pumps is found to be insignificant in comparison to pipe cost and O&M cost and is therefore neglected in the total cost formulation.

The total annual cost of the pumping system is the summary of Y_1 and Y_2, which can be obtained from equations (1) and (5) as follows:

$$Y = Y_1 + Y_2$$

$$= R_1 k_1 D^{m_1} L + \frac{(1 + f)(0.114)\ Q\ 24\ c\ 365}{E_m E_p} \times \left[H_R + \frac{k\ LQ^p}{C^p D^q} \right] \tag{6}$$

By differentiating the total annual cost with the sizes of pipe and equating to zero, optimum pipe diameter can be obtained as:

$$D = \left[\frac{q\ Q\ k_3\ X}{R_1 k_1 m_1} \right]^{1/(m_1 + q)} \tag{7}$$

where

$$X = \frac{k\ Q^p}{C^p}$$

and

$$k_3 = \frac{(1 + f)(0.114)\ 24\ c\ 365}{E_m E_p}$$

Replacing the values of F_2 and X and expressing terms of flow Q, equation (7) can be rewritten as:

$$D = K\,Q^{(p + 1)/(m_1 + q)} \qquad (8)$$

where

$$K = \left[\frac{q}{R_1 k_1 m_1} \times \frac{k\,k_3}{C^p}\right]^{1/(m_1 + q)}$$

Therefore, for a single pipeline, optimum pipe size can be obtained using equation (8) for a known pipe flow value. For a single pipeline system, the optimum size of the pipeline is found to be proportional to $Q^{(p+1)/(m_1 + q)}$.

For p = 1.85, m_1 = 1.27 and q = 4.86, the optimum pipe size is proportional to $Q^{0.465}$.

Branched Pipeline System

The solution of optimum sizes of various sections of a branched pipe system is complicated. Karmeli, Gadish, and Meyers[3] developed a method for design of optimum pipe sizes of branched network system using linear programming techniques. Deb[1] presented a method for the direct least cost solution of pipe sizes of a branched pipe network of known geometry for known water consumptions incorporating various cost functions. However, all these methods are complex and time-consuming. In this paper a simple method has been presented.

Equation (8) derives optimum size of a pipeline for a known flow value, considering the total cost, i.e., cost of pipe, and energy cost. Therefore this simple equation can be used to derive optimum pipe size when flow through a pipe is known. In a branched pipe network of known geometry when water demands at various junctions are known, flows through individual pipe segments are also known. Therefore, using equation (8), a branched pipe network can be solved for optimum cost. In this paper the use of equation (8) which was derived for optimization of single pipeline has been used in determining optimum sizes of a branched network system.

Example

To illustrate the methodology an example of branched network shown in Figure 1 has been used. The network shown in Figure 1 has been solved using rigorous optimization method[1] and simpler method of using equation (8) and both solutions are compared. Table 1 shows various data used in the example problem.

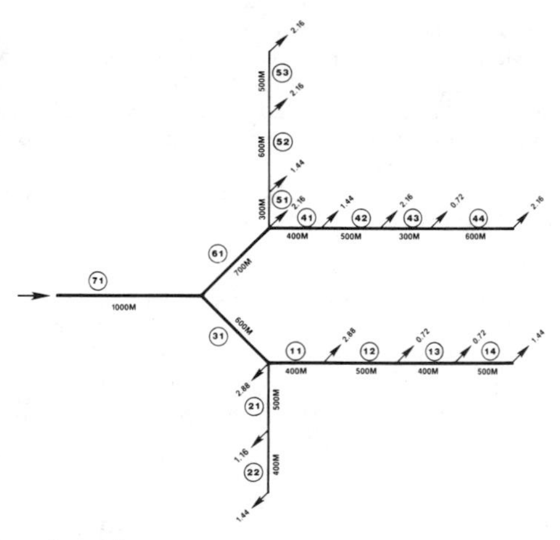

Legend

All Outflows are in 1000 m³/day

Numerical within Circles
Indicate Pipe Numbers

FIGURE 1 EXAMPLE BRANCHED PIPE NETWORK

TABLE 1

VARIOUS DATA USED IN THE EXAMPLE PROBLEM

Pipe Number	Flow in 1,000 Cubic Meters per Day	Pipe Length in Meters	Other Data
11	5.76	400.00	C = 100; P = 1.85; q = 4.86;
12	2.88	500.00	$k = 1.116 \times 10^{12}$, $k_1 = 0.0766$
13	2.16	400.00	$m_1 = 1.27$
14	1.44	500.00	Interest rate = 0.05
			Useful life of pipe, n = 50 years
21	2.60	500.00	Residual head, HR = 20 meters
22	1.44	400.00	$E_p = 0.80$
			$E_m = 0.80$
31	11.24	600.00	Energy cost, c = 0.10 per kwh
41	6.48	400.00	
42	5.04	500.00	
43	2.88	300.00	
44	2.16	600.00	
51	5.76	300.00	
52	4.32	600.00	
53	2.16	500.00	
61	14.40	700.00	
71	25.64	1,000.00	

For a fixed residual head, if the total head loss, H, is increased by increasing the supply head, the least cost pipe sizes for various sections of the branched system will be decreased. With the increase of total head loss, the cost of the pipelines will be decreased and at the same time energy cost will be increased. Thus, for combined total cost of pumping and piping, an optimum total head loss, H, value can be obtained. The branched pipe network shown in Figure 1 is solved using rigorous optimization method developed previously by Deb.[1]

An optimum total head loss of 5.81 meters was obtained. The costs of the systems under various head loss conditions were also developed. Table 2 shows annual costs of pipelines, pumps, O&M, and total costs variation with total head loss in the system. Figure 2 shows the variation of the total annualized costs with total head loss. Figure 3 shows variation of all cost components with a total head loss.

TABLE 2
Total Head Loss Vs. Cost

Total Head Loss in Meters	Annual Cost of Pipe in Thousand Dollars	Annual Cost of Pump in Thousand Dollars	Annual O&M Cost in Thousand Dollars	Total Annual Cost in Thousand Dollars
2.00	83.37	4.32	59.45	147.13
4.00	69.58	4.57	64.85	139.00
5.81	63.13	4.78	69.48	137.66
8.00	58.07	5.04	75.66	138.78
10.00	54.79	5.27	81.07	141.13
12.00	52.25	5.49	86.47	144.21
16.00	48.47	5.92	97.28	151.68
20.00	45.73	6.34	108.09	160.16

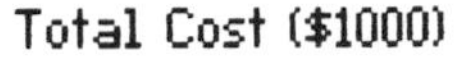

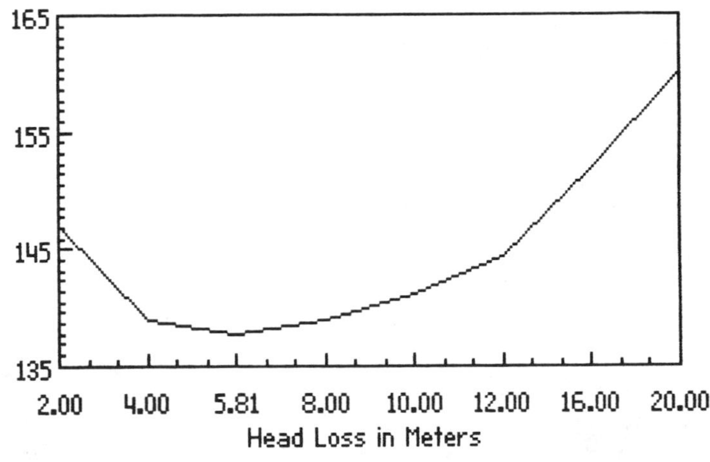

Figure 2. Total Cost Vs. Head Loss

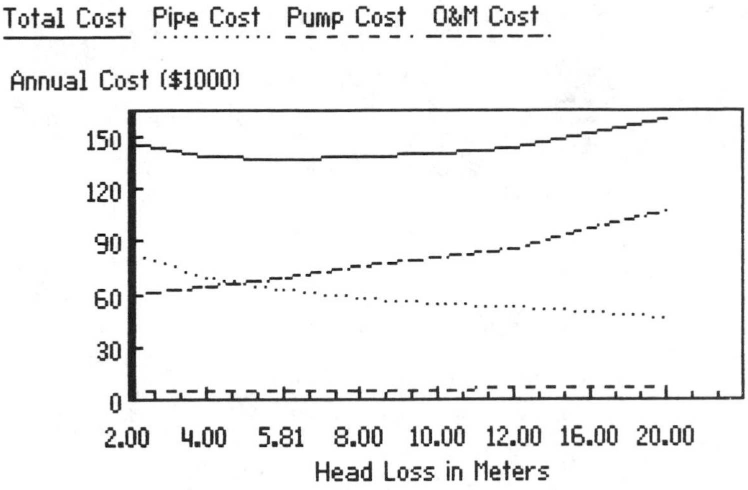

Total Cost Pipe Cost Pump Cost O&M Cost

Annual Cost ($1000)

Head Loss in Meters

Figure 3. Annual Cost Vs. Head Loss

The same branched pipe network (Figure 1) has also been solved with the simplified method using the same data set as shown in Table 1. With known flows through each pipe of the network, optimum pipe size for each branch was calculated using equation (8).

A comparison of optimum pipe sizes obtained by the rigorous method and simplified method is shown in Table 3. The sizes of pipes so obtained are likely to be fractional. In order to obtain a realistic design, one or two alternatives may be tried. In Table 3 theoretical optimum total annual cost of $137,657 is comparable to the optimum total annual cost of $141,969 obtained by simplified method. Variation of annualized total costs of optimum designed system using rigorous and simplified methods are even less: $139,054 versus $140,256. Therefore, for all practical purposes, the simplified method can be used for obtaining an optimum design of a branched pipe network system.

Summary and Conclusions

A new simplified method for optimal analysis of branched pipe network of known geometrical layout and water demands has been developed incorporating various cost functions. The method outlined herein for least cost design of branched water main systems is simple and can be used conveniently with the help of a calculator.

Since branched pipe network is cheaper than a closed loop system, branched network is often used in developing countries. The simplified methodology outlined in this paper will be particularly suitable for designing village water distribution system in developing countries.

TABLE 3: COMPARISON OF COSTS BETWEEN THEORETICAL AND SIMPLIFIED METHODS

	Flow in 1000 cubic meters per day		Theoretical		Design		Simplified Method		Simplified Design	
Pipe Number		Pipe Length in Meters	Opt. Pipe Size mm	Pipe Cost in Dollars	Opt. Pipe Size mm	Pipe Cost in Dollars	Opt. Pipe Size mm	Pipe Cost in Dollars	Opt. Pipe Size mm	Pipe Cost in Dollars
11	5.76	400	354	52868	350	52150	362	54390	375	56926
12	2.88	500	287	50683	300	53597	262	45183	275	47990
13	2.16	400	263	36318	250	34015	229	30508	225	29755
14	1.44	500	233	38871	250	42519	190	30027	200	32026
21	2.60	500	236	39501	250	42519	250	42539	250	42519
22	1.44	400	197	25203	200	25621	190	24022	200	25621
31	11.24	600	455	109167	450	107637	493	120995	500	123047
41	6.48	400	375	56887	400	61788	382	58301	400	61788
42	5.04	500	347	64586	350	65188	340	62841	350	65188
43	2.88	300	294	31279	300	32158	262	27110	275	28794
44	2.16	600	269	56034	300	64317	229	45762	225	44633
51	5.76	300	343	38084	350	39113	362	40793	375	42694
52	4.32	600	314	68224	300	64317	317	68859	325	71199
53	2.16	500	255	43603	250	42519	229	38135	250	42519
61	14.40	700	521	151122	550	162027	554	163358	550	162027
71	25.64	1000	675	300144	700	314415	724	327903	725	328744
Total Capital Cost of Pipes			=	1162573		1203899		1180726		1205470
Annualized Cost of Pipes			=	63128		65372		64113		65457
Annualized Cost of Pumps			=	4783		4747		4922		3837
Annual O & M Cost			=	69746		68935		72934		70962
Total Annualized Cost			=	137657		139054		141969		140256

The cost functions used in this study are valid for the United States only. However, for developing countries, the methodology developed herein can be used in incorporating coefficients and exponents of cost functions valid for those countries.

References

1. Deb, A.K., "Least Cost Design of Branched Pipe Network System," Journal of the Environment Engineering Division, ASCE, Vol. 100, No. EE4, August, 1974.

2. Deb, A.K., "Optimum Energy Cost Design of a Pipeline," Journal of Pipelines, 1 (1981), p. 191, Elsevier Scientific Publishing Company, Amsterdam.

3. Karmeli, D., Gadish, Y., and Meyers, S., "Design of Optimal Water Distribution Networks," Journal of the Pipeline Division, ASCE, Vol. 94, No. PL1, October 1968.

4. Linaweaver, P.F., Jr., and Clark, C.S., "Cost of Water Transmission," Journal of American Water Works Association, Vol. 56, December, 1964.

Problems of Providing Least Cost
Increments of Water Supply Under Urbanizing Situations

Jay M. Bagley* M. ASCE

Abstract

The continuing objective of municipal water managers is to provide a safe, dependable, and least cost supply in accordance with customer demands. This objective must be sustained in a dynamic environment of shifting water use patterns and changing institutional jurisdictions that are inescapable consequences of the urbanization process. The identification, evaluation, and adoption of management options to provide municipal and industrial water service under a policy requiring each increment of additional supply to be least cost may be straightforward so long as options can be independently exploited. Quite commonly, however, potentially attractive supply options entail collaborative arrangements and joint participation in projects that offer scale economies. In many instances, large project solutions to water supply problems are under governmental sponsorship and are characterized by rather long and unpredictable completion schedules. Participation in such projects requires commitments to operating policies, structural design features, and conditions of water delivery that often complicate the efficient meshing of new supply availabilities with the existing water management matrix of individual water supply organizations.

In this paper, the classes of management options commonly available for water planners and managers to consider for adoption in some determined time sequence are identified. These are: 1) appropriation of presently unappropriated waters from surface or subsurface sources; 2) acquisition of water rights currently belonging to others either through voluntary purchase or by eminent domain authorities; 3) purchase, rental, or lease of water on a volume (commodity) basis and for specified durations from present owner, generally a wholesaler such as a special purpose water district; 4) negotiated trades and exchanges of water or water rights that result in advantage to trading parties; 5) implementation of appropriate conservation measures that extend utility of given levels and sources of supply; and 6) joint participation in large water supply projects with opportunity to subscribe for needed supplies. Problems and issues associated with the adoption of each of these measures for meeting incremental supply objectives under urban growth conditions are discussed in this paper and displayed in tabular form.

Specific examples and experience in the incorporation of each of the above kinds of management options are presented. Their suitability and utility under particular socio-economic situations are assessed. The

*Professor of Civil and Environmental Engineering, Utah Water Research Laboratory, Utah State University, Logan, UT 84322-8200.

legal, institutional, and political influences and implications in the adoption of particular management measures are discussed. Lessons learned and pitfalls to avoid in adding new supply increments are outlined.

Perhaps the more significant cautions come out of experiences with capital intensive projects that are joint ventured. Once implementation is set in motion, design configurations and operational plans are difficult to adjust or modify even though projections of demand and utilization may change before project completion. Participants often come to find only at later stages that costs and conditions of delivery veer from original expectations and current desires. Capital budgeting processes of individual participants may be upset. Yet concerns for maintaining financial, institutional, or political integrity of the cooperative project may result in suppression of creative mid-course adjustments and updated examination of alternatives that may be crucial to sustaining a cost effective supply objective. Managers need to beware of collaborative arrangements that require substantial transfer or relinquishment of decision making authorities to new super entities with less accountability to rate payers and taxpayers. They also need to make sure that cooperative involvements do not weaken important lines of informational flow that blurs continuous assessment of the direction and progress toward the standing objective of safe, dependable, and least cost water supply.

Introduction to Technology and Engineering

Charles Morse and Dennis Warner

Presentations include sessions on planning and research in water supply and waste areas. Examples of planning are covered as extensively as would be required for a major national program for a large country or as specifically as would be the case for a Pacific island. There are also sessions on the design and implementation of specific areas of water supply and waste projects and practical field application of systems for the provision of basic services.

A wide range of topics are discussed of interest in the supply of potable water, including windmills, ground water protection, rainwater supply, plastic pipe standards, and the like. In respect of sewerage and wastes, sanitation for small housing projects, experiences in the use of pit latrines on a large scale, and solid waste management are covered.

INDONESIA'S MASSIVE PROGRAMME FOR IKK WATER SUPPLY SYSTEM

BY: Ir. SOERATMO NOTODIPOERO*

Background

Indonesia is one among the nations in developing countries that has commited itself towards the achievement of the goal set out in the International Drinking Water Supply and Sanitation Decade, to provide safe drinking water supply by the year 1990.

The responsibility of providing drinking water supply in urban areas in Indonesia rests with the Directorate of Water Supply, Directorate General of Human Settlements (Cipta Karya) within the Ministry of Public Works.

In line with the fourth National Five-Year Development Plan (Repelita-IV, covering period 1984 - 1989), the national target is to provide 75 % of urban population by the end of Repelita-IV (year 1989). One of the huge efforts towards the achievement of the target is to implement the so called IKK Water Supply Programme, which was initiated in the early eighties.

The programme aims at providing bacteriologically safe water to an average of 75 % of the population of capital of Sub District (Ibu Kota Kecamatan - IKK). The Government planned to complete this programme at the closure of Repelita-IV in 1989. By that time some 1,800 IKK would have water supply system. The size of the IKK town is between 3,000 to 20,000 inhabitants.

Main Features of the IKK-Programme

The National policy with regard to the IKK Programme has been set out by Cipta Karya in the document entitled "IKK Water Supply Programme: Strategy and Scope" (1982).

Three basic premises form the foundation of the programme:

1. Cost effective facilities will be designed and constructed for those IKKs which are considered to be capable of supporting a water supply system.

2. Long term operation and maintenance of these systems is deemed to be equally important as the construction itself.

3. The institutional capacity to ensure both of the above mentioned is to be developed to coordinate physical, human and monetary resources.

*Directorate of Water Supply,JL.R. Patah I/1, Jakarta, Secatan, INDONESIA

Low cost water supply systems with standard production capacities of 2.5, 5 and 10 lps have been developed to serve IKK town population of 3,000 to 20,000. In order to achieve the requirements of minimum cost, simplicity of design, construction and operation and minimum time span, the system utilizes some novel features:

- daily peak flows are eliminated by serving all consumers at a constant rate over 24 hours through pressure independent flow restrictors (the concept of Continuous Flow - CF System). This eliminates the need for water meters;

- no central storage is provided, for public taps a 3.5 cubic meter tank is provided; for house connections use is made of the available water storage normally found in Indonesian homes;

- standarization has been introduced at all levels in the programme, including survey, design, construction and materials.

The above features allow minimum source capacities, pump and pipe sizes and permit the use of sub-professional human resources for a large part of the process from survey to operation.

Review and Modification of the Approaches

On the basis of early experience in the implementation of IKK schemes, the Directorate of Water Supply has already made several changes to policies and programmes. In early 1984 the IKK Review Mission was formulated. The Mission, financed jointly by the Australian Development Assistance Bureau and the Government of the Netherlands, prepared the Report of IKK Review Mission (July 1984).

The factor of people's acceptance, cost recovery and communications were considered to be focal points of concern that was reviewed in the Report. The Mission suggested some recommendations for improvements.

At the end of Repelita-III (1984), after some schemes had been in operation, the Government decided to make overall review on the schemes that has been in operation.

In early 1985 IKK Review and Action Programme was initiated under the bilateral cooperation with the Netherlands Government. For this purpose, 10 IKK schemes located in West Java, North Sumatera and Aceh Provinces were selected for immediate improvement during the "action" period. (Basic information of the IKK scheme under review is shown in Annex 1). The Review was completed in March 1986 and the Report recommended several suggestions for further improvements.

Within the framework of the Review and Action Programme the consultant prepared the investment cost calculation depending on the size of the system and the water source. For a system operating 24 hours a day and a house connection/public hydrant ratio of 50/50 the percapita investment costs are:

System capacity	No. of consumer	Percapita Investment (US $)1)
Spring captation		
2.5 lps	3,600	26
5.0 lps	7,200	19
10.0 lps	14,400	14
Deepwell		
2.5 lps	3,600	26
5.0 lps	7,200	19
10.0 lps	14,400	13
Treatment Plant		
2.5 lps	3,600	43
5.0 lps	7,200	31
10.0 lps	14,400	20

Note: 1). Based on exchange rate of 1 US $ = Rp. 1,077 in 1986 price

The consultant also made detailed financial analyses of the recurrent costs for operation maintenance and replacement, resulting in tariffs ranging from Rp. 1,150 (US $ 1.07 at 1986 price) per household per months for large gravity systems up to Rp. 8,300 (US $ 7.71) per month for small surface water systems.

The suggested modifications on the approach were proposed by the Mission:

- on the process for priority ranking and town selection in a uniform and objective manner through a system of selection criteria, and
- on the design criteria.

New Approach and Design Criteria for IKK Programme

In new approach the priority ranking and town selection is of prime important and has to be done at the early stage before the project is implemented. The selection criteria consisted of population size, need for water and cost of project. In the previous criteria the ability and the willingness to pay by the prospective consumer as well as the attitude of the IKK community towards a CF-system were not included.

A new scoring system for IKK-town selection criteria was developed. Annex 2 shows a flow chart of this new scoring system, maximally 100 points can be scored and half of this total is allocated to criteria for project implementation need and affordability, while other half is allocated to the project implementation effort. The scoring criteria and allocated points for each criteria is shown in Annex 3.

New design criteria was also developed on the basis of the assessment of the functioning of the IKK-system in the Provinces under consideration. The result is shown in Annex 4 that compares the "new" criteria to the "old" one.

The most important changes that can be noted from the modified criteria are:

- the ratio of house connections/public taps is now variable and should be determined on the basis of the socio-economic survey,

- the number of households per public tap is decreased from 20 to 10,

- a shorter daily operation time (e.g. 12 hours instead of 24 hours).

The later modification seems to give a better acceptance of the system by the consumer (the 600 l/day is now supplied through a flow restrictor of twice the capacity). Also the operating agensy is in general in favour of a shorter operation period. Moreover, it adds to the flexibility of the system while production can easily be increased.

As a result, the standarized approach of the system capacities have also been modified to become 5, 10 and 20 lps.

The latest status of the Programme progress (March 1987) shows that some 591 schemes have been completed at various stage of completion. With the modified approach and new design criteria, the IKK schemes that have been in operation seem to have better performance, they are now in good operation and well managed.

Some Aspects on Financial Viability

In new approach the ratio of house connection/public tap has been modified to become in the range of 50/50 to 80/20. By adapting a ratio of 80/20 instead of 50/50 the total population served is reduced with about 17 %, resulting in an increase of investment cost per capita of about 17 %. Furthermore the changes from 24 hours operation with a double capacity restrictor result in an increase in investment cost per capita of about 20 to 40 % depending on the type os system. To achieve financial viability, coverage of recurrent cost is needed, this covers operation & maintenance (O & M) cost plus depreciation of short life assets.

A differentiated tariff structure, taking into the above consideration, has been prepared by the consultant and is presented in tabel below (in US $/month/house connection). The tariffs needed for coverage of of the O & M costs together with the tariffs needed for coverage of recurrent costs are shown for systems operating 12 hours per day with a house connection/public tap ratio of 80/20. The tariffs have been calculated with an assumed billing efficiency of 90 % for house connections and 75 % for public taps.

Type and Capacity of System	Tariff necessary for recovery of: (US $)		
	O & M	Replacement	Total Recurrent Cost
Gravity spring			
5 lps	1.97	-	1.97
10 lps	1.49	-	1.49
20 lps	1.07	-	1.07
Deepwell			
5 lps	2.88	1.86	4.74
10 lps	2.55	1.39	3.94
20 lps	2.14	1.07	3.21
Treatment			
5 lps	4.88	2.83	7.71
10 lps	4.04	1.72	5.76
20 lps	3.48	1.16	4.64

Some Aspects on Operating Costs

A major component (33 to 53 %) in the operation costs of pumped systems are the energy costs. A possibility for reducing the energy costs is the use of State Electricity instead of generating sets. Calculations made by the consultant reveal that the energy costs will decrease up to 50 %. Unfortunately that most of the IKK schemes are located in the remote areas that are still not connected to the State electricity network. With the use of alternative energy sources such as solar, wind and mycro-hydro power, etc. the savings might be even more.

At present pilot projects on the use of solar and wind power as sources of enèrgy are being undertaken by Directorate of Water Supply. The projects are being evaluated and seem to have future prospective.

Conclusion

IKK Programme is indeed the answer to the commitment of the Water Supply and Sanitation Decade. Several reviews and action programmes for immediate improvements have proven satisfactory result, yet further reviews are still needed and immediate action as has been implemented in several Provinces has to be repeated in the other part of the country to lead to the successful IKK Programme.

List of Reference:

1. Report of IKK Review Mission, July 1984
   Report prepared by a joint Team of Government of Indonesia, Australian Development Assistance Bureau and Government of the Netherlands.

2. IKK Review and Action Programme, Summary of Final Report, March 1986.
   Iwaco-DHV-Waseco-Deserco.

STATUS ACTION IKKS PER 1/1/86

Name IKK	Production Unit	Cap. (l/s)	Number of h.c.	Number of p.h.	Start Operation	Operation Hours (hrs/day)	Tariff for h.c. Rp/month	Billing Efficiency	* Surplus Operation Costs
WEST JAVA									
Karangnunggal	treatment	2,5	103	0	September 1985	9	5,000.-	100%	+
Bangodua	treatment	5,0	310	11	September 1985	13	2,500.-	80%	+
Sumber	connection	5,0	250	0	December 1985	24	110.-/m3 **)	87%	++
Waled	deepwell	5,0	being	rehabilitated	by action	team, start	operation in	March 1986	n.a.
Losarang	treatment	5,0	220	3	September 1985	13	2,500.-	70%	+/-
ACEH									
Matangkuli	shallow well	2,5	107	0	November 1985	1	2,500.-	not yet started	+/- *)
Peudada	deepwell	2,5	147	0	November 1985	24	2,500.-	not yet started	+ *)
NORTH SUMATRA									
Tanjung Langkat	river bank	2,5	162	0	November 1985	24	2,500.-	not yet started	+ *)
	infiltration		90	0	January 1986	24	2,500.-	not yet started	- *)
Sei Silau Timur	deepwell	2,5	78	0	January 1986	24	2,500.-	70%	-
Sei Alim Ulu	deepwell	5,0							

*) theorotically calculated with actual number of h.c., tariff and billing efficiency of 90%

**) the status of Sumber has been changed from Ibu Kota Kecamatan to Ibu Kota Kabupaten. As the source allows abstraction up to 15 l/s, it was decided to install watermeters instead of flow restrictors.

IMPLEMENTATION SCHEDULE FOR CF-SYSTEMS IN IKK TOWNS

Annex. 2

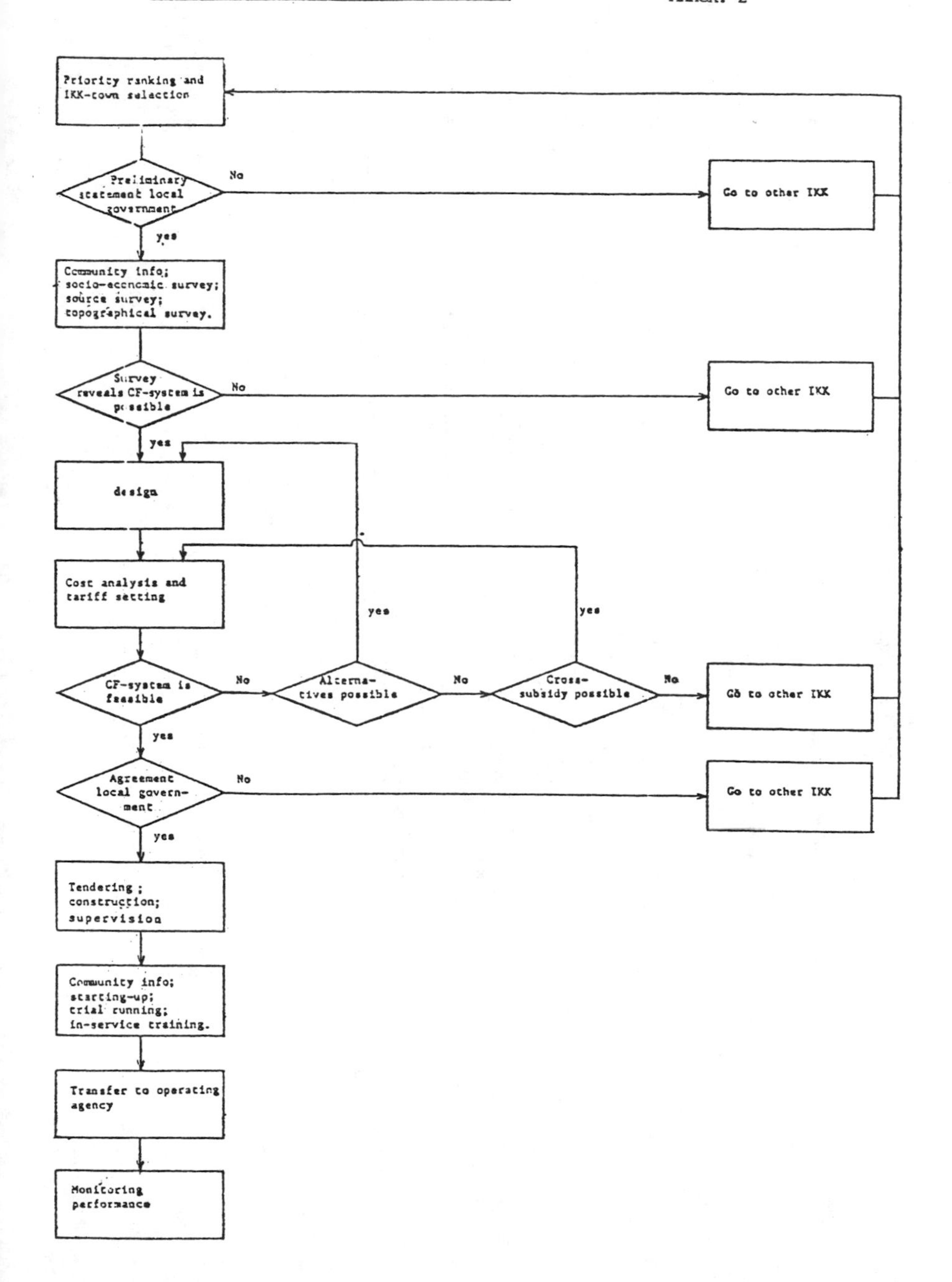

Annex. 3

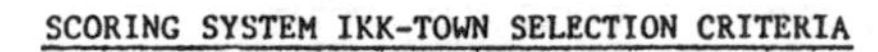
SCORING SYSTEM IKK-TOWN SELECTION CRITERIA

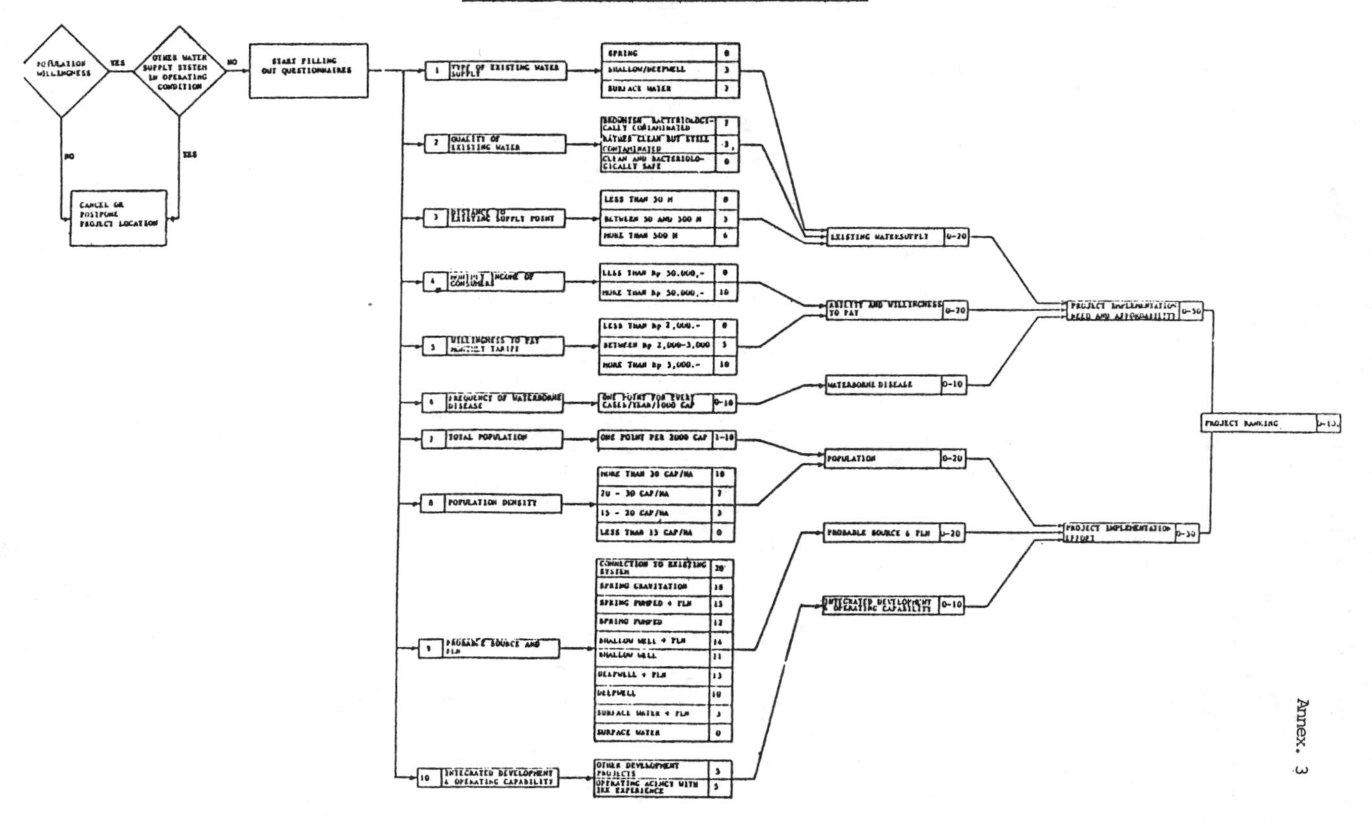
POPULATION WILLINGNESS
YES
NO
OTHER WATER SUPPLY SYSTEM IN OPERATING CONDITION
YES
NO
CANCEL OR POSTPONE PROJECT LOCATION
START FILLING OUT QUESTIONNAIRES
1 TYPE OF EXISTING WATER SUPPLY
SPRING 0
SHALLOW/DEEPWELL 3
SURFACE WATER 7
2 QUALITY OF EXISTING WATER
CLEAN AND BACTERIOLOGICALLY SAFE 0
3 DISTANCE TO EXISTING SUPPLY POINT
LESS THAN 50 M 0
BETWEEN 50 AND 500 M 3
MORE THAN 500 M 6
4 MONTHLY INCOME OF CONSUMERS
LESS THAN Rp 50.000,- 0
MORE THAN Rp 50.000,- 10
5 WILLINGNESS TO PAY MONTHLY TARIFF
LESS THAN Rp 2,000.- 0
BETWEEN Rp 2,000-3,000 5
MORE THAN Rp 3,000.- 10
6 FREQUENCY OF WATERBORNE DISEASE
ONE POINT FOR EVERY CASES/YEAR/1000 CAP 0-10
7 TOTAL POPULATION
ONE POINT PER 2000 CAP 1-10
8 POPULATION DENSITY
MORE THAN 30 CAP/HA 10
20 - 30 CAP/HA 7
15 - 20 CAP/HA 3
LESS THAN 15 CAP/HA 0
9 PROBABLE SOURCE AND FLN
CONNECTION TO EXISTING SYSTEM 20
SPRING GRAVITATION 18
SPRING PUMPED + FLN 15
SPRING PUMPED 12
SHALLOW WELL + FLN 14
SHALLOW WELL 11
DEEPWELL + FLN 13
DEEPWELL 10
SURFACE WATER + FLN 3
SURFACE WATER 0
10 INTEGRATED DEVELOPMENT & OPERATING CAPABILITY
OTHER DEVELOPMENT PROJECTS 5
OPERATING AGENCY WITH IKK EXPERIENCE 5
EXISTING WATERSUPPLY 0-20
ABILITY AND WILLINGNESS TO PAY 0-20
WATERBORNE DISEASE 0-10
POPULATION 0-20
PROBABLE SOURCE & FLN 0-20
INTEGRATED DEVELOPMENT & OPERATING CAPABILITY 0-10
PROJECT IMPLEMENTATION EFFORT 0-50
PROJECT RANKING

Annex. 4

The following table compares the "old" and the "new" criteria :

	Initial	New
1. Supply level at house connections	60 l/c/d	60 l/c/d
2. Supply level at public hydrant	30 l/c/d	30 l/c/d
3. Population served	50% to 100%	50% to 100%
4. Ratio of population served by house connections population served by public hydrants	50%-50%	50%-50% up to 80%-20%
5. Water allocation for non-domestic demand	5%	5%
6. Water allocation for leakage in the system and production losses	15%	15%
7. Peak factor for maximum day	1.1	1.1
8. Peak factor for maximum hour	1.0	1.0
9. Number of people per house	10 capita	10 capita
10. Number of households per public hydrant	20 families	10 families
11. Minimum pressure in distribution system	10 mwc	10 mwc
12. Design horizon for distribution system	5 years	*)
13. Design horizon for transmission	"based upon economic feasibility"	*)
14. System operating period	24 hours per day	12 hours per day

*) for selection of system capacity the present population figures are used; however a 12 hours per day operation leaves some room for future extension.

**Long Island Water Problems -
A Lesson For Developing Countries**

Albert Machlin, Member ASCE *

Abstract

This paper discusses the negative impacts that development of Long Island, N.Y. has had on groundwater, the only source of drinking water for that area. It reviews some water quality and quantity problems encountered on Long Island, and highlights the pitfalls to be avoided by developing countries which are also dependent upon groundwater resources.

Background

Long Island, which is located in the southeastern corner of New York State, is approximately 120 miles long (193 KM) and 25 miles (40 KM) at its widest point, encompassing an area of 1400 square miles (3600 SQ.KM). It is bordered by the East River on the west, Long Island Sound on the north, and the Atlantic Ocean on the east and south (see Figure 1). It consists of four counties - Kings (Brooklyn) and Queens, which are part of New York City, and Nassau and Suffolk which lie east of the City and represent two thirds of the Island's land area. Brooklyn and Queens are highly urban areas, while Nassau and Suffolk have suburban development, with eastern Suffolk still remaining largely rural. Of its approximately 6.8 million population, about 3.1 million (in part of Queens and all of Nassau and Suffolk) derive all of their drinking water from groundwater.

The three major aquifers - the Upper Glacial, the Magothy, and the Lloyd - store trillions of gallons of fresh water (see Figure 2). At present, the Magothy and Lloyd Aquifers are the principal sources of public drinking water supply on Long Island. Although the Upper Glacial is used in eastern Suffolk County for private water supplies, it is not being used in most of Long Island because of poor quality. The total amount of groundwater withdrawn from the aquifers of Long Island is approximately 700 million gallons per day (MGD) or 30.7 cubic meters per second (CU.M/SEC), of which 85% is for Nassau and Suffolk Counties.

*Regional Engineer for Environmental Quality, New York State Department of Environmental Conservation, Building 40, State University of New York, Stony Brook, New York 11794

Figure 1

MAP OF LONG ISLAND, NEW YORK

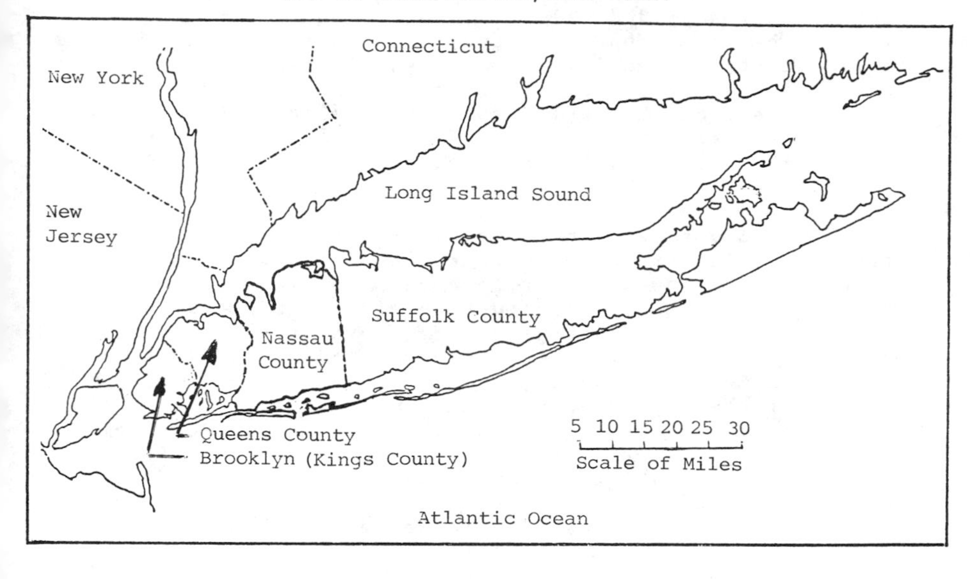

Figure 2

TYPICAL GEOLOGIC SECTION

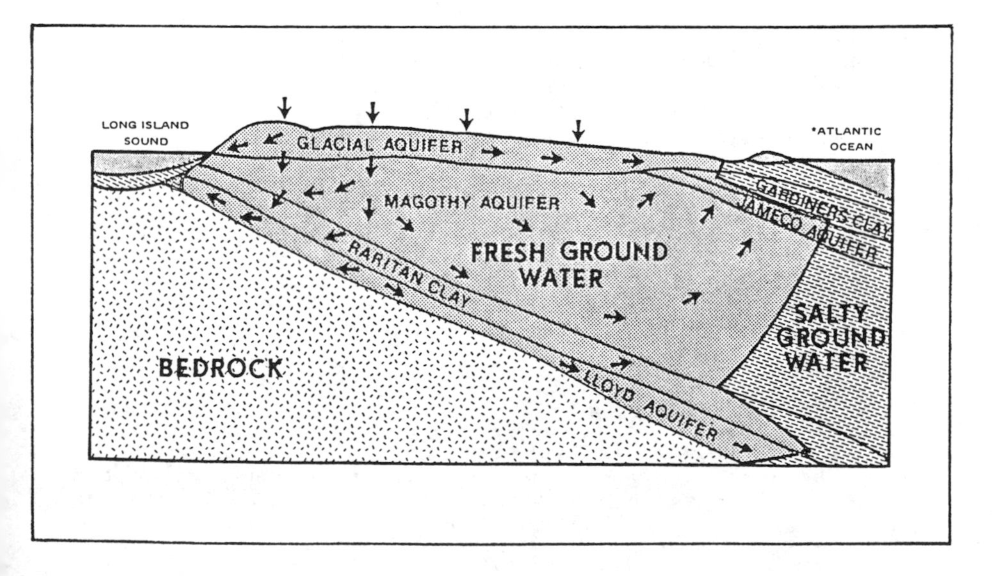

In Nassau and Southeastern Queens, 46 private water companies and municipal water districts supply all the public water. The rest of Queens and all of Kings (Brooklyn) is supplied from off Long Island by the New York City surface water supply system. In Suffolk County, there are 67 private and municipal water suppliers. The Suffolk County Water Authority is the largest of these and accounts for 75 percent of the public water supply pumpage. The most important difference between Suffolk and the other counties is the number of private domestic wells (about 70,000).

Development

The earliest settlements clustered around the harbors since Long Islanders were dependent upon the sea for food and for the transport of farm products and wood to the New York City market. The construction of the railroad during the second quarter of the nineteenth century provided impetus for the establishment of new settlements along its route. Completion of an East River tunnel in the early twentieth century further spurred development of the communities along the North Shore, main line and South Shore branches of the railroad in Nassau and into western Suffolk. As early as the mid 1920's, increased mobility provided by the automobile permitted the expansion of older settlements. However, the Great Depression and World War II brought new construction to a virtual standstill. Commencing in the late 1940's, new household formations and a backlog of unsatisfied housing needs generated high levels of construction activity, first in the central and eastern Nassau, then in western Suffolk. This brought a need for vastly expanded services and provided a skilled labor force for Long Island's growing defense and high technology industries. From 1985 to 1972, the Long Island Expressway was constructed along the spine of the Island from Nassau County to Riverhead - encouraging more development. From 1948 to 1965, the Northern State Parkway, a scenic route, was built from New York City to Hauppauge. The Southern State Parkway was completed in 1959 to ease travel from New York City to western Suffolk County.

The population of Long Island increased almost five-fold between 1900 and 1985, from under 1.5 million to 6.8 million. Of these, about 2.6 million resided in Nassau and Suffolk Counties and were completely dependent on groundwater for their source of drinking water supply. Approximately 500,000 people in Queens County also depend on groundwater sources.

Water Quality Problems

Synthetic Organic Problems

The most significant water quality groundwater contamination problems which have occurred on Long Island are from synthetic organic chemicals - from industrial/commercial solvents and degreasers, from gasoline and petroleum product constituents, and from pesticides and herbicides. Another significant problem is that of nitrate contamination.

Benzene is the only one of these chemicals that is a known human carcinogen - it can cause leukemia. The other synthetic organic chemicals are either suspected human carcinogens or toxic. The New York State Department of Health Drinking Water Guidelines for some of these chemicals are presented in Table I. Water with elevated concentrations of nitrate can reduce the blood's capacity for oxygen transport and transfer - resulting in a condition called methemoglobinemia, which can be hazardous or fatal to infants six months of age or younger. The drinking water standard for nitrate nitrogen is 10 mg/liter.

TABLE I

New York State Department of Health Drinking Water Guideline Levels (ppb)

Chemical	ppb
1,1,1 Trichloroethane	50
Trichloroethylene	50
Tetrachloroethylene	50
Benzene	5
Toluene	50
Xylenes	50
Aldicarb	7

NOTE:

1. Other synthetic organic or pesticide chemicals not listed will have a maximum level of 50 ppb for any one compound or 100 ppb for any combination until a specific evaluation is conducted.
2. The combined level for carbamate pesticides (aldicarb, Carbofuran, etc.) shall not exceed unity in the following formula:

$$\frac{\text{Concentration Aldicarb}}{7} + \frac{\text{Concentration Carbofuran}}{13} + \ldots$$

3. Unless otherwise noted, the combination of compounds should not exceed 100 ppb.

Solvents and degreaser chemicals which are most commonly encountered are trichloroethylene, 1, 1, 1 trichloroethane, and tetrachloroethylene. The main sources of solvents and degreasers are commercial laundries, furniture stripping plants, metal processing plants, machine shops, food and beverage processing plants, and cleaning of transportation vehicles. In addition, there are a number of consumer products containing these chemicals which include fabric and rug cleaners, workshops and auto cleaner solvents, and sewer and cesspool/septic tank cleaners. Improper storage disposal, or accidental spilling of these products can result in groundwater contamination (such as in junkyards, landfills, cesspools, abandoned wells, etc.).

Petroleum Products

Benzene, toluene, and xylene are found in commercial petroleum products (gasoline, etc.) and result in contamination when petroleum spills or leaks occur, or through stormwater runoff. Most leaks are from underground storage tanks at gasoline storage stations, pipelines, and petroleum storage depots. Spills occur as a result of tank truck accidents, overfilling of tanks or trucks, and poor handling. Drainage from areas such as roadways, oil terminals and gasoline service stations, etc., often carry petroleum products to ground and surface water bodies.

Pesticides and Herbicides

Contamination of Long Island groundwater with synthetic organic pesticides has become a serious problem over the last decade. Pesticide and herbicides are widely used on crops, in parks and golf courses, and on lawns, trees and shrubbery. These chemicals generally degrade quickly on the land surface or in the surface water environment. However, degradation processes appear to function differently in the subsurface environment. If the transport time of a pesticide in a soil/water solution to the groundwater table is relatively short, even a non-persistent pesticide may have the ability to contaminate groundwater. Highly permeable Long Island soils are particularly sensitive in this regard. The early pesticides were metal-based with high chemical stability, environmental persistence, and low water solubility. Due to their low solubility, these compounds tended to remain in the upper layers of soil (root zone) and substantial residues may still remain there. The action of microbes in surficial soils can break down some pesticides into non-hazardous forms.

In 1978, wells on both the North and South Forks of Long Island were sampled for the pesticide Aldicarb which had been used extensively on eastern Suffolk County potato farms. Other pesticides detected in groundwater to date include Carbofuran, Dachtal, 1.2 dichloropropane, Dinoseb, Methomyl, Paraquat, Oxamyl, and Carbaryl.

Nitrates

The main sources of nitrate in groundwater on Long Island are human wastes entering the groundwater through cesspools and agricultural and lawn fertilizer. Agricultural fertilizer remains a source of nitrate in eastern Suffolk County where significant farming is still practiced. In most other areas, the increase in population density has resulted in increasing nitrate contamination of groundwater by human waste and lawn fertilizers. Groundwater monitoring data indicate that nitrate nitrogen contamination has occurred throughout Long Island. Nitrate contamination of the aquifer system in Nassau County is widespread geographically and extends to the Magothy aquifer, the major source of water.

Large scale sewering with discharge to sea was recently completed in Nassau County and southwest Suffolk County to prevent pollution from cesspool systems. However,the continued presence of nitrate in the upper layers underlying sewered areas indicates that there are other sources, particularly turf and lawn fertilizer. In areas with no sewers, nitrate pollution has been controlled by requiring minimum building lot sizes to limit population density.

Water Quantity Problems

A groundwater quantity problem occurs when there is too little groundwater (depletion) or too much groundwater (flooding) in a localized area relative to some "normal" groundwater condition or some existing level of human development. Water table elevations on Long Island were first mapped in 1903, which have been used as a benchmark of pre-development conditions.

In Queens County, a cone of depression exists with water table elevations below sea level throughout a large portion of the county - a direct consequence of overpumpage of public water supply wells. This was accompanied by reductions in stream flow and advancement of the salt water interface. There are other areas on the north and south shores and the forks which also exhibit saltwater intrusion.

The expansion of sanitary sewerage systems into areas previously served by on-lot system (cesspools and septic tanks) constitutes an increase in consumptive use of water. Wastewaters that were previously recharged to the groundwater system through on-lot systems are collected and discharged to surface waters. This results in a reduction of water in storage within the aquifers underlying the areas. From the 1950's to the early 1970's, expansion of sanitary sewerage in southwest Nassau County contributed to declines in groundwater levels. Average groundwater levels in a 32 square mile (83 SQ.KM) portion of the sewered area declined 12.2 feet (3.7 M) more than in the unsewered area to the east. In addition, land use practices may have an effect on water table elevations by creating an imbalance of withdrawals and discharge. Runoff from paved developed areas and above-grade structures is diverted to surface waters, reducing the quantities available for recharge. This lowering of the groundwater has caused the water levels to lower in ponds, lakes, streams, and wetlands - often resulting in complete elimination of these valuable resources. High water tables have been caused by cessation of groundwater pumpage. Intensive groundwater development in Kings County resulted in a decline of the water table from 1903 - 1936 of approximately 20 feet (6.1M) - remaining stable until 1946. Following the cessation of pumpage in 1947, the water table rose from a low of about 12 feet (3.7 M) below sea level to about 8 to 10 feet(2.4 to 3.0 M) above sea level in March 1976. This resulted in significant flooding of structures which were installed in dry areas before the cessation of groundwater pumping.

Avoiding Long Island Problems

Under a grant from the United States Environmental Protection Agency, a Long Island Comprehensive Water Treatment Management Plan was completed in 1978. This plan identified groundwater quality and quantity problems, outlined its causes, and recommended future actions to control these problems.

As part of that study, eight hydrogeologic zones were identified - from the most to the least sensitive. The most sensitive zones were the deep recharge areas where surface contamination would most affect the water quality of the principal groundwater aquifers.

After these groundwater sensitive zones were identified, it could be readily seen that the location of industrial and transportation facilities in these areas strongly contributed to groundwater contamination. After the Long Island Expressway was constructed, increasing numbers of industrial and commercial developments occurred close to the expressway, and along several of the major north-south highways - in the groundwater sensitive areas. The study showed that the most significant potential sources of contamination by solvents and degreasers were from industrial facilities as follows:

- Leaks from product storage
- Spills and accidents, and poor facility housekeeping
- Improper Industrial Hazardous or Residual Waste Disposal
- Improper Industrial Wastewater Disposal

The sources of contamination from gasoline and petroleum products were from leaks (product storage), spills and accidents, and facility housekeeping. These often were associated with industries and highways.

Recommendations were also made on actions to be taken for correction of existing problems, and prevention of future problems where feasible. Many of these are included below.

Areas of the world which depend on groundwater as the principal source of water, and have not yet reached the stage of development existing in Long Island, may still be able to prevent some of the same groundwater problems from occurring.

To avoid the water quality problems of Long Island, the following measures should be taken wherever possible:

1. Identify groundwater sensitive areas.
2. Establish strict Groundwater Quality Standards.
3. Establish a strong regulatory groundwater pollution program - promulgate strict wastewater regulations for control of domestic and industrial discharges.
4. Place zoning restrictions on groundwater sensitive areas - restricting use to less polluting facilities.
5. Plan principal transportation facilities so that development in groundwater sensitive areas is not encouraged.
6. Restrict landfills in groundwater sensitive areas.
7. Restrict major chemical and petroleum storage facilities in groundwater sensitive areas.
8. Control use of consumer products which would contribute to groundwater contamination.
9. Prohibit use of polluting chemical septic tank cleaners.
10. Prohibit hazardous waste disposal sites in groundwater sensitive areas.
11. Establish an emergency program to deal with water quality accidents.
12. Promulgate strict standards for chemical and petroleum storage.
13. Regulate transportation of industrial and residual wastes.
14. Establish minimal residential lot sizes in areas where on-site domestic sewerage systems are used (for nitrate pollution control).
15. Place restrictions on fertilizer application rates to prevent excess use of nitrates on lawns and agricultural areas.

16. Establish a pesticide regulation program - prohibit use of certain toxic pesticides and herbicides in groundwater sensitive areas and place restrictions on application rates for specific toxic chemicals.

To avoid the groundwater quantity problems of Long Island, the following measures should be taken whenever possible:

1. Determine maximum groundwater withdrawal rates for areas under consideration.
2. Establish a well permit program to regulate construction, location, and depth of wells, and control withdrawal rates so as not to exceed groundwater replenishment rates.
3. Whenever possible, incorporate water recharge systems to offset effects of lowered groundwater tables caused by construction of sanitary sewerage systems with remote disposal discharge points.
4. If feasible, collect and recharge non-polluting stormwater runoff from paved areas to ground.
5. Insure that construction codes and zoning regulations recognize realistic groundwater conditions to avoid further flooding problems.

Acknowledgement

I would like to thank Lucy Lindstrom for her invaluable assistance in preparation of this manuscript.

References

Baier, J. and Moran, D. (1981). Status Report on Aldicarb Contamination of Groundwater as of Sept. 1981. Suffolk County Department of Health Services.

Div. of Water. (1986). Final L.I. Groundwater Management Plan. N.Y.S. Department of Environmental Conservation.

Katz, B., Lindner, J., and Ragone, S. (1980). A Comparison of Nitrogen in Shallow Ground Water from Sewered and Unsewered Areas, Nassau County, N.Y., from 1952 through 1976. Ground Water 18, 607-616.

Koppelman, L.E. (1978). L.I. Comprehensive Waste Treatment Management Plan. Nassau-Suffolk Regional Planning Board.

National Academy of Sciences. (1977). Drinking Water and Health, 436-439.

Smith, S. and Baier, J. (1969). Report on Nitrate Pollution of Groundwater. Nassau County Department of Health.

WATER RESOURCES MANAGEMENT IN THE CARONI RIVER SYSTEM

G.S. Shrivastava[1], M.ASCE and P.R. Thomas[2]

ABSTRACT

The Caroni River System with its tributaries and underlying aquifers constitutes the largest water resources system in the Caribbean island of Trinidad. However, the utilization and control of the water resources in this river system lags behind the socio-economic developments within its catchment and has only taken place in a fragmented and ad-hoc manner. In this paper the major sources of pollution are identified and measures for the control of the same are outlined. An integrated water resources systems planning in the form of a linear programming model, which incorporates the elements of water supply, flood control, irrigation, recreation and ecology, is also formulated. The paper thus provides the groundwork for a comprehensive water resources planning in the Caroni River System for the 21st Century.

INTRODUCTION

The Caroni River System (Fig. 1) with its tributaries and underlying aquifers constitutes the largest Water Resources System in the Caribbean island of Trinidad. Its catchment boundary also encompasses the largest socio-economic sector of the country in terms of population, urban and rural settlements, educational institutions, infra-structures, industries, agriculture and forestry. However, the utilization and control of the Water resources in this river system lags behind the other developments and has only taken place in a fragmented and ad-hoc manner. The absence of scientific management of the Water resources has given rise to a number of socio-economic and environmental problems. Difficulties are encountered in the river based Water Supply Systems during rainy seasons by the poor water quality, while in the dry seasons by the scarcity of water. Large scale urban developments, quarrying activities, industrial wastes, increased use of pesticides and fertilizers and animal and poultry farming in the recent years have severely polluted the river system. Also, frequent flooding in the lower reaches of the Caroni River and the lack of proper irrigation and drainage facilities in its plains cause wide-spread damages in the agricultural sector and disruptions in social and economic activities. Furthermore, water-logging problems create poor enviornmental conditions resulting in the frequent outbreaks of Water-related diseases.

In this paper the major sources of pollution in the catchment of the Caroni River System are identified and measures for the control

[1]Lecturer in Water Resources Engineering, [2]Lecturer in Environmental Engineering, University of the West Indies, St. Augustine Trinidad and Tobago.

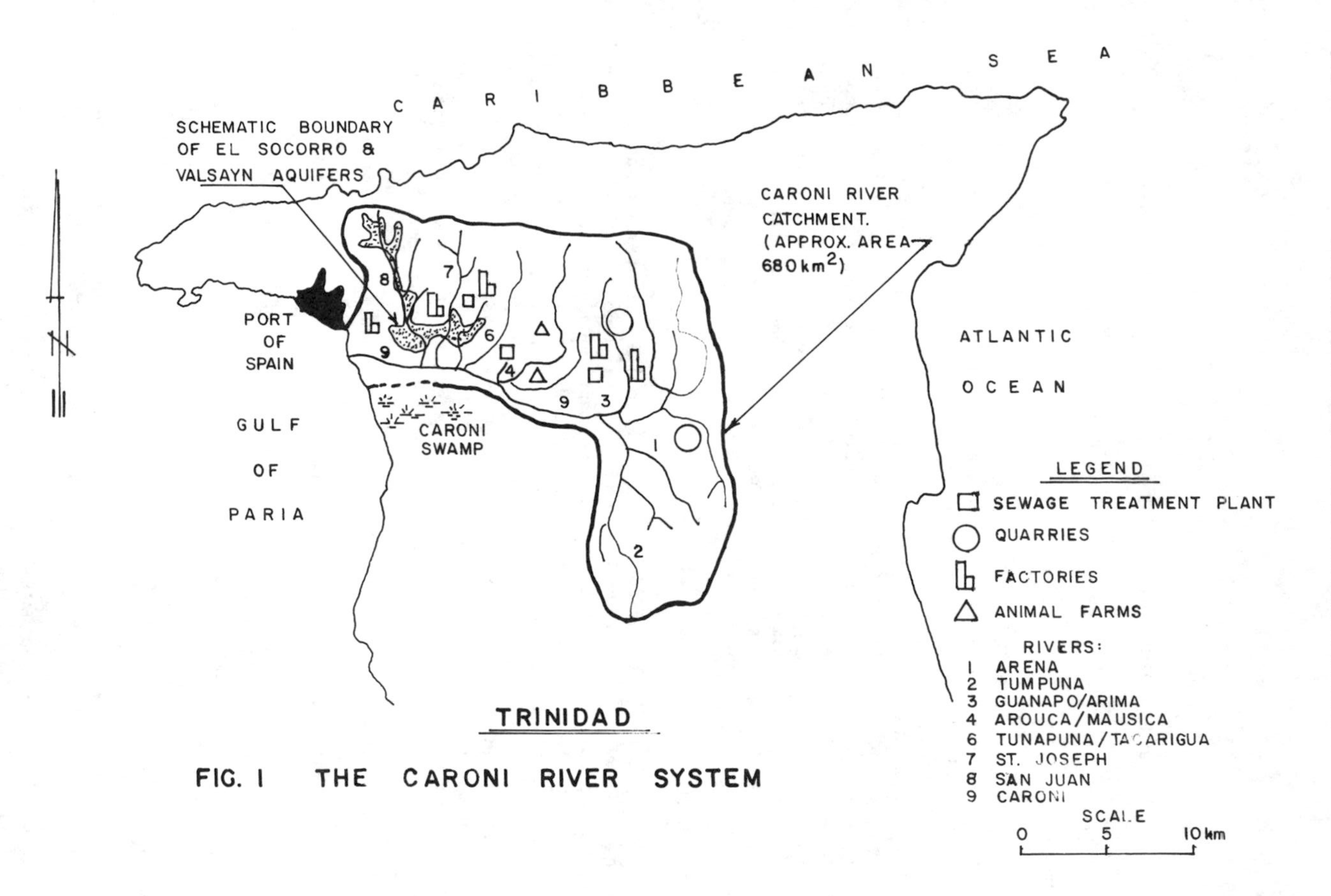

FIG. I THE CARONI RIVER SYSTEM

of the same are outlined. Also, the use of multi-purpose run-off detention basins is proposed for attenuating flood peaks and for improving the quality of river water by storing a number of small floods which have a high pollution load [6]. Such detention basins also improve the reliability of water supply during dry seasons and provide for irrigation as well. Finally an integrated water resources systems planning in the form of a linear programming model which incorporates the elements of water supply, flood control, irrigation, recreation and ecology is also formulated.

WATER QUALITY

The sources of pollution to the Caroni River System come from both the environment and the man. The pollution by the environment is due to the effects of heavy rains resulting in run-off from the land bringing silt, clay and organic matter into the river, thus increasing the turbidity and reducing the quality of water. The pollution of the Caroni River System by man takes the form of physical, chemical and bacteriological impurities and these are contributed by sewage effluents, quarries, industrial effluents and agricultural activities. Figure 1 shows major sources of such pollution.

The tributaries Mausica and Arouca receive sewage effluent, from Arima Sewage treatment plant, with a 5 day biochemical oxygen demand BOD_5 of 40 mg/l and a suspended solids concentration of 50 mg/l. A recent study [5] showed that BOD_5 and suspended solids concentration to the Arouca river from a badly designed stabilization pond system averaged 35 mg/l and 93 mg/l respectively together with a mean faecal coliform concentration of 2.4×10^6/100 ml. Also it has been observed that during dry season the flow in these tributaries were mainly the effluents from the respective treatment plants. Furthermore, samples taken at St. Joseph river, near its confluence with the Caroni River, indicated an average BOD_5 of 50 mg/l and a faecal coliform concentration of 4×10^5/100 ml. Quarrying and gravel-washing activities in the Arima district bring suspended matter and turbidity into the Caroni river via the tributaries. Also, high alkanity has been recorded in the Arima river. In the recent years large industrial estates have been constructed in the catchment of the Caroni River. Table 1 gives a summary of these industries.

TABLE 1: INDUSTRIES AND TRIBUTARIES RECEIVING EFFLUENT

Industry	Tributary Receiving Effluent
(1)	(2)
Meat Processing	Arima River
Animal and Vegetable Fats	St. Joseph
Soft Drinks	Tacarigua
Milk, Solids, Coffee	St. Joseph
Batteries	Arouca
Paints	Arima

THE PROBLEM

The absence of scientific management of Water Resources in the Caroni River Systems gives rise to a number of economic and environmental problems.

Frequent flooding in its lower reaches and the lack of adequate irrigation and drainage facilities in its plains cause millions of dollars worth of damages annually in the agricultural sector. In addition disruptions in social and economic activities, due to frequent flooding, are also considerable. Furthermore withdrawal and storage of water in its upper reaches has altered the pattern and quantity of fresh water inflow into the Caroni Swamp; whose ecological well being depends upon a delicate balance between tidal and freshwater inflows [1]. The ecological decay of the Caroni Swamp, which houses a Bird Sanctuary, becomes painfully obvious to anyone who has regularly visited this swamp over the years. Not only the flora and fauna have withered but the Scarlet Ibis bird population has also dwindled due to the deterioration of the natural environment apparently brought about, inter alia, by a change in the pattern and quantity of freshwater inflow.

It is also significant to note that the possible beneficial use of flood flows in recharging the Valsayn and El Socorro aquifers by the diversion and injection of such flows, and the increased conjunctive use of ground and surface water sources, especially during periods of high turbidity or low flows, have not yet taken place. It may be noted that a favourable time lag of approximately four months between the end of rainy season and recharge to the Northern Range aquifers makes the conjunctive use a practical proposition for improving the reliability of Water Supply and for controlling the waterlogging problem along the East-West Corridor [4].

Finally there is a need to construct detention basins at suitable locations on major tributaries for reducing the flood peaks and to provide for irrigation during dry spells. Such detention basins, apart from being effective sediment traps for reducing the turbidity of river water, can provide idyllic settings for recreation parks; with facilities for fishing, boating, picnics etc. It is to be noted that such parks are sadly lacking along the East-West Corridor.

MATHEMATICAL FORMULATION

For the purpose of mathematical formulation, the Caroni River System can be viewed as a network of streams, reservoirs and other components as shown in Fig. 2. The time unit of one month is used for reducing the magnitude of numerical work. The present problem formulation for the sake of simplicity, however, does not take into account the water quality aspects; which are of equal, if not greater, importance in Water Resources planning. It is therefore envisaged to unify these two aspects during the future enhancements of the mathematical model.

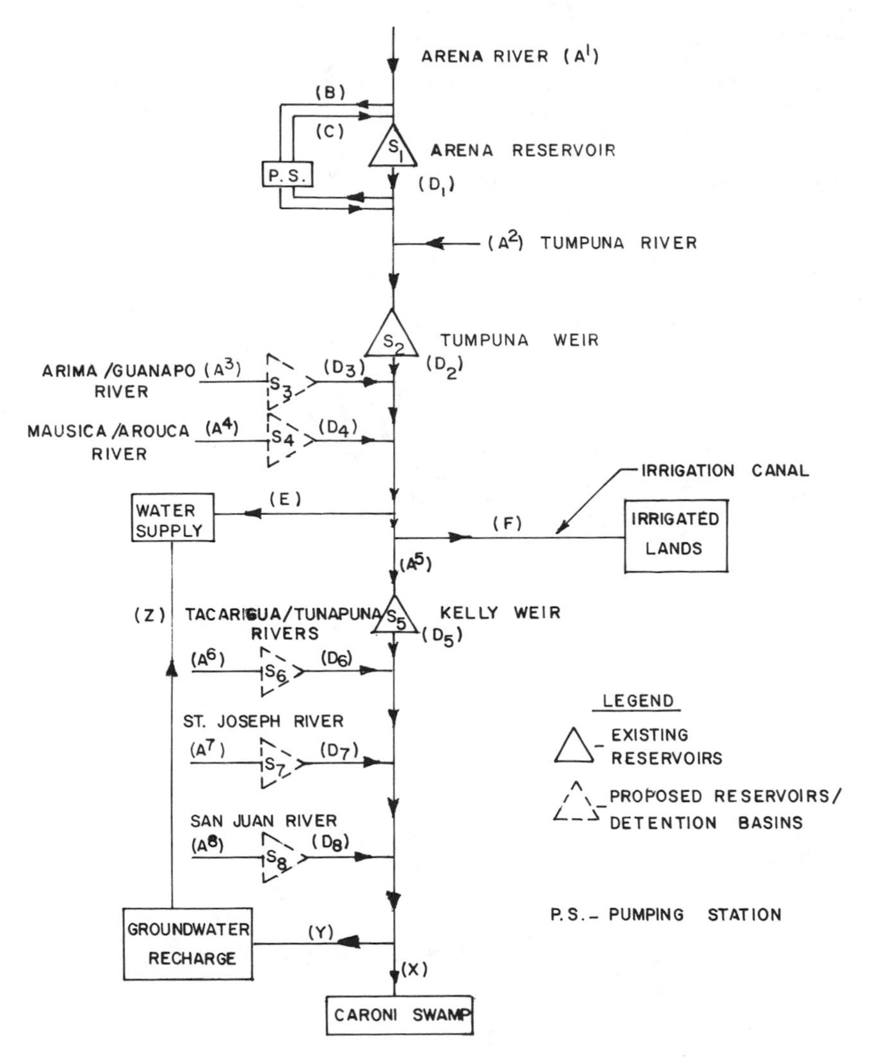

FIG. 2 SCHEMATIC OF THE CARONI RIVER SYSTEM. SYMBOLS IN BRACKETS DENOTE THE MONTHLY STREAMFLOW, PUMPING AND CHANGE IN STORAGE.

Assumptions

(a) The monthly streamflows follow the Markovian property i.e. the probability of any event is dependent only on the outcome of the preceding event [3].

(b) The probabilistic component of stream flows follow Log-normal probability distribution [2].

(c) The inter-relationships between various system parameters are linear.

(d) All system parameters represent time averaged monthly values.

(e) Infiltration and evaporation losses from the reservoirs are not considered for the sake of simplicity.

(f) The proposed detention basin on any tributary may be viewed as an equivalent reservoir representing, if necessary, a number of smaller linear reservoirs placed either in series or parallel - as dictated by the site conditions.

System Parameters

A = Streamflow into a Reservoir
D = Outflow
E = Flow Diversion for Water Supply
F = Flow Diversion to Irrigation
G = Change in Groundwater Storage
j = Month
k = Streamflow Number
l = Reservoir Number
n = Outflow Number
R = Natural Recharge to Aquifers
S = Change in Reservoir Storage
W = Areal Extent of Flooding
X = Outflow into the Caroni Swamp
y = Flow Diversion for Groundwater recharge
Z = Groundwater extraction from the aquifers
α = Economic Return Factor - Ecology/Recreation
β = Economic Return Factor - Irrigation
γ = Economic Return Factor - Water Supply
λ = Objective Function
μ = Economic Return Factor - Groundwater Recharge
ψ = Economic Penalty Factor - Flooding
ϕ = Water Demand for Public Water Supply

Linear Programming Model

$$\lambda = \text{Max} \sum_{j=1}^{12} [\alpha X + \beta F + \alpha E + \mu Y - \psi W] \quad \ldots . (1)$$

Subject to the following constraints:

$$X \geqslant X\ min \qquad \ldots . (2)\ \text{(Ecology Constraint)}$$

$$F \geqslant F\ min \qquad \ldots . (3)\ \text{(Irrigation Constraint)}$$

$$X + Y = \sum_{n=5}^{8} D_n \qquad \ldots . (4)\ \text{(Ground Water Recharge Constraint)}$$

$$E + Y = \phi + G - R \qquad \ldots . (5)\ \text{(Water Supply Constraint)}$$

$$W \leqslant W\ max \qquad \ldots . (6)\ \text{(Flood Control Constraint)}$$

and the non-negativity conditions

$X \geqslant 0$; $F \geqslant 0$; $Y \geqslant 0$; $E \geqslant 0$; and $W \geqslant 0$.

The nature of parameter W is shown by equations 7 and 8.

$$W = f\left(\sum_{n=2}^{4} D_n\right) + f'\left(\sum_{n=5}^{8} D_n\right) \qquad \ldots . (7)$$

$$\sum_{n=1}^{8} D_n = \sum_{k=1}^{8} A^k - \sum_{l=1}^{8} S_l \qquad \ldots . (8)$$

in which f and f' are functional symbols representing elevation - area-storage relationships in flood plains.

In equation 8, A^k is defined as:

$$A_{ij}^{k} = A_{j}^{-k} + e_{j}^{k} \frac{\sigma_j^k}{\sigma_{j-1}^k} A^k_{i-1,\ \overline{j-1}}\ A^{-k}_{j-1}$$

$$+ t_{ij}^{k}\ \sigma_j^k \sqrt{1 - e_j^k} \qquad \ldots . (9)$$

in which equation 9 is a stochastic streamflow generator and:

A_{ij}^k = Streamflow of Stream (K) in a given year (i) and in a given month (j)

σ = Standard Deviation of A^k

e = Serial Correlation Co-efficient of A^k

t = Log Normal Probability Distribution Parameter of A^k

A^{-k} = Mean of A^k

CONCLUSIONS

(a) The solution of the Linear Programming (LP) Model should provide guidelines for the design of detention basins and for the operation of reservoirs in keeping with the most efficient utilisation of Water Resources in the Caroni River System.

(b) It is important to note that the L.P. Model would also provide as a by-product, information on the areal extent of flooding with reference to the frequency of flooding; which is one of the prime requirements for instituting a scheme of flood risk insurance and in the determination of its premium.

(c) The problem formulation immediately suggests the need for a multidisciplinary team effort for the solution of the Water Resources Management Problem in the Caroni River System.

(d) The pollution of the Caroni River continues because of the design, operational problems and degrading performance of the sewage treatment plants, indiscriminate discharge of industrial effluents and increased agricultural activities in its catchment.

(e) The sewage treatment plants need to be upgraded and maintained for improved effluent quality. Also, stringent effluent standards and legislation for the prevention of pollution are required.

REFERENCES

1. Bacon, P.R., "The Ecology of the Caroni Swamp", Ph.D. thesis, University of the West Indies, St. Augustine, 1968.

2. Dillion, M.M., "Trinidad Water Resources Survey-Final Report", London, Ontario, Canada, 1969.

3. Linsley, R.K. et al, "Hydrology for Engineers", McGraw Hill, New York, 1982, p. 391-394.

4. Shrivastava, G.S., "The Optimisation of Pumping Operations in the El Socorro Aquifer", Ph.D. thesis, University of the West Indies, St. Augustine, 1977.

5. Thomas, P.R., and Phelps, H.O., "A Study of Upgrading Waste Stabilization Ponds", Water Science and Technology, 1987, Vol. 19, pp. 77-83.

6. Whipple, W., et al, "Stormwater Management in Urbanizing Areas", Prentice Hall, New Jersey, 1983, p. 151.

DEVELOPMENT OF NATIONAL WATER SUPPLY AND SANITATION PLANS IN AFRICA

Dennis B. Warner, Ph.D., P.E.*

Introduction

Safe potable water and sanitary disposal of human wastes for all were one of the primary goals of the United Nations Water Conference at Mar del Plata, Argentina in 1977. The action plan developed at this conference called for the establishment of an International Drinking Water Supply and Sanitation Decade over the period 1981 to 1990 to vigorously address the water and sanitation needs of over two billion people on Earth. A key tool in promoting the Decade was to be the establishment of national water and sanitation coverage targets and the preparation of country-level Decade plans.

In early 1981, at the start of the Decade, only nine countries had reported the development of Decade plans for their water and sanitation sector. By 1983, this number had grown to 59 country plans with another 31 under preparation. Although the pace of plan preparation was slower in Africa than in other regions, by the end of 1983 WHO reported that 18 countries had completed Decade plans, with another 13 plans in progress in the Africa Region.

This paper describes the role of the U.S. Agency for International Development (USAID), through its technical assistance arm, the Water and Sanitation for Health (WASH) Project, in helping countries in Africa to develop Decade-related plans. USAID was instrumental in developing the first national Decade plan (Sri Lanka) in 1980. This initial plan was the result of a high intensity effort by a large team of outside consultants and contained little direct input by host country officials. The work described in this paper, however, illustrates a highly collaborative approach, whereby the pace of planning activities and the nature of the planning products are directly determined by the capabilities and efforts of host country officials.

This approach and its outcomes are described for the countries of the Central African Republic, Zaire, and Swaziland.

Planning Process

When the WASH Project was asked in 1981 to assist in formulating a detailed Decade water and sanitation plan for the Central African Republic, WASH staff saw an opportunity to improve upon the classical master plan pattern in which outside consultants arrive in-country, quickly prepare a document, and then leave. Through its initial advisory experiences in various USAID-assisted countries, WASH staff saw planning as a process, rather than an event resulting in a document. Decade planning, therefore, should occur over the entire ten-year period and should result in a progressive strengthening of the planning capabilities of host government agencies and their representatives. Numerous examples existed of weak, ineffective national planning agencies

*Deputy Director, Water & Sanitation for Health (WASH) Project, Camp, Dresser, & McKee International, Inc., Arlington, VA

unable to control, or even influence, water and sanitation activities promoted by donor countries. As a result, national water and sanitation development too often was influenced by donor interests rather than national goals.

Because of the problems that arose from top-down planning, WASH saw the goals of Decade planning to be more than the production of a document. In fact, experience showed that the preparation of such a document by outside experts in a country with weak development institutions usually resulted in an inappropriate report, largely ignored by host country officials. The real purpose of Decade planning, therefore, had to be closely associated with local capabilities and local involvement (or "ownership") in the planning process. The objectives of an institutional strengthening approach rather than a document-oriented approach were several:

* to establish a sense of local ownership and responsibility for national planning and implementation

* to increase host government control over the development of its water and sanitation sector

* to attract donor resources to the water and sanitation sector

* to establish a sustainable national planning process

WASH adopted the following principles in its approach to Decade-related planning:

* Decade planning is a process of institutional strengthening rather than the production of one or more documents.

* The process involves assisting host country nationals to understand their water and sanitation needs, to decide what can be done about them, and to take the necessary steps to achieve these ends.

* The process, from the WASH standpoint, is non-directive and only advisory in nature.

* Host country personnel make all major decisions and carry out all planning activities.

* The pace and resulting timetable of the plan development process is dependent upon the interest and efforts of host country personnel.

* The long-term process of institutional strengthening is more important than the immediate production of technically-sophisticated plans and reports.

In applying these principles, WASH consultants were instructed to be advisors to the host governments and not primary authors of planning documents. They were to support and encourage the development of planning activities among host-country personnel but not to carry out the work themselves. Other general guidelines for consultants:

* Draw from experience, especially what has happened in the host country.

* Provide background information on important planning issues and concepts.

* Suggest approaches and, where necessary, alternatives but do not give <u>the</u> (consultant's) correct solution.

* Find ways to support host government decisions.

* Identify methods of strengthening the planning process.

In applying the above principles to Decade planning assignments in Africa, WASH consultants generally followed a similar pattern of planning-related events in all three countries. This pattern, with some variation between countries, generally was as follows:

1. Establish an interministerial national action committee.

2. Formulate national policies and strategies for water and sanitation development.

3. Hold a national seminar on the policies and strategies to inform all parties of official objectives.

4. Prepare a short-term action plan at the conclusion of the national seminar.

5. Establish a planning unit for national water and sanitation development.

6. Carry out longer-term planning activities (data collection, donor coordination, project proposals, sector plans, etc.).

Depending on the state of planning capabilities in a given country, the above sequence of activities has taken from one to four years after the initial WASH visit. Each country has decided its own pace and the rate at which planning activities are to be undertaken. The pace of events, however, is not considered to be as important as the actual institutional development of host country planning capabilities.

Three countries have requested WASH assistance in Decade planning activities. These countries and the dates of initial WASH assistance are the Central African Republic (1982), Zaire (1984), and Swaziland (1985). The following sections summarize the Decade planning process and the corresponding WASH involvement in each country.

Central African Republic

At the start of the International Water Decade in 1980, the Central African Republic was beset by serious water supply and sanitation problems. Only four towns in the entire country had piped water systems, while none had sewerage systems, and improvements in the rural areas were limited to a few wells and springs. In general, conditions were deteriorating for lack of resources and the Government of the Central African Republic (GOCAR) had no effective water and sanitation programs to build upon.

UNDP and other UN-affiliated agencies worked for almost two years to convince the GOCAR to establish an interministerial National Action Committee (NAC) to coordinate water and sanitation activities in the country. By mid-1982, these efforts began to show some success. In July, the GOCAR drew up a draft decree for the establishment of an NAC and requested WASH Project assistance in formulating a detailed Decade water supply and sanitation plan similar to one that USAID had prepared for Sri Lanka in 1980.

USAID and WASH responded to the GOCAR request by sending a consultant to assess the situation and define the nature and extent of subsequent assistance needed to develop a Decade plan. The consultant visited the CAR in August 1982 and concluded that GOCAR planning capabilities were limited by inadequate human, technical, and financial resources and a lack of coordination among institutions having water and sanitation responsibilities. Rather than outline the donor resources needed to prepare a Decade Plan, he recommended the following five-step process designed to strengthen GOCAR planning capabilities and progressively mobilize the water and sanitation sector in the CAR:

1. Establishment of a National Action Committee.

2. Preparation of a water and sanitation strategy document.

3. Official GOCAR adoption of the strategy documents.

4. Water supply and sanitation seminar.

5. Implementation of initial project development activities.

The GOCAR accepted these recommendations and, with expectations of subsequent assistance from USAID, began to implement them. In September 1982, the NAC was formally established by presidential decree. Four months later (January 1983), a Technical Subcommittee was appointed to work with a WASH Team on the preparation of draft policies and strategies for the water and sanitation sector. The policy and strategy document was completed in February and formally approved by the Council of Ministers in April. In May, the NAC held a national seminar on water and sanitation strategies for technical and administrative officials from different parts of the country. The purpose of this seminar was to inform all relevant GOCAR officials and representatives of donor organizations of the new policies and to solicit their participation and support in future planning and implementation activities. At the conclusion of the seminar, the Technical Subcommittee prepared a short-term action plan to guide subsequent activities and set three-year coverage targets for the period 1982-1985.

To assist the NAC in establishing the necessary planning organization, a WASH consultant spent November 1983 in the CAR the Technical Subcommittee helping to plan a secretariat and a permanent Planning Bureau. This bureau was authorized by the GOCAR in January 1984 and staff were appointed in September 1984. Meanwhile, UNDP provided funds for the construction of offices; a new State Secretariat for Hydraulics was created to be responsible for water supplies, and the NAC began to hold coordination meetings with WHO, the World Bank, and other donors. WASH consultants returned in September 1984 to review progress and determine further needs.

In January 1985, the Planning Bureau of the NAC began to collect field data in anticipation of an international donors' meeting. A WASH consultant spent the summer of 1985 working with the Planning Bureau and with the State Secretariat for Hydraulics preparing project proposals for donor support and in outlining planning activities for the remainder of the Water Decade. These proposals were presented at the donors' meeting held in Geneva in November 1985. A year later, in October 1986, UNDP and the planning bureau began formulating the terms of reference for a detailed rural water and sanitation master plan. This plan will be developed over an 18-month period beginning in mid-1988

Zaire

National planning is relatively new in Zaire. The first attempt at an overall national economic plan was the Mobuto Plan, which covered the period January 1981 to September 1983. This attempt was followed by another intermediate economic plan for the period October 1983 to December 1985. In 1985, the Government of Zaire (GOZ) began work on its first Five-Year Development Plan for the period 1986-1990.

National planning for water supply and sanitation is also very new in Zaire. Until recently, little was done to coordinate the GOZ organizations working in the sector. Urban water supply is under the jurisdiction of REGIDESO, an autonomous public enterprise whose origins date back to 1939. In 1984, REGIDESO began preparing a national urban water supply plan as input to the impending Five-Year Development Plan of 1986 to 1990.

Rural water supply, on the other hand, is the responsibility of the National Rural Water Service (Service National d'Hydraulique Rurale, or SNHR), which began operating in 1978 in a limited number of rural areas with material support from UNICEF. Between independence in 1960 and 1978, rural water activities were almost non-existent, and only some missions and non-governmental organizations (NGO's) were able to maintain their efforts. SNHR was formally reconstituted as a national service in September 1983 under the Ministry of Agriculture and Rural Development and given responsibility for the provision of potable water in rural areas. One of the first activities of the new SNHR was the preparation of a list of projects for a five-year program expansion over the period 1983 to 1988.

Sanitation responsibilities are divided between the Department of Public Health and the Department of the Environment, Conservation of Nature, and Tourism. The National Sanitation Program (Programme National d'Assainissement, or PNA) was created in 1981 to be responsible for planning, coordination, execution, and evaluation of sanitation activities affecting public health. In practice, the PNA has concentrated on urban centers and has had little impact on rural areas, although in 1985 it did prepare a national program of activities. The Department of Public Health has limited responsibilities for rural sanitation. In 1982, a Division of Primary Health Care (Direction des Soins de Santé Primaires, or DSSP) was established within the Department of Public Health, with authority to integrate water and sanitation activities within rural health zones. At present, the DSSP is most active in those rural health zones where it can work closely with the USAID-financed SANRU primary health care project.

In February 1981, the GOZ created an interministerial National Action Committee (Comité National d'Action de l'Eau et de l'Assainissement, or CNAEA) to be responsible for all planning, coordination, and execution of development and rehabilitation programs in the overall water and sanitation sector. The CNAEA immediately set a target of 35 percent rural and 70 percent urban coverage, respectively, for potable water supply by the end of the Water Decade in 1990.

Between 1981 and 1984, the CNAEA was involved in a number of planning-related activities, including the organization of an international donors' conference in Kinshasa in February 1983. A background report, prepared by WHO for this meeting, was the first attempt in Zaire to compile overall data on existing resources and projected needs.

In early 1984, the CNAEA requested assistance from USAID in preparing a national rural water supply plan. A WASH Project team visited Zaire in September 1984 and recommended a process of institutional development and planning-related activities leading to the eventual preparation of a national plan.

The process recommended by WASH consisted of five major steps, as follows:

1. Formulation of rural water policies and strategies.

2. Presentation of policies and strategies at a national seminar.

3. Establishment of a permanent planning unit.

4. Preparation of a national rural water plan.

5. Preparation of an overall national water supply plan.

In October 1984, the CNAEA accepted the WASH recommendations, but broadened the area of concern to include rural sanitation as well as rural water supply. Step 1 of the foregoing process was begun in January 1985 when a two-person WASH team assisted a technical subcommittee of the CNAEA in preparing a draft of national policies and strategies in the rural water and sanitation subsector. Step 2 occurred when these policies and strategies were formally reviewed by all branches of the GOZ at a national seminar held in Kinshasa in May 1985. A WASH consultant assisted with the design and organization of the seminar. The revised policies and strategies developed in the seminar were officially approved by the CNAEA in October 1985.

Step 4 of the process was initiated in July 1985 when the GOZ requested USAID/WASH assistance in preparing the rural water and sanitation plan. (Step 3, the establishment of a permanent planning unit, has not yet been implemented, although the GOZ has decided to establish such a unit in SNHR. The planning activities undertaken to date on the rural water and sanitation plan have been under the general direction of the CNAEA.) To develop the rural water supply and sanitation plan, the CNAEA, in mid-1985, set up a subcommittee of senior experts from the various GOZ departments concerned with rural water and sanitation. This subcommittee was assisted in its work by representatives of WHO, UNICEF, and the SANRU project.

With financial assistance from USAID, the CNAEA subcommittee carried out a series of field investigations in all eight rural regions of Zaire from October to December 1985 in order to obtain basic information for the rural plan. A WASH team then visited Zaire in January 1986 and again in April 1986 to assist the CNAEA in defining the elements of the plan and in reviewing preliminary plan drafts.

By June 1986, the committee had completed the plan for the rural sector. This document was promptly reviewed and approved by the National Action Commitee (CNAEA) and then combined with an already-existing urban water and sanitation plan to form a national water and sanitation plan for both rural and urban areas. By the end of 1986, the national water and sanitation plan was formally incorporated into the Five-Year Development Plan for 1986-1990. These planning efforts, culminating in the incorporation of the water and sanitation plan into the Five-Year Development Plan, marked the establishment of the first official national plan ever developed in Zaire for the water and sanitation sector.

Swaziland

Swaziland has shown interest in national water and sanitation planning since March 1977 when it participated in the United Nations Water Conference at Mar del Plata. In March 1979, following a WHO/SIDA-sponsored mission, it was proposed that a National Action Group (NAG) be established and administered by the Rural Water Supply Board (RWSB). This proposal was requested in a Cabinet paper in July 1979 and approved later that year.

During its first few years, the NAG presided over a number of studies, including the 1980 Decade Country Report, a survey of manpower needs and a plan for manpower development, and several technical and health-related investigations. Between late 1981 and mid-1984, however, a series of ecological, political, and administrative events caused a decline in Decade-related activities of the NAG. Nevertheless, the NAG was able to initiate some sector planning activities, including an analysis of projected costs for water and sanitation targets for both rural and urban areas of Swaziland through the year 2001, a draft plan for the water and sanitation sector, and a review of the National Plan for Zambia by various Swaziland officials.

In August 1984, a working subgroup was established to be responsible to the NAG and to assist the Senior Engineer of RWSB, who serves as the Secretary of the NAG, in gathering and analyzing background data and in drafting a sectoral policy and a national action plan. In November 1984, the Government of Swaziland requested assistance from the WASH Project to assist in developing a sector plan.

Although the NAG had not met for some time, by August 1985 government officials had prepared a draft outline for a revised Decade Plan. In addition, partial sections of the Fourth National Development Plan 1983/84 - 1987/88, were available from the appropriate ministries along with other planning documents.

A two-person WASH team visited Swaziland in September 1985 and worked with the Technical Subgroup on the design of a program of national water supply and sanitation planning. The consultants recommended the following steps as input to the development of a Decade plan:

1. Formulation of national policies and strategies for water supply and sanitation development.

2. Presentation of policies and strategies at a national seminar.

3. Preparation of a short-term action plan.

4. Preparation of the water and sanitation component of the Fifth Five-Year National Development Plan, 1988/89 - 1992/93.

These recommendations were accepted by the GOS. The Technical Subgroup then began to collect background information on existing policies, strategies, and programs in the water and sanitation sector.

In February 1986, a WASH team was sent to Swaziland to assist the Technical Subgroup with the preparation of a national policy and strategy document. Rather than submit the document directly to the Cabinet for approval, the NAG decided to first review and discuss the proposed policies and strategies at the national seminar. A WASH consultant returned to Swaziland to assist the Technical Subgroup with the design and operation of the seminar, which was held in June 1986. Over sixty participants and observers from government, the private sector, NGO's, and the donor community attended the seminar.

Following completion of the national seminar, another WASH consultant worked with the Technical Subgroup during July 1986 on the preparation of a Two-Year Action Plan for the period for 1987/88 - 1988/89. This document contained proposed water supply and sanitation activities for the urban, peri-urban, and rural subsectors. It also recommended that a five-year sectoral development plan be prepared during the two-year period. The Action Plan was reviewed extensively by government and revised several times by the Technical Subgroup over the next four months. One major revision to the plan was that the Technical Subgroup take responsibility for the preparation of rolling three-year capital development plans for the sector. In April 1987, a final version of the Two-Year Action Plan, along with the policies and strategies document, was sent to the NAG for official approval.

At present, the Technical Subgroup continues to meet on a regular basis, as it has since July 1986. It intends to begin the development of a Four Year Master Plan in the near future with assistance from USAID, the World Bank, and UNDP. The overall experience of the NAG and the Technical Subgroup in Decade planning activities already has provided several favorable results, including greater coordination between GOS agencies, increased donor funding of water and sanitation programs, and growing interest among GOS officials in copying the interministerial Decade planning mode in other development sectors.

Conclusions

The Decade planning approach adopted by the WASH Project puts primary emphasis upon involvement and decision-making by host country officials in all planning-related activities.

Based on experience in the Central African Republic, Zaire, and Swaziland, this approach has been enthusiastically accepted by national officials responsible for water and sanitation development. The overall time necessary to develop specific planning documents takes longer with this approach than the conventional method of producing plans with outside consultants, but local institutions are stronger, a sense of ownership of the plans is imparted, and the long-term sustainability of the planning process is enhanced as a result of this process. Other factors which sometimes arise from the above approach are increased donor involvement in the water and sanitation sector (due to better organization of the sector and more efficient management by the host government) and restructuring of the water and sanitation sector (in terms of the creation of new organizations for water and/or sanitation development).

The most important outcome of this approach, however, is that the host government gains greater control of water and sanitation development in its own country. This is not to say that external donor involvement is no longer needed. Indeed, the establishment of effective Decade planning processes should provide more opportunities for useful donor cooperation, but in areas and for reasons determined by the host government. Thus, strengthening national planning capabilities in host governments provides them with corresponding power and responsibility over the health and welfare of their citizens.

Appendix - References

World Health Organization, International Drinking Water Supply and Sanitation Decade: Briefing Document, EHE/80.8, 27 February 1980.

United Nations, Progress in the Attainment of the Goals of the International Drinking Water Supply and Sanitation Decade: Report of the Secretary-General, A/40/108,E/1985/49, 6 March 1985.

Central African Republic

WASH Field Report No. 53, Recommendations for Initial Water and Sanitation Decade Planning Activities in the Central African Republic, September 1982.

WASH Field Report No. 72, Formulation of National Rural Water Supply and Sanitation Strategies in the Central African Republic, March 1983.

WASH Field Report No. 97, A Seminar on Water Supply and Sanitation Strategies in the Central African Republic, July 1983.

WASH Interim Report No. 106-1, Assistance to the National Committee for Water and Sanitation: Central African Republic, December 1983.

WASH Field Report No. 137, Progress in the Establishment of Water Decade Committees in the Central African Republic, January 1985.

WASH Field Report No. 158, Central African Republic: Identification and Formulation of Water Supply and Sanitation Projects, December 1985.

Zaire

WASH Field Report No. 135, Recommendations for National Rural Water Supply Planning in Zaire, January 1985.

WASH Field Report No. 142, Formulation of National Rural Water Supply Policies and Strategies in Zaire, May 1985.

WASH Field Report No, 150, National Seminar on Rural Water Supply Policies and Strategies in Zaire, June 1985.

WASH Field Report No. 171, Design of a National Rural Water Supply and Sanitation Plan in Zaire, February 1986.

WASH Field Report No. 184, Preparation of a National Rural Water Supply and Sanitation Plan in Zaire, May 1986.

Swaziland

Unpublished Reports, WASH Project Files, 1985-1986.

CAN THE WATER DECADE STILL REACH ITS TARGET BY 1990 ?

B. Z. Diamant *

Abstract

The International Drinking Water Supply and Sanitation Decade (1981-1990) is approaching its last stages with quite moderate results and humble prospects. Investigations have shown that more people are living now without safe drinking water, and in particular, without proper sanitation, than those who lived so at the start of the Decade. The blame for this falls on the wrong approach adopted in the execution of the Decade during the last 6 years, mainly in 2 major issues : 1) Too much attention has been allocated to the small privileged urban sector, while neglecting the large rural sector, and 2) most solutions suggested and practised for the Decade's problems, were not based upon appropriate technology adjusted to existing conditions in the developing countries, for which the Decade has been mainly meant. Nevertheless, it is not yet too late to reach the Decade's target of providing safe water and proper sanitation to all by 1990, by immediate shifting from costly and non-practical solutions to appropriate traditional local practices, such as improved dug wells equipped with manual pumping means, development of roof catchments, using simple purification systems like clay-pot chlorination, and embarking on mass-production of simple, but efficient, human waste disposal means, made of local cheap materials, that rural people can afford to obtain and construct. Most of the Decade's activities must concentrate in the rural areas, where more than three quarters of Third World people live. Along with these urgent short-term activities, the planning of long-term more sophisticated solutions can be continued for future development.

Introduction

The poor state of drinking water supply and sanitation in the developing countries has been a known fact, but the alarming extent of the situation has been revealed for the first time by the findings of the Global Survey on Water Supply and Wastewater Disposal carried out by the World Health Organisation during the previous decade. An Interim Report on the Survey, covering the first half of the decade (1970-75) was released in 1976 (WHO,1976). The Report revealed that in 1975

* Professor of Environmental Health Engineering. 24 Barkai Street, Ramat-Gan, Israel.

3 out of 4 developing countries people had access to safe drinking water in the urban areas, but less than 1 out of 5 in the rural areas. The sanitation situation was found to be by far worse, when less than 3 out of 4 urban people enjoyed proper sanitation facilities, against less than 1 person out of 8 enjoying similar facilities in the rural areas. These were very severe findings, considering the fact that nearly 3/4 (72 %) of the developing countries people were living in rural zones, which in the average left 2/3 and 3/4 of the Third World population without safe water and adequate sanitation respectively in 1975. The Interim Reportcontained also optimistic predictions for the end of the Survey period (1980), according to which 9 out of 10 urban people would have by then safe water while only 1/3 of the rural population would be covered at the same time. In sanitation the privileged urban people would enjoy a 9/10 coverage, while the rural people - less than 1/4 coverage.

But even these humble predictions did not come true, and survey carried out in 1981 (World Water 1981) showed for 1980 a 3/4 only of urban water coverage and a 1/4 rural water coverage. In sanitation the 1980 urban coverage was about 1/2 and the rural coverage only 1/7. In the average these findings showed that in 1980 less than every second person and lees than every fourth person in the Third World were living without safe drinking water and proper sanitation respectively. Quite disappointing achievements against the predictions.

The WHO Global Survey findings had a strong impact on the U.N.Water Conference held in La-Plata, Argentina, on March 1977, and in its resolutions the Conference requested "the commitment of all national Governments to provide ALL PEOPLE with water of safe quality and adequate quantity and basic sanitary facilities by 1990." This resolution was the basis for the International Drinking Water Supply and Sanitation Decade 1981-1990 (IDWSSD), approved and launched by the General Assembly of the United Nations, in a special session held on 10 November 1980. The goal of the IDWSSD was "the provision of clean water and adequate sanitation FOR ALL by 1990."

The Set Back

The humane goal of the IDWSSD known as the Decade, of providing safe water and proper sanitation for all people, was soon found to be too ambitious. A short time after the Decade's activities have started, the WHO being the U.N. coordinator for these activities, has announced a modified more humble target, according to which the following rates of coverage were expected for 1990 : No change in the urban water supply coverage, which would still be standing on 100 % coverage. Urban sanitation would be reduced to 80 % coverage, and the biggest cuts were meant for rural water supply and sanitation, both reduced to 50 % rates of coverage. The rural areas of the developing countries, where the real core of the problem lies, has had the worse set-back. (NCE,1982). But this was not the end of set-back. In his address to the 39th World Health Assembly, held in Geneva on 21 March 1986, the Director-General of WHO, had set up new targets for the Decade, as indicated in Table-1. (IRC,1986).

Table-1. New Coverage Rates for the Decade

Year	Water Supply		Sanitation	
	Urban	Rural	Urban	Rural
1980	72 %	32 %	54 %	14 %
1983	76 %	33 %	58 %	15 %
1985 *	77 %	36 %	60 %	16 %
1990 *	79 %	41 %	62 %	18 %

* Predictions

Rural sanitation has been mostly affected by the new coverage rates, it has suffered severe set-backs, from 100 % coverage in 1980, to 50 % covergae in 1982, and again to only 18 % coverage in 1986. It should be noted that lack of proper rural sanitation affects nearly a third of the whole world's population. Poor sanitation has a direct affect on the quality of drinking water. The Decade's problem in the Third World is very seriously severe, as can be realised from the following example : According to WHO records (WHO,1980), out of the 13.6 million children up to age 5, that died in the world in 1979, 13.1 million were from the developing countries. Most deaths were caused there by water-borne diseases and "could have been, therefore, potentially prevented."

The irony involved in the whole matter lies in the sad fact that although statistically improvements can be noticed in respect of percent-coverage, in real figures these immaginable advances turn to be quite retreats. Investigations have shown that the number of people lacking safe drinking water has risen by 100 million, between 1975-1981. The number of people living without proper sanitation has risen by 400 million during the said period (Agarwal,1982). This dangerous trend has been confirmed by a recent WHO press release (Water World,1987), stating that "slightly more people will be without water in 1990 than at the start of the Decade, and approximately 200 million more will have no access to appropriate means of disposal of faecal wastes."

The Wrong Approach

Water and sanitation are considered to be major corner stones in the structure of the human environment. No development of any kind can be accomplished without these two primary factors being properly treated and adjusted to human needs. The set-back in achieving the Decade's target affects, therefore, the whole development of the human environment. This set-back can be attributed to the wrong approaches practised in the various stages of executing the Decade. These approaches include first of all the separation of the water aspect from the sanitation and the constant allocation of inferior attention to the latter, although the two are closely interrelated and complement each other (Diamant,

1985). The gap between the water coverage and the sanitation coverage is constantly growing along the periodic development stages, in particular in the crucial rural sector, as indicated in Table-1. This separation has been strongly criticized by the Director-General of WHO at the very early stages of the Decade. He was "very startled to read that safe drinking water has been separated from sanitary waste disposal and that the population coverage target for waste disposal was only half that of water." (Mhaler,1981). Despite this criticism the gap still exists and even growing wider. It seems that efforts must have been made to try and reduce this gap, but they must have failed, because, as the director of the WHO Environmental Health Division in Geneva, said "sanitation occupies a secondary place in targets set by developing countries and drinking water still commands by far the greatest attention." (Dieterich,1982). This attitude must and can be changed, whether by better health education practices, or rather by re-directing the allocation of international funds towards this end.

The small urban sector in the developing countries is by far more developed and privileged than the huge rural sector. Nevertheless the majority of the Decade's efforts and funding are being invested in the former. This wrong approach can be justified by various reasons, all of which can not stand a fair trial. It is true that the influential people live in the urban areas, it is easier to operate in these areas, projects in these areas are more "bankable" etc etc. If the Decade is expected to reach its early targets, all efforts must be shifted as soon as possible to the rural sector, because here lies the core of the problem awaiting the coverage. This shifting can and must be done by the relevant U.N. agencies involved in the Decade's development. A glance at the World Bank's list of loan allocations for water and sanitation, reveals instantly that most funds were approved for "bankable" projects, located ofcourse in the urban areas. This approach must be entirely revised towards the final stages of the Decade.

The small Decade activities that take place in the rural areas of the Third World, have been also affected by the wrong approach. Most solutions here are practised and executed, while lacking the important principles of appropriate technology. Therefore, although being technically sound and efficient, these solutions are disconnected from reality and from the prevailing conditions in the areas of development. In the water supply field, international organisations with the support of national local bodies, are operating sophisticated costly neans of development, such as rigs and mechanical pumps, that require costly spares which are not available, as well as fuel and power which are also not available, in particular in the rural areas. Due to these reasons, the actual coverage is almost negligible, and the extremely costly operations are restricted to the performance of "demonstration projects". But these are, ofcourse, of no value, because no rural person or even a community in the developing countries, can afford to follow the demonstrated example and drill a borehole with a rig, or purchase and maintain a diesel operated mechanical pump. On the other hand the huge funding necessary for mass solutions of this kind, will never be available by any national or international organisations. A complete change of approach is, therefore, required in this field if the Decade's target is to be reached at all.

Appropriate solutions for rural sanitation have to be more simple and economical than similar solutions for rural water supplies. The latter are normally meant for a whole community or a group of houses, where more efforts and contributions can be collected. A human waste disposal device should not be shared by more than one family, in order to prevent maintenance problems. Solutions must be, therefore, cheap and simple so that a common rural person can afford to build one. Unfortunately, rural sanitation solutions adopted in recent years, do not fall in line with the appropriate requirements. Here again,the adopted solutions have been developed upon a demonstration basis. But due to their costs and complexity, ordinary rural people can not afford to build them.

Two major types of human waste disposal devices have been developed in recent years, mainly by international organisations. The more common one is the Ventilated Improved Pit, known as the VIP. This device incorporates, in addition to its main purpose as a human waste storage, also a smell and fly control means. The other type is more costly and complicated and is known as the Permanent Improved Pit, or the PIP. It operates like the VIP, but it can be also emptied. Both devices are technically sound, but they are costly and can not be built by the future owner without hired labour and costly materials. A cost analysis has been carried out on a UNICEF assisted VIP and PIP project in Imo State in Nigeria (Iwugo,1982). The analysis showed that a VIP unit cost US $ 352 and a PIP unit - US $ 522. According to the above-mentioned recent coverage rates for water and sanitation (IRC,1986), some 1,500 million rural people were living without proper sanitation facilities. Assuming 6 people per family, then this figure is related to 250 million families that need 250 million human waste devices. If the VIP is selected for the solution then a sum of US $ 88 billion will be needed for the performance of the solution. if, however, the PIP is selected for this purpose, then the required sumwill increase to US $ 130 billion. These are, ofcourse, not appropriate solutions at all. Nevertheless, dozens of demonstration projects of VIPs and PIPs are carried out in numerous developing countries. Each project is able to build several hundreds or even thousands of units, when actually many millions are required.

Some efforts are being invested even in more sophisticated solutions such as the Aqua Privy, where costly impregnated concrete structures are incorporated in the device. Such solutions are even less appropriate than the VIP and the PIP.

Appropriate Rural Water Supplies

Demonstration projects can be carried out by local Governments or by international organisations. When mass production of devices for millions of people is involved, then only the people's efforts can raise, with the aid of the authorities, this heavy load. In view of the relatively short time left for the end of the Decade, and considering the tremendeous needs for billions of people, two appropriate solutions are suggested by this Paper for the rural water supply problem - the dug well and the roof catchment. Both can easily be geared into mass-production and serve as an urgent immediate solution at present.

The digging of wells has been an old tradition in all developing countries. Most dug wells have a circular shaped cross-section with a diameter of 2 metres and a depth of about 10 metres, though some dug wells can reach double that depth. Dug wells are considered to be shallow wells, and as such are liable to easy contamination. There are, however, some economical preventive means to protect the well, that are listed below. A protected dug well can produce relatively safe drinking water if the maintenance procedure is properly followed, which will depend mainly on efficient health education among the users. Since the know-how of digging wells is available in the rural areas, and the costs are low and mainly in kind rather than in cash, it is possible to embark on mass production of these water supplies, in respect of hundreds of thousands and even millions.

The protective means, that can be introduced also in existing badly maintained shallow wells, include the digging of a small drainage trench around the well to protect it running rain water ; the building of a low fence around the well, for protection from surface contamination and to protect children from falling in ; the use of one rope and bucket only. No other private ropes and buckets should enter the well ; a wooden pulley on a wooden structure can be fitted above the well to facilitate the pulling of the rope. A trough can be fixed under the pulleyand a catch that turns over the bucket into the trough when it reaches the top, leading hence the water to the user's private container without anyone touching the well's bucket ; the disinfection of the well by means of clay-pot chlorination. This exists of a small clay-pot filled with a mixture of sand and bleaching powder and closed with wax. The pot is lowered in the water with a connected rope and the chlorine penetrates slowly through the clay in the water in small quantities sufficient for periods up to 2 weeks, when the pot has to be replaced. Later stages of improvement can include the cover of the well with a concrete slab and the installation of a hand pump. There are numerous kinds of hand pumps, and the appropriate choice has to rely on simplicity and on the use of improvised spares, when such are not available, for example, a possible replacement of a broken metal handle with a locally made wooden handle.

The dug well campaign has to be organised by the relevant local authority, such as the local health office, and can be assisted by international organisations. The organisers have to provide the administration, small running expenses, for the purchase of digging tools and for the common rope and bucket, and most important of all - the provision of health education activities that have to precede and then follow the construction and maintenance stages.

Roof catchment water supplies comprise small quantities, usually beyond the regular domestic needs, and can be sufficient, therefore, only for drinking purposes. Being clean rain water, it is entirely pure and safe for drinking. The development of roof catchments requires, quite often, the replacement of an existing straw roof with a corrugated sheet roof. In such cases two targets are achieved simultaneously : the provision of safe drinking water and the improvement of housing (Diamant,1982). The organisation of a roof catchment campaign is similar to that of the dug well, including the incorporation of health education.

Human Waste Disposal

Wastewater disposal consists in the rural areas of developing countries almost entirely on human waste disposal, due to lack of running water in the houses. A human waste disposal device for these areas is, as a matter of fact, a storage for the waste, preventing it from being carried away by the rainwater to the water courses. Every family must have its own device, because sharing these devices raise serious cleaning and maintenance problems. Therefore, the device must very simple and economical, made of locally available materials and can be built by the user without hiring labourers. The cost of the device has to be mostly in kind so as to enable a full-scale mass production. The most applicable human waste disposal device, in this respect is the Ishara Pit Latrine which was developed in the rural location of Ishara in Nigeria (Diamant,1978).

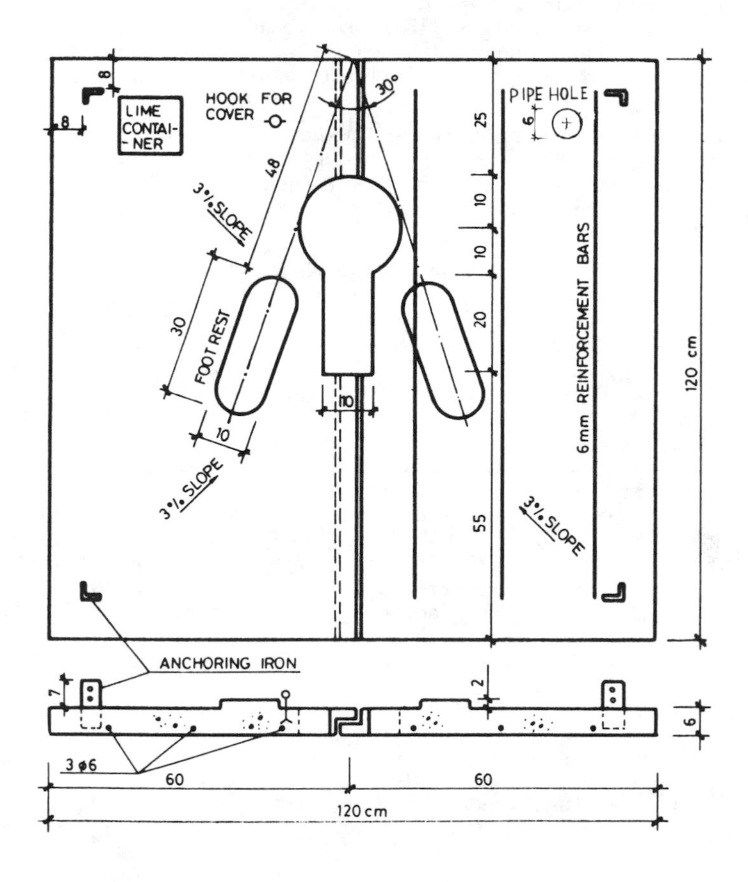

Fig.1 The Ishara Pit Latrine concrete slab

The device consists of 3 components : the dug pit having a square cross-section of 1 x 1 m and a depth of 2.50-3.00 m. ; the super-structure, made of a skeleton of raw wood covered with home-made woven mats. The purpose of the structure is to provide privacy for the users, most of whom live in hot climates. The cost of the structure is, hence, just in kind; the third and most important component is the concrete slab. The design of the slab includes considerable improvements which do not increase its price, since the improved mould is to be provided by the organisers of the campaign. The features include a key-shape outlet, slopes towards the outlet, elevated foot rests, two half slabs connected by tongue and groove system, anchors in the 4 corners for binding the super structure etc. (Fig.1). All expenses are in kind apart from the cost of half a bag of cement required for the casting, and probably few reinforcement rods, though bamboo canes can be used instead.

Due to its simplicity and low cost the Ishara device can be mass-produced by organised campaigns. It is recommended to standardized metal moulds for mass production purposes. The concrete slab is long-lasting and can be removed from a full pit to a newly dug one.

The above mentioned suggestions are meant as short-run emergency measures to meet the Decade's target. Along with these urgent activities, long-term planning can continue with sophisticated more costly devices for future development stages.

References

AGARWAL, A. et al, 1982. Water, Sanitation, Health - For All ? Earthscan Publication, London.

DIAMANT, B.Z.,1978. Pit Latrines. In : Sanitation in Developing Countries, (Aditor A. Pacey).Chap.6. Wiley & Sons, Chichester, UK.

DIAMANT, B.Z.,1982. Proc. Intern. Conf. on Rainwater Cistern Systems. Univer. of Hawaii, 15-18/6/82. Honolulu. pp 276-83.

DIAMANT, B.Z.,1985. The Plight of Rural Third World in the Decade Era. Journal AQUA, No.2 1985. London. pp 70-76.

DIETERICH, B.H.,1982. The Water Supply and Sanitation Decade. Proced. Intern. Water Supply Assoc. Congress. Zurich, 6-7/9/82.

INTERNATIONAL REFERENCE CENTRE,1986. Newsletter No.163 (Sep.'86. The Hague.

IWUGO, K.O.,1982. Recent Developments in Dry Excreta Disposal Systems. Proc. '82 African Water Technology Conf.Nairobi 30/11-1/12/82.

MHALER, H.,1981. Partnership for Health For All. WHO Chronicle. 35(6).

NEW CIVIL ENGINEER,1982. Intern. Drinking Water Supply and Sanitation Decade. N.C.E., March 1982. p.4

WATER WORLD, The Overseas Newsletter of the IWES, 1987. The Water Decade. p.3

WHO, 1976. Community Water Supply and Wastewater Disposal. WHO Chronicle, Vol.30(8). Wld Hlth Org., Geneva.

WHO, 1980. The Less Developing Countries. WHO Chronicle. Vol.35(6).

WORLD WATER,1981. D-Day for the Water Decade. World Water, Liverpool.

Water Treatment for Apia, Western Samoa

P. M. Berthouex* and C. Potthof**

Summary of The Design

Apia, the capitol of Western Samoa, will construct three water treatment plants, new treated water storage facilities, and renew the reticulation network. This paper will deal mainly with the treatment plants and some pre-design investigations of treatment options.

The 1981 population of Apia was estimated as 39,000. Population growth rates of 1.6% per annum up to the year 2000 and 2.2% between 2000 and the planning horizon of 2010 were used. The resulting design populations were 52, 860 in the year 2000 and 65,710 in 2010. The present average domestic consumption was estimated as 273 L/day per capita. The principal factor creating the high domestic consumption is the traditional methods of laundry and dishwashing, with water flowing freely throughout the washing operation. A education program for water conservation, coupled with metering of household use, is expected to decrease the per capita consumption. Accordingly the demands used for design were 190 L/day per capita at year 2000 and 180 L/day per capita at year 2010. Adding industrial, commercial, and public sector use gave total average production capacities of 17,026 m^3/ day in year 2000 and 20,211 m^3/day in year 2010. A maximum day of 1.3 times the average day, with a peak hour factor of 2 was adopted.

The design recognized that construction for the ultimate 2010 demand must be done in stages. The first design/construction phase provides the following capacities: intakes and raw water mains were designed for the maximum day production in year 2010, treated water mains for maximum day production at year 2000, reservoirs were designed for 75% of average day production at year 1995, and the reticulation system for hourly demand at year 2000. The primary and secondary reticulation system comprises about 98 km of new mains and the rehabilitation of about 51 km of existing mains. The service area is divided into sixteen pressure zones. The need for this is apparent from elevations given below. A schematic of the proposed system is shown in Figure 1.

Treatment plants, which consist of sedimentation (0.75 m/hr overflow rate) up-flow roughing filters (1 m/hr filtration rate), slow sand filters (0.125 m/hr filtration rate), and chlorination (1.0 mg/L average dose), are designed

* Dept. of Civil and Envir. Engineering, The University of Wisconsin-Madison, Madison, Madison, WI 53706, USA
** GKW CONSULT, 1600 Mannheim, Gottlieb-Daimler Str. 12a, Fed. Republic of Germany

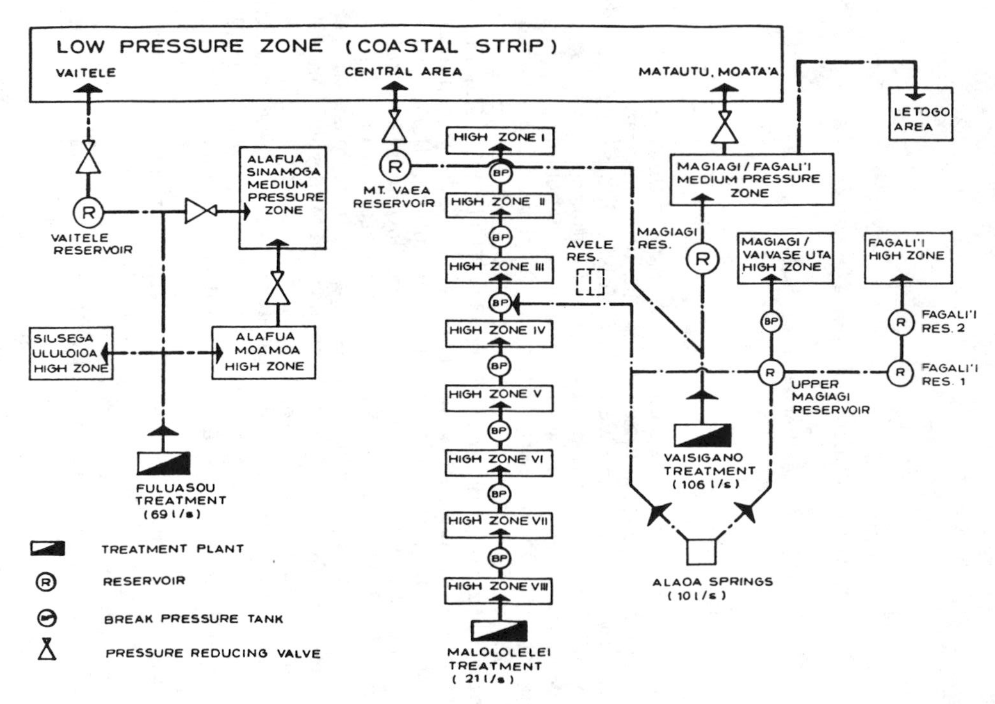

Figure 1. Schematic of the Apia Water System

for average day production at year 1995. Flow from all intakes to treatment plants and then to storage reservoirs is by gravity.

The ultimate demand of 20,211 m^3 /day, equivalent to an average flow of 234 L/s, will nominally be supplied from four sources, with treatment plants at the Alaoa Tailrace, Malololelei, and Fulu'asou:

(1) Alaoa Springs - ultimate yield =10 L/s. This spring water needs no treatment. Two spring water collection boxes will be built and the pipeline will be rehabilitated.The Malololelei system has been designed with capacity to serve the area normally supplied from Alaoa if the springs should fail.

(2) Alaoa Tailrace - ultimate yield = 127 L/s taken from intakes on the East Vaisigano River and on the Alaoa Power Station Tailrace.; average 1995 design flow = 9125 m^3/day. Treated water is directed to an existing 4500 m^3 reservoir at Mt. Vaea and a new 2250 m^3 reservoir at Magiagi. The Alaoa Tailrace intake elevation is 118.86 m (all elevations given with reference to mean sea level); East Vaisigano intake = 113.81 m; water elevation in slow sand filters = 105.8 m; water elevation in plant storage reservoir = 100.4; water elevation Mt. Vaea reservoir = 84.1 m.

(3) Malololelei - ultimate yield = 15 L/s taken water from the West Vaisigano River; average 1995 design flow = 1860 m^3/day. Treated water distribution will be integrated with three existing break pressure tanks; four new break pressure tanks will be constructed. The intake elevation is 528.05 m; water elevation in slow sand filters = 492.5 m; water elevation in storage reservoir = 490.5 m.

(4) Fulu'asou - ultimate yield = 82 L/s taken from the East Branch of the Fulu'asou River; average 1995 design flow = 5915 m^3/day. Treated water goes to the 2 new 2250 m^3 reservoirs at the treatment plant and Viatele. The intake elevation is 152.9 m above MSL; water elevation in slow sand filters = 135.8 m; water elevation in plant storage reservoir = 133.8; water surface elevation Viatele reservoir = 96.0 m.

Four sources and three treatment plants were needed because the the safe yield of the rivers is very low. Because of the mountainous topography of the Apia area, separate extraction and treatment sites did have some hydraulic advantages, in particular elimination of pumping.

Raw Water Quality

The three rivers that provide the Apia water supply are naturally clean. Their chemical and bacteriological properties are excellent for drinking water. In dry weather the water is virtually free of turbidity. The Fulu'asou's turbidity was always observed to be less than about 20 NTU, even during heavy rains. The Vaisigano River is different. A heavy rain in the catchment area can increase the turbidity turbidity within minutes from below 10 to more than 100 NTU. At times the turbidity will exceed 200 NTU. The turbidity drops soon after the rain stops because the porous volcanic soil absorbs a high percentage of the rainfall.

Figure 2 shows how the intake at the Alaoa Treatment Plant will use a mixture of water from the Alaoa Power Station Head Pond and the East Branch of the Vaisigano. The East Branch was expected to be turbid during storms, while the tailrace would discharge relatively clear water, because the hydropower statation head pond would function as a clarification basin. Figure 3 shows that the head pond does not effectively clarify the water that passes through it. Instead, it stores turbid water which is released for hours after the East Branch has flushed itself clean. The turbidity level and the pattern of turbidity fluctuations were both important. They indicated the importance of making field studies during the season when raw water quality is at its worst.

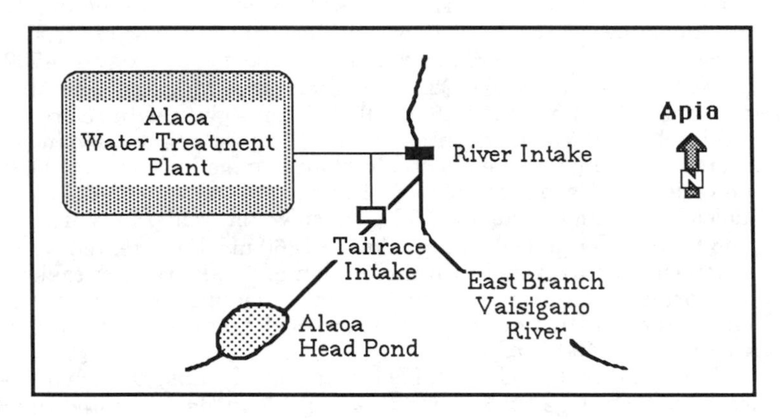

Figure 2. Primary raw water intake for the Alaoa Plant

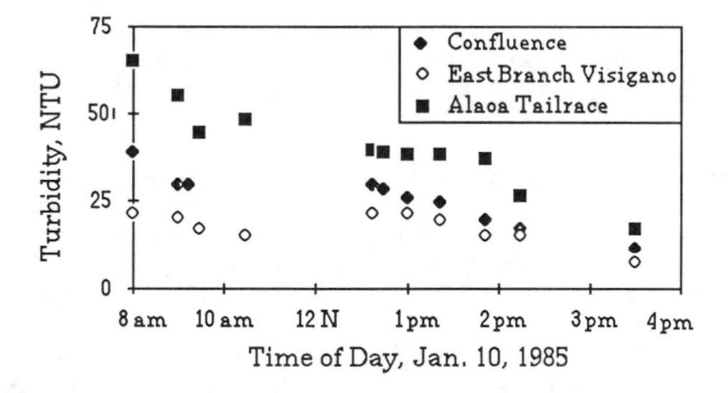

Figure 3. Turbidity measurements at the Alaoa Intakes

Treatment Plant Design

The design goal was to use technology that was reliable, simple, and had minimal use of energy or chemicals. The final design uses no energy or chemicals. Treatment will consist of sedimentation, vertical flow roughing filtration, slow sand filtration, and chlorination. The lowest bid was for circular settling tanks, roughing filters, and slow sand filters. The layout of the Malololelei Plant is shown in Figure 4. This is the smallest of the three plants. Multiple slow sand filters were used to increase flexibility of cleaning and to increase the total system reliability.

The design was based on pilot plant and field tests of water treatability characteristics. This paper explains the field tests that were used to guide the design. Investigations of this kind, which are not difficult or expensive, have great value for selecting appropriate technology and optimizing process design. They can also serve as a training exercise for local staff who may become the treatment plant operators.

Slow Sand Filtration

A pilot plant was operated from June 1984 to late-January 1985 in order to evaluate local sand filter media and to test performance at different filtration rates. The study showed that treated water quality would be excellent and the interval between cleanings would be 6 to 10 weeks if the raw water turbidity is in the usual range of 2 to 20 NTU. Treating water with turbidity of 50 to 200 NTU will block the slow sand filters in a matter of hours. The high turbidity episodes that occur in the rainy season could cause catastrophic failure of the slow sand filter system.

The design is based on only one filter at a time being out of service. If more than one is down for cleaning, the operating filters would be overloaded so much that they would clog rapidly even if the turbidity remained low.

Sand is a scarce resource in Western Samoa. Coral sand is not suitable as filter media because it is weak and chemically reactive (95% loss in a standard acid test).. Basalt sand, which can be prepared by crushing basalt rock or dredged from a few beach areas, performed well in the pilot tests and met the criteria for acid solubility and strength. The size distribution of the sand from each source was measured. None had an ideal gradation, but acceptable filter media can be prepared by washing.

Sedimentation

When the river level rises during a storm the turbidity, the amount of leaf debris and other large particles increases. Plain sedimentation was effective for removing the large particles but it would not reduce the turbidity enough to eliminate other pretreatment. Figure 5 shows that three hours detention time accomplishes most of the removal that is possible. A more detailed analysis of the settling test data showed that an overflow rate of 0.75 m/hr was appropriate.

These tests changed a preliminary design in two ways. The final

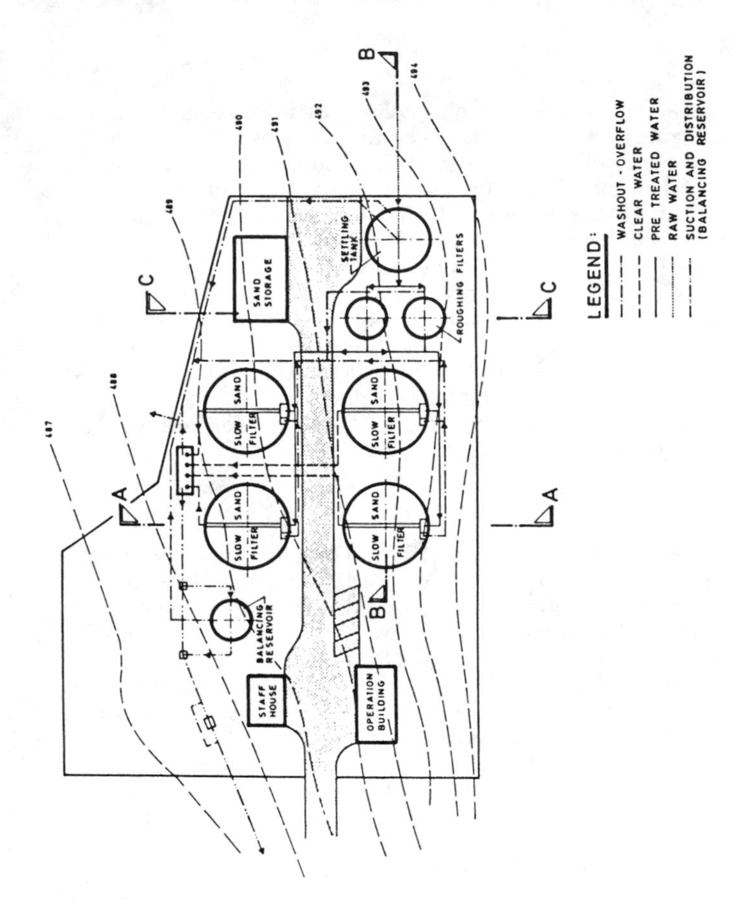

Figure 4. Layout of the Malololelei Plant

design used 0.75 m/hr rate, which was about four times larger than the rate used for preliminary design. Also, the preliminary design had assumed that sedimentation would accomplish the necessary degree of pretreatment, which was demonstrated to be false. It also became evident that, even thought some roughing filtration would be needed, sedimentation should also be used to remove heavy material and floating matter.

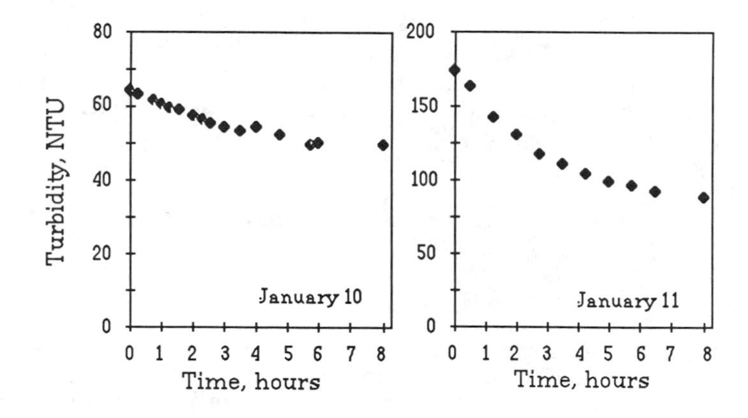

Figure 5. Batch sedimentation column test results.

Chemical coagulation with alum would enhance sedimentation enough to reduce the turbidity to less than 10 NTU. Adding too much or too little chemical, however, would give a poor result so very precise control of the chemical addition would be needed. The turbidity changes so fast, however, that precise control of the chemical dosage would be impossible. Chemical addition destroyed the alkalinity of the water and decreased the pH. Also, it would be expensive. For these reasons, chemical addition was not used.

Roughing Filtration Studies

The inadequacy of plain sedimentation and the rejection of chemical coagulation caused other pretreatment alternatives to be investigated. Roughing filtration, with either horizontal flow or vertical flow, was effective.

Figures 6 shows some test results for vertical flow roughing filter, which were nearly equally effective in removing turbidity as the horizontal flow filters (data not shown). Vertical flow filters were selected because they were better suited to the space and hydraulic constraints of the treatment plant sites. They will be cleaned by flushing.

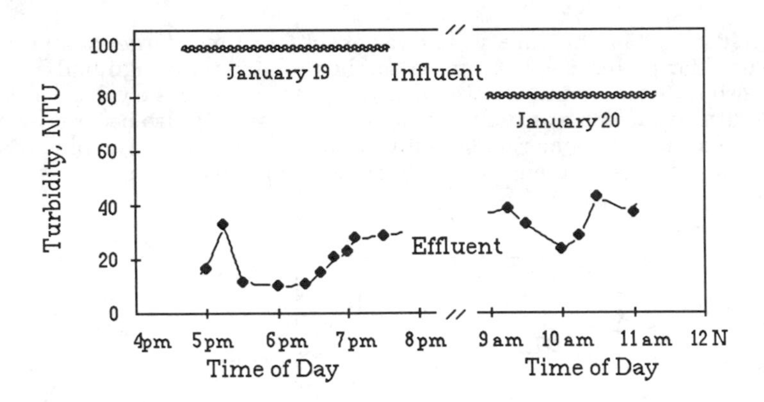

Figure 6. Vertical flow roughing filter test results, rate = 1 m/hr

Summary

Pilot plant studies of slow sand filtration verified that this process would give excellent effluent quality and that the interval between cleaning would be satisfactory for an average design flow rate of 1.25 m/hr. Local basalt sand was shown to be a suitable filter medium.

The studies also indicated that pretreatment would be needed to protect the slow sand filters from rapid clogging when the turbidity increases during storms. Experiments were done to simulate various pretreatment methods. Plain sedimentation and vertical flow roughing filters were included in the design.

The cost of pre-design tests such as these is reasonable and will be repaid in several ways. They can be essential for selecting the most appropriate treatment technology and for optimizing the system design. They also serve as a training exercise for local staff who may become treatment plant operators.

Acknowledgements

The Public Works Department in Apia was appointed as the Project Supervisor by the Government of Western Samoa. Funds for the project were provided by the governments of the Federal Republic of German, New Zealand, and Saudi Arabia. The design was done as a joint venture between GKW Consult of Mannheim, Federal Republic of Germany, Murray-North Partners Limited of Aukland, New Zealand, in association with G. M. Meredith & Associates, Apia, Western Samoa. Mr. Karl Wasserman and Mr. Jim Hodges were the project managers for GKW Consult and Murray- North Partners, respectively.

Checks to Building Station Water Systems in Sudan

David T. Higgins, M.ASCE*

Abstract

Construction of water supply systems for four agricultural research stations was incomplete after five years of effort. With some exceptions every element in the systems--the boreholes, pumps, elevated tanks, and pipelines--proved to be unbelievably difficult to accomplish. This paper describes some of the difficulties and also asks some questions about potentially beneficial applications of newer water supply technology in Sudan.

Introduction

For five years I served as civil engineer for a USAID/World Bank-sponsored development project in Sudan. I was based in Khartoum to assist in the construction of four agricultural research stations in Sudan's West. When I arrived in June of 1980 I expected to accomplish my activities within two years. Station construction was to be completed in three. Yet five and a half years later neither was done.

Some of the reasons for the slow pace of accomplishment are quite astonishing. To share them with those who have not worked under similar conditions is the primary purpose of this paper. A second purpose is to let me ask some questions about potentially beneficial water supply technology not presently applied in Sudan.

Description of the Project

The Western Sudan Agricultural Research Project (the Project) was designed to support scientific activity for improving traditional rain-fed farming practices in western Sudan. Rain-fed agriculture is distinguished from irrigated agriculture which was purposely excluded from the Project.

The support included building the research stations, equipping and staffing them, training agricultural scientists, and partially

*Assoc. Prof., Dept. of Civil and Environmental Engineering, Washington State University, Pullman, WA 99164.

supporting operations for a number of years. Station locations are shown on the next page.

The stations were to provide a modern living and working environment for the research scientists, most of whom would be Sudanese, and for support staffs. Central to this were reliable supplies of electricity and running water, as well as good quality housing, offices, laboratories, motor pools, and the like.

The largest station, at El Obeid, includes a two-story administration building; a two-story research laboratory/ library/office building; senior, middle, and junior housing; guest housing; motor pool and machinery maintenance structures; warehouses; a diesel-electric powerhouse; ground and elevated water storage tanks; a pump house and water distribution system; a sewerage system with stabilization ponds; and roads and street lighting. The three smaller stations are basically the same.

Project Engineering

Engineering for the Project was separated into two parts: that provided for the main construction programs by architect-engineer consultants and that provided by the Project's engineer. The former, in essence, designed the stations, oversaw the bidding and contracting, and served as the Project's agent in station construction supervision. The project engineer, again in essence, advised the project director about engineered facilities that would be required for Project operation but were not included in the station main construction contracts.

Main Construction Program

Of the four stations the first to be completed was Kadugli. Its construction had been started by the Russians but came to a standstill when the political winds shifted in the 1970s.

To get the Project moving quickly in 1979 a Khartoum A/E redesigned the Kadugli station, incorporating what they could of the earlier construction. A Lebanese contractor executed the works.

At the same time a Portland, Oregon A/E, with experience in Sudan, designed the stations for Ghazala Gawazet, El Obeid, and El Fasher. A Sudanese joint venture won the internationally tendered contract for the three stations. They, in turn, found it necessary to subcontract much of the work to a mainland China construction force. The Chinese provided a badly needed skilled labor component.

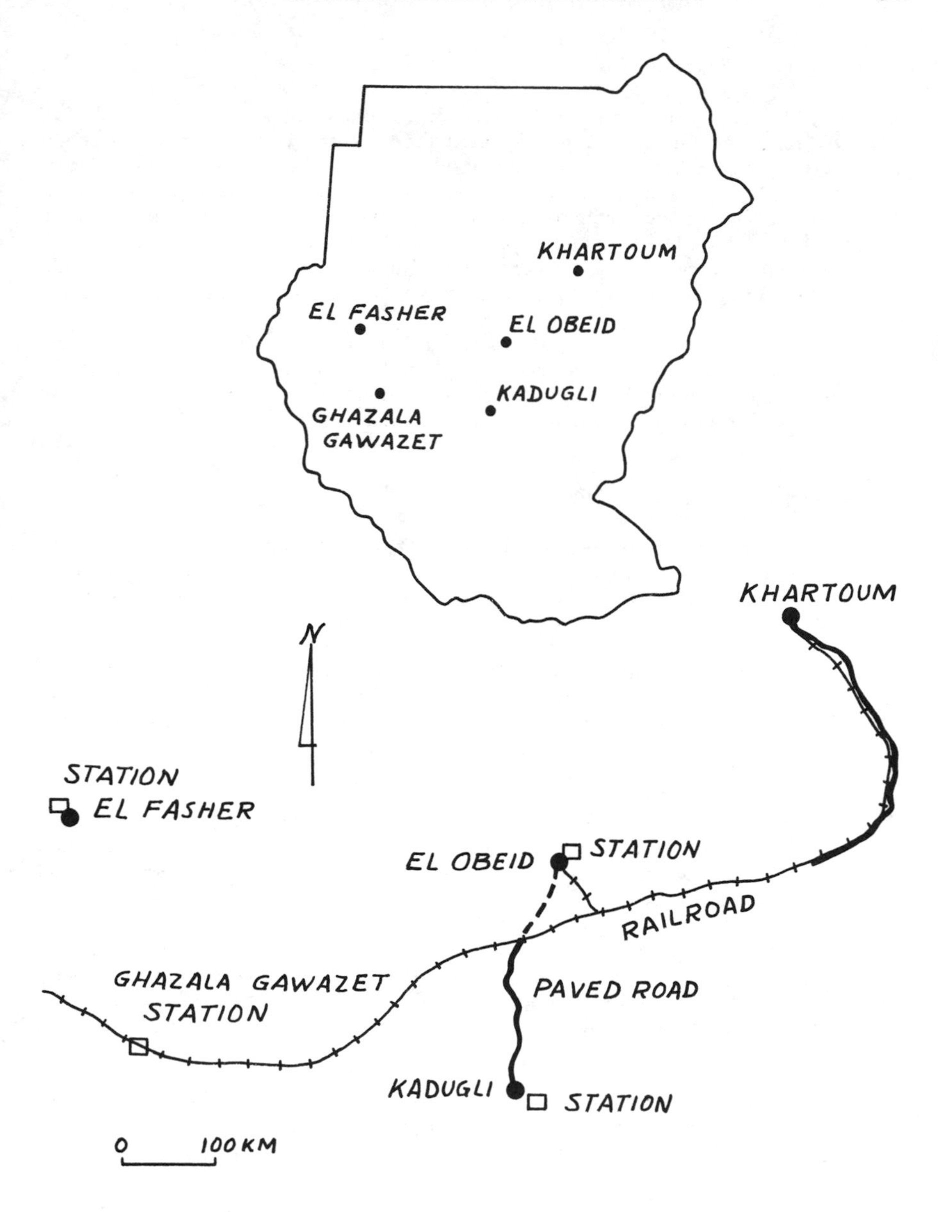

STATION LOCATIONS

Force Account Engineering

Falling outside the main construction program was a need for the design and construction of facilities not fitting easily within the main program or not anticipated at the time of main program design. Among these were roads, airstrips, fencing, drainage structures, building rehabilitation, farm facilities, and water supply systems. It is the latter that this paper considers. This engineering was referred to as force account. It was not an apt name; the Project initially had no internal construction capability.

Over the years it did hire a young Sudanese engineer and develop three foremen for simple construction jobs. But the majority of the force account construction was done by Sudanese government agencies or by private Sudanese contractors.

Water Supply in Western Sudan

Having a dependable, 24-hour-a-day, pressurized water supply would make the Project stations almost unique among Sudanese communities. Not even the better districts of Khartoum are so favored.

In general Sudan is a Sahelian country. In the Project area rainfall averages vary from about 200 mm at El Fasher to 700 mm at Kadugli. Annual departures from the average can be substantial. Furthermore, precipitation occurs in the region only during the rainy season, roughly June into September.

Much of the Project region's farming is by families in small villages centered in a cropped area extending a few kilometers around the village. Availability of drinking water, (duration and proximity) is a limiting factor in farm size.

Typical sources of rural water supplies are the seasonal streams, shallow dug wells, boreholes and hafirs. In many areas only the borehole supplies water year around, and then only if it and its pump are maintained, and, if the pump is engine driven, fuel is available.

The hafir is a basin excavated by scrapers in clay soils adjacent to a stream and filled by diversions from the stream. It generally supplies water through the rainy season and perhaps into January.

Both hafir and borehole water sources built during ambitious rural development programs in the 1950s and 60s have deteriorated in the last decade and a half.

Water supply systems for the larger population centers such as Kadugli, El Obeid, and El Fasher may be more sophisticated but are in equally poor condition.

Kadugli's water supply is a city field of ten wells. The failing pumping, storage and distribution systems deliver water to many areas of the town for only an hour or two a day and at pressures insufficient to fill roof tanks. The municipal water supplies of El Obeid and El Fasher are surface impoundments located a few kilometers from town and augmented by boreholes within the city limits. Impounding volume at El Obeid is on the order of three million cubic meters. As in Kadugli the condition of the systems is such that water is generally available for only a very few hours a day and at low pressures. In 1985 the El Obeid reservoirs were exhausted in March and did not begin to fill until July. Very limited water supplies came from town boreholes and from railroad tank cars hauling water from the Rahad reservoir to the southeast.

Station Water Supply

The stations at Kadugli and Ghazala Gawazet are dependent on their own wells. Neither is close enough to a municipal supply system. However, the stations on the outskirts of El Obeid and El Fasher were to be served by the municipal systems. In their planning, the A/E anticipated low municipal system pressures and designed an on-station, in-ground sump to accumulate municipal flow as it occurred, and a booster pump to an elevated tank. In addition they provided a separate landscape irrigation distribution system to be served by a supplementary borehole source provided by force account. But, as we learned during station construction, the municipal systems often could not supply even construction water. The supplementary wells were then seen as the only reliable source of water for all aspects of station operation. Yet, except at Kadugli, there was almost no progress during the five years in developing station water sources.

Why the Work Went Slowly

Sudan has faced endless troubles since independence in 1956: civil strife, coups and counter-coups, the 1973 oil embargo, inflation; the devisive effect of the Islamic Sharia law imposed in 1983; the burden of famine and war refugees from Ethiopia and Chad; its own recent drought; the loss to emigration of many of its professionals, technicians, and skilled craftsmen, etc.

National communications and transportation facilities during the five years were poor. Telephone service between station sites either did not exist or only sporadically worked. There were only about 160 km of paved roads in the Project region. During the rainy season overland vehicle travel was at best uncertain. The railway system was of occasional service to the Project but required arrangements to be made months in advance.

The Project contract was referred to as "host country." The significance of this, to me, is that a Sudanese national holds the position of project director. And he, then, must see that the Project follows Sudanese government (GOS) regulations with respect to the disbursement of Sudanese currency, which funded most force account activity. Other ramifications of having to follow Sudanese government regulations will be seen in the following list of what I thought of as constraints to speed of progress.

Constraining Conditions

1. GOS funds of Sudanese currency for force account construction were slow in appearing and slow in flowing throughout the five-year period.

2. Particular requirements of the GOS for procurement and for awarding construction contracts that led to delays were the following:

   a. need for special approval of purchases in excess of L.S.1,000 (roughly $US 300 to 500). Approval could take weeks.

   b. need for a month's advertising for construction bids or large material supply contracts, then a week or two in the Project office determining the winner and advising the GOS, then sometimes months before their approval was given--in which time materials went to someone else, or prices increased, or fuel disappeared, or the rains came.

   c. unexplained interference with contract awards.

   d. unexplained rules that forbade the Project from selling unneeded materials to other GOS units, yet allowed the Project to buy materials from other GOS units. This policy prevented the occasional purchase of usually hard to get materials before their need had been fully confirmed.

   e. excessively involved store keeping and release-from-stores requirements.

3. Project management was such that there were never enough vehicles, or supplies, or personnel to implement an aggressive force account engineering program.

4. The GOS would confiscate Project vehicles, already in short supply, for electioneering.

5. Periodic nationwide fuel shortages would slow or stop most vehicle activity.

6. Many of Sudan's professionals and skilled craftsmen have emigrated.

7. GOS-controlled wage limits did not generally attract good quality construction labor. Often, only boys, rather than grown men, were available at the allowable wage. Their work was noticeably inferior to that of men who could have been attracted by a higher wage.

8. Vacations for government workers can last some six weeks or more. Disruptions of normal activity during such absences occur frequently. Expatriate leaves of a month or so could be equally disruptive.

9. The Islamic holy month of Ramadan is a time of inefficient field work, especially when it falls in the hot season. Men do not work well without food and water during the day, and adequate sleep is not assured during the night.

10. Often a working day is no longer than four or five hours. The Sudanese breakfast hour begins at about 9 am. Not a lot of useful work occurs before then, and by 2 pm preparations for transportation home begin.

11. The GOS refused to allow the Project to employ capable private Sudanese well drillers even when the government water agency failed, in some places for years, to provide wells.

12. The GOS required the Project to depend upon Sudan Airways for aircraft maintenance. The Project airplane was sometimes unavailable for service for weeks or months at a time.

13. GOS telecommunications officials delayed licensing the Project radio communications system until December of 1981. Crystals could not be cut until frequencies were allotted.

14. Surface travel to the West and sometimes to Kadugli was severely limited during the three or four month rainy season. If materials, fuel, and men were not in place before the rains, work delays were inevitable.

15. Boundary and topographic maps were often difficult or impossible to find. In some cases GOS survey units would not cooperate in furnishing services.

Without doubt a lack of vehicles, fuel, and spare parts made it most difficult for the government water and survey units to maintain morale and to function in a timely and efficient manner. The government water unit at Kadugli cooperated most effectively when the Project supplied fuel.

Potentially Beneficial Water
Supply Technology

During the five years a number of engineering-related questions arose for which I still do not have answers. Perhaps others at this conference do.

1. Polyethylene pipe appears to be most suitable for small water systems. Project force account installed several thousand meters of 2-inch polyethylene pipe. Additional thousands are installed annually by others in Sudan. The pipe is manufactured in Khartoum in diameters up to 4 inches. I tested samples of the 2-inch pipe to 90 psi. The Project was able to exploit this strength because it could import the nylon insert-hose clamp systems designed for this pipe. But Sudanese polyethylene pipe users have to cut threads in it in order to use locally available galvanized pipe fittings. The threads weaken the pipe. Working pressures are generally held under 10 psi. Are the insert and stainless steel hose couplings available in the markets of other African countries? Could they be manufactured in Sudan and sold at a competitive price? Is there a more practical, alternative coupling system?

2. Plastic membrane linings may solve some of Sudan's potable water storage problems. I looked in detail at lined ponds with floating covers for the stations at El Obeid and El Fasher. Among the questions that arose are: Will they be damaged by termites seeking water? What will be the effect of wind blown sand on the floating cover? What installation skills are required for some of the smaller units?

3. Although many kinds of borehole pumps are installed in the West of Sudan, the reciprocating piston pump is much used in wells in the 1000 GPH range. EDECO and Schoeller-Bleckmann pumping systems are common. The Mono pump (helical rotor) is also relatively common especially in wells of 2000-3000 GPH capacity. Has anyone comparative maintenance information for both kinds of pump in sandy wells? At Ghazala Gawazet piston leathers were replaced at 3-week intervals. How would we have faired with a helical rotor pump?

Conclusion

This paper barely hints at Sudan's burden of economic, social, and political problems. It then presents a one-sided view of some of the difficulties facing those who would build a seemingly small project in this country with all its problems.

I could, as readily, have told a very different story: one about the Sudanese with whom I worked or with whom I did business--their ability, their willingness to work under trying conditions, their initiative, wisdom and intelligence, and friendship.

Rainwater Catchment Systems: Advancements in an Appropriate Technology

Richard J. Heggen,* A.M. ASCE

Within the past ten years, rainwater catchment for domestic water supply in both developing and developed nations has emerged as a technology worthy of scientific and engineering analysis. Advancements in both design and construction have brought about decreases in costs and increases in efficiency. Computer modeling has allowed evaluation of catchment systems under stochastic conditions of rainfall and demand, a major improvement over conventional sizing for average conditions. This paper reviews rainwater catchment usage and selected recent advancements in technology and analytic methodologies.

Introduction

Rainwater catchment (RWC) systems constructed to intercept, divert and store rainwater for domestic use are employed in virtually every temporate, tropical or arid nation in the world. RWC's are employed across the spectrum of developemnt, from such industrialized nations as Japan and the United States to such sustenance economies as those of rural Togo or Haiti. More than 100 million people employ RWC's for water supply. Of nations where English is a language of technical publication, over half are currently invovled in RWC research and development. In the broad field of water supply planning, however, RWC potential is often ignored. Certainly RWC systems are not suited for many applications, but where they are suited, they may be an efficient alternative for water supply.

This paper addresses three issues related to the incorporation of RWC systems into potable water development. These issues are 1) the dilemma of "appropriateness", 2) technical developments and directions, and 3) the contribution of systems analysis.

A note is in order regarding the references following this paper. There are currently more than 500 published papers on RWC. The several referenced publications each summarize multiple papers and together contain an extensive bibliographic list.

*Assoc. Prof. Civ. Engrg., Univ. of New Mexico, Albuquerque, NM 87131.

Appropriateness - Diverging Criteria

RWC systems conventionally are of small scale, traditionally servicing but a single household. RWC systems generally can be designed by nonprofessionals, frequently following no more than nomographs and illustrated drawings. RWC systems commonly can be built of inexpensive materials, often having galvanized iron roofing as the most costly item. RWC systems typically can be constructed with common labor, usually requiring no more than hand tools.

From the perspective the "Small is Beautiful" aesthetic, RWC's are classic in appropriateness. Unfortunately however, even those who share a philosophical bent toward intermediate technology must address the pragmatic necessities of large-scale development in the modern world. Is the continued development of RWC practice part of the planning solution, or are RWC's part of the problem? Both sides can be argued.

On the positive side, RWC systems may provide improved water supply in areas such as rural Thailand where more technical alternatives (say wells) are costly or difficult to maintain. RWC technology can be modified to a variety of hydrologic conditions, be it the near-daily rainfall of tropical islands or multi-month drought seasons in Australia. Even a poorly designed RWC (place a series of buckets under the eaves) may provide a reasonable percentage of that which might be obtained from an optimal design, while an inadequately-engineered well or dam may yield nothing. RWC's can be easily modified in response to changes in demand or technology. In short, RWC systems can be an immediate, affordable and forgiving technology.

On the negative side, RWC systems (especially if government funds are consumed in their development) may represent but a diversion in the long-term objectives of sufficient, hygenic public water supply. In Malaysia, such perception has been noted. RWC water supply will rarely meet anything above personal requirements; there are virtually no reports of RWC employment for community use other than schools. If water is an agent in economic development, catchment yield may not suffice for the rate of growth desired by ambitious planners. Whereas RWC systems are generally shown to be less prone to pollution than streamflow or shallow aquifers, treatment, if required, is likely to be costly when economies of scale are considered. In some cases where RWC's and an alternative water source exist in proximity, consumers will turn toward the alternative, an indicator of socio-preference that cannot be ignored.

As in most "objective" decision processes, a subjective decision framework can exist. RWC systems can look bad or good, depending upon the criteria selected. As with any technology, tradeoffs are made. It is not the thesis of this paper that RWC systems are necessarily appropriate in a given instance, but rather that such judgement be made from a position of scientific assessment. Such assessment should not be based upon vague impressions of rainbarrels, but rather upon knowledge of current developments and potential for technological advancement.

Technological Developments and Directions

The reference list at the end of this paper provides a broad spectrum of research and practice reports of current developments. It would be impossible to summarize that literature here. Rather, to illustrate the degree of ongoing work, a single component of the RWC system, the storage container, will be discussed.

The reservoir or cistern element of RWC systems can be divided into two categories: excavated and elevated. Traditionally excavated cisterns at times illustrate a remarkable degree of structural sophistication. Iranian catchments a century old employ arch, shell and dome roofs with tapered walls. Modern excavated cisterns increasingly are making use of ferrocement or impermeable membrane liners. PVC has been extensively employed, but improved liners offer additional resistance to weather and wear. Butyl rubber has been employed in Africa. Twenty mil CPE has been shown to be generally adequate for domestic systems having a planning life of 15 years. Reinforced Hapalon has been used for more durable facilities. Synthetic lining material has been incorporated into floating covers for large cisterns.

Soil cement grouting in a trench holding an inflated balloon-like form has been studied in Portugal. In India and California, RWC water is recharged to the groundwater to augment that conjunctive-use resource. Lined, rock-filled artificial aquifers have been proposed for RWC storage in Japan. Compartmentalized reservoirs have been developed in the Caribbean, reducing abstractions and facilitating maintenance.

Elevated cisterns have traditionally made extensive use of containers salvaged from other uses: oil drums, aircraft fuel pontoons, rum barrels, jars and buckets. Historical practices for larger cisterns have followed reservoir practices in general: reinforced concrete, steel, masonry or wooden stave structures. Current developments are more directed toward material efficiency. Interlocking reinforced mortar blocks have been developed in Thailand to take advantage of shear strength. Adobe blocks have been tried, but the results are yet unreported. Ferrocement is a technology gaining wide acceptance throughout much of the developing world. Containers exceeding 100 m^3 have been successfully built. Technology exists for both interior panel-molded and nonmolded construction. An innovative Indian technology involves a sawdust-filled sack mold which is simply emptied when the ferrocement cures. An Indonisian campaign for rural water supply has produced a manual for 10 m^3 ferrocement tanks and a 12-day course in their local construction. Similar technical-organizational education is being carried out in Thailand, Bolivia, East Africa and the South Pacific.

Bamboo-cement has been promoted for storage containers, but in Thailand, where bamboo reinforcement has been used for up to 20 m^3 vessels, field testing now indicates that the bamboo deteriorates in a matter of years. Palm fiber and sisal reinforcement have been used as well. The Ghala basket, mortar applied over a basket frame, is being used for up to 6 m^3 cisterns in Kenya. Concrete ring tanks are in use over much of Southeast Asia. An inward-curved compression shell panel prefabricated ferrocement design has been developed in Australia.

Fiberglass and rubberized bladder tanks provide an alternative for transportable RWC systems. The Water and Sanitation for Health Project has developed a 12x27 decision matrix for cistern selection.

Emphasis in Australia and Papua New Guinea is now on marketing commercial cisterns, illustration of a change from technical to infrastructure focus as development progresses.

Systems Analysis

Runoff simulation and reservoir operation are areas in Civil Engineering where operations research and systems analysis have had large impact. As both the physical and information structure of RWC systems are akin to those of runoff-reservoir systems, it seems appropriate to transfer the analytic insight to issues of RWC design and analysis. Again, it is beyond the scope of this paper to review all the impacts of systems analysis that have been applied to RWC analysis. One aspect, that of operation, will be discussed to provde illustration.

RWC water demand may be specified in two manners, as a fixed target or as a dynamic parameter influence by system states. Fixed target demands are generally unrealistic, and while employed extensively in crude RWC cistern sizing in scores of countries, merit little discussion. The Rippl mass-curve method has been used for tank sizing in India; more data-efficient stochastic critical period mass-curve analyses have been devised in Canada and Australia. Correlation has been employed in Papua New Guinea to relate the extent of dry season to the average daily rainfall in the driest month, another technique useful for data-sparse operational investigations.

Dynamic demands, on the other hand, are both realistic and amenable to econometric specification. Rationing has been incorpated into probably less than ten percent of RWC design development, but where it has been included, the results have been positive. Rationing for RWC systems has been taken as stepwise or logistic functions relating consumption as a percent of unrationed target to remaining storage. As the rain barrel empties, daily withdrawals are curtailed. Work in Canada has taken rationing to be a 25 percent reduction if less than a month's demand is stored. In Nova Scotia, withdrawals are halved when the tank is 14 percent full. In Austrailia, reduction is 40 percent when less than 10 days of normal supply remain. In Hawaii, demand is reduced by 33 percent when cisterns are half empty. Continuous rationing schemes have been evaluated in Australia, Hawaii and New Mexico. In Hawaii, a multi-stage optimization of the rationing rule has been solved by dynamic programming. In all cases, an improved analysis is made of the system's ability to provide water during dry periods. RWC design can reflect realistic water usage.

The yield-before and yield-after-spill operating rule alternatives are the same as those used for reservoir simulation. With RWC's the split is roughly 50-50 between the methods. For households where water use can come in quick response to new rainfall ("real time" operating), the yield-before-spill model is perhaps the more appropriate. In California, system reliability for both operating rules has been statistically evaluated. Differences were negligible when daily rainfall was used, but became significant when monthly values were employed.

Optimization of RWC systems is generally done by a cost effectiveness criteria. Rough calculations relating average rainfall to average demand have been done in at least two dozen nations to match cistern size to fixed (generally the roof size) catchment area. Improved RWC design by simulation employing real or synthetic rainfall traces and time-varient (sometimes with rationing capability, sometimes without) demand has been carried out in Canada, Australia, Japan, Hawaii, Micronesia, Singapore and Great Britain. Catchment and storage parameters have been optimized in Yugoslavia and varied through expansion path analysis in Micronesia. Stochastic influences on RWC system behavior has also been explored explicitly by the Gould method.

Conclusions

RWC is a technology that reflects current developments in engineering materials and design. RWC systems play a major role in potable water supply for a significant number of people. In planning for development, especially in rural settings, RWC may be a viable alternative for domestic water provision.

Appendix I. - References

Fujimura, F., Proceedings of the International Conference on Rain Water Cistern Systems, Honolulu, HI, 1982.

Keller, K., Rainwater Harvesting for Domestic Water Supplies in Developing Countries, Working Paper No. 20, Water and Sanitation for Health Project, Arlington, VA, 1982.

Pacey, A., and Cullis, A., Rainwater Harvesting, the Collection of Rainfall and Run-Off in Rural Areas, Intermediate Technology Pub., London, 1984.

Smith, H. H., Ed., Proceedings of the Second International Conference on Rain Water Cistern Systems, St. Thomas, VI, 1984.

Vadhanavikkit, C., Ed., Proceedings of the Third Internatinal Conference on Rain Water Cistern Systems, Khon Kaen, Thailand, 1987.

WASH, Rainwater Roof Catchment Directory and Bibliography, Water and Sanitation for Health Project, Arlington, VA, 1986.

Women, Water, and Windmills

An Approach to Third World Water Supply

Harry S. Bingham, F.ASCE*

Abstract

The plight of the women water-bearers of the Third World has been the motivating force to develop a strategy for constructing and servicing village water supply centers based on the use of the time-tested water-pumping windmill. Inexpensive developments near the center of a village in wind-swept environments are seen as an investment in health and nation-building in contrast to the debilitating trips to distant and contaminated streams. First-hand knowledge of the water-bearers' habits in Ethiopia and Zaire prompted the development of a prototype village water supply center in Pennsylvania. The prototype consists of a water-pumping windmill on a 40-foot tower situated over a 140-foot drilled well and a wooden-stave water tank with multiple faucets conveniently situated at the bottom of the tank. A resting bench was built into the foundation of the tank at a convenient sitting height where the water-bearer could rest his/her vessel during the filling operation. Variations on the design and use of the prototype are discussed relative to appropriate technology criteria for Third World countries where 200 to 500 persons might be serviced by the simplistic technology of such a water supply system. An organizational scheme is set forth to guide a host country in developing a program of village water supply centers. The implementation of the program must be well-planned to include comprehensive programs of education and communication that start with the host country ministries and extend down to the lowest villager to be affected. A well-defined plan which accounts for a system of preventive maintenance is essential to avert the abandonment of the windmills when a problem develops. Several decades of successful experience with water-pumping windmills by the farmers and ranchers of the developed nations can provide a model for village water supply development for the developing nations if the old pitfalls of technology transfer are avoided.

Figure 1: Prototype of a Third World village water supply center

*Director of Facilities and Land Management, Rider College
2083 Lawrenceville Road, P. O. Box 6400, Lawrenceville NJ 08648-3099.

Introduction

Between 1966 and 1968 while I was living on the outskirts of the Ethiopian capital of Addis Ababa I was intrigued almost daily by the procession of women who would come past my door with their water jugs on their backs going to and from a central watering station in my neighborhood. There were few homes in this "village" of a half a million people which had the luxury as I did of a central water supply. The women (the traditional water-bearers of the Third World) made their daily treks to a location a few hundred yards from my home where they could purchase from the local water supplier a jug full of water during certain hours when an attendant was available to turn on the valve to this precious supply of water.

Unfortunately it was my experience to observe, and for the large majority of the Ethiopian women to experience, a much different set of circumstances out in the small villages beyond the limits of the big city of Addis Ababa. There it was very common for women to walk several miles in pursuit of an even more questionable water supply. Ten years later when I had a consulting assignment in Zaire I observed a similar set of circumstances, the only difference being that the women and children carried the water in vessels on their heads rather than strapped to their backs. I will never forget watching a pregnant woman carry approximately 5 gallons of water balanced on her head up a hot dusty road which led from a spring easily 2 miles away over a course that required a change in grade of over 400 feet. The footprints of that barefooted woman are still imprinted in my mind. Therefore, the question has always lingered, "What would be a simple means of providing a safe and reliable water supply at village centers to eliminate the time and drudgery associated with the labors of these women?" It was obvious that capital-intensive water supply systems with diesel pumps were not the answer.

A Simple Revelation

One day I received a post card from a niece in California with a picturesque scene of an old fashioned water-pumping windmill. A storage tank connected to the mill was serving the water needs of a whole herd of cattle. At the time I was uninformed about these wind machines, but I wondered why we couldn't make wider use of this time-tested but simple technology to serve the water needs of men, women, and children around the world. As a sanitary engineer I was aware of the World Health Organization's estimates that 80% of all diseases are water-related and that approximately 2 billion men, women, and children in the world are without reasonable access to a safe and adequate water supply.

The thought of using the water-pumping windmill became exciting to me. In 1975 I attended a two-week course in windmill technology at New Mexico State University. I learned that these simple machines require little more than a quart of oil a year to keep them running. Of course, you have to have a favorable environment with prevailing wind and a source of groundwater that can be accessed with a dug or drilled well. I was pleased to learn that a typical installation might be expected to pump a thousand gallons of water a day with an average of only four hours of favorable wind. At the rate of four gallons per capita such a supply could sustain a village or neighborhood of 250 persons.

After talking and corresponding with several groups of people interested in the plight of the Third World nations, I found that there was a great amount of interest in the concept of using the windmill and a storage tank for a village water supply. However, there was always the suggestion that the principle could be best demonstrated and eventually accepted if a prototype could be constructed that would be accessible to interested groups.

A Prototype Is Born

It takes an unusual set of circumstances to create the proper conditions where a Third World village water supply system can be constructed at an eastern Pennsylvania boarding school. Nevertheless, it came to pass as a result of my involvement with George School, a Quaker high school in Newtown, Pennsylvania, where three of my four children have attended.

Several years ago I offered to provide some consulting services for an Alternative Energy Center being planned by Mr. Dale Miller, Chairman of the Science Department. Over the years a large solar greenhouse has been constructed to function as a laboratory/classroom for a number of energy-related activities. During the planning of the Center I proposed that they erect a water-pumping windmill with a low head storage tank that could serve the greenhouse and some exterior garden and nursery plots through trickling (drip) irrigation. With minor variations the scheme has been modified to serve as a prototype of a Third World village water supply. (See Figure 1 for a picture of the prototype.)

The George School installation, which consists of an 8-foot diameter Aermotor windmill atop a 40-foot galvanized steel tower, an 8-foot diameter 5-foot deep cedar tank, and all related plumbing was purchased and installed for a total cost of approximately $6,500 (not including the well) in 1984 dollars. Most of the labor was furnished by students, staff, and volunteers, although the tower and the windmill were erected by three Amish farmers at a nominal cost. Several of the costs, including an elaborate foundation for extra height would not be incurred in a typical Third World installation. Also, provisions had to be made in Pennsylvania for protection against winter freezing conditions. It is interesting to note that the windmill, tower, and all of the piping inside of the well casing were installed or erected by the three Amish farmers in 8 hours without the benefit of any power equipment.

Lessons Learned in Zaire

In my early planning on the subject, and as reflected in the prototype, I was thinking almost exclusively of prividing a wooden stave storage tank adjacent to the windmill as both a storage and a dispensing facility. But, after doing a consulting assignment in 1978 on a joint Peace Corps/UNICEF project in Zaire, I started to think of some alternatives. The most promising option is a modification of the typical spring development scheme as practiced in Zaire and elsewhere. In the case of a spring development, a natural spring is "captured" with the construction of a low dam on the hillside at the spring site. Further construction efforts render the source protected from human and animal contamination. Then a pipe is introduced at the collection site to convey the water a short distance away to a distribution site around which persons can gather to fill their water vessels.

The variation of the spring development technique, which should be much more cost effective than using wood stave or other substantial above-grade water tanks, is illustrated in Figure 2. Chart 1 compares the cost of several tank alternatives.

Chart 1: A COMPARISON OF STORAGE TANK ALTERNATIVES

TYPE	1) Wooden stave, cedar, 8' diameter, 5' in height with galvanized steel hoops and wooden cover	2) Above ground, residential type swimming pool, 12' diameter, 3' deep with PVC liner and wooden cover	3) Collapsible steel modular tank, 7' x 7' x 4' with 20 mil PVC liner, steel cover	4) PVC liner on earth base utilizing concrete units made on site (see Figure 2)
Capacity	1800+/- gallons	2500+/- gallons	1500+/- gallons	1300+/- gallons
Approx. cost of tank and cover materials FOB New York City 1987 $$$	$2,250	$1,000	$1,250	$ 500

Windmills Versus Handpumps

Much good work has gone into the rural water-supply handpump project directed by the United Nations Development Program (USDP) and the World Bank. Through that project there has been an identification of more reliable and economical hand pumps that can be supplied for Third World applications. It is my recommendation that windmill installations for village water supplies should always be accompanied with a nearby hand pump. In the George School prototype there is a hand pump at the top of the well serving the windmill as a supplementary pumping source. I now question the advisability of including this feature for Third World use as it encourages tampering with the pump rod as the system is changed from a wind-driven to a hand-operated mode. A more reliable system would be a dual installation utilizing two separate wells which would provide a back-up system to preclude many emergencies resulting from inoperative equipment.

A Strategy Is Imperative

I would be naive to think that we could commence to spot windmills at random across the prairies and hillsides of Third World nations in the somewhat random manner by which they sprang up in the United States. Historians tell us that something like six and a half million windmills were put in place between 1880 and 1930 in the United States. The

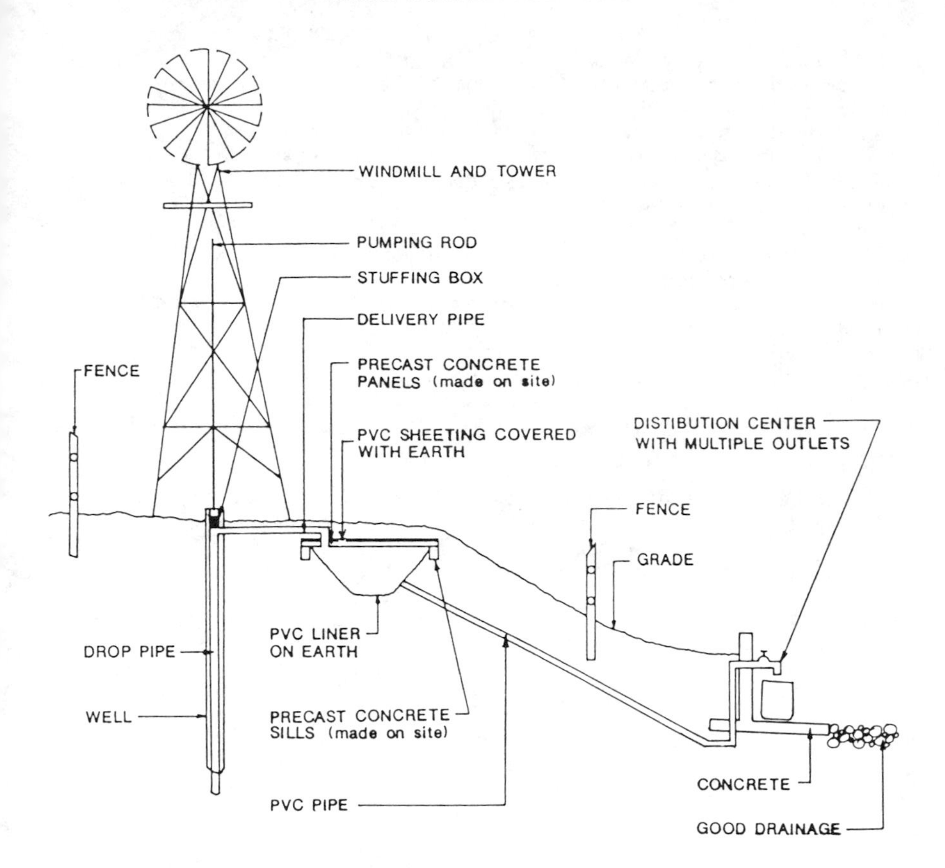

NOTE:

1. A TRAPEZOIDAL STORAGE TANK WITH A DEPTH OF 3 FEET, A TOP WIDTH OF 5 FEET AND A BOTTOM WIDTH OF 1 FOOT CAN PROVIDE OVER 1300 GALLONS OF STORAGE IN A 20 FOOT LENGTH.

2. IF THE NATURAL TERRAIN DOES NOT PROVIDE FOR A CHANGE IN GRADE, THE STORAGE TANK CAN BE SITUATED ON A MAN-MADE MOUND ADJACENT TO THE WINDMILL.

3. COST OF ALL MATERIAL FOR AN 8 FT. DIA. MILL, 40 FT. TOWER, ALL ASSOCIATED PIPING AND TANK MATERIAL IS ESTIMATED AT $5,000 DELIVERED TO A TYPICAL THIRD WORLD LOCATION. COST OF WELL AND THE INSTALLATION OF ALL OF THE ABOVE COMPONENTS IS NOT INCLUDED.

Figure 2: VILLAGE WATER SUPPLY CENTER WITH PROTECTED STORAGE

(not to scale)

rapid advance of civilization in the nineteenth and twentieth centuries brought many needs for water in the United States by many different groups. For example, the railroad needed a supply of water at out-of-the-way locations to service their steam locomotives, then there were the rural folks who wanted running water in their new bathrooms, and also the farmers and ranchers who needed irrigation water and a source of water for their livestock.

For the Third World I have in mind a much more controlled effort that is subsidized and supervised by a national or provincial government with possible foreign aid or World Bank Support. The host country would have to be convinced that such a program would help achieve a higher standard of health for its people and be willing to commit its personnel and other resources to making it work. In some economies it may be feasible to think in terms of amortizing the cost through payment of taxes and/or use costs. Cooperatives should also be considered.

To introduce the use of the windmill in Third World nations on a grand scale would require a comprehensive program including extensive education and communication in the early going. A strategy must be worked out through a Ministry of Interior, for example, and its Division or Department of Water Resources and/or Community Development. Chart 2 shows a proposed organization for a program of rural water supply development within an existing governmental structure. As indicated in the chart there are four major areas of responsibility: 1) a Community Liason Section; 2) a Technology Section; 3) a Finances and Economics Section; and 4) a Long Range Planning and Manufacturing Section. The duties given for each in the chart are self-explanatory.

It would be my recommendation that such a program should be started on a pilot project basis with 6 to 10 villages within a radius of 100 kilometers of an existing and appropriate host country agency office. Of course, it goes without saying that the host villages would have to be supportive of the program, and the physical location of the sites would have to have favorable wind conditions which have been carefully confirmed by on-site monitoring.

The most important ingredients to such a development program are the training and equipment support of the installation and maintenance forces. There is no question but what Peace Corps-type volunteers could be taught the basic skills to carry out most or all of the assignments listed in Chart 2. A training program of 8 to 10 weeks would probably be required. However, the commitment must be there on the part of the host country to train some of its own personnel concurrently and commence that difficult process of technology transfer - be it ever so simple. Certainly an important part of the training process is going to be the familiarization with the tools and procedures necessary to install and maintain the system. For a successful program it seems almost certain that one well-equipped installation and maintenance vehicle would be required for a program of any size. One vehicle could service approximately 100 installations after erection depending on the dispersion of the sites and the terrain.

Chart 2

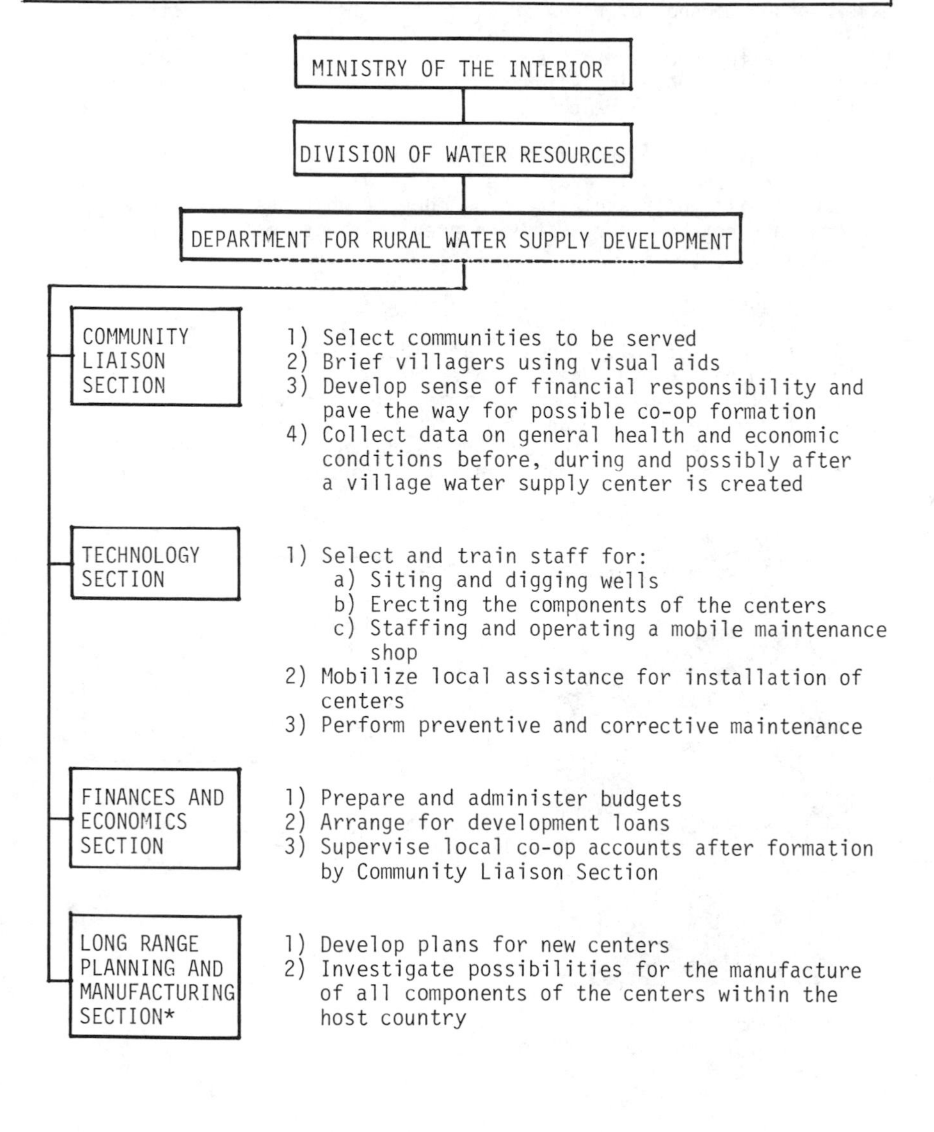

*Only after a successful pilot project

Lessons Learned

The virtues concerning the use of water-pumping windmills have been extolled in connection with the historical records for the developed nations. Unfortunately, some of the Third World applications have not enjoyed the same success. The missing ingredients are the ones that have plagued technology transfer since the dawn of the industrial era - maintenance, spare parts, and regular follow-up visits. Any new venture, no matter how simple the technology, is doomed to failure without a commitment to these ingredients with even greater intensity than the initial project development.

The opportunities for providing improvement to world health through the increase in more readily available potable water sources are limitless if we will diligently apply the knowledge of what has been successful and what has not been successful in previous ventures utilizing windmills.

Lest We Forget

In 1987 the cost of fully-treated water delivered to a residence in the northeastern part of the United States is in the vicinity of 30 cents per 100 gallons. As I contemplate this luxury, how can I or anyone else so fortunate not make an effort toward helping realize the goal which has been set for our decade by the United Nations - an accessible, clean water supply and adequate sanitation for everyone in the world by 1990.

Woman sitting
at prototype of
village water supply center

METRICS
1 quart = 0.000945 m^3
1 gallon = 0.00378 m^3
1 foot = 0.3948 m
1 mile = 1609.344 m

Micro-Water Projects in Swaziland

David M. Kemp * and David R. Maidment, Member, ASCE **

This paper presents the concept of micro-water projects and how these projects are conceived and implemented in the Southern African nation of Swaziland. Micro-water projects are water supply projects which are built, funded, and maintained primarily by the community they serve. The projects serve between fifty and five hundred persons and have initial capital expenditures of between $1,000 and $15,000 depending on the size and type of system built. Micro-projects arose out of the need for water projects which were less prone to the problems of vandalism, neglect, community dissension, and lack of maintenance that plagued larger, government-built projects in Swaziland. In the late 1970s, the government of Swaziland responded to this need with the inception of micro-project crews. These crews assisted eligible communities by providing engineering design, skilled artisans, transportation, partial funding, and links to governmental agencies.

The paper highlights the basic concepts of micro-projects for water supply in Swaziland. An example is presented of a micro-project in the community of Ekuphakumeni. In this project, animal traction is used to power a high speed pump which delivers water to the community. The members of this community not only provided a large percentage of the funding and labor required, but they were also responsible for routine maintenance and arrangement of the pumping schedule. The process for choosing a given community for a micro-project is discussed along with the number and types of micro-projects which have been built to date. The advantages and limitations of the micro-projects concept are outlined.

Introduction

The idea that community input and participation are crucial ingredients to a successful rural water supply project is a concept which has acquired much support in the last twenty years. However, there are not many successful programs which have fully utilized this vital resource. The government of the southern African nation of Swaziland has developed large water projects which have experienced continued problems with operations and maintenance. Recognizing the link between community participation and the success of a rural water project, the government in 1979 initiated a pilot program of micro-water projects.

---

* Department of Civil Engineering, The University of Texas at Austin, Austin, Texas 78712.

** Associate Professor, Department of Civil Engineering, The University of Texas at Austin, Austin, Texas 78712.

The main goal of these projects is community participation in the planning, design, construction, operation and maintenance of rural water schemes. By emphasizing community participation through the micro-water projects, the Rural Water Supply Board of Swaziland (R.W.S.B.) hopes to improve the success rate of rural water projects. This program is a mechanism by which the government and local communities work together to construct and operate rural water systems. The first author spent three years in Swaziland, from 1981 to 1984, working as a volunteer for the R.W.S.B. During this period, he was responsible for supervising the design and construction of 20 micro-water projects in the Manzini district. Each of these systems provided water to between fifty and five hundred people. This paper describes the concept of micro-water projects, describes how these are undertaken in Swaziland, and assesses the advantages and limitations of this concept.

The Need for Micro-Water Projects

The micro-water project concept arose out of the need for water projects which were less prone to the problems of vandalism, neglect, community dissension, and lack of maintenance that plagued larger, government-built water projects in Swaziland. Furthermore, a method to better utilize the very limited resources of the R.W.S.B. was sought. The government of Swaziland responded to these needs with the inception of micro-project crews. These crews assist eligible communities in the planning, construction, operation, and maintenance of water projects. Micro-water projects emphasize community input of labor and funds in a water project. In this manner it is hoped that the community will feel more protective of and responsible for the project. The sense of pride and ownership which accompanies this type of project is also expected to carry over into other forms of community development.

Micro-Water Projects Defined

Micro-water projects typically serve rural communities with populations of between fifty and five hundred people. The initial capital expenditures for a micro-water project are between $1,000 and $15,000. The micro-water project crews attempt to locate, through community meetings, those communities which possess the desire and motivation to participate fully in the development and operation of a community based water system. In this manner, the government obtains community input into the type, size, location, and cost of the project. While it is anticipated that these types of projects will encourage community based organizations to attempt solutions to local problems, it is also hoped that links between the local community and various governmental organizations will be established during the project. In particular, the involvement of district health workers is encouraged. In addition, since most of these projects are located in relatively remote areas where the ability of the R.W.S.B. staff to maintain them is limited, it is anticipated that local communities will be able to perform routine maintenance. Lastly and most importantly, the government hopes to encourage communities to become less reliant on the national government and more inclined to attempt solutions to local problems.

Micro-Water Projects in Swaziland

Swaziland is divided geographically into four districts and each district has a micro-projects crew. Each micro-projects crew consists of an engineering technologist, a pipe fitter, a builder, two assistants, and a driver/laborer. The crews work out of the district headquarters and, with the exception of the periods when they are living on site, they commute to their various projects from this location. The engineering technologist is directly responsible to the District Clerk of Works, who is in turn responsible to the Senior Engineer of the R.W.S.B. Although the micro-water project crews are somewhat autonomous, they are required to submit project proposals, design drawings, and complete project applications to the design office of the R.W.S.B. The micro-water projects crews have use of the R.W.S.B. equipment and vehicles, at the discretion of the District Clerk of Works. The Clerk of Works, although primarily involved in the direct supervision of larger systems, also serves as a technical resource for the micro-water projects, and site visits by the Clerk of Works to the various micro-water projects are frequent.

Initial Meetings and Project Responsibilities

Communities wishing to participate in the micro-water projects must first request a meeting either through the national or district headquarters of the R.W.S.B. After an initial site investigation by the R.W.S.B. design office and the micro-water crew, a meeting with the community is arranged. At this meeting, which has the micro-water projects crew and the majority of the community members in attendance, the expectations, needs, and assets of the community are meshed with the abilities and policies of the government. The possible source of water, the type of system, the estimated cost, the time of construction, the location of standpipes, and the community contributions of both labor and funds are discussed. It has been found that this initial meeting is the most critical to the success of a given project, so questions from community members are dealt with thoroughly. If the project is both technically and economically feasible, it is important at this stage that the community fully understand what its responsibilities are and what the responsibilities of the government are.

In order to ensure that both parties understand the nature of the agreement, the duties of both the community and the government are presented to each community in a detailed and straightforward manner. The community is informed that it will be responsible for all labor associated with the project as well as partial funding for construction, operation, and maintenance. The amount of funding required for each community is dependent on the type of system and the size and resources of the community. The community is also responsible for selecting five to ten community leaders to be members of a water committee. This committee is responsible for collecting funds, organizing work parties, and delegating responsibilities for performing operations and maintenance tasks. In addition, the community selects three or more individuals who are willing to master basic maintenance and operations techniques.

Since most micro-water systems utilize hand pumps or gravity distribution systems, maintenance usually involves the replacement of leaking taps and the repairing of broken distribution lines, tasks which can be performed by properly trained community members. The individuals selected for the maintenance committee work closely with the R.W.S.B. pipe fitter during the construction of the project and spend at least one day replacing taps and repairing leaking distribution lines under the direct supervision of the R.W.S.B. pipe fitter. These individuals are supplied with the basic tools and materials needed to perform their duties. Finally, these individuals are responsible for reporting the R.W.S.B. the nature and location of any major maintenance problems. The members of this committee are reimbursed by the community for their services and supplies.

The R.W.S.B. design staff is responsible for performing a preliminary site investigation, locating a source of water, providing partial funding, and designing the system. The R.W.S.B. also supplies technical expertise, transportation, materials, and continued technical support for the project. The micro-water project crew establishes links between the community and various other governmental and nongovernmental agencies which can assist the community. Finally, R.W.S.B. is available on a limited basis to assist communities with operations and maintenance.

If, at the conclusion of the initial meeting, the project is deemed feasible by both the community and the R.W.S.B., the community is requested to raise a certain amount of funding and to begin some of the preliminary excavations. The community is informed that R.W.S.B. personnel will begin design and construction activities once the required fund raising and excavation are completed. It has been found that communities which perform these initial tasks are very likely to participate fully in the rest of the project. Once the community has raised the necessary funding and performed the required excavations, the R.W.S.B. staff begins the design phase of the project. The micro-water project technician produces the initial design and then obtains a design approval from the R.W.S.B. design office. The technician produces the design knowing that community members will do most of the maintenance on the system and designs the system accordingly.

Construction of Micro-Water Projects

Once the initial meeting and site investigation are completed, the construction stage of the project begins. Since the micro-water crews are only needed for technical assistance and guidance, the crew can have as many as five or six projects in progress at one time. This requires a well-planned schedule, but results in productive utilization of precious resources. The micro-water crews survey an excavation site or a pipeline and then provide the community with depth gauges and markers so that work can continue without the assistance from the micro-crews for five to ten days. The micro-crews attempt to visit each site at least once every two weeks to check the progress of the work. When the construction of reservoirs, the laying of pipelines, or the construction of standpipes is anticipated, the micro-water crews schedule all such work for a specific block of time. During this period, the micro-water crews live on site and travel to other nearby projects as needed. In this way the community has direct contact with the crews and has the expertise of specialists at hand when they are needed. However, even during periods such as reservoir construction, the community assumes primary responsibility for the construction activities and relies on the R.W.S.B. staff for technical assistance and advice.

Problems Associated With Micro-Water Projects

One of the major problems that the micro-projects experience is communities which slow down or stop work on the project. There are many causes of these slow downs and most projects experience a lull in the work after the initial phase is completed. The micro-water crews have developed a method of dealing with this type of problem. After two or three site visits with little or no progress, the micro-water crew technician will inform the community leaders that until progress is seen on the project, the R.W.S.B. will not participate further in the project. Since the community has a substantial amount of funds and labor invested in the project at this time, the community usually resolves whatever problems were hindering the progress of the project. The R.W.S.B. personnel ask the community to contact the micro-water crews when they are ready to begin activities again. This has proved to be a very effective method of assuring the completion of a project. The key is requiring the community to contribute funds and some labor before R.W.S.B. contributes anything more than the preliminary site investigation.

Maintenance of Micro-Water Projects

As has been the experience in other areas of the world, maintenance is the largest problem which rural water projects face. Not only is maintenance a very large portion of the cost of a rural water program but it is also a never-ending drain on the resources of the nation. In addition, since rural water projects are by nature often located in remote areas, the time required to report a maintenance problem and the cost of transportation to the site can both be substantial. If members of a community are unfamiliar with a system, the exact nature of a problem can be unclear, and in some cases trips to the site to resolve a simple maintenance problem are necessary. In the micro-water systems, community members acquire a working knowledge of how the system functions so many routine maintenance tasks can be performed by the community. Also, since the community has contributed a significant amount of funds and labor, the community has a vested interest in the continued operation and maintenance of the project. This results in a lower cost of maintenance to the central government, as well as systems which are better cared for by the community and therefore longer lasting.

A Typical Project

The community of Ekuphakameni is one of many communities which has participated in the micro-water projects program. This community, which consists of 150 people, is located on a hill 70 meters above a spring.

The spring has been a traditional source of water and was open to contamination by both people and animals. A delegation from Ekuphakameni met with the Manzini micro-projects crew and a site visit and meeting was arranged. After a positive site investigation and several community meetings, it was decided that the spring would be protected and water would be pumped to the community. Since the community was located in a remote section of Swaziland and the ability to obtain fuel was limited, the use of a diesel pump was not feasible. After four meetings with the community, it was decided that a pump which could utilize animal traction should be located and tested. A pump which met the requirements was located at the Rural Industries Innovation Center in Botswana. After inspection of the pump by a mechanical engineer and the micro-water technician, it was determined that the pump met the requirements of ease of maintenance, ruggedness of construction, and low cost of operation. The community of Ekuphakameni provided all labor for construction of the project, as well as $2,800 for piping, and an agreement to perform required maintenance and operations. The total cost of this project was $15,000 and the difference was provided by the R.W.S.B. and the Canadian International Development Agency. The time required to develop this project was seven months. The community has organized members to perform routine maintenance and to provide oxen for pumping. Three hours of pumping are required to fill a 9,000 liter reservoir, which is sufficient for three days of water for the community. The R.W.S.B. micro-water crew visits the site routinely and are assisting the community with the establishment of pumping and maintenance schedules.

Conclusions

The choice by the Swaziland government to encourage community members to participate in the planning, design, construction, operation, and maintenance of small rural water projects has proved to be a wise decision. Micro-water projects have increased the feeling of pride within communities as well as better utilizing scarce resources. In addition, communities have been exposed to the various agencies which can provide technical, educational, and financial assistance. Communities have also become more protective of their water systems. This attitude has lead to a decrease in the expenses associated with both the operation and maintenance of projects. Finally, the communities have gained a sense of power having had a direct input in the development of their projects.

Water Supply Planning in Developing Areas:

Lessons from Puerto Rico

Dr. Gregory L. Morris *

INTROUCTION

Puerto Rico offers an instructive environment for examining water planning issues pertinent to developing areas.

* Physically, the island consists of 8,890 square km of generally mountainous topography with an extremely diverse hydrologic system, onto which a 1985 population of 3.3 million and a GNP of $15 billion has been superimposed along with their respective water demands and environmental impacts.

* Historically, the island has experienced rapid economic development and it now provides a level of water supply service that developing areas will find enviable. For example, the P.R. Aqueduct and Sewer Authority now supplies municipal water to 96% of the population.

As a consequence of rapid development and intense utilization of resources, Puerto Rico is today addressing water management issues which may not become important in other developing areas until well into the next century or even beyond.

Puerto Rico also shares another pattern in common with many developing areas: water supply development has been relatively unhindered by an integrated planning processes. This approach is typical of developing areas where the pattern and rate of economic performance is volatile, an engineer's political affiliation is important, and technical resources are scarce. It is somewhere between difficult and academic to apply sophisticated planning techniques and formalize the planning process in such an environment. Nevertheless, this does not eliminate a developing area's need to use limited economic and technical resources efficiently.

This paper outlines several basic planning concepts, illustrated using examples from Puerto Rico, which can be applied to help avoid severe errors during project conceptualization and planning.

* Morris & Assoc. P.O. Box 5635, San Juan, PR 00905

Concept #1: WATER SUPPLY IS NOT A RENEWABLE RESOURCE

Contrary to popular misconception, although rainfall is a renewable resource, water supply is not. This apparent contradiction is understandable when it is reccalled that rainfall arrives at irregular intervals producing streamflows that may vary over three orders of magnitude during the course of a year. The firm yield of streams is a small percentage of the mean annual runoff; storage is required to increase water supply (Table 1).

TABLE 1: STORAGE CAPACITY AND FIRM YIELD COMPUTED FOR SIX STREAMGAGE STATIONS IN PUERTO RICO (Morris, 1984).

	Storage Capacity	Firm Yield
	(Percent of Mean Annual Streamflow)	
No reservoir	0	5
Small reservoir	1	15
Large reservoir	75	71

Storage is not a renewable resource unless specifically designed and operated with that objective in mind. However, most surface water impoundments are neither designed nor operated as renewable resources due to the problem of sedimentation.

The susceptibility of different reservoir to sedimentation varies dramatically. The useful life of some reservoirs may exceed 1,000 years, as exemplified by the Proserpina and Cornalbo reservoirs which were constructed by the Romans during the 1st and 2nd Centuries, and which today supply water to the City of Mérida, Spain (Ordóñez, 1984).

However, long-lived reservoirs are the exception rather than the rule and many important reservoirs are subject to rapid sedimentation. Of the 26 reservoirs in Puerto Rico, only Carite is expected to have a life approaching 1000 years based on its high ratio of capacity to watershed area (DNR, 1984).

Carraizo reservoir, constructed in 1953 and the principal source of water supply for San Juan, will have a useful life of only about 60 years if sediment control measures are not implemented. In only three decades approximately half of the reservoir volume has been lost to sediment accumulation and sediments are approaching the level of the crest gates. Sedimentation of this reservoir, which controlls the island's largest watershed, is an acute problem. Downstream there are no potential reservoir sites, upstream the river branches into several small tributaries before any sites suitable to reservoir construction are encountered, and the dam crest cannot be raised because the City of Gurabo would be flooded.

Today we accomodate reservoir sedimentation by providing dead storage, leaving future generations to solve the problem of sedimentation. However, for the City of San Juan the future has now arrived and it is unpleasant. Sediment control strategies are being examined with particular emphasis on changing operating rules to route sediments through the reservoir (Morris 1986). Fan Jihua (1985) provides a good overview of sediment control techniques that have been employed successfully around the world.

It is important to remember that municipal water supply projects are long-term investments in economic and environmental infrastructure. The long-term consequences of water supply development should be carefully considered, and development strategies which result in a non-renewable water supply should be avoided even though short term (eg. 50 year) economic analysis indicate they are "cost effective". Future generations will be as dependent on water resources as we are today; do not leave our children's children's children a legacy of sedimented reservoirs, contaminated aquifers, and long pipelines to tap increasingly remote sources of water supply.

Concept#2: YOU CAN'T PREDICT THE FUTURE

Projecting water demand is perhaps one of the most difficult and dangerous tasks in developing areas. Good projections are difficult to make because they require more insight than analytical sophistication, and there is often a wide disparity of opinions concerning probable or desirable patterns of future development.

Projections are dangerous because are the basis for financing large and costly infrastructure projects which may never achieve payback if the projections are not realized. The problem of projecting water demand for industrial and agricultural uses is particularly acute since these sectors may expand or contract rapidly in response to their changing competitiveness in the international marketplace.

The dangers inherent in uncertain projections, and development strategies to minimize these problems are illustrated using the 5-stage Toa Vaca project. This project was proposed in the early 1960s as a system of reservoirs and tunnels to divert water from Puerto Rico's moist northern watersheds to the drier south coast to supply a new heavy industrial sector, which was based largely on petrochemicals and growing rapidly. There was great pressure to develop a major new source of water supply because the existing south coast supplies were already appropriated for irrigation (sugarcane) and little local water supply remained to support industrialization.

The Toa Vaca project suffered three basic conceptual difficulties which plague projects in developing areas all too frequently.

1. Half of the demand (and most of the income) was projected to come from growth within the new heavy industrial sector. However, there were serious problems with the projection of industrial water demand, as suggested by Figure 1.

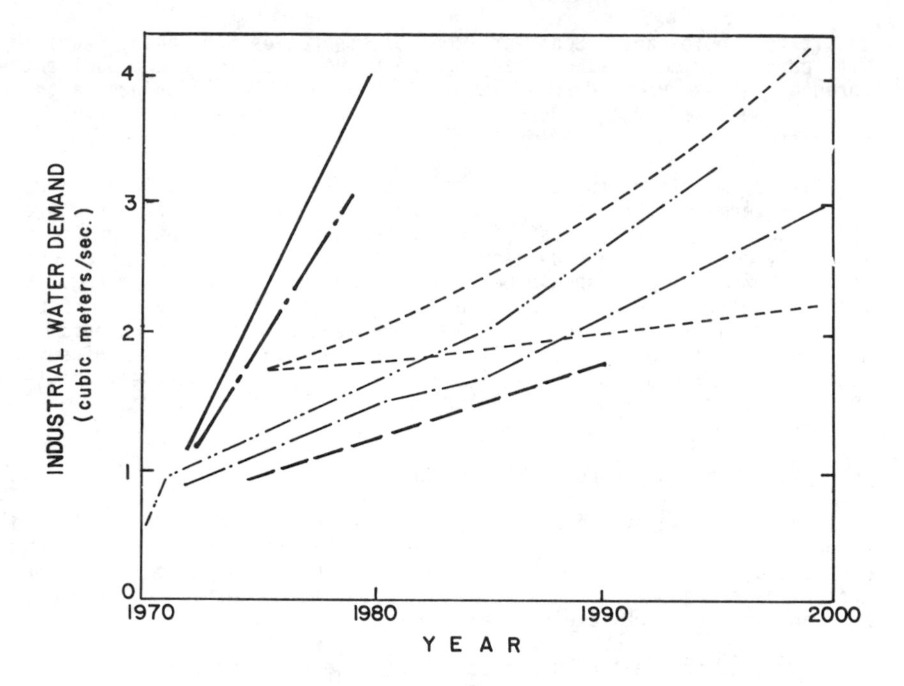

Figure 1: Projections of industrial water demand on the south coast of Puerto Rico prepared during the late 1960s and early 1970s (Morris, 1979).

2. Given the uncertainty in demand it would be logical to look for a project that could be conveniently implemented in several phases, each of which could pay its own way. However, the project was designed with extremely high early costs (due to a 14 km tunnel in Stage II) and could not achieve break-even until completion of Stage V (Table 2).

3. Viable alternative water supply schemes including other reservoir sites, wastewater reuse, and improved management of groundwater were not considered in the project analysis. In the economic analysis the Toa Vaca project was compared to only one alternative: seawater desalination. With this type of comparison it was predictably the "least cost" alternative.

TABLE 2: SUMMARY OF COST IN 1970 DOLLARS AND YIELD FOR PROPOSED TOA VACA PROJECT, PUERTO RICO (Development & Resources Corp., 1970).

Stage	Cost Per Stage ($ millions)	Yield Increment (m^3/sec)	Cost Per m^3/sec of Yield Increment ($ millions)
I	$17.9	23	$30
II	53.5	20	62
III	59.4	60	23
IV	32.1	25	30
V	50.6	100	11
TOTAL	213.5	228	

A subsequent analysis of potential reservoir sites Islandwide by the U.S. Army Corps of Engineers (1980) revealed a number of alternative projects which could have been implemented at much lower cost and with better staging that the Toa Vaca alternative.

Only Stage I of the Toa Vaca project was constructed, thanks to the oil crisis of 1974 and the rapid stagnation of the industrial boom on Puerto Rico's south coast.

In summary, given unreliable demand projections, an unreliable supply of foreign exchange, or other unknown factors that can frustrate project performance, it is not necessarily best to search for projects that will provide the "optimum" long-term economic performance. Rather, it is logical to look for projects that can be conveniently implemented in reasonable increments and over a flexible timetable, without creating an undue financial burden if the projections are incorrect. Whenever the failure to carry a multi-stage project to completion will have severe economic consequences, implementation should be seriously questioned and alternatives sought.

Concept#3: PLANNING AND MANAGEMENT COSTS LESS THAN CONSTRUCTION

There is a certain favorable auroa that surrounds new construction. It has often been suggested that this is because politicians like to attend project inagurations against the backdrop of thousands of fresh tons of concrete and steel and have their names engraved in bronze plaques.

Less widely recognized, but I believe far more important to understanding the new project syndrome, is the reality that it is administratively easier to construct a large new project than to optimize the management of existing systems. New projects don't change the status quo of inefficient work habits. In contrast, the efficient management and optimization of any water supply, treatment and delivery system requires efficient and daily attention to numerous details.

Yet, there do exist some opportunities to achieve a dramatic increase in water use efficiency without significantly re-ordering work priorities. In Puerto Rico one such approach is to require that all new toilets installed be of certified water-conserving design through regulation or law.

There is very limited lawn irrigation in Puerto Rico, and without lawn irrigation the water use by conventional toilets can approach half of the total residential water demand. In addition to reducing consumer demand:

* Water conservation reduces the required deliveries (or reduces their rate of growth) which should produce a corresponding reduction in water loss. Thus, in computing the water savings it appears appropriate to count both the water not used in the toilet PLUS the water not lost in the distribution system. This can be an important consideration in developing areas where distribution losses are typically high.

* The enlargement of transmission pipelines and construction of new supply and treatment facilitiies can be postponed or eliminated.

The economic savings that can be realized from water conservation are large (Table 3), and conservation is so important that it should be given at least as much emphasis as new project development.

TABLE 3: SAVINGS IN WATER PRODUCTION AND THE COST OF NEW WATER SUPPLY FACILITIES RESULTING FROM THE INSTALLATION OF WATER-CONSERVING TOILETS IN PUERTO RICO AS A FUNCTION OF DISTRIBUTION SYSTEM LOSES (DNR, 1984).

	Percent of Water Lost	
	17%	40%
Economies in Water Production (m^3/sec)		
Reduction in water demand	2.02	2.02
Reduction in water production	2.54	3.59
Economic Savings ($ millions)		
Reduction in construction cost of required new facilities	$145	$205
Reduction in interest payments on new facilities not constructed	238	336
Reduced operation & maintenance expenditures each year	5.2	7.4

Best of all, an informal survey of building suppliers in Puerto Rico revealed only a $2 difference in cost between the "economy" models of water-conserving and conventional toilets. Thus, the conversion to water-saving technology does not necessarily entail economic burden.

CONCLUDING REMARKS

Major errors during project conceptualization can often be avoided by applying basic planning concepts. Examples similar to those from Puerto Rico cited in the article abound in all areas of the world. It is up to conscientious water professionals to look at the development history in their own area to understand past mistakes that should not be repeated and to uncover new opportunities to be exploited.

REFERENCES

Development & Resources Corp. 1970. Economic Feasibility of the Toa Vaca Complex. Report to P.R. Aqueduct and Sewer Authority, San Juan.

Fan Jiahua. 1985. Methods of Preserving Reservoir Capacity. In: Bruk (ed) Methods of Computing Sedimentation in Lakes and Reservoirs. UNESCO, Paris.

Ordóñez, José A. Fernández. 1984. Catálogo de Noventa Presas y Azudes Españoles Anteriores a 1900. Comisión de Estudios Históricos de Obras Públicas y Urbanismo, Madrid.

Morris, Gregory L. 1979. Regional Development and Water Resource Management: Implications of a Changing Agricultural Sector on Puerto Rico's South Coast. Ph.D. Diss., Univ. of Fla., Gainesville.

Morris, Gregory L. 1984. Water Supply Augmentation Using Small Reservoirs. pp 125-142. In: Proceedings, First Caribbean Islands Water Res. Conf. Water Res. Research Inst., St. Thomas.

Morris, Gregory L. 1986. Controlling Reservoir Sedimentation: The Billion Dollar Opportunity. pp 33-37. In: Anuario 1986, P.R. Section, AWWA-WPCF. San Juan.

Puerto Rico Dept. of Natural Resources. 1984. Plan de Agua Para Puerto Rico (draft). San Juan.

U.S. Army Corps of Engineers. 1980. Island-Wide Water Supply Study for Puerto Rico. San Juan.

A STRATEGY FOR CARIBBEAN WATER SUPPLY PLANNING

Henry H. Smith, Ph.D.*

ABSTRACT

In most of the smaller islands of the Caribbean, there is no single natural source of water that may be used to satisfy the ever rising demands for consumption and sanitary purposes brought on mainly by increasing standards of living and visitor arrivals. Mountainous terrain makes buildable land dear and along with high evaporation rates, makes large scale surface water impoundments impractical. Ground water supplies are limited due to high runoff rates and little opportunity for recharge. Salt water intrusion is generally a problem of major concern. Rainfall harvesting has traditionally been the major and preferred source of water for residential use.

Desalination of sea water is an attractive option to islands where the raw resource is readily available, free, abundant and of high quality. Experience has shown that while desalination may theoretically be a solution to water shortage problems, there is a high potential for the creation of many other problems which may preclude acheiving the true goal - provision of water where needed, as needed, and of desirable quantity and quality.

A model is developed to allow the simulation of an integrated system which includes desalination, rainfall harvesting and ground water withdrawal. This model facilitates productive interaction between the research staff, the decision maker and the systems analyst who may not all have the same objectives, constraints or may have difficulty communicating with each other for reasons such as differences in expertiste, training, etc. A solution strategy is proposed for workable and acceptable systematic step-wise development of conjunctive use system configurations that incorporates input from all involved in the decision making process.

* * * * *

*Director, Water Resources Research Center, University of the Virgin Islands, St. Thomas, VI 00802.

The attractiveness of the smaller lesser developed islands of the Caribbean as places to live is constantly increasing. This has resulted in high demands being placed on water distribution systems which may have been served from already severely taxed ground water supplies.

Ground water in these islands, though potentially least expensive of all water, for the most part is limited due to the steep volcanic relief as well as limited land area to serve as catchments areas for the spurious, brief, intense, tropical showers. What limited ground water there is must be cautiously managed to prevent sea water intrusion.

The steep relief and high evaporation rates restrict surface water supplies to ephemeral streams. Surface dams tend to be small ponds for stock watering and generally only of significance during the wetter periods of the year. Dams for potable water supply are uncommon.

Rainfall in these islands is seasonal with a large portion of the rain falling between August and November. Rain water has been harvested on a large scale on some islands through the use of hillside catchments made by using corrugated galvanize, concrete or compacted limestone. The large areas, high storage volumes required and questions of the quality of water derived from these system have restricted their use.

More popular is rain water harvesting by individual homes and subsequent storage in cisterns. In fact, cisterns meeting particular size requirements are required by law in several of the islands. These cisterns generally form an integral part of the foundation of residences and can be very costly. They are often estimated as costing approximately one quarter of the cost of the house.

From the water manager's vantage, cisterns are attractive for costs associated with distribution of water are reduced. At the same time there has been a traditional preference for cistern water. It has been suggested that this may be a result of a feeling of independence by users of these sytems. Users are more comfortable knowing that they have °control' of the quality and quality of their water supplies.

On the other hand, when cisterns are predominant and a central distribution system is not in place, fire fighting is major problem. Additionally without distribution systems, when cisterns fail to meet demands, distribution of water to where it is needed is a major undertaking. Additional problems associated with cistern systems are the maintenance of water quality in these systems, the dependance of these systems on the highly variable rainfall and the inadequacy of these systems for use with multistoried structures where demands on them are multiplied but the associated catchment area is not.

Sea water desalination is an alternative that is appearing to be more and more attractive in the Caribbean. Desalination makes use of a plentiful readily available free resource to produce a potentially unlimited supply of superb quality water. Particularly attractive is that this water may be produced at will.

Unfortunately there are several costs that are associated with desalination systems. There are energy, capital and site specific costs which must be met. Increasing the size of desalting plants do not necessarily reduce unit costs of water as is commonly the case in other water projects. Plant capital costs are greatly affected by desired plant efficiency and availability. Additionally, the operation and maintenance of desalination plants require highly trained personnel.

In the setting described above increasing standards of living coupled with rapidly increasing visitor arrivals as well as permanent populations have resulted in spiralling water demands. To make matters worse, high water demand periods coincide with the periods of lowest rainfall. Reliance on single sources of supply is proving to be unwise and moreover unsatisfactory. Crude attempts at utilizing multiple sources of water have often been unecessarily costly and in many instances have completely failed leaving consumers worse off than they were before. There is then need for systematic development of planning and management strategies for attainment of demand satisfactions. Such strategies should incorporate to as high a degree as possible the merits peculiar to each water source while similarly minimizing to as high a degree as possible the lesser desirable aspects. This is the essence of proper management of water supplies in conjunctive use strategies.

In a conjunctive use scheme in which cisterns and sea water desalination is to be used there is much potential. Each of these sources have attributes not common with the other source. Cisterns may provide water at a relatively lower cost but become less reliable as demands increase and rainfall becomes more variable. Desalination systems on the other hand can provide high quality water in great quantities as needed but is comparatively more expensive.

From a strictly mathematical approach, a conjunctive use strategy for desalination systems and cistern systems may be developed. Such a strategy would determine the optimal cistern size for use in these situations. The cisterns would be large enough to serve as main water supplies during periods of heavy rainfall but not so large that there is invested capital in cisterns whose full benefits are seldom realized. On the other hand the cisterns would not be so small that they continually overflow and provide a reliability that would essentially be no greater that what in surface water hydrology would be termed °run of the river'.

In the strictly mathematical approach the desalination plant would be designed to meet demands during the periods that the cisterns 'fail'. A design based solely on volume requirements would not be adequate. Instead, desalination plant costs may be reduced if the desired plant reliabilty is also incorporated in the plant design considered.

In developing the conjunctive use strategy, it was considered of critical importance to not end up with a product that is based only on the physical parameters but equally important, on the non-commensurate factors that are very often the crucial ones in many of these situations.

What was desired was a product that is also useable and acceptable to its potential users. The strategy then includes a component that allows input to the model developer from the research staff as well as the decision maker.

The conjunctive use model developed was structured around simulation of a system utilizing ground water, cistern systems and a desalination plant. Among the critical required input for the model are a weekly rainfall record, an anticipated demand, the water harvesting area available, the catchment efficiency of this area, and the volume of cistern storage and supplemental storage in the system. The physical characteristics of the available aquifer must be specified as well as the capacity of the desalination plant and its governing operating rule.

The conjunctive use model first determines the amount of water that may be harvested during the first week of the simulation as well as the water which may be obtained from wells. Since ground water in this model is considered transient, demand is first placed on this source. If there is insufficient ground water then the excess demand is placed on the rain harvested. If demands are satisfied the excess is added to cistern storage. If the demand is not satisfied then the cistern storage and supplemental storage are tapped in that order. Whether or not the desalting plant is producing water during that simulation period is determined principally by the volume of water in supplemental storage and the operating rule for the plant which was specified as an input.

This process is repeated for each week in the simulation period with the state conditions in each week directly related to the occurences in the previous week. Model output at the end of the simulation include the duration and amounts of shortages and surpluses, desalination plant operation history, etc. A sample of a portion of the tabulated output is presented in Table 1.

Table 1. Selected Tabulated Output From Conjunctive Use Model

Plant Capacity (Mgal./Week)	Cistern Capacity (Gal./Ft of Roof)	Supplemental Storage Needed (Mil. Gals.)
1.75	2.5	a
	5.0	22.4
	7.5	6.3
	10.0	2.9
2.00	2.5	a
	5.0	18.1
	7.5	2.5
2.50	2.5	a
	5.0	5.6
	7.5	0.0
5.00	2.5	3.8
	5.0	0.0
	7.5	0.0

a. Excessively large storage volumes required.

The output of the conjunctive use model then is used to determine if the desalination plant used in the original simulation is suitable for the application intended. A long record of shortages as well as extended plant running times might be indicative of a plant capacity that is too small. Few shortages and some spills but having either very frequent plant running times or very extended running times may point to an inadequate operating rule.

In the strategy developed the simulation model is run for a wide range of conditions and several consequence curves developed. These curves are the vehicle through which the decision maker can provide input to what has previously been a purely technical development. The input of the decision maker is critical not only from a relations standpoint but also because of the unquatifiable input that the knowledgeable and experienced decision maker may provide. The decision maker is aware of political expediency, people's customs and traditions and several other non-technical factors on which the success of a project may depend. At the same time, the decision maker may not have the technical expertiste to evaluate a proposal based on its technical merits. Consequence curves developed using technically sound reasoning provide decision makers the opportunity to view the range of consequences of decisions they may make.

In the procedure, after review of the consequence curves, the decision maker may choose his preference or suggest other system configurations. The availabilty of a range of consequence curves allows the decision maker to develop a sensitivity to the effects of varying physical parameters that he normally would not have. The suggested configuration would then be simulated again by the analyst along with closely related configurations to develop another series of curves.

This process of simulation and revision provides the decision maker with the opportunity to learn the technical operating environment of the system by experience and similarly the analyst gets a feel for what the priorities, objectives and constraints of the decision maker are. By iteration between the analyst and the decision maker, a system configuration should result that is acceptable to both parties and with proper implementation should achieve the desired objectives.

SELECTED REFERENCES

Dracup, J.A. 1966. The Optimum Use of a Groundwater and Surface Water System, A Parametric Linear Programming Approach. Contribution No. 107, University of California, Berkely, California.

Mawer, P.A., and Burley, M.J. 1968. "The Conjunctive use of Desalination and Conventional Surface Water Resources." Desalination, Vol. 4.

Spiegler, K.S. and Laird, A.D.K. (ed.) 1980. Principles of Desalination. Academic Press, New York.

Young, R.A. and Bredehoeft, J.D. 1972. "Digital Computer Simulation for Solving Management Problems of Conjunctive Groundwater and Surface Water Systems." Water Resources Research, Vol.8, No. 3, June.

Rainwater Collection and Treatment

at a

Small Industrial Plant in Puerto Rico

Carlos E. Pacheco, P.E. *

Gregory L. Morris, Ph.D **

INTRODUCTION

The "Federación para el Desarrollo Agrícola" (FEDA), a cooperative serving small farmers in the central mountainous region of Puerto Rico, elected to construct their new 950 m^2 (10,000 ft^2) fruit processing facility in Barrio Carite. The plant site, at an elevation of approximately 600 meters above sea level, offered location advantages to the nearby farmers. However, this area had deficient water supply infrastructure.

The Barrio Carite potable water supply operated by the Puerto Rico Aqueduct and Sewer Authority (PRASA) consisted of a stream diversion which provided no treatment other than chlorination. In addition to producing highly turbid water during periods of wet weather, the system also suffered from an unreliable pumping and chlorination system. In short, the existing system could not provide water of the quality and reliability required for a food processing supply.

The factory is located on the crest of the divide between the La Plata and Patillas watersheds which ruled out the use of either a nearby stream (there were none) or groundwater. This left the project with only two alternatives: (1) providing storage and treatment for the existing PRASA supply, and (2) use of rainwater. A system was designed incorporating both alternatives.

* Pacheco, Carlos & Assoc., Sagrado Corazón 521, Santurce, PR 00915
** Gregory L. Morris & Assoc., P.O. Box 5635, San Juan, PR 00905

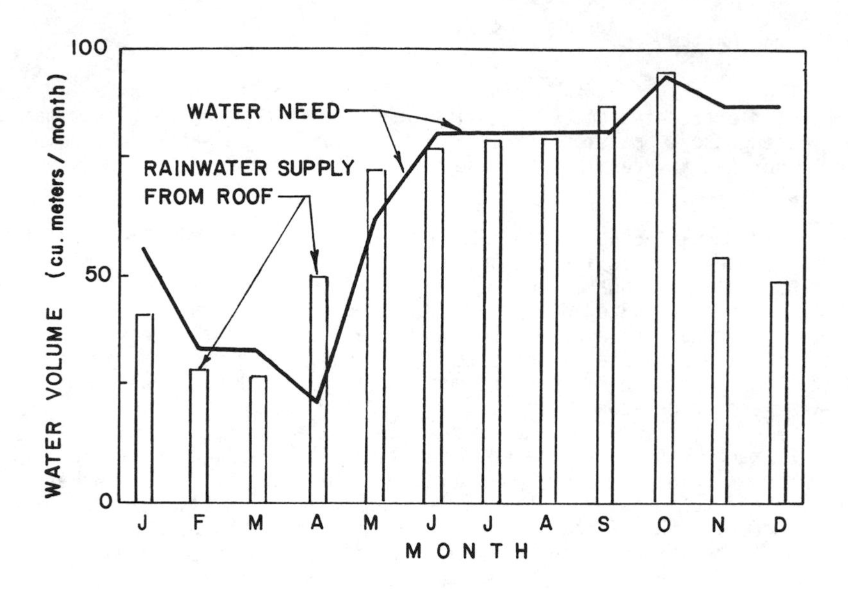

Figure 1: Seasonality of projected water need and average rainwater supply from the available roof area, FEDA food processing facility, Barrio Carite, Guayama, Puerto Rico.

BASIS FOR DESIGN

The factory's water requirements were estimated based on the projected processing activity. The seasonality of the factory's major processing crop, passion fruit, plus the different water requirements for other fruits, resulted in the seasonally variable water need shown in Figure 1.

Based on 1,455 sq m (15,652 sq ft) of roof area, 1828 mm (72 in) average annual rainfall and a projected annual water need of 3140 cu m (828,000 gal.), rainwater collection could theoretically provide up to 85% of the total water demand with the provision of adequate storage. However, actual rainwater use will be significantly less than that theoretically possible since this would require a very large storage volume to compensate for differences in the seasonal patterns from year-to-year and large storm events.

Storage was sized based primarily on PRASA water availabilty, providing a rainwater storage volume equivalent to 38 mm (1.5 inches) of roof runoff to supplement the PRASA supply. The utilization of rainwater, despite the limited rainwater storage, is enhanced because months of high water demand generally correspond to months of high rainfall (Figure 1).

The major components of the water system are diagrammed in Figure 2. A 322 cu m (85,000 gal.) tank provides both storage and plain sedimentation of PRASA water prior to filtration. A gravity sand filter was constructed and valved so that all rainwater was filtered but PRASA water could bypass the filter when its quality was acceptable. Chlorination was provided following filtration. A 57 cu m (15,000 gal.) tank provides rainwater storage capacity equivalent to 38 mm (1.5 in.) of roof runoff and also serves as the sump for the pressure water pump.

The gravity sand filter (Figure 3) was provided with 3.35 sq m (36 sq ft) of surface area, with a nominal filtration capacity of 7.9 lps (126 gpm). Roof runoff first passes through a 20 mesh stainless steel screen on top of a 1.9 cu m (500 gal.) stainless steel rinse tank salvaged from another factory. The screen prevents the entry of leaves and twigs into the system. Referring to Figure 3, the pipe at the bottom of the rinse tank (#6) diverts the initial runoff out of the system and water enters the filter only when the level rises to pipe #1. During a long storm or during periods of continuously wet weather a valve on pipe #6 (not illustrated) can be closed to limit rainwater losses.

The filter consisted of 24 inches of sand over 21 inches of gravel supported by an underdrain system consisting of pre-cast elements with 9.5 mm (3/8 inch) orifices. Backwash supply is provided by gravity from the 85,000 gal. tank. A pellet chlorinator was used with pellet discharge proportional to flow as driven by a water meter.

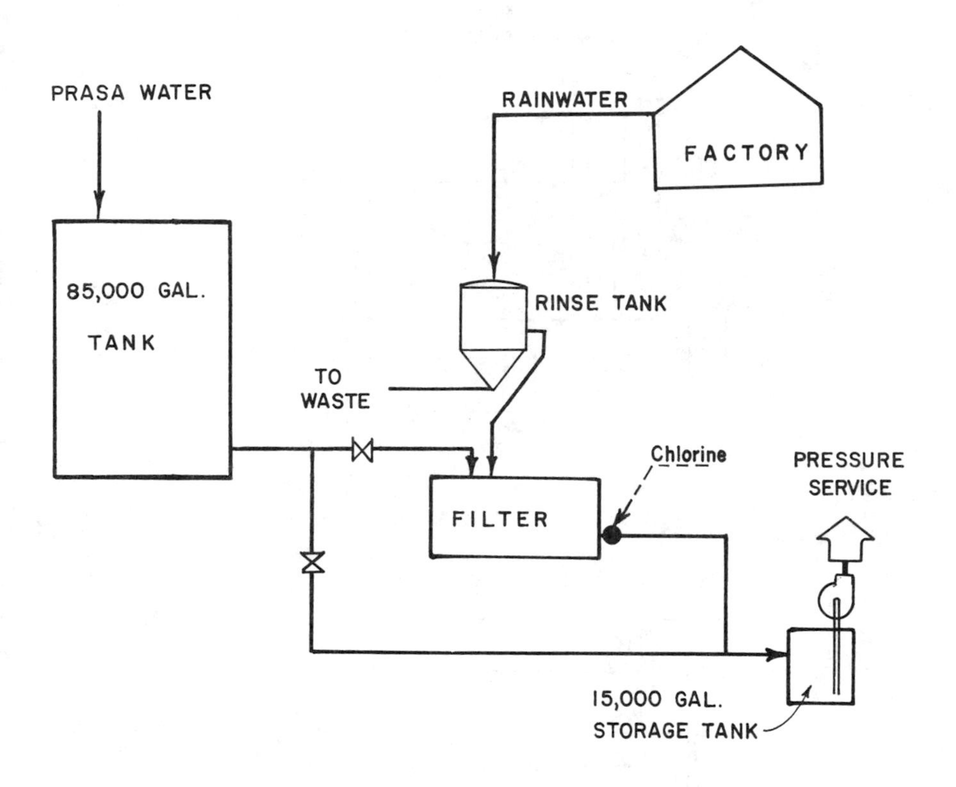

Figure 2: Schematic diagram of the water supply system using both purchased water and rainwater, FEDA food processing facility, Barrio Carite, Guayama, Puerto Rico.

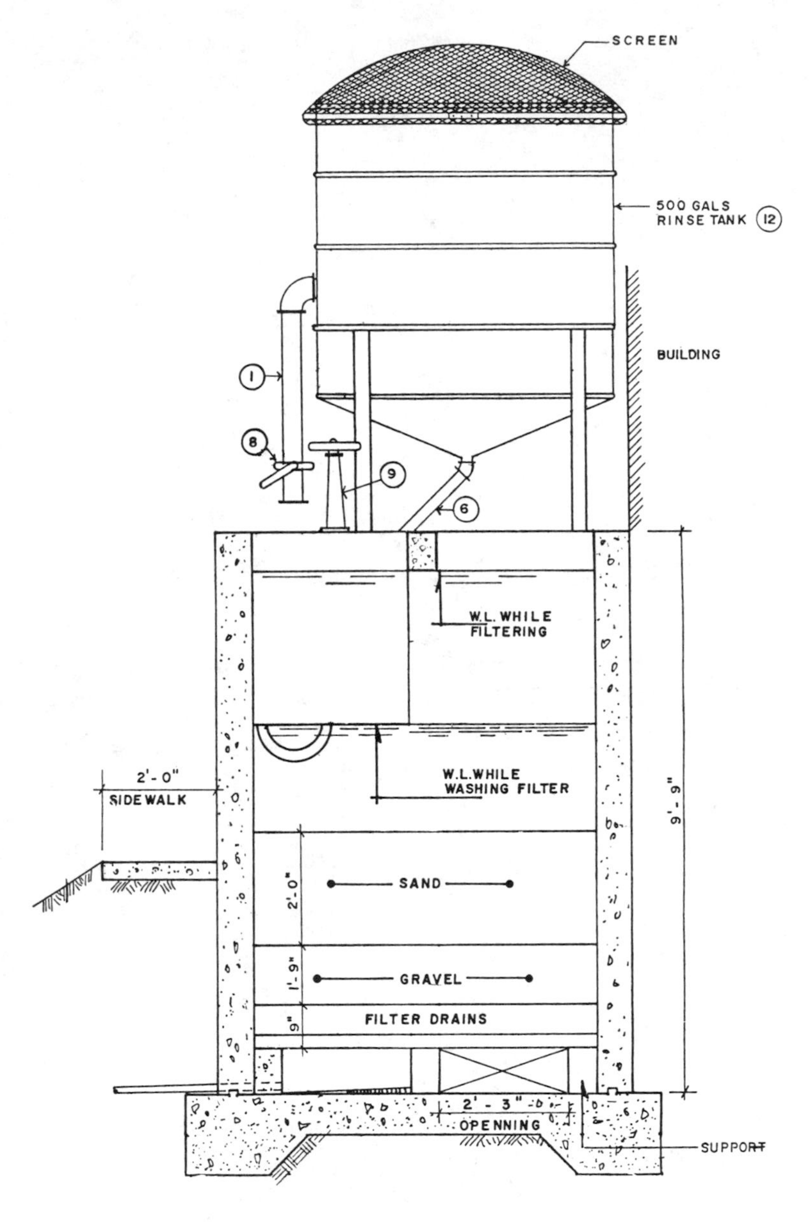

FIGURE 3

RAPID SAND FILTER

SYSTEM COSTS

The entire system (collection, storage, treatment and pumping) cost $ 90,000 to construct as summarized in Table 1.

TABLE 1: CONSTRUCTION COST OF WATER SUPPLY SYSTEM (1983 costs).

Component	Construction ($ US)	Cost (% of Total)
STORAGE TANK (85,000 gal.)	$ 53,204	59 %
FILTER & CHLORINATOR	14,816	17
STORAGE TANK (15,000 gal.)	13,480	15
RAINWATER COLLECTION SYSTEM	8,500	9
TOTAL	$ 90,000	100 %

The purchase of 3,134 cu m/year (828,000 gal./year) of PRASA water would cost $2,864, equivalent to a present worth value of $28,119 using an 8% annual discount rate. The price of PRASA water does not represent the true cost of the Barrio Carite water supply, since their billing structure is based on average water supply costs island-wide.

Rainwater collection and treatment is not an economical alternative to the purchase of water from a central water supplier in this instance. Morris et. al. (1984) have previously shown that a rainwater collection and storage system costs substantially more than a conventional water supply in Puerto Rico. However, given the need to install a storage and treatment system due to deficiencies in the purchased water supply, the addition of a rainwater collection system was highly cost effective in this instance.

SUMMARY

Even when a rainwater collection and treatment system is not an economically attractive alternative to the purchase of water from a central water supplier, under certain conditions a rainwater collection system can be successfully incorporated into an industrial facility to generate cost savings.

Given the tendency for industrial facilities to have multiple types of water use with a variety of quality requirements, the use of rainwater in an industrial environment can be facilitated by directing it to non-sanitary uses to eliminate the need for treatment. Alternatively, rainwater could be separately stored for use or blending in processes which require water with a low mineral content.

The greatest impediment to rainwater use is the requirement for large amounts of storage to take advantage of intermittent rainfalls. The use of rainwater is facilitated in situations where water need exhibits a seasonal pattern similar to rainfall and where rainfall can be used to supplement an existing water supply.

REFERENCES

Morris, Gregory L., Raymond Acevedo-Pimentel and Gaddial Ayala. 1984. Yield and Cost of Water Supplies from Rain-Fed Cisterns: Puerto Rico. Proc. 2nd Intl. Conf. on Rain Water Cistern Systems, Caribbean Research Inst., College of the Virgin Islands, St. Thomas.

PLASTIC PIPES FOR DEVELOPING NATIONS

B. Jay Schrock,* Member ASCE

ABSTRACT

The proper use of Plastic Pipes for water supply for developing nations is paramount. Short term economics should not supplant long term performance criteria. It is essential that the Engineering Consultant and the user Agency understand plastic pipe physical, mechanical and chemical characteristics, and that proper design and construction specifications be administered.

The International Standard Organization (ISO), the American Society for Testing Materials (ASTM) and the American Water Works Association (AWWA) Standards should be understood and utilized as a minimum for pipeline project activities. If the need arises, the project design and construction may require upgrading the Standards to meet local conditions.

The proper enforcement of installation criteria must be performed in order to meet long term operational features. Certain plastic pipe limitations have been reasonably well documented world-wide. This information should be reviewed by the Consultant and Water Supply Agency. These items should be understood and provisions enacted in order to capitalize on the many advantages of plastics and avoid certain disadvantages.

The manufacture of plastic pipes must meet certain long term performance testing criteria. Various testing procedures have been developed during the past thirty (30) years. Pipes manufactured by different companies can look the same, however, long term performance can vary many fold. Therefore, proper certification of pipes manufactured and delivered to the project must be obtained.

INTRODUCTION

The growth of plastic pipe use during the past forty (40) years has been tremendous. This has transpired despite the lack of a coherent analytical design technology, in particular, the mode of prediction of performance under stress. The use of plastic pipe has been developed almost exclusively by trial and error. Better technologies and un-

* Consultant, JSC International Engineering, Sacramento California, USA.

derstanding has significantly improved during the past 15 to 20 years. The introduction and use of long term pressure testing, and improved jointing systems, has contributed greatly toward better performance of plastic pipe under pressure.

The various AWWA, ASTM and ISO Standards have established minimum levels of product performance criteria. It should be noted that these Standards should not be interpreted broadly and/or followed blindly. It is necessary for the Design Engineer and the User Agency to evaluate the Standard and make necessary adjustments for local conditions, sometimes on a case-by-case basis.

The need for developing rational formulas when using plastic pipe has compounded in recent years due to emphasis on product reliability and the need for substituting plastics for metal due to the energy crunch.

The historical use of various stress-strain formula for elastic materials such as metals cannot be used directly on plastics due to their viscoelastic nature. Metals have constant Stress-Strain below their yield point, where plastics not only are not constant but vary greatly with time under load, with rate of loading, with small changes in temperature, and with details of testing. The stress-strain relationship with plastics is a moving target in that the strain properties accelerate relative to an increase in stress.

The plastic pipe industries are constantly changing by upgrading their product standards and using improved quality materials. The industries are developing better and more meaningful testing methods and procedures that simulate actual field operating conditions.

With the foregoing in mind, it is imperative that the Design Engineer and User Agency follow Standards and require manufacturer certification to these Standards in detail.

PIPE MATERIALS

The ASTM Standards for long term pressure testing states that for a given resin compound or formulation and reinforcement orientation, certain tests must be conducted in a specific fashion and a specific formula must be adhered to and used. What this essentially means is that any changes in the product materials or ingredients does require a new long term test program. These various long term tests identify certain properties such as creep, fatigue, and other paramenters that are inherent in the materials.

It is necessary when designing pipes for potable water

use that the product and product materials are in conformance to the criteria published in the National Sanitation Foundation (NSF) Standards. Standard No. 14 requires that the plastic piping system components and related materials meet certain features that bear upon public health, safety, and performance-oriented requirements.

Manufactures of materials for use in plastic pipes for pressure applications shall provide evidence of design stress conformance. Thermoplastic materials must comply with the policy and procedures for developing recommended hydrostatic design stresses (HDS) for their thermoplastic pipe materials when tested in accordance to ASTM D 2837. Thermoset materials must be tested in accordance to ASTM D 2992 for obtaining similar information.

AWWA C 900 Standard for PVC thermoplastic pipe establishes a hydrostatic-design-basis (HDB) rating of 4000 psi for water at 23C. A design service factor 2.5 is applied and provides a HDS value of 1600 psi for attaining the needed wall thickness for a specific pressure class rating. The factor of safety provided for sustained working pressure in the class ratings is 3.0 or greater. When temperatures of the fluid conveyed is greater than 23C, the same allowable design stress is derated according to a table provided in the AWWA C 900 Standard.

Polyethylene (PE) and polybutylene (PB) pipes, ½ inch through 3 inch, for potable water conveyance adhere to AWWA C-901 and C-902 Standards, respectively. The Standard for PE covers low (Type I), medium (Type II), and high (Type III) density materials. The Standard for Pb is Type II. These polyolefins each have their own respective HDB with a design service factor of 2.0 for determining their required wall thicknesses. It should be noted that normally smaller pipes are subjected to more frequent surge and water hammer than larger pipes. The designer should take this into consideration and may consider a higher pressure rated pipe than normal system pressure in order to minimize the potential problem of fatigue in plastics.

The AWWA C 950 Standard for Fiberglass established an HDB rating with a design factor of 2.0 for long-term. This will determine the required wall thickness. This includes a 40% allowance for surge pressure or water hammer for transmission mains. When designing distribution piping or when pressure surges or water hammer is known, the pipe pressure rating should be increased to include these transient conditions.

INSTALLATION

Acceptable installation results from the proper choice of materials and construction techniques that recognize the factors that contribute to good long term performance.

The pipe zone embedment material consists of bedding, haunching below the springline and the initial beckfill to over the top of the pipe.

The pipe zone material should receive a positive compaction, and a density of 90 percent (Proctor Density), or higher and shall have the requirement listed in the project specification.

Vertical deflection control is an important paramater for plastic pipes and is often used for determining installation quality control. It is common practice to reference installation quality to an allowable initial deflection limit. Various user agencies measure initial deflections during installation and have established a specified allowable limit.

The installed pipes are inspected by using a mandrel having a fixed dimension for small diameters, (6-inch and larger), or by man-entry and measurement on the larger diameters. Common practice in the USA is to limit initial deflection on thermoplastic pipe to a maximum of 4 or 5 percent. The practice in the USA, for the larger diameter thermosetting resin pipes, is to set a limit of initial deflection of somewhat lower levels which is dependent upon the pipe design. In any event, the purpose of setting the initial deflections much lower than the various piping materials stress and strain limits is the knowledge that the long term pipe and soil strengths must be controlled.

A study prepared for the Plastics Pipe Institute (2) determined that kinks in PB service lines formed during coil unrolling and/or installation have resulted in later pipe wall rupture, therefore this should obviously be avoided. Also, significant number of leaks have been attributable to fittings; a chief problem is localized pipe failures at the tips of insert stiffners used in compression fittings: other sources of failures include such deficiencies as corrosion of hose clamp screws and cracks within non - PE or PB plastic fittings.

Thermosetting resin has much lower ultimate strain resistance than most thermoplastic resin piping materials. This also varies with the type of thermosetting resin and the reinforcing fiberglass wall design and build-up.

There are many parameters which may affect pipe deflection. Listed below are the more significant parameters that govern long term pipe response: 1 - Pipe stiffness, 2 - Soil Stiffness, 3 - Soil load (Dead and live loads), 4 - Trench Configuration, 5 - Haunch Support, 6 - Non-elliptical Deformation, 7 - Effect of Moisture on Soil and Pipe, 8 - Construction Methods, 9 - Installation Variability and 10 - Time.

Several laboratory and field research projects conducted during the past 10 years have clearly indicated that pipe stiffness can have tremendous impact on initial and long term pipe deflection; the most recent conducted by Bishop (1), Lang (3), and Moser (4). The lower the stiffness the more variable and greater the pipe deflection, as addressed in the Plastic Pipe Overview 1983 (5).

When referencing and using AWWA C 900 as the material specification or the higher pipe stiffnesses listed in AWWA C 950, the deflection and non-elliptical deformation problems are reduced. Generally soils consolidate and get stronger with time, however, during that time period the pipes have a tendency to deflect more moving to fill the provided void caused by this consolidation. This is commonly known as the time deflection lag factor.

When proper haunch support is provided during installation the maximum long term soil load imposed on the pipe can be calculated by the prismatic loading formula.

When the trench wall insitu soils are 90 percent (Proctor Density) or greater, the width of the trench normally has little impact on the pipes performance. When insitu soils have lower densities, trench widths should be wider, with pipe zone material compacted to 90 percent (Proctor Density) or more.

High ground water can reduce the soil loading on the pipes, however, it will also reduce the pipe zone compaction and lateral support. This should be evaluated during the design stage of the project.

The construction methods along with installation variability can be kept within acceptable levels when the Owner Agency and/or the Engineering Consultant provide full time competent inspection.

OPERATION AND MAINTENANCE

Plastic pipes should be stored on uniform non-point loading supports. It should not be exposed to ultra-violet exposure for extended periods of time. It should not be dropped or impacted during transport or handling. Stones larger than 25 millimeters in size should not impact the pipe.

When making service connections to the pipe it is important to use a double-strap saddle and a sharp tapping tool. Plastic pipes can be tapped, however, a slightly different procedure should be encorporated in a User Agency Training program.

APPENDIX - REFERENCES

1. Bishop, R.R. and Lang, D.C. "Design and Performance of Buried Fiberglass Pipes - A New Perspective", ASCE National Convention, October 1984.

2. Chambers, R.E., "Performance of Polyolefin Plastic Pipe and Tubing in the Water Service Application", October 1984.

3. Lang, D.C., and Howard, A.K., "Buried Fiberglass Pipe Response to Field Installation Methods", ASCE International Conference in Underground Pipeline Engineering", August 1985.

4. Moser, A.P., Bishop, R.R., Shupe, O.K., Bair, D.R., "Deflection and Strains in Buried Pipes Subjected to Various Installation Conditions", 64th Annual Meeting, TRB, January 1985.

5. Schrock, B. Jay, "Plastic Pipe Overview - 1983", Europipe "83 Conference, Basel, Switzerland, June 1983.

APPROPRIATE DESIGN OF WATER SYSTEMS IN MICRONESIA

By: Shahram Khosrowpanah[1]
Leroy F. Heitz[2]

ABSTRACT: Operational failures of municipal water distribution systems remain a chronic problem in the Micronesian Islands. The water systems in Kolonia, Pohnpei, capital of the Federated States of Micronesia, and Moen Island, the capital of Truk State, are prime examples of island center systems experiencing continuing problems. In spite of seemingly adequate supplies delivered to the system, water is available to the customers only a few hours a day.

The water systems of these two islands were analyzed by reviewing their design, construction, maintenance and operation. These reviews were accomplished through field investigations and discussions with individuals holding key roles in system operation and maintenance. Low system maintenance and high consumption rates are the problems that were found to be common to the two systems. Low maintenance was found to be caused by the complexity of the design not matching the level of training of those responsible for maintenance of the system, and climatic conditions not adequately considered in the original designs. High consumption rates are primarily caused by cultural and social factors that were not considered in the original designs.

Appropriate design considerations which mesh the hydraulic aspects of the system with the local climatic conditions, local maintenance and construction capability and the social and cultural aspects of water use are introduced in this paper. An analysis scheme which considers the effect of failure of various components is suggested. This scheme forces designers to create a system that can remain operational under the given environmental conditions and considering local maintenance and construction capabilities.

The paper emphasizes that appropriate design is really an attitude rather than a new design procedure. The hydraulic considerations remain the same no matter for what part of the world the system is designed, but the final design must reflect the ability of the local infrastructure to construct, maintain, and operate the system.

---

[1]Asst. Prof., Water and Energy Research Institute, University of Guam, UOG Station Mangilao, Guam 96923.

[2]Director of Water and Energy Research Institute, University of Guam, UOG Station Mangilao, Guam 96923.

INTRODUCTION

The Micronesian (Western Pacific) area lies between the equator and the tropic of cancer and extends from longitude 130 E to 180 E. More than 2,000 islands are spread across this area. Of this total, almost 100 are inhabited. The islands range in size from less than 100 ft^2 (9 m^2) to 212 sq miles (549 km^2). Annual rainfall ranges from 80 to 400 in. (2,032 to 10,160 mm) per year, with humidity readings of 90 to 100 percent and temperatures of around 80 °F (27 °C) occurring all year round.

The water distribution systems serving Pohnpei and Moen Islands were investigated by the authors. Pohnpei is located approximately 3,100 miles (5,000 km) west-southwest of Honolulu, Hawaii, and 2,300 miles (3,700 km) southeast of Tokyo, Japan. This island is 129 sq miles (334 km^2) in area with peaks of volcanic origin exceeding 2,400 ft (732 m) above sea level. Moen is located approximately 3,500 miles (5,500 km) west-southwest of Honolulu, Hawaii, and 2,100 miles (3,500 km) southeast of Tokyo, Japan. This island has an area of 35 sq miles (91 km^2) with mountain peaks ranging to 1,460 ft (445 m) above sea level. The climate of these two islands is similar to the rest of the islands of the Micronesian group with uniformly warm and humid conditions prevailing year round. The average island rainfall is 350 in. (8,890 mm) and 126 in. (3,200 mm) per year for Pohnpei and Truk, respectively (7).

The water distribution systems of these two islands were designed and constructed according to modern design specifications in the early 1970s. In spite of the physical existence of these modern systems with plentiful surface as well as groundwater supplies, the people are not served on a dependable 24 hour per day basis. This condition is locally known as "Water Hours". This unreliable water service contributes to sanitation-related problems as well as a lack of fire protection. The problems which exist with these two water systems are typical problems to those experienced with other water systems in the Micronesian area. These problems are: system leakage, low system maintenance, components of the water system not functioning as originally designed, water usage by consumers much higher than the design values for the system, and non-existence of trained personnel for operation purposes.

In this paper, the Pohnpei and Moen water systems have been analyzed in terms of how their design, construction, operation, and maintenance affect the problems which exist with these two systems. The items on which water system designers should place special emphasis have been introduced.

DESCRIPTION AND OPERATION OF THE POHNPEI WATER SYSTEM

The present water system on Pohnpei have developed over two distinct administrative periods:

The period of Japanese administration (1914-1940). The first modern water distribution system was built during this period. The system included: diversions on several streams and rivers, a small treatment plant, several concrete storage tanks, and a transmission network of mainly galvanized iron pipes.

The period of administration by the United States under the Trusteeship of the Pacific Islands after World War II. Beginning in 1968, under several contracts administered by the United States Navy's Officer in Charge of Construction (OICC), an expanded and modernized water distribution system consisting of intake, transmission, treatment plant, storage, and distribution facilities was developed. Several of the facilities constructed by the Japanese were retained and integrated into the system. These tasks were completed by OICC by 1976. Since then, the Pohnpei District Department of Public Works (DPW), under the district administration of the Trust Territory Government, became involved in constructing new water lines, for providing service to areas not covered by the Japanese and U.S. Navy constructed systems (2).

The existing water supply and distribution facilities include: a concrete diversion dam and primary settling basin located on the Nanpil River and a three mile transmission line which carries raw water by gravity from the diversion dam to the Water Treatment Plant (WTP). The Water Treatment Plant consists of a settling tank, five sand bed filters, two 40 Hp (30 kw) pumps, a chlorination system, and a clear well. Four steel storage tanks, three with capacities of one million gallons each and one with a capacity of one half million gallons, are located at various points in the system. The water from the WTP can flow to three of these tanks by gravity. The fourth tank is fed by two existing pumps at the WTP. The distribution system consists of galvanized iron and asbestos cement pipes with sizes ranging from three to twelve inches. There are three pressure relief valves in the system. Neither of these valves is presently operating properly.

Two million gallons (7,600 m^3) of water are delivered daily from the Nanpil river to the WTP. The raw water discharges into a primary settling tank at the WTP and then is fed to five sand bed filters. Chlorine is added to the water which then flows into a small clear water well. Two 12 in. (300 mm) pipes carry the water to the main distribution system.

The water service is from 5 to 9 o'clock in the morning and evening periods, after which water service is curtailed to most users. Closing and opening the system invariably involves manipulating various valves in order to divert water from the users to one of the storage tanks and vice versa. Providing a longer water service period for the hospital is also a major operational consideration. This mode of operation has been the norm for years and recently the "Water Hours" have become shorter. Sometimes the DPW has been unable to fill the storage tanks sufficiently to provide water during emergencies.

DESCRIPTION AND OPERATION OF THE MOEN, TRUK SYSTEM

The development of the modern water distribution system on Moen, Truk, had its earliest beginnings during the Japanese administration of the island. The Japanese constructed several small dams and storage tanks and a simplified distribution system. Part of this system was destroyed during WW II. Up to the late 1960s the system had developed to the point where water from two surface water catchments and several deep water wells was being used to supply a relatively small area of Moen. The water distribution system consisted of 2, 2 1/2, and 4 in. (50, 60, and 100 mm) pipes laid on top of the ground (8).

From the early 1970s to the present a series of U.S. sponsored Capital Improvements Projects were carried out to upgrade the water system. Improvements included installation of 8 and 12 in. (200 and 300 mm) lines to replace existing lines and to provide new service to areas not previously served by the existing system. System storage capacity is supplied by a total of seven storage tanks with a combined storage capacity of approximately 8 million gallons (30,300 m^3). New wells have been drilled bringing the total to 30 wells available for water production. The design of the previously described improvements was carried out by reputable and competent engineering design firms under contract with the U.S. Navy. Construction inspection was handled by an OICC office which was located on Moen during the period of construction.

Many of the sophisticated automatic controls that were designed for the system are inoperative. Automatic level control valves for the supply tanks are rusted and vandalized to the point where they no longer function, and many of the chlorination stations are inoperative. Parts of the new system are rumored to have severe problems with leakage and the wells experience a high rate of off-line time due to maintenance problems.

Daily system operation consists mostly of turning on and off control valves in order to supply most of the service areas with at least two to four hours of water per day. Some areas are only serviced once or twice a week.

SUMMARY OF PROBLEMS

Water supply and distribution can be analyzed in the simplest fashion by looking at the simple water balance equation shown below:

SUPPLY = DESIRED DELIVERY + LEAKAGE

In this simplified equation the storage in the systems is neglected so long term averages for all components of the equation must be used. It is the duty of the water system design engineer to find a way to make the two sides of the equation balance. When looking at the Pohnpei and Moen water systems it is very easy to see that the left hand side of the equation (the supply) is always less than the right side of the equation (desired delivery plus leakage). The end result of this unbalanced situation is water hours to restrict the consumption of water by the users.

In order to analyze the two systems it must be determined what problems lie in the components of the water balance equations and why the problems exist. On Pohnpei and Moen, the supply side of the equation comes from surface and ground water sources. Some problems that have been observed with surface supplies in Pohnpei are poor maintenance of the intake works and supply lines that has lead to reduced supply capacity. The water delivered to the WTP was far less than the design capacity of the system. In Moen a severe supply problem exists partly because of pumps being off-line due to a need for maintenance. Some wells are also inoperative because of high salt content. These high chloride levels are due to location and/or poor management of the wells to prevent salt water intrusion. These problems can be categorized into two areas, poor maintenance and poor resource management.

The first term on the right side of the water balance equation deals with water usage. In reviewing design documents for the systems it appears that adequate but not excessive (by U.S. standards) water consumption figures were used. It appears that little consideration has been given to cultural or social aspects of the local population that could lead to higher consumption rates (5). Another problem that occurs is really a self defeating cycle that is triggered by the use of water hours. Since water hours limit the time when water is available, people try to store enough water to last through the period when the water is off. Therefore, the rate of demand is very high during the water hour period. In some cases taps are left on continuously to be sure that storage containers are filled even if unattended. This can lead to wastage and in some cases extremely unsanitary conditions. Sometimes when the lines are shut off, negative pressure appears at certain points in the system. This causes water to back flow into the systems from poorly maintained local supply containers such as pans and drum cans.

Leakage, the second term on the right hand side of the equations exists in both the Truk and Pohnpei systems. Leakage is quite understandable in some of the older portions of the two systems. Leakage problems in the newer portions of the systems are due to poor construction practices and poor inspection during construction or designs that were not adequate for existing conditions. Problems with leakage could be summarized as maintenance problems in the older sections of the systems and design or construction problems with the newly constructed portions of the system.

APPROPRIATE DESIGN CONSIDERATIONS FOR TROPICAL WATER SUPPLY SYSTEMS

The designers of new systems or improvements to existing systems play a key role in helping to eliminate the many problems that exist with tropical area water supply systems. First and foremost the designer must realize <u>where</u> the system is to be installed. Even though the laws of hydraulics work exactly the same in the tropical Pacific as in other parts of the world, the non-hydraulic factors which do depend on location can be the key to the success or failure of a system.

The first non-hydraulic factor that should be considered is, who are the people to be served. In the case of the tropical Pacific, the Micronesian people are not at all similar to the typical American population in water use and in many other ways. Their culture, politics and social customs are quite unique, and it is important for the designer to understand this uniqueness in order to decide what design strategies to use.

One important aspect of this "people orientation" is to understand that water consumption rates and patterns may vary drastically from those that might be typical in a North American or even on a typical military installation in a tropical area. An example of this can be seen on Pohnpei where water consumption rates have been estimated as high as 100 gallons per capita per day (gpcd) (380 liters per capita per day, l pcd) (5) as compared to 50 gpcd (190 lpcd) for typical U.S. applications (6). Native populations may have a completely different attitude toward "WATER" than those represented in normal textbooks or design guides. The designers should look to basic water use studies completed by water research centers or other university organizations as valuable sources of information on this important issue. In this case a familiarity with cultural considerations is just as important as knowing the proper pipe friction loss factor to apply.

Another major design consideration is the skill of those who will be constructing the project. There is a good chance that the capability of the local construction force may be below that of a typical North American contractor. Available on-island construction machinery and tools may be far less sophisticated than those available in other parts of the world. If the contract is let to off-island firms there is a good chance that part of the construction force will be non-English speaking and may encounter some difficulty in interpreting plans and specifications and in communicating with others working on the project. The key in this area is to keep designs as simple and fail-safe as possible. The design should include only a minimum of special construction techniques, and all construction drawings and specifications should be very clear, concise and to the point. Cut and paste designs from those previous used in North American or more developed area should be avoided. The designer should ask himself at every step along the way. Will the available work force be able to build this design?

Another important consideration is the skill of those that will be maintaining the system after construction. The design should be as maintenance-free as possible, and the required maintenance should be kept at a level requiring a minimum of skills. "High tech" automated gadgetry will most likely fail due to a lack of maintenance and is simply a waste of money to include in a design. Any specialized maintenance equipment required should be specified and included in the contract if possible. The final design should include a maintenance manual that specifies exactly what needs to be done and specifies a set schedule for all required maintenance. Training should be scheduled for local maintenance people especially if specialized equipment is to be installed.

Another important consideration under maintenance is the availability of spare parts. Backups should be included for all essential system components as part of the original design. It can take many months to get spare parts to many tropical locations. Therefore, if a particular component could fail and incapacitate the system, then that part should be available on the island for immediate replacement. Another important maintenance item deals with the availability of local funds for maintenance. In most cases only a minimum of funds will be available. The designer should strive to keep maintenance costs to an absolute minimum. High maintenance cost items will probably not be maintained properly and will likely lead to eventual system problems.

Another factor that must always be considered is the environment to which the design will be exposed. While the tropical islands are some times called paradise, they are far from paradise when considering the environment's effects on various components of a water distribution system. High humidity coupled with the salty sea air present corrosion and rusting problems that must be seen to be believed. Almost any component of the system that has a potential for corrosion or rusting will, if left exposed, be rendered useless in very short order. Special precautions must be taken in choosing materials, providing protection, and in specifying maintenance schedules in order to guard against damage from the environment.

A final consideration is really to bring all the above factors into play in what is termed a failure mode analysis. In this analysis the system is reevaluated in light of user, maintenance and environmental considerations to see what effects various system component failures might have on the operation of the total system. Again, recognizing that maintenance levels might be relatively low, and environmental conditions will be rather harsh, the designer should evaluate the system's capability to serve its design purpose over the project life. This analysis might identify certain key areas that might require redundancy or at least on the spot backups so that the system can function at a minimal level until more permanent repairs can be made. In this case the designer is asking himself how best to provide a system that performs as designed for the longest period of time.

SUMMARY AND CONCLUSIONS

The water distribution systems for the islands of Pohnpei and Moen, Truk, were examined. Various problems dealing with the planning, design, construction and maintenance of these systems were identified. Factors identified included: maintenance and resource management under the water supplies heading, unanticipated high consumption rates and shortage amplification due to water hours under the delivery category, and design, construction and maintenance considerations under the leakage heading.

In order to eliminate or at least minimize the possibility of similar problems existing in new designs, an elevated level of design awareness is advocated by the authors. An enlightenment to the culture of those to be using the system, an awareness of the skills and resources of those who will be constructing and maintaining the system, and a knowledge of the effects of the tropical environment on equipment are all factors which the designer must be keenly aware.

Finally the authors advocate a failure analysis study be made on the system to evaluate the effects of failure of key components in the system. This type of analysis will indicate areas where back up redundancy might be required or areas where more design work is necessary.

A successful design for a tropical water supply system requires the designer to have an attitude of awareness of "Where" the system will exist, not on a drawing board but in the coral sands of a remote tropical island.

REFERENCES

1. Austin, Smith and Associates. "Bid, Specifications, Contract and Bond for the Construction of Ponape Water and Sewerage Systems, Project No.2." For the Trust Territory of the Pacific Islands, Honolulu, Feb. 1971.

2. Austin, Tsutsumi and Associates, Inc. "Water Resources Development Plan for Moen Island, Truk ECI." Pre-final report to the Naval Facilities Engineering Command, Honolulu, 1980.

3. Barret, Harris, and Associates. "Leak Detection Service, Kolonia Water System, State of Ponape, Federated States of Micronesia." Report to Pohnpei State Government, Honolulu, 1983.

4. Heitz, L. "Development of a Computerized Distribution System Model of the Moen Island Water Distribution System." Technical Report No. 62, University of Guam Water and Energy Research Institute, Mangilao, Guam, 1986.

5. Khosrowpanah, S., and R. Hunter-Anderson. "Socio-Political Impacts on the Water Distribution System of Kolonia, Pohnpei, Capital of the Federated States of Micronesia." Paper to be presented at the 8th Annual College of Art and Science Research Conference, University of Guam, Mangilao, April 1987.

6. Lindsley, R.K., and J.B. Franzini. "Water Resources Engineering." McGraw-Hill Book Company, New York, 1979.

7. NOAA. "Local Climatological Data, Annual Summary with Comparative Data, Ponape Island, Pacific." Asheville, NC 1983.

8. Van der Burg, O. "Water Resources of the Truk Islands." U.S. Geological Survey Water-Resources Investigations Report 82-4082, Honolulu, 1983.

9. Van der Brug, O. "Water Resources of Ponape, Caroline Islands." U.S. Geological Survey Water-Resources Investigations Report 83-4139, Honolulu, 1984.

Slow Sand Filtration of Tropical Source Waters

Joy M. Barrett*

ABSTRACT

The setup and operation of pilot-scale slow sand filters for tropical applications are discussed. Considerations unique to tropical climates, including temperature and influent water chemistry, are examined. Results of preliminary investigations are presented, which show that slow sand filtration is highly effective in reducing coliform indicator organisms and turbidities of warm, nutrient-rich, turbid source waters.

INTRODUCTION

It is estimated that two billion of the earth's inhabitants (nearly 50%) do not have access to safe drinking water (Huisman and Wood 1974). Most deaths in developing countries are due to diseases from water-borne pathogens. That means they are preventable. Potable water need not be a privilege restricted to residents of developed and/or affluent countries. Because it is non-energy intensive and requires only a small initial capital output, slow sand filtration for the production of potable water has worldwide applications.

Slow sand filtration is a combination of processes whereby physical straining, microbial predation, and biochemical conversion of contaminants all serve to render raw water potable. Three crucial components of a slow sand filter are: the raw water reservoir directly above the sand bed; the "schmutzdecke", an organic film on the top surface of the filter-bed; and the filtration medium itself. The reservoir functions as a settling basin. The schmutzdecke, by its sticky nature, provides an adhesive surface for suspended solids, and can also trap organics and attract dissolved organic matter (DOM). Water is fed to the sand bed at a rate of 0.1 to 0.3 m/hr, or 2 to 7 m^3/m^2/day. Unlike multi-media rapid filtration beds, the slow sand filter consists of just one grain size, the optimum being in the range of 0.1 to 0.3 mm. This sand bed actually acts as a substrate for a myriad of microbial processes.

After an operating period of 1 - 6 months (which depends largely on climate and raw water quality) the

* Department of Civil and Environmental Engineering, University of Colorado, Boulder, Colorado 80309 U.S.A.

filter must be cleaned. Generally, some quantity of head loss, or fluid flow impairment, is used as an indicator of the need for cleaning. The most common method for cleaning slow sand filters is to drain the water to below the sand surface, and scrape off the schmutzdecke. After cleaning, the filter must be recharged, or allowed to ripen, before it can again filter effectively. This process of ripening is simply the development of a new schmutzdecke, and the period of time required varies from 1 - 30 days, again depending on the climate, and raw and desired effluent water qualities.

Extensive research has been performed on the efficiency of slow sand filters for treatment of water in temperate climates. When source waters are relatively low in temperature and turbidity, this process has been found to be effective in the removal of organics, certain ions, bacteria, pathogenic protozoa, and viruses (Bellamy et al. 1985, Poynter and Slade 1977, Slezak and Sims 1984). Slow sand filtration plants currently serve as sources of potable water in many tropical countries, including Columbia, India, Jamaica, Kenya, Sudan, and Thailand (Visscher et al. 1986). This water treatment process has also been recommended for schistosome cercariae control in north Cameroon village water, having been found an effective barrier to cercariae at very low hydraulic loading rates (Kawata 1982). Although slow sand filters are not uncommon in developing nations, evaluations of their performance under tropical conditions are few. Yet, filter effectiveness is very sensitive to variations in influent water quality, which is frequently determined by climatic conditions. In addition, this technique of water treatment relies heavily on microbial predation and degradation of contaminants within the filter, which are temperature-dependent processes.

Research is being conducted at the University of Colorado on the characterization and performance of slow sand filters in the purification of warm, turbid waters. A simulated tropical source water was used in the study, based on data from a compilation of chemistry of the world's rivers and lakes (Thurman 1985). In the current investigation, labile organic solutes were not added to the feed water. An appreciable fraction of dissolved organic matter (DOM) consists of such humus-like compounds, which are not readily available nutrients for aquatic organisms. It has been estimated that no more than 10% of total dissolved organic carbon (DOC) is usable substrate for microorganisms (Seki 1982). Therefore, it is crucial to distinguish between usable and non-usable components of DOM.

An influent temperature range of 25 to 30°C was chosen, which corresponds with the annual range of 26 to 29° C observed in the Orinoco river in Venezuela (Nemeth et al). Raw water turbidities in this study were varied to simulate source waters during both dry and rainy seasons.

MATERIALS AND METHODS

Influent Chemistry and Column Inoculation

The chemistry of the source water for the slow sand filters is summarized in Table I. The feed water was formulated so as to yield a total DOC content of 1 mg/l, consisting of 30% glutamic acid, 10% salicylic acid, and 60% glucose. Glutamic acid, which is the most prevalent amino acid in South American rivers, was added at a concentration equal to that of total dissolved amino acids in such large tropical rivers as the Parana and the Zaire (Thurman, 1985). Salicylic acid and glucose were added to allow greater species diversity, with respect to metabolic pathways of microorganisms, than a single organic nutrient would provide. Quantities of ammonium carbonate, potassium phosphate dibasic, and magnesium sulfate (heptahydrate) were added so as to yield C:N:P:S ratios of 30:4:1:0.5, which is slightly carbon-rich relative to the ratios in cell material (Grady and Lim, 1980). The carbonate and phosphate salts also served to buffer the system, though sodium hydroxide was required to raise the pH. The desired ionic strength was achieved by both sodium hydroxide and calcium chloride. Source water batches of 757 liters were prepared each day. Due to this large volume, chlorinated tap water was used. Dechlorination was accomplished by reaction with sodium thiosulfate, as prescribed in Standard Methods (1985).

Ripening of the schmutzdecke and biofilm with warm-climate organisms was enhanced by inoculation with South Florida pond water. This inoculum was slightly alkaline, with a pH of 7.6. Therefore, daily batches of source water were adjusted to a pH of 7.6 ± 0.2.

Physical Set-up and Hydraulics

The pilot-scale slow sand filtration system is depicted in Figure I. Each column is constructed from two cylinders, the lower of which is completely filled with the filter medium. The sand bed is 105 cm in height, and is supported by a 25-cm underdrain. The sand used had an effective size of 0.30 mm and a uniformity coefficient of 1.76. The space above the sand bed functions as a reservoir. The water level rises as the schmutzdecke and biofilm develop, restricting fluid flow. The connected cylinders of each column can be disassembled for inspection and cleaning of the schmutzdecke.

Feed rate was controlled by orifices in a constant-head box above the filters. Flow was maintained at 243 ml/min per filter, which corresponds to a hydraulic loading rate of 0.2 m/hr. Head-loss compensation was accomplished by adjusting valves in the effluent lines leading to spill cups.

Two ports on each column, located 16 cm above the sand bed, allowed reservoir sampling and temperature monitoring by stationary thermocouples. An additional 90-cm thermocouple was used to measure the feed tank temperature. All thermocouples were connected to an Omega model DSS-650 digital multi-channel thermometer. The feed tank and filters were insulated. A feed tank water temperature of 31°C maintained filter temperatures at 26 ± 1°C, so that warm tap water precluded the need for auxiliary heaters.

Two pump discharges allowed simultaneous supply to the constant-head box and recirculation to the feed tank. Recirculation kept the feed tank water well-mixed.

The time required for flow from reservoir sample ports to effluent spill cups was calculated to be approximately 3.5 hrs. This period was therefore used as the time interval between reservoir and effluent sampling.

Microbiological and Physical Techniques

Background levels of coliforms were determined through rigorous sampling. Coliform indicator organisms were then added to the columns from pure cultures of E. coli maintained in the laboratory. After a mixing period, samples were obtained from sample ports. These samples provided influent counts of coliforms. Corresponding effluent samples were acquired from spill cups 3.5 hours after collection of influent samples. All sample containers, glassware, and buffered dilution water were sterilized before using. Total coliform counts were achieved using techniques from Standard Methods (1985).

Turbidities were measured of the source water in the feed tank, and of samples taken from reservoir ports and effluent spill cups. A Hach model 2100A nephelometric turbidimeter was used for analysis. A turbid suspension was prepared by addition of four parts silt and one part kaolin. Silt-clay additions ranged from 0.053 g/l to 0.115 g/l.

Coliform and turbidity studies were performed separately and all tests were performed at least nineteen days after filter startup.

RESULTS AND DISCUSSION

Table II summarizes the performances of filter columns 1 and 2 with respect to coliform reductions. A series of forty samples prior to coliform addition yielded a background coliform count of 3/ml. A fraction of the post-inoculation samples was discarded for showing fewer coliforms per ml than the background levels, or no coliforms at all. Subtracting the background level from the remaining sample counts yielded an average reduction in column 1 of 6 out of 9 coliforms per ml or 66.67%, with a standard deviation of 5.51%. Likewise, filtration through column 2 decreased the average coliform count from 9.5 per ml to zero, or 100% reduction, with a standard deviation of zero.

A noteworthy observation is that effluents were consistently contaminated with organisms other than E. coli, whereas influent samples produced no such colonies. Apparently, organisms residing in the filter penetrated the sand bed, and appeared in the coliform test. Because these results are from the first filter runs on new sand, it is possible that additional time is required for complete maturation of the biofilm.

Turbidity reductions are summarized in Table III. Contrasting the source water and reservoir turbidity values (after silt/clay additions), significant settling occurred in the water column above the sand bed. Turbidity values of the constant-head box effluents for the most turbid solutions were included to demonstrate that some settling did occur in the relatively stagnant box. The data indicate that regardless of source water turbidity, slow sand filter effluent turbidities were uniformly low.

Initial evaluations of slow sand filter performance under tropical conditions have demonstrated that this method of water treatment significantly reduces coliform indicator organisms and turbidities of warm, nutrient-rich source waters. Results of coliform reductions in this study indicate that the initial ripening period may need to be longer than predicted from nutrient content and temperature considerations. The existence of parallel filters in the pilot-scale facility will allow parameters to be varied in future work. For example, comparisons can be made of filter efficiencies for different influent turbidities and coliform concentrations. Additionally, the rate of biofilm and schmutzdecke development as a function of temperature can be examined. Finally, the impact of humic materials may be investigated, as such labile substances comprise a significant portion of DOC in many of the world's lakes and rivers.

SUMMARY

Preliminary investigations have shown that slow sand filtration is effective for the production of potable water in tropical environments. With respect to initial capital output, materials for construction, energy requirements, and operation and maintenance, slow sand filtration is an appropriate technology which can dramatically reduce the incidence of water-borne diseases in developing countries.

Aknowledgements

The author is grateful to the following persons for their contributions and/or advice: Dr. David Hendricks, Dept. of Civil Eng., Colorado State Univ., for the loan of the filter columns; Dr. JoAnn Silverstein and William Hogrewe, Dept. of Civil and Environmental Eng., Univ. of Colorado; Dr. Steve Schmidt, Dept. of Environmental, Population, and Organismic Biology, Univ. of Colorado; and Dr. William Cooper, Water Quality Research Center, Florida International Univ., for the S. Florida pond water sample.

Table I
Simulated Tropical Source Water for Slow Sand Filtration

Chemical	Concentration (μg/l)
Glutamic Acid	736
Salicylic Acid	164
Glucose	1502
Ammonium Carbonate	217
Potassium Phosphate Dibasic	188
Magnesium Sulfate 7-Hydrate	128
Calcium Chloride	5548
Sodium Hydroxide	1519

Table II
Average Coliform Reductions in Slow Sand Filters

	Influent coli/ml	Effluent coli/ml	Numerical Reduction	Percentage Reduction
Column 1	12(9)*	6(3)*	6/9	66.67
Column 2	12.5(9.5)*	1(0)*	9.5/9.5	100

*Numbers in parentheses represent the actual count minus the background level of 3/ml, except for Column 2 effluent, where post-background adjustment was considered to be zero.

Table III
Turbidity Reductions in Slow Sand Filters

	Turbidity (NTU) No Add'n.		Turbidity (NTU) Add'n. of 0.053 g/l		Turbidity (NTU) Add'n. of 0.115 g/l	
	Col 1	Col 2	Col 1	Col 2	Col 1	Col 2
Feed Tank	0.38	0.38	17.00	17.00	13.00	13.00
Constant-Head Box	---	---	---	---	11.00	11.00
Reservoir	0.38	0.33	2.15	2.80	7.50	7.40
Effluent	0.20	0.26	0.22	0.23	0.22	0.29

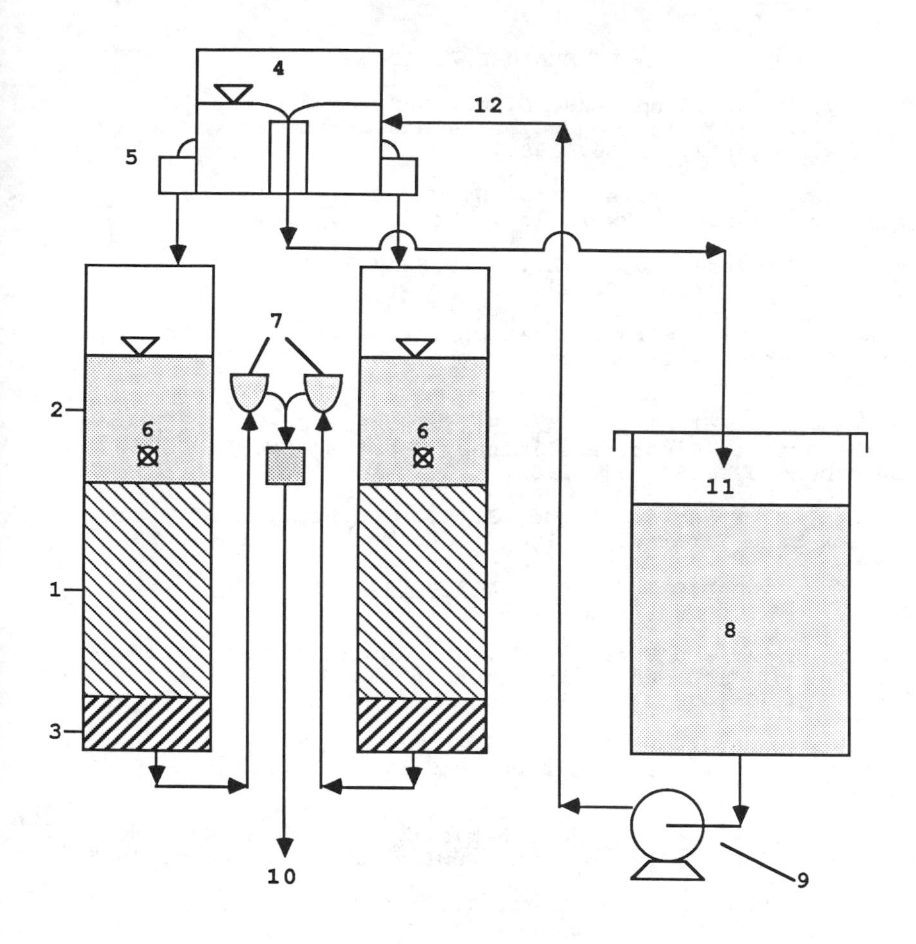

Figure I. Pilot-Scale Slow Sand Filtration Facility.

1. Sand bed
2. Reservoir
3. Underdrain
4. Constant-head box
5. Orifices
6. Thermocouple and sample ports
7. Effluent spill cups
8. Feed tank
9. Pump
10. Effluent
11. Overflow
12. Feed

LITERATURE CITED

Bellamy, W. D., Hendricks, D. W., and Logsdon, G. S.. Slow Sand Filtration: Influences of Selected Process Variables. J.A.W.W.A. 77:12, 62-66, 1985.

Grady, C. P. L., Jr. and Lim, H. C.. Biological Wastewater Treatment. Marcel Dekker, Inc., New York, 1980.

Huisman, L. and Wood, W. E.. Slow Sand Filtration. World Health Organization, Geneva, 1974.

Kawata, K.. Slow Sand Filtration for Cercarial Control in North Cameroon Village Water Supply. Wat. Sci. Tech. 14, 491-498, 1982.

Nemeth, A., Paolini, J., and Herrera, R.. Carbon Transport in the Orinoco River: Preliminary Results. SCOPE/UNEP Sonderband 52, 357-364, 1982.

Poynter, S. F. B. and Slade, J. S.. The Removal of Viruses by Slow Sand Filtration. Prog. Wat. Tech. 9, 75-88, 1977.

Seki, H.. Organic Materials in Aquatic Ecosystems. CRC Press, Inc., Boca Raton, 1982.

Slezak, L. A. and Sims, R. C.. The Application and Effectiveness of Slow Sand Filtration in the United States. J.A.W.W.A. 76, 38-43, 1984.

Standard Methods for the Examination of Water and Wastewater. Sixteenth Edition, A.P.H.A., Washington, D. C., 1985.

Thurman, E. M.. Organic Geochemistry of Natural Waters. Martinus Nijhoff/ Dr. W. Junk Publishers, Dordrecht, 1985.

Visscher, J. T., Paramasivam, R., and Santacruz, M.. IRC's Slow Sand Filtration Project. Waterlines 4:3, 24-27, 1986.

Optimal Design of Rapid Sand Filters for Water Supply Treatment

Fazal H. Chaudhry and Luisa F. R. Reis*

Abstract

The operation of a filtration system in a water treatment plant involves fixing a number of parameters that must be considered at the design stage due to their economic importance. Thus the design of a filtration system should consider the diverse aspects of operation besides the various details of the processes in a abridged manner. This study presents a systems analysis view of the decision process and generates designs by optimization. The design problem was expressed as that of unconstrained minimization of a nonlinear cost function and was repeatedly solved by a quasi-Newton search algorithm. Examples of optimized solutions for the decision variables and the minimized costs are presented graphically. The present study demonstrates the feasibility of design through optimization of the complex process of filtration.

Introduction

Although there is need for new technologies appropriate for developing nations, the cities even in such countries will continue to require large scale production of high quality drinking water supplies. It be hooves engineers to design the treatment facilities such that their scarce resources are put to optimum use. Whether one is considering conventional treatment of water or otherwise, granular media filtration is generally present in the recommended schemes. It is also one of the most expensive processes in terms of both the investment and maintainance costs. In spite of the fact that filtration parameters which have been successfully employed elsewhere usually form the basis of design of new units, their relevance to a particular situation, specially from the economic point of view, can not be assured.

The difficulty of designing filtration units through cost minimization lies with the complexity of the system in terms of the physico-chemical, geometrical and operational aspects involved in them (3,4,5). There is a proliferation of mathematical models, some requiring numerical solution of differential equations for each set of filter varia-

* Professor and Assistent Professor, Dept. of Hydraulics and Sanitary Eng., San Carlos School of Eng., San Carlos, San Paulo,Brazil 13560.

bles, but there is no effective utilization for them. Such lack of use of models, even for predictive purposes, makes it impossible to consider design as a cost minimization problem.

It is the purpose of the present paper to show that it is possible to organize all the diverse economic and technological information related to sand filtration units in terms of a cost objective function and analyse it to obtain combinations of design variables corresponding to minimum total cost. Three distinct steps were followed to obtain the sets of decision variables. Technological functions based on a mathematical model of filtration were developed in explicit form and an objective function in the form of total annual costs established for four filter layouts. Six decision variables were identified, namely, filtration rate, medium depth, number of filters, box length to width ratio and backwash and filtred water pipe diameters.

Technological Functions

The technological functions describe the relationships between the design and operational variables like the filtration rate v, bed depth x, run time t, influent and effluent concentrations, c and C_o and hydraulic gradient J for a given combination of pretreatment and sand size. Of all the models of rapid sand filtration in literature, the one recently presented by Adin and Rebhun (2) is most appealing both for its conceptual description of kinetics of filtration and its wholesome experimental verification. However, its predictions must be obtained for each set of operating conditions. Given that the optimization procedure proposed for design must analyse a very large number of alternative conditions, it would be very time-consuming if each condition were evaluated numerically one at a time.

It is desirable to work with Adin-Rebhun model in dimensionless form as below,
continuity Equation:

$$\frac{\partial c'}{\partial X} + \frac{\partial \sigma'}{\partial T} = 0 \qquad (1)$$

Kinetic Equation: $$\frac{\partial \sigma'}{\partial T} = c' \ (1 - \sigma') - R \ \sigma' \ (1 - \sqrt{\sigma'} \ )^{-3} \qquad (2)$$

to be solved under the following boundary and initial conditions,

$$c' = 1 \qquad X = 0 \;,\; T \geqslant 0$$
$$\sigma' = 0 \qquad X \geqslant 0 \;,\; T = 0 \tag{3}$$

Here $c' = c/C_o$, $\sigma' = \sigma/F$

$$T = k_1\, C_o\, vt \;,\quad X = k_1\, Fx \tag{4}$$

and $R = k_2/k_1\, C_o\, K_o$

Besides the variables mentioned above, the formulation introduces the parameters k_1 and k_2 which describe respectively the attatchment and detatchment mechanisms, the filter deposit capacity F, clean filter permeability K_o and the second dependent variable σ, the specific deposit. R is called the process parameter as it characterizes the particle--flocculent-media interaction.

The above system of equations was solved numerically by the method of characteristics as suggested by Adin (1), to obtain the dimensionless c' and σ' profiles in space and time for each R. As the systems are usually designed to produce a desired quality of water, the run lengths T_ℓ corresponding to breackthrough concentrations $c' = c'_\ell$ were obtained for various bed depths X = L or

$$T_\ell = T_\ell\ (L;\ R,\ c'_\ell\ ) \tag{5}$$

Fig. 1 shows an example of the technological function in Eq. (5).

Further the specific deposit, σ', profiles at $T = T_\ell$ were integrated over L to obtain the average deposit $\bar{\sigma}'$ in the bed and consequently the terminal head loss gradient J_ℓ, corresponding to the imposed c'_ℓ, by the Sekhtman's equation as,

$$J = J_o\ (1 - \sqrt{\bar{\sigma}'}\ )^{-3} \tag{6}$$

Thus the second technological function for $G_\ell = J_\ell/J_o$ in the form,

$$G_\ell = G_\ell\ (L;\ R,\ c'_\ell) \tag{7}$$

is determined. Fig. 2 presents an example of this function for the same c'_ℓ as in Fig. 1.

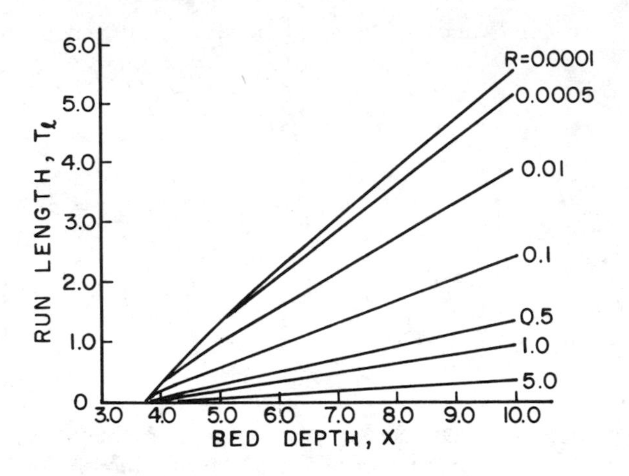

Figure 1. Dimensionless Run-Length as Function of Bed-depth and Process Parameter R for c'= 0,025

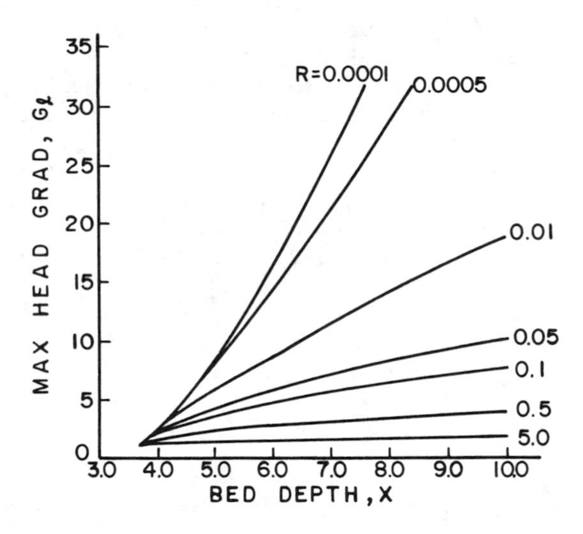

Figure 2. Maximum Relative Head Gradient as Function of Bed-depth and Process Parameter R for c'= 0,025

The technological functions obtained numerically were represented mathematically by fitting the generalized solution of Eqs. (1) and (2) for $R = 0$.

Objective Functions

The objective function was composed of the annual investment and maintenance costs. The capital recovery factor was based on 7% annual interest rate and variable lives for different elements considering 30 years as the economic life of the project.

For the formulation of the costs, the filtration system was considered to be composed of N traditional constant-rate filters of which only N-1 are operational at a given time. Individual filters are rectangular placed side by side along the pipe gallery, their number, area and aspect ratio considered as the decision variables. Four different layouts were formulated separately depending on the choice between single or twinned units and whether the units are arranged in a single or double row along the gallery. The system operation is controlled from the covered control room above the gallery where the command consoles are placed.

The settled water headers are provided on the sides opposit to the gallery.

The filter medium is sand of size medium or coarse for filtration respectively of suspentions treated with alum or cationic polyelectrolyte. These combinations are considered in order to make use of the experimental information on various model parameters obtained by Adin and Rebhun(2). The medium sand is supported on a sequence of gravel sizes placed on a false floor with orifices. The backwashing was considered to be that with water and air, each having a separate dispensing system. It consisted of first introducing water only for ten minutes at a rate corresponding to 15% bed expansion and later with air at half that rate for 6 more minutos. The wash water is to be collected through concrete troughs.

Each filter is provided with individual piping connecting the effluent side to the backwash and filtered water pipelines and a discharge controller. The filter operation is effected through hydraulically controlled influent, effluent and backwash valves.

The cost functions of all structural elements, equipment, energy and chemicals were based on market prices in Brazil. Some simple structural and hydraulic considerations were necessary in the design of filter box, foundations and the pipelines. The maintainance costs consisted of those related to low head pumping during filtration, backwash water and air pumping and to the treatment of water lost during backwashing.

Optimization and Results

The cost functions are too elaborate to be included in this paper. Only eight independent variables were necessary to quantify the various components of the objective function F. These are: x_1 = filter length to width ratio; x_2 = the bed depth; x_3 = filtration rate; x_4 = diameter of backwash pipe and fittings; x_5 = diameter of filtered water pipe and fittings; x_6 = the number filter units in the system; x_7 = run length and x_8 = healdloss across the medium.

The filtration system design problem then is to find the decision variables $x_1, x_2, ..., x_8$ such that $F(x_1, x_2, ..., x_8)$ is minimized and technological functions in Eqs. (5) and (7) are respected. These functions in dimensional form are:

$$x_7 = f_1 (x_2, x_3) \quad (8)$$

$$x_8 = f_2 (x_2, x_3) \quad (9)$$

where the parameters C_o, c'_ℓ , k_1, k_2, F and k_o are not included either being fixed by the problem or being functions of filtration rate x_3. The best way to satisfy technological restrictions in Eqs. (8) and (9) is to eliminate x_7 and x_8 by direct substitution from the objective function $F(x_1, x_2, ..., x_8)$. Thus the problem reduces to one of unrestricted optimization as,

$$\min_{x_1, x_2, ..., x_6} FC (x_1, x_2, ..., x_6).$$

The cost function FC for our problem was highly nonlinear so that the optimal solutions was found by direct search. The subroutine E04JAF of the NAG Library which is quase-Newton type was employed to determine the optimal combinations of variables for given sets of input data on C_o, c'_ℓ , suspension-medium combination, filter layout and the plant capacity Q. The search was greatly facilitated by the well-behaved nature of the response surface. The minimum cost region was found to be well--defined which could be reached from diverse initial positions. Optimal designs were determined at C_o = 5, 10, 20, 40, 80 mg/ℓ , c'_ℓ = 0,05, 0,025, 0,01, Q = 250 m^3/h - 32.000 m^3/h for two suspension-medium combinations and four plan layouts. As an example, we present in Figs. 3 and 4 the optimal designs respectively for the combinations (c'_ℓ = 0,025, alum treated suspension-medium sand, single row of filters, twinned units) and (c'_ℓ = 0,025, cationic polyelectrolyte treated suspension-

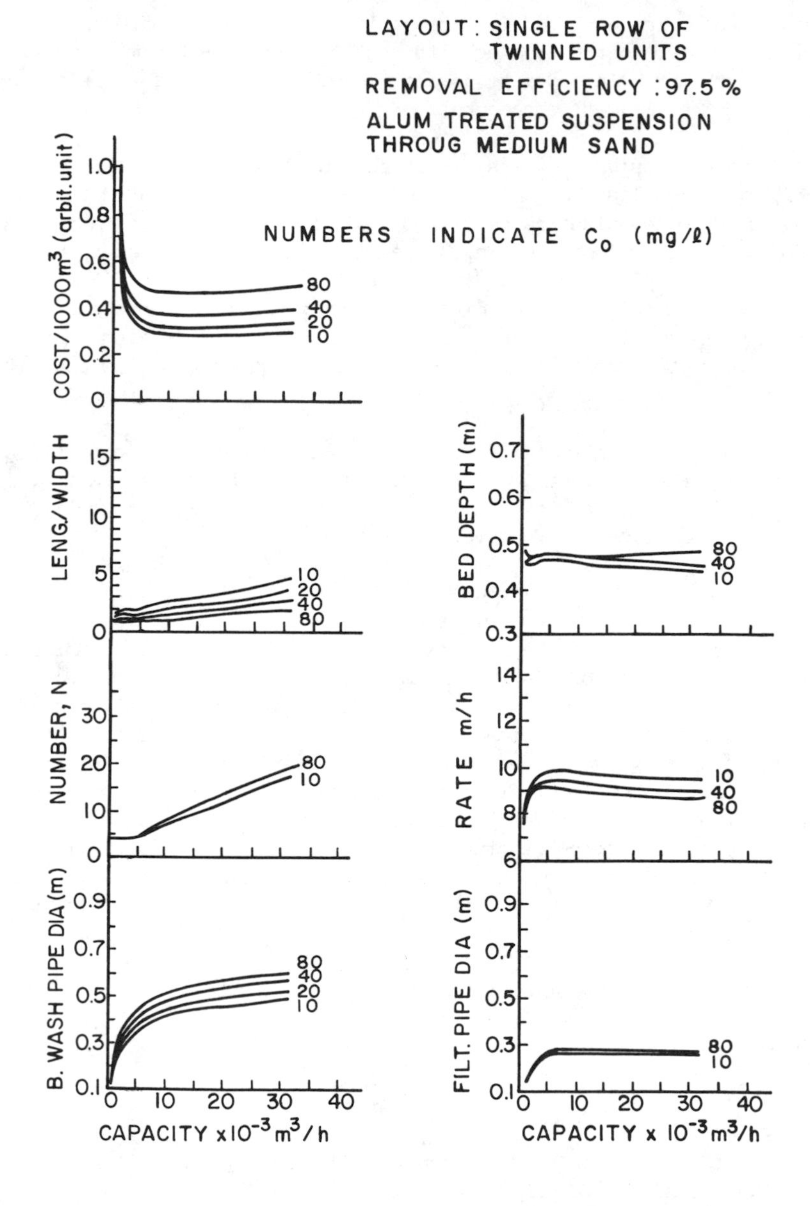

Figure 3. Optimal Designs for Alum Treatment

-course sand, single row of filters, twinned units). Also presented in these figures are corresponding unit cost of treating 1000 m^3 of water in arbitrary units in order to assess the effects on it of the plant capacity and the influent concentration.

The optimal design requires, as is obvious from the partial presentation of results that the structural and operational parameters need be adjusted to the plant capacity and influent concentration. A rather profound change is indicated in the length to ratio and backwash pipe diameter as the plant capacity increases. The bed depth and filtration rate in the case of cationic polyelectrolyte pretreatment are shown to be more sensitive to plant capacity and influent concentration than the alum one. The unit cost of treatment presents a minimum in the capacity range of 10-15 x 10^3 m^3/h.

This cost is shown to increase sensibly with influent concentration. This result should help assess the feasibility of direct filtration. A comparison of unit costs of alum and polyelectrolyte treatments confirms the latter is more cost effective.

Conclusions

It is concluded that the filtration systems can be designed through direct consideration of its cost by application of systems analysis concepts. This procedure permits evaluation of various design alternatives from the point of view of total economy at the same that determines all the relevant design variables. The most important conclusion is that the mathematical models of filtration expressed in adequate form are a powerful tool for filter design. It is necessary, however, that the models are applied to pilot study data to produce enough parametric information relevant to various media-suspension combinations.

Acknowledgments

The authors are indebted to San Paulo State Foundation for Research Supports - FAPESP for supporting the present research through research scholarships.

Appendix I - References

1. Adin, A. - "Solution of granular bed filtration equations" - Journal E.E.D., ASCE, 104 (EE3): 471-485. Jun., 1978.

2. Adin, A. & Rebhun, M. - "A model to predict concentration and head-loss profiles in filtration". Journal A.W.W.A., 69(8): 444-453; Aug., 1977.

3. Huang, Y.C. & Bauman, E.R. - "Least cost sand filters design for iron removal". Journal S.E.D., ASCE, 97(SA2): 171-190, Apr., 1971.

4. Letterman, R. - "Economic analysis of granular bed filtration". Journal E.E.D., 106(EE2): 272-291, Apr., 1980.

5. McNaughton, J. - "Cost reduction while maintaining performance during in - depth filtration for potable water treatment". Filtration and Separation, 15(3): 253-261, May/Jun., 1979.

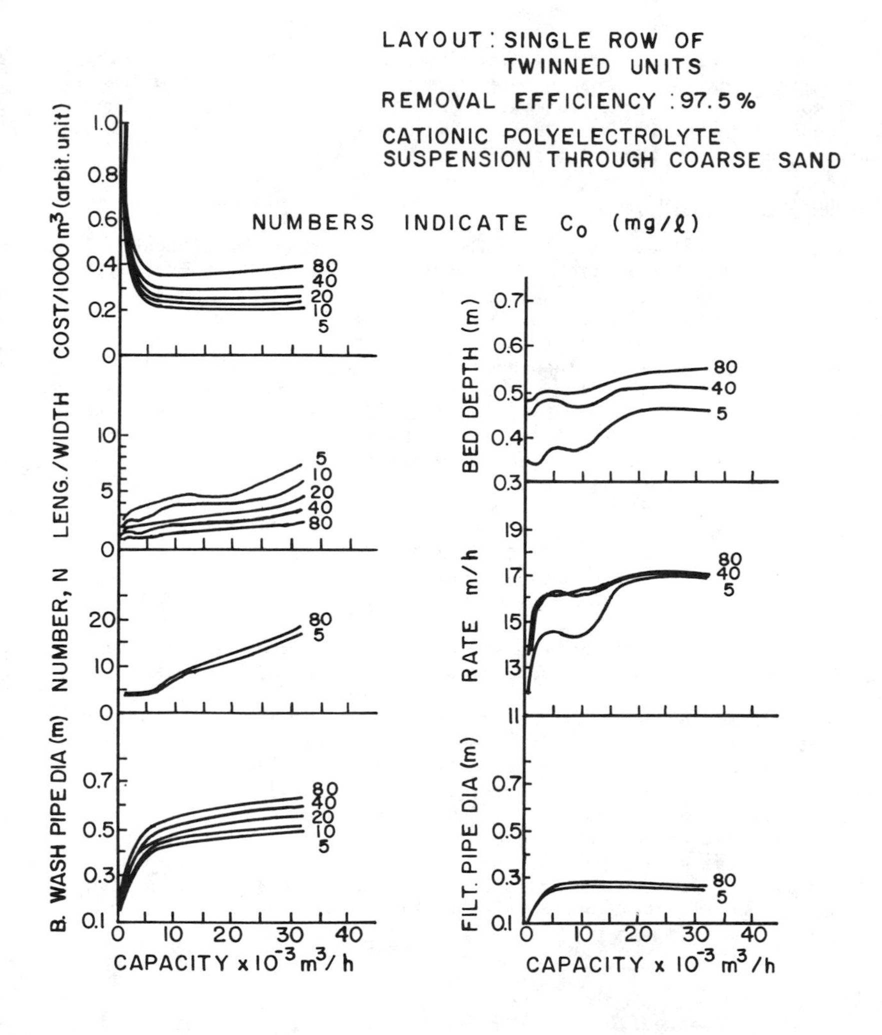

Figure 4. Optimal Designs for Cationic Polyelectrolyte Treatment

Drinking Water Treatment Using Slow Sand Filtration

Kim R. Fox*

Associate Member, ASCE

Problems faced by developing countries in providing safe drinking water directly parallel those of small communities in the United States. Recent work by the U.S. Environmental Protection Agency and others on slow sand filtration for small water systems may be applied in developing countries to assist in providing safe drinking water.

Introduction

The United Nations General Assembly has designated this decade (1981-1990) as the International Water Supply and Sanitation Decade, stressing the need to provide safe, reliable and adequate community water supplies on a world wide basis by the year 1990. The World Health Assembly (emphasizing primary health care of which a safe water supply is a major component) has adopted a goal of "Health for All by the Year 2000." Both of these goals placed emphasis on providing safe drinking water to developing countries where estimates have shown that 1.5 billion people lack reasonable access to safe water. Reasonable access was defined as having a public fountain or standpost located not more than 200 meters from a house in an urban area and implied for a rural area that "the housewife or members of the household do not have to spend a disproportionate part of the day in fetching the family's water needs." Safe water was defined as "supplies including treated surface waters or untreated but uncontaminated water such as that from protected boreholes, springs and sanitary wells" (McJunkin 1982).

In order to help achieve the safe water goals, treatment technologies for drinking water that are effective, economical, and easy to operate and maintain in developing countries must be used. One such treatment technology that has been recently re-examined is that of slow sand filtration. Slow sand filtration (SSF) is not a new process, but one that has been used to effectively treat drinking water since the early 1800's. The recent re-interest in slow sand filtration was brought about by the needs of small communities to install treatment technologies that are effective, less costly, and easier to operate and maintain than the more sophisticated rapid sand filters. These simpler technologies for small communities can easily be adapted and used in developing countries to provide safe drinking water.

---

*Environmental Engineer, U.S. Environmental Protection Agency, 26 W. Saint Clair Street, Cincinnati, Ohio 45268, USA.

The United States Environmental Protection Agency (EPA) has reinvestigated slow sand filtration as a simple technology for treating drinking water. New analytical techniques such as, particle counting, improved turbidimetry, improved growth media for microbiological analysis, and advanced techniques for measuring organic constituents, allow for a more detailed study than was possible in the early 1900's. Two pilot scale filtration systems were set up, operated and monitored at EPA laboratories under controlled conditions. In addition to the laboratory studies, EPA funded several field projects to further evaluate the effectiveness of slow sand filters and to examine labor requirements and operation and maintenance costs. The work that EPA and others have done will be described in this paper to illustrate that slow sand filtration is an effective treatment technology that can be applied in developing countries.

Slow Sand Filtration

Some of the first references to the forerunner of the slow sand filter were to a system known as the Lancashire filter developed around 1790. Almost forty years later (1827 and 1829), slow sand filters were installed in Greenock, Scotland and London, England. The design of the filters in London by James Simpson became the model by which modern day slow sand filters are designed (Baker 1948).

A slow sand filter (Figure 1) consists of an open-topped box, partially filled with sand (or other filtering media) and a water collection system beneath the sand. During slow sand filtration, water enters the reservoir above the filter media and is retained there for 3-12 hours. Some sedimentation and biological action takes place in this reservoir. The water flows downward (due to gravity) and passes through a thin layer that builds on top of the media known as the schmutzdecke. As the water enters the schmutzdecke, biological action breaks down some organic matter and inert suspended particles may be physically strained out. The water then enters the top layer of media where more physical straining and biological action occurs, as well as some attachment of particles onto the sand grain surfaces. Some sedimentation also occurs in the pores between the media grains.

Because the operation of slow sand filters is relatively simple, only moderate amounts of time are needed for daily maintenance, which entails checking flow rates and head loss, measuring filter water turbidity levels, monitoring disinfection equipment, conducting miscellaneous sampling, and recording appropriate data.

When the schmutzdecke builds up to a point at which it clogs the top sand layer and causes headloss through the filter to reach a predetermined level, the filter must be drained and the top layer of sand removed. This period of scraping is the only time during normal operation that additional personnel may be required to maintain the filter. The length of time between scrapings depends on how much material is filtered out of the water; high turbidity shortens the cycle length.

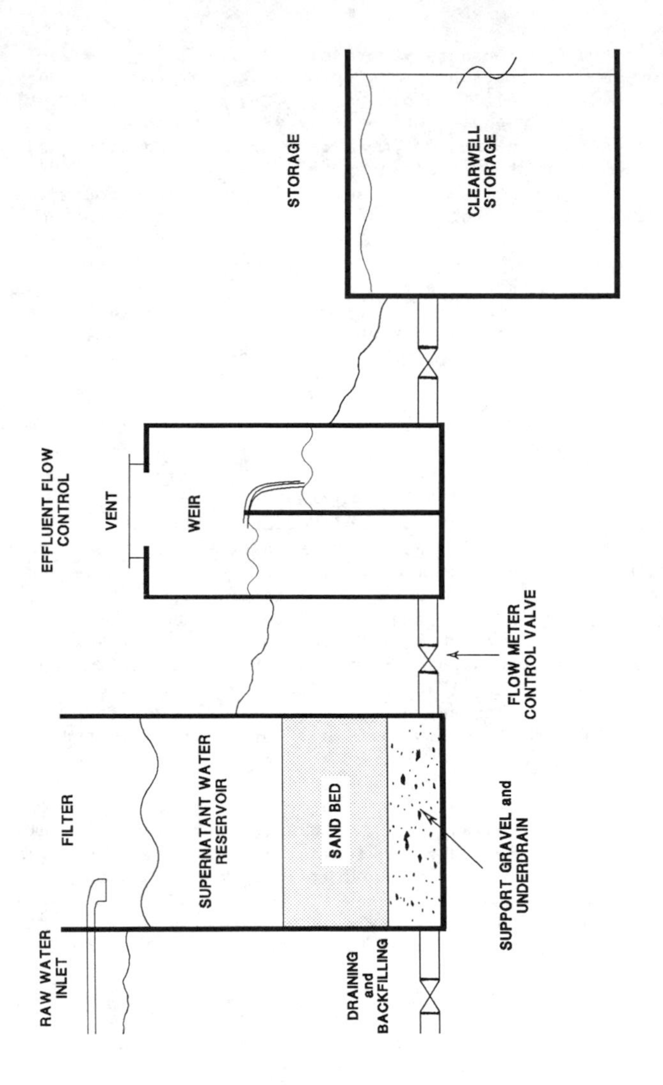

FIGURE 1. TYPICAL SLOW SAND FILTER

Slow sand filtration offers several advantages over conventional filtration techniques: 1) single step treatment that can improve the physical, biological and chemical quality of a low turbidity water, 2) ease of operation, 3) lower operating costs with operating costs being principally labor rather than energy or chemical, 4) ease of construction and maintainance with few mechanical parts, 5) proven performance over decades of use, and 6) reproven performance with the re-investigation of slow sand filters with modern analytical measurements. Disadvantages of slow sand filtration include: 1) higher capital costs incurred by the larger area of land required for construction, 2) the inability of slow sand filers to treat high turbidity waters, and 3) the inability of slow sand filters to treat waters subject to many types of industrial wastes.

Pilot Plant Research

The Drinking Water Research Division (DWRD) of the U.S. EPA constructed two pilot-scale, slow-rate filtration systems and during the first phase of the research, operated a slow sand filter (Filter A) for 800 days and a slow carbon filter, (slow-rate filter containing granular activated carbon (GAC)) for 200 days. The slow carbon filter (Filter B) was operated in parallel with a slow sand (Filter B) filter to compare the effluent quality between the sand and the carbon filter. The characteristics of the pilot-scale filters are shown in Table 1. Both filter systems were housed in thermoplastic and the filter media supported by torpedo sand and gravel. The underdrain systems were designed for collecting the filtrate and for backfilling the filter after scraping (backwashing was not a design factor). A complete description of the pilot filter systems can be found in the literature (Fox 1984). The second phase of the slow sand work is under way and EPA is investigating the removal of inorganic contaminants by slow sand filtration.

The sand used in filter A was described by the supplier as a "clean builder's sand." This sand was used to take advantage of local materials and to determine if readily available sand would perform satisfactorily in slow sand filtration or if specialized sand would be neces sary for proper operation. The work by EPA showed that the builder's sand was very successful as a slow-rate filter media. Other researchers and designers used local sand (sized according to design specifications) to take advantage of availability and cost savings (Seclaus 1986 and Paramasivan 1977).

Both filtration systems were subjected to a variety of source water conditions. Typically, low turbidity surface water was spiked with raw sewage (a sewage-to-water volume ratio of 1:2500) to produce a water with a high level of coliforms. The sewage spike created conditions such that the influent coliform levels to filter A ranged from 15 to 10,000 colony forming units (CFU)/100 mL. After the initial ripening period (with the exception of one period during the 800 days of operation) the effluent coliform counts were less than 1 CFU/100 mL. Heterotrophic bacteria reduction was typically 2 to 3 logs in magnitude on filter A. Turbidity levels for the influent to sand filter A ranged

Table 1.

Characteristics of Pilot-Scale Slow-Rate Filters

Parameter	Sand Filter A	Sand Filter B	GAC Filter B
Inside Dimension $-m^2$	0.21	0.1	0.1
Media			
Type	Sand	Sand	12 x 40 GAC
Effective Size-mm	0.17	0.29	0.55 - 0.65
Uniformity Coefficient	2.1	1.8	1.6 - 2.1
Depth -m	0.76	0.82	0.82
Mean Height of Reservoir above media - m	1.25	1.09	1.09
Flow Rate-m^3/d	0.58 0.87*	0.29	0.29
Loading Rate-m/h	0.12 0.18*	0.12	0.12
Mean Detention time in Reservoir -h	10.6 7.07*	9.21	9.21
Filter empty Bed contact time -h	8.69 5.79	6.87	6.87

*Flow rate was increased 50 percent during study.

from 0.2 to 10 ntu. Once the filter passed the ripening stage, the effluent turbidity never exceeded 1.0 ntu and for the last 400 days of operation were routinely below 0.35 ntu. Particle counting was also used to measure the effectiveness of the filter. A particle counter* capable of quantifying size and distribution of particles in the range of 2.5 to 150 um diameter in an aqueous sample was used to analyze both the influent and effluent. Filter A displayed one to two log reductions of Giardia-cyst-sized (7-12 um) particles and tended to show greater reductions as the filter became clogged. Counts of other particle sizes exhibited similar reductions. Profile sampling showed only minor reductions in particle counts below the top few centimeters of sand. In order to compare organic removals, sand filter A was sampled during the first 200 days and displayed average removals of 19 and 18 percent for total organic carbon (TOC) and seven-day trihalomethane formation potential (THMFP), respectively.

---

*HIAC Particle Size Analyzer, Model PC-320, CMH-150, Pacific Scientific Co., Montclair, Calif., USA. (Mention of commercial products does not imply indorsement by USEPA).

The GAC used in carbon filter B was a commonly available water-grade mix (12 x 40 mesh coal base) with an effective size ranging from 0.55 to 0.65 mm. Because this size is much larger than that typically used for slow-rate filtration media, particulate penetration was anticipated. Carbon filter B was subjected to a source water with a turbidity ranging from 0.4 to 23 ntu. Even with the high influent turbidities and the larger effective size of the carbon media, carbon filter B lowered the turbidity to below 1 ntu and routinely to below 0.5 ntu. After the initial ripening period for carbon filter B, one to two log reductions were observed in heterotrophic plate counts and one to three log reductions in coliform densities. Comparison of sand filters A and B showed that bacterial control improved as the effective size of the sand decreased, as previously reported by Hazen (1913). Control of organics by carbon filter B was very effective but, as expected, was dependent on time. Mean TOC and seven-day THMFP removals were 88 and 97 percent, respectively over the first 250 days of operation. Periodic profile sampling showed a time-dependent wavefront movement of organics through the adsorbent (Fox, 84).

Field Research

In conjunction with the pilot-scale research conducted at EPA laboratories, EPA initiated several field studies to evaluate slow sand filtration. The studies were done utilizing both pilot-scale systems and a full-scale slow sand filter that was off-line. The principal objectives of these field studies were to evaluate slow sand filters for the removal of *Giardia lamblia* cysts, bacteria, turbidity, and trihalomethane formation potential. Operational data were collected during the course of these projects to define the hours and costs associated with operating and maintaining slow sand filters. An evaluation also was made of filtered water quality both before and after a filter was scraped to determine how a filter's efficiency was affected by scraping.

Giardia cyst removal was shown to be affected primarily by the biological maturity of the filter and by low temperaures (Bellamy 1985 and Pyper 1985). *Giardia* cysts were spiked into the influent water in two of the field studies. At both locations, removal rates were related to the hydraulic loading rate, temperature, sand size, sand bed depth, and to schmutzdecke removal. *Giardia* cyst removal was very dependable with removal exceeding 99.9 percent seen during warm water and biologically mature filter conditions. Removal decreased during cold weather ($< 7°C$), but still exceeded 99 percent during non-optimum treatment conditions. Sand size (below 0.278 mm effective size), sand bed depth, and hydraulic load rate had little or no affect on *Giardia* cysts removal; greater than 99.9 percent removal was observed under varying conditions. The high removal under most conditions was due in part to the large size of the *Giardia* cysts (7 - 12 um).

Coliform and heterotrophic plate count bacteria removal was not as efficient as *Giardia* removal and was affected by varying operating conditions (Bellamy 1985). Coliform removal by slow sand filtration was approximately 99 percent during optimum operational conditions.

The coliform removal was affected by cold temperatures, increased hydraulic loading rates, larger effective sized sand, shallower bed depth, sand scraping, and resanding. Under various adverse conditions, total coliform removal ranged from 80 percent to over 99 percent under optimum conditions. A new slow sand filter was capable of removing only 60 percent of the total coliforms at start-up until the filter biologically matured.

One of the pilot-scale filters contained a sand with an effective size of 0.615 mm (Bellamy 1985). Total coliform removal by this filter was lower than by filters with a smaller sand. At ambient water temperatures of 17°C, this filter achieved 92.8 percent removal of total coliforms. This removal was lowered to approximately 83 percent when the source water temperature was reduced to 5°C. The filters containing the smaller sand also showed a decrease in total coliform removal with reduced source water temperatures, but still maintained an average of 90 percent removal at 5°C. A shallower bed depth also lowered the total coliform reduction. One of the test filters housed a bed depth of only 48 cm, as compared to 97 cm in the other filters. This filter demonstrated an average of 95.3 percent removal for total coliforms. Hydraulic loading also affected coliform reduction; when the flow rate was tripled from 0.12 m/hr to 0.40 m/hr, coliform removal decreased from 99 percent to 95.7 percent.

At the full-scale plant that was experimentally used, coliform removal was not as efficient as in the pilot units (Bellamy 1985). Total coliform removal averaged around 80 percent for the test periods (various intervals over two years of operation). The source water quality had extreme variations in coliform counts. The effluent from the full scale slow sand filter did not show an immediate improvement in water quality when low coliform counts were present in the source water. Thus, low removal occurred when influent water of good quality was applied following application of influent water of poor quality.

Heterotrophic plate count bacteria removal by slow sand filtration was more erratic than the coliform removal (Bellamy 1985, Letterman 1986 and Pyper 1985). This was due in part to the fact that a slow sand filter is also a biologically active community and that sloughing of bacteria from an active filter can occur. Removal averaged 89 percent for heterotrophic plate count bacteria. When high levels of heterotrophic plate count bacteria were present in the source water (5 X 10^5 CFU/mL), removal exceeded 99 percent. Removal was not as high when the influent heterotrophic plate count was lower, but effluent counts generally ranged from 8 to 200 CFU/mL.

Good reduction in turbidity and particle counts were seen in most of the field studies (Bellamy 1985, Letterman 1985 and Pyper 1985). Poor reduction in turbidity was seen at one test site, which was attributed to an abundance of extremely small particles in the source water. Slow sand filters were capable of lowering the turbidity to below 1 ntu when the source water quality did not routinely exceed 15 ntu and was not comprised of extremely small colloidal material. Under optimum conditions, effluent turbidities were routinely below 0.3 ntu at the experimental full-scale plant.

Trihalomethane formation potential was also evaluated at one field site and reductions similar to the EPA laboratory studies were observed (Pyper 1985). The full-scale unit showed a reduction of around 10 percent in formation potential. This removal is low and in areas where trihalomethane formation is a problem, slow sand filtration is probably not an acceptable treatment technique.

During routine operation, slow sand filters do not need much manpower for proper operation. The daily operational procedures such as monitoring flow rates, measuring head loss, analyzing turbidity samples, and recording appropriate operational data take a relatively short period of time. The labor requirements rise dramatically when a slow sand filter must be taken out of service for schmutzdecke scraping. Data was collected from seven operating slow sand filters to attempt to quantify the labor requirements for both routine operation and for filter scraping (Letterman 1985). An attempt was also made to relate the frequency of filter scraping to various operating conditions such as raw water quality, hydraulic loading rate, and sand size.

At one of the seven plants, one inch of sand was removed from the top of the filter manually during filter scraping and then transported from the filters with motorized buggies or by hydraulic transport. The man power required under these conditions was approximately 5 man-hours per 1000 ft^2 (92.9 m^2) of filter surface area. More man-hours would be required where the sand would be transported by hand drawn carts. Resanding, where new sand is placed in the filter to replace several scrapings of sand, normally required approximately 50 man-hours per 1000 ft^2 (92.9 m^2) (replacing 6 - 12 (15 - 30 cm) inches of clean sand).

After resanding or routine scraping, a reripening period is normally necessary before the slow sand filter is effective. In both the pilot-scale laboratory work and pilot-scale field studies, this ripening period was not evident using turbidity, particle count data, or bacteria data as the water quality measurement. The data collected from the full-scale plants did show a deterioration in water quality following filter scraping or resanding (Letterman 1985). This discrepancy in results is probably due to the fact that a full-scale filter is disturbed more during scraping than a pilot unit. In a full-scale filter, workers are walking across the sand and possibly driving vehicles across the sand during the maintenance operations, while in a pilot scale unit the maintenance operations can be performed with minimal disturbance to the filter.

The length of the ripening period in the full-scale plants following resanding or scraping ranged from 6 hours to 2 weeks in duration (based on turbidity and particle count data) (Letterman 1985). The predominate water quality factor enfluencing the length of the ripening period was the nature of the particulate matter in the raw water. The data collected did not show any evidence that source water temperature or scraping methodology had an effect on the presence or absence of a ripening period.

Costs Data

DWRD has conducted several studies and has acquired a large data base to explain the costs associated with slow sand filters. Cost equations and computer modeling have been completed and are available from EPA (Fox 1985). These data may not be suitable for use in developing countries as construction practices, labor requirements, and materials available may not be the same as used in the United States. These data can be used as guidelines in predicting labor hours for operation and maintenance of a system.

Summary

The pilot-scale studies and the field research projects have shown that slow sand filtration can be a relatively easily operated and effective treatment process for small systems requiring a treatment facility for a low turbidity surface water. Further investigation by EPA and others is ongoing to determine the role of slow sand filtration to remove inorganic contaminants from drinking water and to find ways of improving the organic removal by slow sand filtration. Although slow sand filtration is not a panacea for water treatment, it can be utilized in developing countries to help in providing safe drinking water to the world's population.

Appendix - References

Baker, M.N., "The Quest for Pure Water," American Water Works Association, Volume I, 1948.

Bellamy, W. D., Silverman, G. P., and Hendricks, D. W., "Filtration of *Giardia* Cysts and Other Substances, Volume 2. Slow Sand Filtration." EPA-600/2-85/026. U.S. Environmental Protection Agency, Cincinnati, Ohio, 1985.

Fox, K. R., Miltner, R. J., Logsdon, G. S., Dicks, D. L., and Drolet, L. F., "Pilot-Plant Studies of Slow-Rate Filtration," Journal American Water Works Association, Vol 76, pp 62-68, December 1984.

Fox, K. R., Clark, R. M. and Logsdon, G. S., "Slow Sand Filtration for Drinking Water Treatment: U.S. Experiences," 2nd World Congress on Engineering and Environment, New Delhi, India, NTIS PB86-128725/AS, November 1985.

Hazen, A., The Filtration of Public Water Supplies. John Wiley & Sons, New York, 3rd ed., (1913).

Letterman, R. D. and Cullen, T. R., "Slow Sand Filter Maintenance Costs and Effects on Water Quality." U.S. Environmental Protection Agency, Cincinnati, Ohio, NTIS PB85 - 199669/AS, 1985.

McJunkin, F. E., Water and Human Health, U.S. AID National Demonstration Water Project, AID/DSAN-C-0063, July 1982.

Paramasivan, R. and Gadkari, S. K., "Slow Sand Filtration, Project Report Phase I." National Environmental Engineering Institute, Nagpur, India, December 1977.

Pyper, G. D., "Slow Sand Filter and Package Treatment Plant Evaluation: Operating Costs and Removal of Bacteria, *Giardia* and Trihalomethanes," U.S. Environmental Protection Agency, Cincinnati, Ohio, NTIS PB85-197051/AS, 1985.

Seelaus, T. J., Hendricks, D. and Janonis, B. A., "Design and Operation of a Slow Sand Filter," Journal American Water Works Association, Vol. 78, pp 35, December 1986.

THE COMBINED CONVEYANCE/SEWAGE TREATMENT CONCEPT

Ahmad H. Gaber* & Said S. El Nashaie*

1. Introduction

Over the last few decades, sanitary engineers have experimented with various innovative techniques for increasing the contact between oxygen, bacterial mass, and organic wastes as a means of enhancing wastewater treatment. Increasing the mass transfer area has been tried using methods like extended aeration, packed beds, and rotating discs.

Recently, Evens (1) suggested a new method that is based on creating a large area of microbial film (500 m^2) in an aerobic ecosytem by providing a conduit of approximately 1 meter diameter and 12 meter length with partitions of polyamide sheets. The air is induced into the sewage stream flowing under pressure at the Venturi throat effect.

Motivated by the problem of the Scarcity of land available for treatment facilities in many villages in Egypt, the authors of this paper wished to consider the feasibility of using force mains themselves as wall film biological reactors (WFBR) for sewage treatment in small communities (5,000 - 15,000 population). If this proves to be feasible, it will be a good example of task integration, as shown in Figure 1.

This paper reports the initial assessment of the proposed concept by which the "Force Main" is used as a "site" for the biological treatment of sewage where sufficient air can be introduced in the system. The initial investigations examines kinetic data available in the literature and information obtained from the Evens Process in order to evaluate feasibility of the process. In this preliminary investigation only the residence time, concentration of the microorganisms on the wall of the pipe, and the rate of substrate consumption in BOD units are considered. Hydrodynamic problems associated with this multiphase flow system are not investigated, except with respect to the cross sectional area occupied by each phase which is taken to be proportional to the flow rate of each phase. Suspended solids are considered a part of the liquid phase. The efficiency of oxygen transfer to the reaction site is assumed to be 100%, and to be in excess of the oxygen necessary for the biological reactions. The effect of changing the oxygen transfer efficiency (η_c) is investigated without trying at this stage to correlate oxygen transfer efficiency and operating conditions. Oxygen transfer efficiency (η_c) is related to operating conditions, especially the air-to-fluid ratio and the temperature, which affect the value of dissolved oxygen concentration. However at this stage of investigation we will consider η_c to be 100% in most of the computation.

* Faculty of Engineering, Cairo University.

2. Mathematical Model for the Wall Film Biological Reactor for Sewage Treatment (WFBR)

Figure 2 shows a schematic diagram for the type of wall film biological reactor under consideration, which consists of a circular pipe of diameter d_t, length L, where the sewage is fed at the entrance with a volumetric flow rate G_f (m^3/day) and the Venturi system described by Evens (1) introduces the air at a flow rate of G_A (m^3/day). The microorganisms grow on the wall, reaching a steady state condition in about 8-30 days of start-up. The concentration of microorganisms on the wall may be expressed as Xw (mg of microorganisms / m^2 of exposed surface of pipe). The sewage BOD_5 is taken to be 106 mg/m and the required output BOD_5 is taken to be 5×10^4 mg/m^3.

The specific rate of growth of microorganisms, μ, is,

$$\mu = \frac{\mu_{max} . S}{Ks + S} \quad day^{-1}$$

where S is the substrate concentration in BOD_5, (mg/m^3) and Ks is a kinetic constant (mg/m^3).

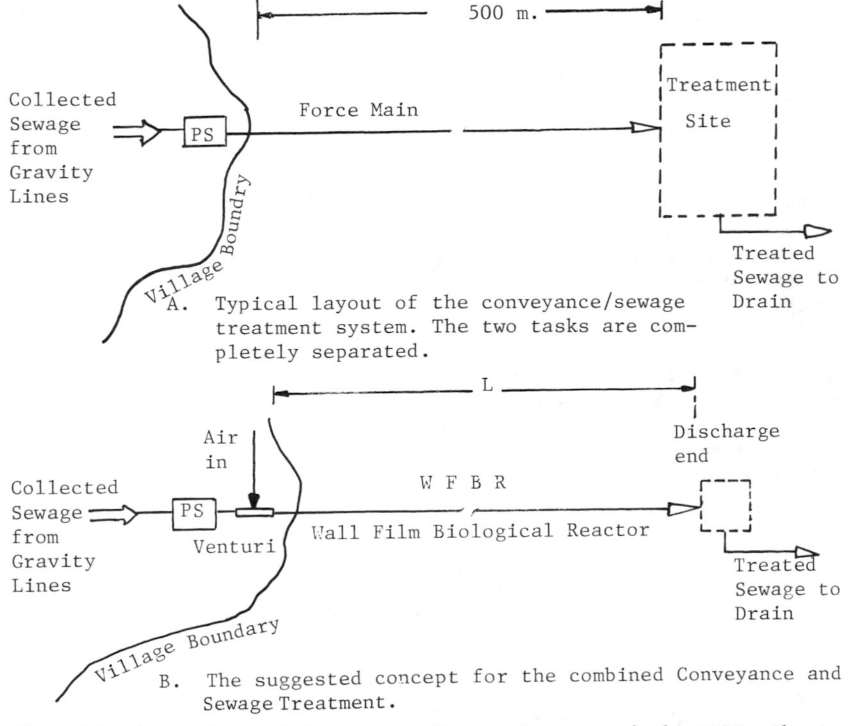

A. Typical layout of the conveyance/sewage treatment system. The two tasks are completely separated.

B. The suggested concept for the combined Conveyance and Sewage Treatment.

Fig. (1) Comparison of Conveyance Sewage System and the WFBR, Showing integration of Conveyance and Treatment Functions in the WFBR.

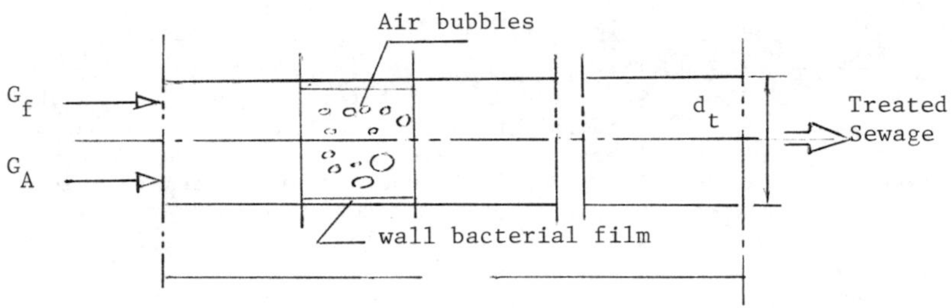

Fig. (2) Wall Film Biological Reactor (WFBR) for Sewage Treatment

A differential material balance over an element of the length of the reactor for the produced microorganisms gives

$$G_f \frac{dx}{dl} = \mu (X A_f + Xw \; d_t) - A_f . K_d . X \qquad (1)$$

where X is the concentration of microorganisms in the fluid (mg/m^3), Xw is the concentration of microorganisms on the wall (mg/m^2), and A_f is the cross sectional area occupied by the fluid,

$$A_f = \emptyset . A$$

where A is the cross sectional area of the pipe and $\emptyset$ is given by,

$$\emptyset = \frac{G_f}{G_f + G_A}$$

K_d is the rate constant day^{-1} for endogenous respiration working on the microorganisms produced in the bulk of the fluid. We consider the concentration of microorganisms on the wall to be constant, and l is the length coordinate along the length of the pipe.

In equation 1 the first term on the right hand side accounts for the reaction in the bulk of the fluid, the second term accounts for the reaction at the biological film attached to the wall, and the last term accounts for endogenous respiration as described earlier. We assume the feed to be free of microorganisms, X(o) = 0; this assumption gives a lower rate of reaction than if X(o) > 0, thus it affects the results in the direction of system overdesign rather than underdesign.

The substrate consumption is related to the rate of production of microorganisms through the yield factor, which is defined as:

$$Y = \frac{\text{gms microbial solids produced}}{\text{gms } BOD_5 \text{ removed}}$$

thus,

$$G_f \frac{ds}{dl} = -\frac{1}{Y} \mu (X . A_f + Xw . d_t . \pi) \qquad (2)$$

with S = So at l = 0

3. Estimation of Parameters for the WFBR

In this section we obtain values for the parameters of the WFBR. We found in the literature two sets of kinetic parameters. We will differentiate between them using data published on the Evens process.

3.1. Kinetic Parameters

The two sets of values found in the literature are as follows:

1st set of parameters (2)

$Ks = 60.0 * 10^3$ mg/m^3, $\mu max = 3.2$ day^{-1}, $K_d = 0.09$ day^{-1},
$Y = 0.4, 0.57$

2nd set of parameters (4)

$Ks = 16.8 * 10^3$ mg/m^3, $\mu max = 0.331$ day^{-1}, $K_d = 0.035$day^{-1},
$Y = 0.56$

3.2. The Concentration of Microorganisms on the Wall (Xw)

The parameter Xw is estimated from the Work of Evens which gives Xw for polyamides to be 1.5×10^6 mg/m^2. If we consider the growth of the microorganisms on the wall of the pipe to be 10% of that for polyamide (a very conservative estimate since on glass, which gives the lowest growth, it is 1/6 of that for polyamides), then Xw for our case is taken as 15×10^4 mg/m^2. This value can also be shown to be a conservative value if we consult the work of Characklis and Trulear(5). They found that the thickness of the biofilm formed is in the order of magnitude of 0.1 cm, and that the biofilm density is about 7.5 mg dry cell/cm^3. From these data and Evens' (1) estimation that dry matter concentration of the aquatic organism is 1% and that active microorganisms compose 50% of the total, thus,

$$Xw = 37.5 \times 10^4 \text{ mg/m}^2$$

This value is more than twice as much as the value of $Xw = 15 \times 10^4$ mg/m^2 that we are using, so it can be safely assumed that the value of X is not overestimated.

Use of a low value of Xw at this initial stage of investigation is justified by the fact that we don't have accurate estimates and by the fact that at this stage of investigating the feasibility of the process it is preferrable to avoid overestimating the rate of reaction (underdesign) than to underestimate the rate of reaction (overdesign). Low values of Xw in fact underestimate the rate of reaction in the direction of overdesign.

3.3. The Air Flow Rate (G_A)

For a sewage feed of BOD_5. $So = 10^6$ mg/m^3 (1000 mg/lit) and the exit water having BOD_5 $S = 50 \times 10^3$ mg/m^3 (50 mg/lit). Simple calculation shows that the necessary volumetric flow rate of air is given by:

$$G'_A = 3.454 \; G_f \; \text{m}^3/\text{day}.$$

However G_A computed on the basis of BOD_5 should be corrected for BOD_u as follows (6):

$$G_A = G'_A \times \frac{1}{1 - 10^{-5K}}$$

where K = 0.1 day^{-1}, thus,

$$G_A = 5.07\ G_f\ m^3/day$$

4. Evens Process

The system of Evens (1) is similar in principle to the WFBR except that the microorganisms are grown on the polyamide sheets immersed along the length of the canalization. For this system the total area of polyamide sheets is distributed along the total length of the system.

The model equations are slightly different than those for WFBR because the surface for microorganisms is polyamide sheet and not the walls of a pipe.

The growth equation is:

$$G_f \frac{dx}{dl} = \mu.X.A_f + Xw.\ a_1.\ \mu - Kd.X.\ A_f \qquad (3)$$

Where a_1 is the area if active surface (polyamide sheets) per unit length of the system, m^2/m.

The equation for the substrate is,

$$G_f \frac{ds}{dl} = \frac{-1}{y} \left(\mu.X.A_f + X_w.a_1.\mu \right) \qquad (4)$$

5. Parameters for Evens Process

Evens gives:

$$X_w = 15 \times 10^5\ mg/m^2$$

and,

$$a_1 = \frac{500}{12} = 41.66\ m^2/m\ \text{length}$$

- The wastewater flow rate varies within the range of 200-500 m^3/hr, and the system can introduce air in the range of 350-900 kg air/hr.
- The diameter of the "canalization" is 1 meter and its total length is 12 meters. In the original design the canalization was a cylinder, the frame of which was made of PVC tubing covered with a strong nylon texture (0.5 mm mesh) and itself protected with a netlon netting of 1 to 2 cm mesh.

If we take the BOD_5 for the feed wastewater to be 1000 mg/lit., and that for the exist to be 50 mg/lit., then to be able to achieve at least the hypothetical air required for the system for the above limits of wastewater flow rate and air supply we must take.

$$G_f = 4800 \ m^3/day \quad \text{(lower limit for } G_f\text{)}$$

$$G_A = 16494.3 \ m^3/day \quad \text{(upper limit for } G_A\text{)}$$

Thus: $\frac{G_A}{G_f}$ = 3.44, which is within the range of hypothetical air requirement.

6. Results and Discussions

In this section we solve the model equations using the given parameters and obtain the pipeline length necessary for the required degree of treatment.

6.1. Evens Process

Using the Evens model equations (equations 3,4) together with the specific parameters given in Section 5 and the two sets of kinetic parameters given in Section 3.1, we obtain the length required to treat wastewater of 1000 mg/lit to output water of 50 mg/lit. Using the first set of kinetic parameters gives a length of 10.8 meters (for Y = 0.4), while using the 2nd set of kinetic parameters gives a length of 131.6 meters. From these results it seems that the second set of parameters underestimates the rate of reaction giving the required length as over 100 meters while the first set of parameters gives a required length which is quite close to the 12 meter length of canalization. Therefore it seems reasonable to use the first set of kinetic parameters in our computations rather than the 2nd set of kinetic parameters.

Table 1 shows the effect of changing "a" (air to fluid ratio) and Y (the yield factor) on the required length of the pipe. We notice that "a" does not have a significant effect on the required length, while increasing Y from 0.4 to 0.5 causes the length to increase from 10.8 meters to 15.6 meters. This is due to the fact for the same rate of growth the rate of substrate consumption increases with the decrease of Y. The insignificant effect of "a" is due to the fact that our model does not include any allowance for the effect of "a" on dissolved oxygen concentration. Therefore "a" only affects A_f which only affects the rate of reaction in the bulk fluid, which is negligible.

The results of this rough simulation of the Evens process shows that even with the very approximate nature of the parameters and operating conditions we can confidently reject the second set of kinetic parameters and accept the first set of kinetic parameters as a first approximation of the rate of reaction at this preliminary stage of the investigation.

6.2. WFBR Results

For the WFBR model we used the first set of kinetic parameters, as justified in the previous section, and the rest of parameters given in Section 3. For a 15 cm diameter pipe and a wastewater flow rate of 250 m^3/day and an air to fluid ratio of 3.454 we obtained a length of about 500 meters (for Y = 0.4) (see Table 2).

Because of the uncertainty at this stage regarding many of the parameters, Table 2 gives the effect of various parameters on the required length of the WFBR. It is clear from the table that the length is not affected by "a", this is of course due to the fact that the model used does not relate "a" to the dissolved oxygen concentration. However the rest of the parameters affect the required length of the pipe considerably. It is clear from the results in Table 2, that the required length can be corellated to the various parameters according to the following very simple relation,

$$L = C_1 \frac{G_f \times Y}{d_t \times X_w} \quad (5)$$

Where all the variables in equation (5) have the values shown in Table 2, and where C_1 is a constant equal to 1.125×10^5.

Table 3 shows the steady state response for changes in the feed flow rate on BOD_5 at the exit for fixed pipe length. The BOD_5 at exit increases with the increase in the feed flow due to the decrease in the residence time.

Table 4 shows the effect of oxygen transfer efficiency on the length of the pipe needed to achieve the required treatment. The dissolved oxygen concentration is not related at this stage of investigation to the design and operating conditions. If we denote the efficiency of the oxygen transfer to the reaction site and its effect on the rate of reaction in comparison with the case of excess oxygen in the reaction site by η_c, then equation (5) will be modified to become;

$$L = C_1 \frac{G_f \quad Y}{d_t \quad Xw} \frac{1}{\eta_c} \quad (6)$$

The relation between η_c and operating/design parameters will be the subject of the next investigation.

7. Conclusions

A simple plug flow model for a wall film biological reactor for sewage treatment has been developed. The kinetic parameters for the reaction system were obtained by checking available data in the literature against the performance of the Evens Process. The density of the microorganisms at the wall was estimated conservatively from the results of the glass rod and polyamide sheets of Evens, and the value of Xw obtained was checked against other published data (5) to make sure that the errors in Xw evaluation will be on the side of underestimation rather overestimation of the rate of reaction.

The results obtained suggested that it is possible to use the force main as a biological reactor provided sufficient air is introduced into the system for aerobic digestion to take place and that the microorganisms developed on the wall of the pipe is within our estimated values. A length varying between 500 and 700 meters for a 15 cm diameter pipe seems to be sufficient to reduce the BOD_5 of a 250 m^3/day sewage flow from 1000 mg/lit to 50 mg/lit.

This investigation represents the first phase towards the final design of this system. Subsequent stages consist of:

i. Determination of the dissolved oxygen concentration and its effects on the rate of reactions.
ii. Experimental verifications of the kinetic parameters in a prototype.
iii. The hydraulic characteristics of the system including analysis of the energy requirements.
iv. Engineering design of the full scale system.

Acknowledgments:

The first author would like to thank Chemonics International Consulting Division for the time and opportunity to develop these ideas. The authors would also like to thank Professor Kazuyoshi Kawata for providing several of the sources used in this paper and for his encouragement, and Mr. Anthony Stellato for editorial assistance.

References:

1. Evens, F., "The Treatment of wastewater and the Rehabilitation of Rivers and Lakes by the "Phallus Process". Academic Royale Des Sciences D Outre-Mer, Extrait Du Bulletin des Seances 1976. pages 112-136.
2. Sherrard, J., Lawrence A., "Design and Operating Model of Activated Sludge". Jour ASCE., Env. Engng. Div., 99, 1973.
3. Servisi, J. and Bogan, R., "Free Energy as a Parameter in Biological Treatment", Proc. ASCE, 89, 1963.
4. Arceivala, S.J., Wastewater Treatment and Disposal,Marcel Dekker Inc., p. 604, table 12-1, 1981
5. Characklis, W.G. and Trulear, M.G. Dynamics of Microbial Film Process, Part V "Biokinetics", 1980.
6. Arceivala, S.J., Wastewater Treatment and Disposal, p. 620, 1981.

Table 1. Effect of some parameters on the computed length for Evens Process.

a	Y	Length (meters)
3.44	0.4	10.8
5.07	0.4	10.8
3.44	0.57	15.6
5.07	0.57	15.6

Table 2. Effect of various parameters on the required length of (WFBR).

G_f m^3/ day	a	d_t meter	Y	Xw mg/m^2	Length (meters)	Velocity of Liquid cm/sec
250	3.454	0.15	0.4	15×10^4	500	72.8
250	3.454	0.15	0.57	15×10^4	710	72.8
250	5.07	0.15	0.57	15×10^4	710	99.3
250	5.07	0.3	0.57	15×10^4	350	24.8
500	5.07	0.3	0.57	15×10^4	705	49.6
500	5.07	0.15	0.57	15×10^4	1430.0	198.5
500	5.07	0.3	0.4	15×10^4	495	49.6
500	5.07	0.3	0.57	30.0×10^4	355	49.6

Table 3. Effect of changing fluid feed flow rate on exit BOD for (WFBR).

L = 350.0 meters, a = 5.07, Y = 0.57, d_t = 0.3 meters, X_w = 15×10^4

G_f	BOD_5 at exit mg/lit.	Velocity of fluid cm/sec
250	36.57	24.81
250 x 0.9	5.45	22.33
250 x 0.8	0.279	19.85
250 x 1.1	92.130	27.29
250 x 1.2	155.250	29.77

Table 4. Effect of Oxygen supply efficiency.

G = 250 m^3/day, Y = 0.57, d_t = 0.3 meter, a = 5.07, X_w = 15×10^4 mg/m^2

Oxygen supply efficiency	Length (meters)
100 %	350
95 %	370
90 %	390
85 %	415
80 %	440

SYSTEMS ENGINEERING SOLUTIONS TO DROUGHT: THE TANZANIAN CASE

Athumani R. Mfutakamba*
Member of American Society of Civil Engineers
(Student Chapter)

ABSTRACT

This paper proposes a drought model that would help overcome the suffering of the millions of people in the world who live in abject poverty, consequent to both climatological, political, economic and social factors. The model is then applied to the Tanzanian case with profound results in the water supply, agricultural and energy systems.

Experience contends that among the factors which strangle economies of most developing countries is the absence of the necessary technological capabilities to offset drought. Various resource allocation models are discussed and critically reviewed and the proposed model is developed in a Systems Engineering approach.

Great emphasis is given to the need to approach the problem in a multi-disciplinary manner. All expertise in the natural, pure, applied, and the social sciences and engineering are required for a successful formulation, implementation, monitoring and evaluation of the model.

Active participation of the population is also stressed to achieve the preset objectives. Risk analysis, reliability and uncertainty, utilizing decision theory and systems analysis are discussed in the choice of a viable alternative for the period of water scarcity. Different sources of water are identified and the best source is selected systematically.

Procedures in model building, verification, diagnostic checking and updating as new information becomes available are demonstrated. The interplay between statistics and probability, hydrology, engineering and economics in water resources is established to help planners and decision makers manage the resources effectively.

The author calls for social and environmental impact studies for the projects. He also underscores the need for regional and inter-regional integration of water resources, sanitation, agriculture and energy systems through good political will and right decision making within and among African countries. This is due to the merits of comparative advantage, economies of scale, and right investments at the right place and time.

*Chairman and Director General New Tech International Institute, P. O. Box 35153 Dar-es-Salaam, Tanzania, East Africa; P. O. Box 2271, Stanford California 94305, U.S.A.

It is concluded that the effective implementation of the model could best be achieved if countries in Africa would consider the formation of a joint African Water Resources, Sanitation, Agriculture and Energy Council under the auspices of the Organization of African Unity (OAU) to plan, design, and implement projects to cater for their needs.

The paper outlines various recommendations to designers, planners and decision makers who are dealing specifically with the role of water economy sector in the development of nations.

INTRODUCTION

The drought model is presented as published elsewhere (Mfutakamba, 1983) and this paper applies it in critically evaluating the Tanzanian experience on rural water supply, irrigation and sanitation systems.

The case study critically assesses a followup to a published report on the earlier work conducted in the summer of 1979 by the Office of Evaluation to examine the effectiveness of rural portable water projects assisted by the USAID and other donor agencies.

The case study is preceded by an overview of current research and implementation efforts in water supply, sanitation and irrigation systems in developing countries.

Successful implementation of the model requires a sectoral approach in deriving the model parameters which are reflected as constraints in the dynamic programming format. Analysis of principal component to determine the contribution (weighing) of each constraint required in the formulation of the model is proposed using decision matrix.

The complete model is presented at the end of the text preceded by the case study. The complexity of model formulation and evaluation is demonstrated in determining qualitatively and quantitatively three of the nine constraints mainly: social, technological and policy (administrative and political) constraints in the Tanzania case.

General recommendations for Africa and particularly for Tanzania are given in the model and case study respectively. Further research work is required in determining the specific values of the constraints and other approaches like simulation studies in determining precise corrective remedial solutions to economies faced with drought problems.

Salient features of the model are reflected in the discussion of the case study. The development of the drought model and closing remarks is attached in the Appendix.

RURAL WATER SUPPLY, SANITATION AND IRRIGATION SYSTEMS

In the developing world, the need for integrated development projects is becoming more important.

The role of water resources in the development of economies is very central. Due to the global power structure the emphasis has been to develop water resources to cater for the needs of the urban centers. The focus has never been to the main foreign exchange earners: The rural poor. Even the budget structure continues to favor urban areas not only in health services, roads, housing, education, but also in water supply and sanitation.

Recently some developing countries, while maintaining the urban emphasis of water supply and sanitation have also turned focus to the rural areas. International, multilateral, governmental agencies and local governments drew plans and strategy to tackle the issue.

It would, therefore, be interesting to evaluate the recent study by USAID on the social, technical and administrative (policy) issues of one case study in a developing country--Tanzania on rural water projects and inject a few lessons from elsewhere in developing countries.

Let us first examine broadly the question of water supply and particularly research efforts now underway in developing countries.

Water Pollution Control in Developing Countries

About 80% of all known diseases are related to water misuse (WHO 1976). Lack of portable water plus three billion people or 72% of those living in developing countries in perpetual danger of these killer diseases. Inadequate nutrition also leaves more than 150 million people in Africa, the majority being women and children in the rural areas starving.

Nationally and internationally, efforts are being made to arrest the situation. Water supply and sanitation facilities in developing countries have to draw on existing knowledge and experience in water supply and pollution control, which the developed countries did not have when facing similar problems.

These alarming finding were among the main reasons leading the United Nations Water Conference held in Mar-del-Plata, Argentina in 1979 (Gust, 1979) to declare the decade 1981-1990 as "The International Water and Sanitation Decade."

Best Technology

The appropriate technology for water supply and pollution control in the developing countries has to be based, therefore, on careful assessments and adjustment of suggested solutions from other countries to local existing conditions and circumstances. This technology has to be applied mainly in the fields of urban and rural water supply, waste water disposal, water pollution control legislation, professional manpower development, the design and construction of both small and large scale irrigation systems and the use of as much as possible of local material, financial and human resources.

This enormous and ambitious task requires a vast commitment of resources. Providing clean water requires political will, technical expertise, financial and material resources all supported in a management structure capable of realizing plans as programs of action according to the Health Science Division of IDRC-Canada. In areas where ground-water is readily available, the handpump is the simplest and least costly method of supplying safe drinking water. Peat latrines and septic tanks have to be well designed and well protected to avoid source pollution.

By the year 1990, some 1833 million (1.8 billion) people in the developing world (excluding China) will require new, clean water supplies. Almost 1400 million (1.4 billion) of those will be living in rural areas. More water supplies and sanitation facilities will be required. Approximately 20 million or more hand pumps may be needed by the year 2000. Replacement pumps will be needed for at least 500 million people during this same period, adding another 2.5 million pumps to a total requirements.

COST OF PUMPS, IMPORTS, VOCATIONAL TRAINING AND TECHNOLOGY TRANSFER

The manufacturing, installation and maintenance of water pumps has to be done by the rural people themselves if any significant headway is to be made in supplying safe water for drinking, livestock and adequate water for small scale irrigation in developing countries in general and Africa in particular.

Pumping devices need to be developed at rural vocational training centers to withstand use and abuse, of a large user group. Rural vocational training centers require resources: material, human and the desired infrastructure and again, a good management structure as well as good planning and forecasts.

The development of a reliable hand pump that can be locally produced, installed and maintained at a reasonable price would be a major step toward providing reliable safe drinking water and other needs in rural communities.

Other renewable forms of energy like solar, wind, geothermal to pump water and provide electricity for other uses could be investigated for their economic and technical viability.

The cost of a hand pump varies from a low of about Tanzania Shillings (TAS) 800 for a simple shallow-well pump to a high of around TAS 80,000 for a heavy duty, deep-well pump. A reliable source indicates that, for Tanzania, one of the East African countries committed to developing rural water supplies, the average maintenance cost per pump per year is approximately TAS 16,000. The International Reference Center for Community Water Supply and Sanitation puts the minimum cost to bring safe water and adequate sanitation to all by 1990 at an estimated TAS 12,000 billion--even using low-cost technologies and community self-help, it is clear that governments and aid agencies alone will not be able to support the bill. Some financing strategies need to be adopted. For instance, in Costa Rica, a revolving fund system was established, whereby urban water users pay more to subsidize the rural water system. This is proportionate to the funds invested in the former to improve the quality and quantity of the water and sanitation systems.

Governments must reduce reliance on imports by researching, testing and using locally available materials. Strategies must be developed to promote community acceptance and self-reliance at the village level. Programs for effective transfer of the technology to the villagers themselves must be developed and implemented. Participation of all the relevant institutions dealing with the technology to be transferred of the recipient country should be fully integrated.

Authorities in developing countries should now press for agreements with donors which stipulate the manufacturing of pumps locally rather than importation. Fortunately materials for making pumps are readily available in Africa and elsewhere in the developing world. Through widespread introduction of plastics technology much use is now focused on polymer resins, specifically polyvinyl-choloride (PVC) and polyethylene (PE) for making pumps.

One of the most salient bottlenecks in rural water supply programs is the high failure rate of conventional manual pumps. Failures occur mainly because pumps were not designed for the level of stress and abuse they routinely receive in the rural areas of developing countries. This observation must, however, be qualified by the fact that there are cases and forms of vandalism even in the developed world of different category. Some pumps breakdown before they were even commissioned; the case of CIDA and SIDA sponsored rural water supply project in Kenya is one example.

Because the materials from which conventional pumps are made, mainly cast iron and steel, are not only expensive, but are not readily available locally, many developing countries must rely on imported pumps and parts supplied by international and bilateral donors. These factors presented difficulties in terms of costs, and maintenance requirements and problems in the procurement of spare parts.

THE TANZANIA EXPERIENCE

Tanzania has an area of about 945,087 sq. km. (364,900 sq. miles) with a population of nearly 20 million people. The economy is agriculturally based with leading industries involved in agriculture processing, mining, manufacturing and tourism.

Most of mainland Tanzania is an infertile plateau, 1,000 to 4,000 feet high bisected from north to south by the Great Rift Valley. On the more fertile margins of the plateau are lake Nyanza (Victoria), Tanqanyika and Malawi. With several permanent rivers and the lakes, the country has a great potential for irrigation and improved rural water supply and sanitation systems.

Based on economics of scale, the moving of over 9 million Tanzanians into 6,900 villages by mid-1975 helped a great deal in supplying them, among other social amenities, safe drinking water and sanitation systems.

There are other advantages in people living and working together for the common good. Education, both formal and informal, health services and an increased level of political awareness and defense are some of the examples.

There are several studies done on various sectors of the Tanzania's economy in general and the role of the water economy in particular. These studies have been carried out by local and foreign experts or as a joint venture by foreign assistance donors and host country officials as they prepare to meet the challenge of delivering a basic human requirement--safe drinking water.

A critical assessment will be made to a study conducted during the summer of 1979 by the office of evaluation to examine the effectiveness of rural portable water projects assisted by the USAID.

The study focuses on technical, administrative and financial viability issues of rural water systems in Tanzania. The study was considered useful because it was based on field observations in one of the least developed countries which has made a serious commitment to provide safe water to its rural citizens, it examined the variety and appropriateness of the technological approaches for supplying water to rural areas; and it contained a number of findings and proposals worthy of serious consideration by donors and recipient countries.

The Study

The team which did the evaluation consisted of three core members: a geographer with USAID, and economist with Development Alternatives and an engineer. A local senior hydrologist with the Ministry of Water, Energy and Minerals (MAJI) was the head of the Tanzanian group that participated, which included Regional Water Engineers, interpreters, and support staff.

The team visited 20 villages that were selected after discussions with MAJI officials and donors. The main criteria for selection were:

1. that the systems were completed;
2. that they represented the technological diversity of rural water supply systems in the country; and
3. that they represented regions with different natural water resources.

The sites visited represented the four major technologies in use in dry regions (Shinyanga and Singida). Wet regions (Iringa and Mbeya), and mixed regions (Morogoro). The team did not attempt to provide a statistically representative sample of the water program. At each site, a standardized interview schedule was used to collect data. The group responding, consisted of the village chairman or secretary, and a group of village elders. Women were not usually part of the group interviewed, but there were exceptions.

It is clear at this point that the team overlooked the important subjects of the study--women, who are primary drawers of water in any developing country. This point is stressed earlier by drawing experience in Sri Lanka and is documented by researchers in the early seventies in Tanzania, and East Africa, for example, Tschannel and Warner and Internationally by Dan Okun (1981), among others.

The Setting

The team identified part of the key question in developing, implementing, and administering a rural water system to be based on local conditions which are unique for each village. However, a representative sample of a small number of villages was necessary to limit the study in terms of costs, scope and time constraints.

An argument could be extended further on the history of past investments and the constraints imposed by climate, economics, and the institutions in competing for the use of the simplest, least expensive technology. The level of technological advancement in a Tanzanian village is still very low for, again, historical reasons. There is of now a high potential of developing such capabilities especially when we realize that the attained level of literacy is even higher than that of the USA as reported by New York based Africa News.

Climate

Because of the weather diversity, rainfall is a major factor in the choice of technology. In part of the country, rainfall is confined to two periods.

During rainy periods, rural residents can collect water from a number of sources that are not available in dry periods. This is critical for planning water systems, since anticipated benefits frequently assume that the improved source of water supply will be used exclusively. The team urges it is difficult to convenience a rural resident to collect water from an improved source that is further than a traditional source close at hand.

The counter argument here is that public health awareness is very important. Through media, seminars, meetings and films which demonstrate the advantages of using improved sources people will avoid using the contaminated water. Some areas with a high level of fluoride content were abandoned when better alternative sources were identified. Rural water supply and sanitation should form an integral part of political literature in Tanzania now that people are literate.

From the personal knowledge of the author and studies and tests done by IDRC-Canada, the in-country design of a foot or hand pumps made of locally available PVC or PE material will be more convenient. This can avoid difficulties during the rainy period when delivery of fuel or spare parts become impossible due to impassable roads and washed out bridges.

The handpumps can be installed close to their homes, the villagers could save a great deal of time and effort in not having to fetch water from the river or other unprotected sources. The plastic pumps are also lighter and easier to operate than metal pumps, a boon for women and children especially.

More acceptability and satisfaction in the quality of water can be increased using PVC pumps. Water from metal handpumps cannot be free from rust and yellow color. Piped water has higher levels of chlorine which families find hard to tolerate. Also problems of turbidity and foul-smell are common in piped water.

Villagers should be discouraged from using traditional sources of water such as streams and rivers which are highly polluted. Simple inexpensive but effective treatment of polluted water must be instituted. System reliability and performance has to be improved to guarantee supply qualitatively and quantitatively. Research on the source and integrating users' rights from planning stages are the necessary prerequisites.

Settlement Patterns

In mid-1977, 12 million people out of a total of 14 million lived in 7300 villages. Projections were that by 1981, virtually all the 15.7 million people of the country will be living in 8000 communities.

The team observed that there are advantages of supplying water to grouped communities as a criterion for village location. However, poor or improper planning and funding caused many villages to have no convenient source of water.

A careful assessment of demand and supply of domestic and irrigation water is necessary before Tanzania can complete its capital movement from Dar-es-Salaam to Dodoma. There are cases in history such as in India where a powerful king built a beautiful city in which no one could live because there wasn't enough water for the residents. Water engineers, therefore, have a responsibility to offer professional advice to decision makers.

Government Policy

The team suggested that government policy and government actions conflict; the result is that national program lacks financial support and is inconsistent in the technologies it uses and the obligations it puts on the communities. It (the team) further argues by advancing an historical review of policies to furnish a perspective for understanding the present situation.

The discussion of the Evaluation team was structured in this manner: the policy of the government in rural water supply has undergone a number of changes. The initial goal, set forth in the Second Development Plan in 1969, was to provide water for everyone in rural areas within 40 years. In 1970, the time period was shortened and interim goals were set:

- to provide a source of clean dependable water within a reasonable distance (less than 4 km) of each village by 1981 as a free basic service; and
- to provide a piped water supply in the rural areas by 1991 so that all people will have access to a public water supply within 400 meters.

The Evaluation report states further that by 1976, the emphasis had been changed. The Third Five-Year Development Plan (1976-1981) included the following guidelines:

- Village water systems should consist of small and simple systems at low costs;
- Various methods of water collection should be used, e.g., rainwater collection;
- Suitable water lifting devices include handpumps and windmills;
- Every 100 villagers will have at least one water source; and
- The village program is not primarily to provide piped water.

Recently, the argument (in the report) continued, the Government changed its policy in two ways: it dropped the 1981 interim goal that each village have at least one source of water; more significantly, it required users to pay for water. The 1991 goal still stands, although the requirement for piped supply conflicts with the 1976 guideline that states that the village program is not primarily to provide piped water.

To conclude, the report states that this inconsistency would not be a problem if a central ministry controlled the formulation and implementation of programs. In Tanzania, this is not the case. There are 20 regions, each with its own rural water program. In some of the regions, diesel-pumped water systems with extensive distribution network are used for rural water supply, while other regions rely on hand-dug shallow wells with hand pumps. In some regions, users are asked to pay during periods of the year and in other regions they are never asked to contribute.

Whereas the facts may be very well stated, the analysis does not reflect the causes of what the team terms as "inconsistencies" and "conflicts" in the Government policy on rural water supply. Rather, these are measures taken by a young economy faced with choosing between essentials amidst decline in world trade, drought, war and an increase in the price of oil.

Planning is basically an art and a science, to have programs which one can allow flexibility and offer an economy breathing space. Any water resources systems engineer dealing with real-time operation of a system knows the benefit of revising an unattainable objective function as new information becomes available or constraints get more defined and clear. Economic austerity measures have to be taken by any country once sign of a declining economy are observed.

On the ability of users to pay although the team did not reveal which villages or regions pay and which ones do not, one may suspect possible reasons as either willingness, ability or how the projects were introduced to the community. Were the systems government funded, self-help, church sponsored, donor or voluntary organization?

Let me, at this juncture, inject an overall comment on the study. First, the team's composition was very much skewed towards water professionals and neglected the whole aspect of the interdisciplinary nature of the water-economy. Thus, the study though very commendable, was very much left with a host of questions rather than answers to the problems of rural water supply.

Second, among other local counterparts, there should have been a Tanzanian political economist or at least someone from the Ministry of Economic Planning. Also, in each region visited, an official of the party CCM, should have been included in the team. That way, a project of any nature can easily get mass acceptance.

Third, village water supply and sanitation are just as important as agriculture is in the rural areas. So, there is an institutional integration need for MAJI and KILIMO (agriculture and Livestock Ministry). The study, therefore, should have had not only regional water engineers but also regional agricultural officers and irrigation engineers. Such focus helps in improving rural integrated schemes (projects).

Fourth, studies such as this one are of practical benefit to local institutions of higher learning such as the University of Dar-es-Salaam and Sokoine University of Agriculture. Some interested faculties and students both graduate and undergraduate should have been invited to participate. Future projects may improve enormously if these experiences can be included in the planning, implementation and management aspects.

Finally, let me identify some general constraints that are likely to be encountered in studies like this one. The design of such a study could have such factors in mind and could, therefore, have some interim guidelines of identifying participants as researchers and subjects too. This may help to attain even higher national objectives with the village as a nucleus (planning unit). The drought model (Mfutakamba, 1983) is attached in the Appendix. The salient feature of the model reflects an optimization scheme:

- Resource inventory constraints and the available and affordable technology
- Physical constraints (if wells, dams, canals, pumps, and pipes are to be built)
- Economic constraints
- Social constraints
- Legal constraints
- Religions, cultural and traditional constraints
- Political and institutional constraints
- Financial constraints

- Environmental constraints

Following the report further, the lack of national program has resulted in wide variation in projects. Complex technologies predominate in one region while simple systems are emphasized in another. Donors chose technologies on the basis of their interest and traditional approach, rather than on the most appropriate solution for the physical and social setting. The guidelines of the Third Five-Year Development Plan have almost completely been ignored or contradicted. In Shinyanga, a rural water system means hand-dug shallow wells with Shinyanga handpumps. In Singida, with much the same physical and social setting, the technology used is predominately deep wells with dual pumps: a wine-mill and a standby diesel unit. In Iringa, they emphasize gravity systems. Morogoro emphasizes hand-drilled shallow wells with Kangaroo handpumps. Tanga is served by an extensive gravity pipeline. In the West Lake area, systems are diesel-pumped.

The examination of the reports' conclusion that there are three conditions that would ensure that rural water systems are and will continue to be successful in Tanzania, need to include a clause on irrigation, especially small scale.

The shallow wells projects in Shinyanga and Morogoro, are both successful because the technology is particularly appropriate for social and environmental conditions, based on systems reliability studies, but neither project has satisfactorily resolved the problem of local financing.

Tanzania is not receiving all the possible benefits from donor programs because the activities of the Regional Water Engineers and the donor projects are not always closely integrated. There are three possible ways to provide the integration, provide counterpart personnel who are trained in the methods and techniques of the project; periodically review and integrate activities of both the donor and Regional Water Engineers to prevent duplication of efforts; and finally, to adopt or adapt the innovative methods of donor projects into the regional program.

Rural water supply and sanitation systems should be integrated into the national and regional water master plans. This will avoid leaving out the very people the water master plans are intended to serve. The same applies to energy and agriculture master plans.

The need for uniformity of components was stressed in the report. The rural water supply program in Tanzania is less effective than it could be because components of water systems are not uniform. The Singida project tried to use pumps that were available in the country, but when none were available, they introduced a different pump this increased the complexity of maintenance. Diesel engines used for rural water systems are made by a number of different manufacturers. British, Italian, Indian and American diesel engines were encountered.

The same situation applies to almost all the components in water supply systems. It is difficult for the country to maintain adequate inventories of all makes and models of water components. The Government should assess the suitability of water systems components; establish an approved list and insist on compliance by donor agencies.

The success of the shallow wells projects compared to the performance of other technologies suggests that, where possible, shallow well should be the technological choice. This only assumed water needs for domestic and livestock, however, when irrigation is incorporated may be another alternative could be arrived at. With community support to assure their continued operation, these systems can serve much of the rural area of the country with a dependable supply of water at low cost.

REFERENCES

Buras, N. (1972). Scientific Allocation of Water Resources.

Dan-Okun (1981). The Proceedings of the International Conference on Water Supply and Sanitation Decade, Civil Engineering Department University of Ottawa, Ottawa, Canada.

Fiering, M.B. and Jackson, B.B. (1977). Synthetic Streamflows, Water Resources

Gust (1979). Proceedings of the United Nations Water Conference Mar-del-Plata, Argentina.

Mfutakamba, A.R. (1983). Systems Engineering Solutions to Problems of Drought in Developing Countries. MAWAZO, Volume 1, No. 2, Stanford University, California, U.S.A.

UK-Africa (1982). Water and Sewage Publication.

United Nations Economic Commission for Africa (1980, 1981, 1982), Quarterly Bulletin on Rural Progress, Vol. 1, No. 4, July; Vol. 2, No. 2, January, respectively.

USAID (1979). Evaluation of Rural Water Supply Projects in Tanzania.

WHO (1973). Report on Water Pollution. Pollution control in developing countries.

WHO (1976). Global Survey on Water Supply and Waste Water Disposal in Developing Countries.

APPENDIX
DROUGHT MODEL

(Mfutakamba, 1983 in Mawazo Publication)

Thirty African countries were affected by drought in 1981, twenty-four of them very seriously with resultant acute water and food shortages. This resulted in both human and livestock losses of astounding proportions. Two years earlier there were ten countries affected by drought. This year drought warnings have been sounded for countries in the Southern and in Eastern Africa regions. There are numerous accounts of social and economic problems caused by adverse weather conditions--a combination of long periods of drought followed by floods. It is pertinent, therefore, to furnish some statistics on various water demands which will underline the apparent need for having a systematic approach in arriving at some viable solution to sustain economies which are partly strangled by natural disasters like drought.

To fully understand how drought can cripple an economy, consider the following statistics: it takes more than 30,000 gallons of water to produce a ton of steel; 200 gallons of water to make a pound of synthetic rubber; 184,000 gallons for a ton of high quality book paper, and one gallon to brew a pint of beer. It takes 4,000 gallons to raise 1 pound of beef; 1,300 gallons to grow one pound of cotton and 500 gallons to grow a pound of rice. One medium size per mill, turning out 1,000 tons of paper per day, uses as much water as all the people in the city of one million. The implications for the establishment of a well-planned water economy and continued industrial expansion and for a rising standard of living are ominous (Forbes, Aug. 1979).

Conservation and preference to projects in which water isn't diverted for long periods of time from its natural channels are inadvisable. However, agriculture versus Hydropower tradeoffs need to be considered in order to optimize between food production and energy generation. Indeed, how do nations set out an industrial and/or agricultural strategy, develop and survive without paying dual attention to the water economy, and hence, preparedness against imminent frequent droughts, and in some areas, cases of floods.

As a conclusive demographic remark, the Lagos Plan of Action for Economic Development of Africa (April 1980) points out clearly that "the population of Africa is approximately 469,359,000 of which 334,183,000 or 71.2% live in rural communities. No development program for Africa--or indeed for any individual African countries can fail to take into account the rural factor. . ."

In keeping with Africa's demographic distribution and the predominant rural emphasis, the plan stresses first food production and agricultural development and then industry. The industrial sector is further designed to provide the supply of the bulk of industrial inputs required for agricultural production, processing, storage and transportation.

The linkage between the water economy, agriculture and industrial sectors, and other economic activities therefore, needs no emphasis. To tackle the afore-mentioned challenges of drought, systems engineering which employs a multi-disciplinary approach is used to propose a model for countries in Africa and elsewhere.

Causes of Drought

Drought is a climatological phenomenon that leads into conditions of water stresses. It is difficult to establish a general definition of drought since various disciplines have different indicators to interpret it. Agriculturists and botanists consider permanent wilting of plants as a signal of drought. Meteorologists look into the amount of rainfall over a period of time.

Hydrologists on the other hand are concerned with water levels in rivers, lakes and reservoirs. Definitions of drought are therefore numerous, but one thing common to all of them is the fact that they all center on the demands of water not met due to adverse weather conditions causing water stress.

Proposed Model for Drought Solution

Basic Approach: The proposed drought model is a decision model with an objective function and given constraints. The objective function will offer standards for judging various design alternatives (Ferring et al.). Both costs and benefits of the structure must be expressed in comparable quantitative terms. The following are categories of objective functions:

1. Maximize expected net benefits, the difference between expected benefits and costs.
2. Maximize the ratio of expected benefits to cost; this objective function measures the return in benefits per unit of money invested but does not tell how the benefits are distributed among the members of the society.
3. Minimize the costs subject to certain cost constraints.
4. Maximize benefits subject to certain cost constraints.
5. Maximize societal and regional goals, political influence subject to prevailing socio-economic and political conditions.

These different objectives do not generally suggest the same optimal design and only one objective can be used or an optimal combination. The selection is frequently influenced by political and social processes.

Having defined the objective function of the model, we then qualitatively and quantitatively determine the decision variables. These will be benefits, existing state of nature (water--its location, quantitative availability in time, quality and organization), desired state through a transformation process, investment for the project, a transformation matrix and cost of removing the undesired byproducts.

Constraints

We optimize the objective function subject to the following constraints:

1. Resource inventory constraints and the available and affordable technology
2. Physical constraints (if wells, dams, canals, pump, pipe, etc. are to be built)
3. Economic constraints
4. Social constraints
5. Legal constraints
6. Religious, cultural and traditional constraints
7. Political and institutional constraints since river basins do not normally follow political boundaries.

In order to obtain a "unique" solution both the objective function and the constraints have to be expressed in quantitative terms. However, some of the constraints like the social and political ones are hard to quantify and therefore, we need to resort to other methods to incorporate them into the model. Attempts in this direction are proposed by using decision matrices (Buras, 1972).

Outcomes

Proposed decision model for drought identifies the alternative course of action which will satisfy the criteria.

Evaluation

The consequences that follow from each alternative are based on six feasibility tests: 1) Engineering, 2) Economics, 3) Financial, 4) Political, 5) Social, and 6) Environmental.

Selection of Preferred Alternative

Selection of preferred alternative is based on the six established criteria. Having passed all the above tests, we weigh preferences as to what level we can achieve the regional and interregional objectives.

Sometimes the objectives are mutually exclusive and totally exhaustive and, thus, competing and conflicting. Under such circumstances, we optimize one objective and set the rest as constraints.

On the basis of value judgement, public reaction (or response) and optimal regional objective, we select the preferred alternative. Here, the complete decision model is utilized again.

Experience from similar projects locally and elsewhere should be closely examined in the implementation of the current projects since water resources planning is not immune to failure.

Implementation

With funds available for the selected project (say ground water exploitation) detailed design and specs of components (e.g., wells, distribution systems, pumps, etc.) are made and specifications for bids are drawn, once aquifer yield studies are completed and various demands of water exploitation is more flexible in both planning, design and operation. Thus, stages of the project could be done in phases as funds become available.

Monitoring

Continuous monitoring of the system performance will assist in improving its mode of operation and produce better similar designs in the future. It is pertinent here, to mention the necessity of utilizing local expertise right from initial planning and design stages.

Ecological, biological, and chemical changes should continue to be monitored. The same should be done with the geomorphological, meteorological, and hydrological regimes throughout the different stages of development of the project. This will help to see if any environmental or health problems associated with the project have resulted and ways to mitigate them should be sought.

Confirmatory studies by social scientists concerning relocated people displaced by the project should continue. Social and environmental impact assessment reports would assist planners and decision makers in evaluating the project.

When the project is fully operational, the regional and inter-regional objectives should be re-examined to see if they have been fulfilled. For ground water projects modifications of the design can be made to achieve the preset goals. Dams, on the other hand, are difficult to modify, if not impossible, once they are built, unless we need to elevate the crest.

Some elaborate standards for water and environmental quality and sanitation should be set or observed for the project.

Closing Remarks

- Available rural technologies could be developed to meet needs or copied from elsewhere using local materials and resources to increase agricultural output (especially of food crops) and water supply.

- Technical or financial aid alone cannot solve development problems. Instead, any development goal can only be realized through the determination, interest, and active involvement of the community whom the project is meant to benefit (UN Water Decade--The Tanzania Experience, Wangingombe Village Water Project, 1982).

- More research and study should be done in the areas of ground water recharge, recycling of treated waste water, rainfall augmentation, ocean and sea water desalination, and even dams and wells before an alternative is selected.

- Democratic public participation is vital for planning of water resources to facilitate cooperation from the public. Thus, water conservation measures, proper waste disposal, and less water pollution will be realized. Utilization of economies of scale by both municipalities and industries could also be realized. Multiplicity of environmental concern in the planning of water resources are vital and should be taken into account.

- Drought being a hydrologic random event, it requires synthetic hydrology for its treatment and understanding (determinism and stochasticity are, in this respect complementary rather than being alternative of each other). This is one aspect of uncertainty requiring investigation. The difficulties in estimating future demands for water, further technological developments or future political decisions calls for criteria to be formulated to optimize flexibility and minimize costs.

- No matter how sound a project may be physically, no matter how profitable it will be economically, it will come about only if political leaders would champion its cause in the right way at the right time.

- An important factor in estimating surface water supply is the storage capacity of the various reservoirs. Due to the mismatching in time between supply and demand, and because of the stochastic characteristics of the hydrologic phenomena storage facilities are necessary for the development of surface water resources.

- Drought is a result of weather systems but desertification is caused by the action of man. Thus, improved farming and grazing techniques, reduction of deforestation by continued replanting of trees for firewood may help the situation. Provision of newer and less conventional forms of renewable energy in rural areas such as bio-gas, windmills, solar power, etc. is also advisable.

- Current and future studies of water deficit, agriculture and energy interaction should include analysis and evaluation of policy alternatives available in the various regions of Africa. These options would include the development, allocation, and use of water resources. One such study, for instance, could focus on the possibilities of the formation of an African Water Resources, and Sanitation Council to work with an Agriculture, Food and Energy body.

Removal of Heavy Metals From Polluted Water
Sources Through Bioadsorption

Roque A. Román-Seda, Ph.D.*

Population growth and economic development all over the world are stressing the capability of our available water sources to satisfy the ever growing demand for it. In some places of the world, such as Israel and even some parts of the United States, it is already not posible to fulfill current needs with the locally available water.

One solution to this problem has been to transport water over long distances from places of oversupply to places with scarcity of water. This solution, however, will not be available for long, since each region is slowly moving towards saturation of demand. This condition is forcing us to take a hard look at our wastewaters, and to devise ways of making them available again as a way to alleviate the growing water deficit.

Reutilization of wastewaters has several problems, among which is the presence of dissolved, toxic, hard to remove constituents. Among these are included nitrate, viruses toxic organics, inorganic poisons such as cyanide and arsenic, and heavy metals.

Ion exchange represents the technology of choice for the removal of heavy metals from water. Yet it is an expensive technology. There is a need to find alternative ways of effectively removing heavy metals from water and wastewater at a lesser cost. This study represents an initial attempt to inquire into such a technology.

An investigation was performed to study the potential of inert biological sludges to serve as adsorbers of heavy metals in wastewaters. The study showed that the adsorption capacity of the sludge is enormous, even when their metal content prior to deliberate use as adsorbers is considerable.

This study provides indirect confirmation that the cell wall is responsible for whatever adsorption takes place. Maximum removal was observed at neutral pH for Lead, Cadmium, and Nickel, after accounting for hydrolisis effects. Chromium was removed best at pH 4.5. Contact times between 2 to 4 hours and sludge concentrations between 2 to 8 percent were shown to be sufficient to perform the observed removal. Correlations with traditional adsorption models did not describe the observed removal behavior adequately. An ion exchange mechanism is thought to be responsible.

Introduction

This work tried to determine what ordinary secondary sludges can do to remove dissolved heavy metals from wastewaters.

The study tried to assess the feasibility of developing a new technology for the removal of heavy metals from wastewaters utilizing biological sludges as adsorbers. Another motivation for the study was to create a useful application for sludges.

*Assistant Professor, Civil Engineering Department, University of Puerto Rico

Another goal of the project was to discover the potential of biological sludges for the removal of heavy metals at neutral pH under no flow conditions. The avoidance of the use of chemicals was perceived as desirable in order to make the technology economically attractive.

Literature Review

Microorganisms have been shown to concentrate cations from their aquatic environment and, specifically, to concentrate heavy metal ions. This phenomenon has been shown for Bacillus subtilis in particular (Beveridge et. al., 1976), and for bacterial cell walls in general (Beveridge, 1980). The uptake of heavy metals by microbes has been documented (Chiu, 1972). The relationship between cupric complexes and cell wall related structures has also been illustrated (Elsabee, 1976). Also, the relationship between biosorption of lead and a specific kind of microorganism has been documented (Paskins-Hurlburt, 1976).

Volesky has shown (Volesky et.al, 1971) that biosorbent activity is not associated with living organisms but that it is present in dead biomass as well to perhaps an even greater degree. Biosorption has been specifically linked to the cell wall.

Several investigators (Isezos et.al, 1981; Polikarpov, 1966; Shumate et.al, 1979; Strandberg et.al, 1980) have gone beyond and demonstrated the feasibility of concentrating uranium, thorium, actinides and other valuable heavy metals from dilute radioactive wastes or seawater by using pure cultures of microorganisms that selectively remove these metals.

Previous work suggests that this biosorption property should exist in wastewater sludges and that these may be useful in removing heavy metals from wastewaters. It is the purpose of this work to investigate such a property and to establish some of the parameters that control the biological adsorption phenomenon.

Purpose of the Study

One of the objectives of the study was to determine whether there exists in biological sludges a significant adsorption capacity for the removal of Cadmium, Lead, Nickel, and Chromium. The other objective was to determine the effects of contact time, sludge concentration, and initial metal concentration on overall removal.

Procedure

A source of dried, stabilized sludge was selected from a municipal wastewater treatment plant of known operational characteristics. The dry sludge was hand ground and passed through a sieve with 425 mm openings to insure uniform size particles. Duplicates of each kind of experimental treatment were run.

Solutions of the desired metal were prepared in redistilled deionized water to obtain the desired metal concentrations. Solution pH was adjusted as needed. Two hundred milliliters aliquots were added to the sludge samples.

Blanks were utilized consisting of redistilled deionized water with the same initial concentration of metal as in the treated samples. These blanks were utilized to correct for changes in solubility due to pH effects.

Isothermal experiments were performed, utilizing a constant temperature water bath equipped with a shaker.

Sludge Characteristics

The sludge came from a Publicly Owned Treatment Works in the town of Añasco utilizing a trickling filter. Añasco is located in the western coastal plain of the island of Puerto Rico and it is a town of about 40000 people, with a very small industrial input going into its domestic wastewaters.

The Añasco plant is a 0.22 m^3/s (0.5 MGD) trickling filter plant with an average influent BOD_5 pf 200 mg/l. The plant has primary sedimentation and a trickling filter with rock media. Effluent is recycled.

Primary and secondary sludge is anaerobically digested and then dried in sand beds, from which can be obtained a well stabilized product. The sludge volume index ranges between 135 to 190 ml/l. The sludge was analyzed for metal content and found to contain an appreciable range of metals and quantities.

Description of Experiments

The independent variables were: 1) pH, 2) contact time, 3) sludge concentration, and 4) initial metal concentration. The dependent variable was dissolved metal concentration at the end of the contact period.

Controls were kept for each experimental treatment. These consisted of solutions with identical initial metal concentrations as the treated samples, but without sludge. At the end of the contact period samples were taken from both controls and samples, filtered with a 0.45 micron membrane and acidified. Standard metal analyses were performed on filtered controls and samples. The dissolved metal concentration in the controls was utilized as the initial metal concentration in the samples, thus accounting for any hydrolysis effects that could have taken place at that pH during the given contact time.

Results

pH and Hydrolysis:

Figure 1 shows the equilibrium dissolved metal concentration for Cadmium at various pH's after a contact time of 4 hours. The upper curve shows the equilibrium concentration for the controls, which was then assumed to be the initial concentration of the samples, thereby taking into account any hydrolysis effects that could take place during that period. For Cadmium, removal was observed to be a maximum around pH 7. Similar results were obtained for Nickel (Fig. 2). Lead, however, showed some deterioration in removal capability due to a marked increase in precipitated metal at pH 7 (Fig. 3). The observed equilibrium concentrations, however, were still near water quality standards. Chromium, on the other hand, showed too great a deterioration in overall removal at pH 7. Maximum removal was observed near pH 4.5. (Fig. 4)

Contact Time:

Figure 5 shows that for Cadmium most removal takes place between 2 and 4 hours. Longer contact times are not warranted. For Nickel, however, 2 hours seem to be more than sufficient. The results observed for Lead and Chromium in relation to contact time were two to four hours.

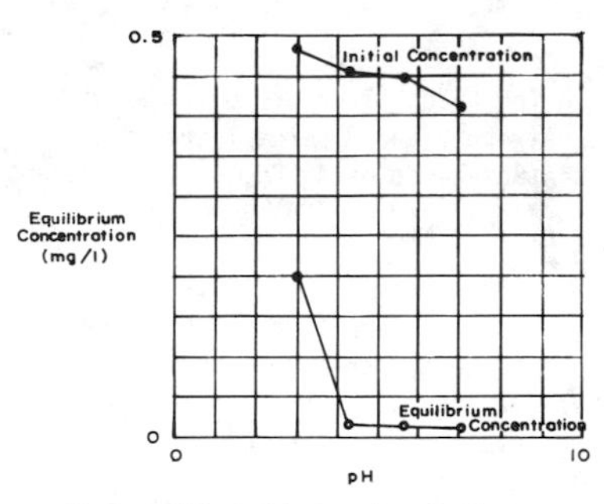

Figure 1. Effect of hydrolysis and pH on the equilibrium concentration of cadmium.

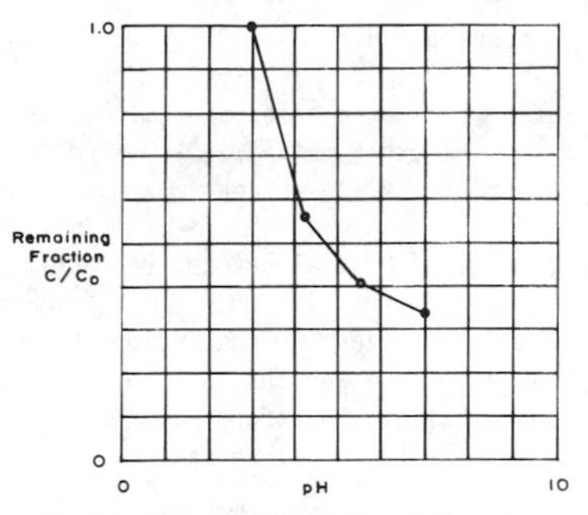

Figure 2. Effect of hydrolysis and pH on the remaining fraction of nickel.

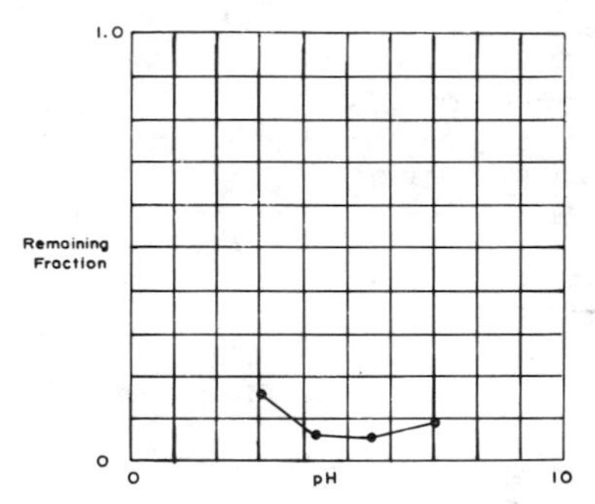

Figure 3. Effect of hydrolysis and pH on the remaining fraction of lead.

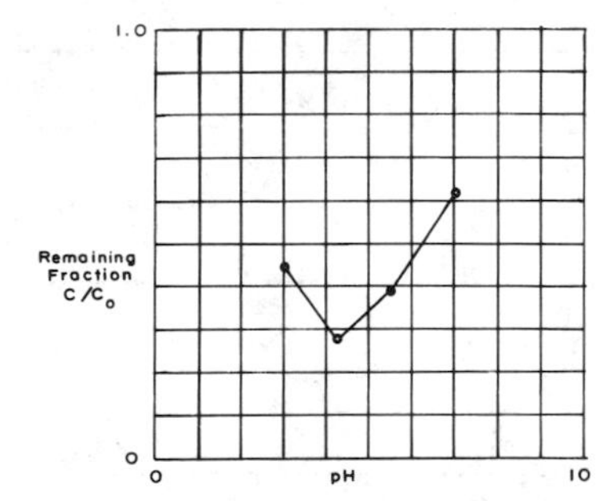

Figure 4. Effect of pH on the remaining fraction of chromium.

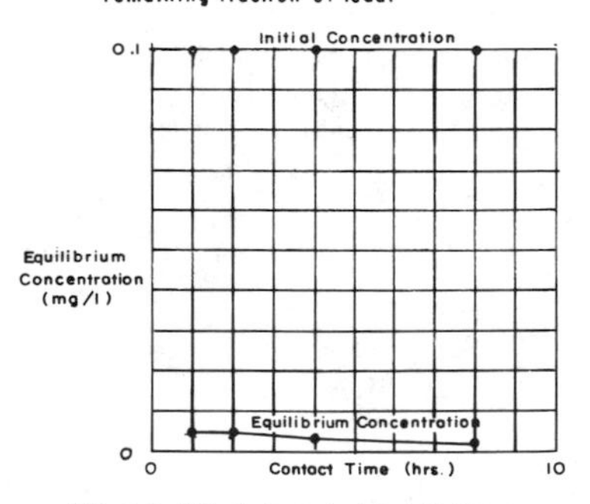

Figure 5. Effect of contact time in the removal of dissolved cadmium (pH = 7.0).

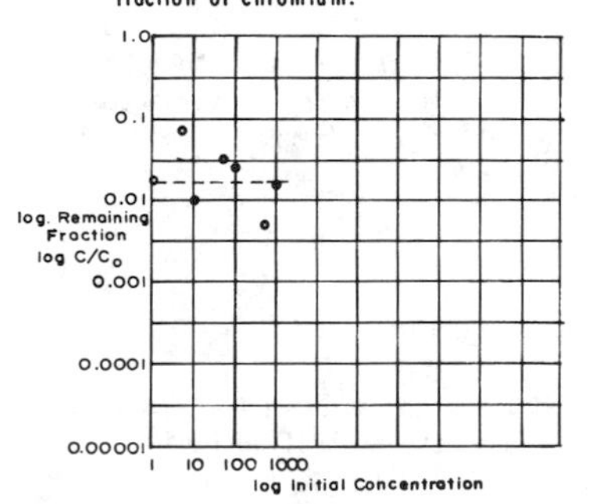

Figure 6. Effect of initial dissolved metal concentration on the remaining fraction of cadmium.

Sludge Concentration:

Table 1 shows the effects of sludge concentration for Lead. These results are typical for the other metals. Sludge concentrations between 2 and 8 percent are sufficient to effect maximum or near maximum removals at contact times of 4 hours.

Correlation with Physical Adsorption Models:

Correlations made for the various metals tested using the Freundlich and Langmuir adsorption models showed that none of the data sets showed good, consistent correlations using these models. Therefore, it appears that a physical adsorption model cannot adequately describe what is happening at a microscopic level.

Table 1: Effect of Sludge Concentration and Contact Time on the Remaining Fraction of Lead

Sludge Concentration (%)	Remaining Fraction, C/C_0		
	Contact Time, Hours		
	1	2	4
1	0.38	0.17	0.18
2	0.19	0.03	0.09
4	0.25	0.13	0.06
8	0.25	0.20	0.03

Initial Metal Concentration:

The observed data suggest that the remaining fraction tends to remain constant in the range of initial concentrations tested. This result suggests that perhaps a mechanism of an electrical nature, perhaps akin to ion exchange, is responsible for the observed removals.

Conclusions

The conclusions from this study were:

* Maximum or near maxi mum removals are obtained for Lead, Nickel and Cadmium around the neutral pH region.

* Maxium removal for Chromium is obtained at pH 4.5 to 5.5.

* Cadmium and Lead can be reduced to levels close to the water quality standards.

* Nickel was reduced 60 percent of its original amount.

* Optimum contact time seems to lie between 2 and 4 hours for the metals tested.

* Optimum sludge concentrations show a spread between 2 to 8 percent.

* The remaning fraction seems to remain constant with increasing initial metal concentration over the ranges tested.

* Conventional adsorption models do not fit the data adequately.

* Adsorption may be caused by electrical interactions rather than by physical adsorption.

* The adsorptive capacity of the sludge is very high.

Acknowledgments

This study was made possible through funding from the Puerto Rico Water Resources Research Institute based in the Engineering School of the University of Puerto Rico and from the Department of the Interior through grant No. 14-34-0001-2141. Partial results of this study were presented at the 40th Purdue Industrial Waste Conference at Purdue University, in May, 1985.

List of References

1. Beveridge, T.J. and Murray, R.G.E., "Uptake and Retention of Metals by Cell Walls of Bacillus subtilis", Journal of Bacteriology, pp. 1502-1518, Sept. 1976.

2. Beveridge, T.J., "Metal Uptake by Bacterial Walls", Paper presented at the Second Chemical Congress of ACS, August 24-29, Las Vegas, 1980.

3. Chiu, Y.S., "Recovery of Heavy Metals by Microbes", Ph.D. thesis, University of Western Ontario, London, Ontario, 1972.

4. Elsabee, M.Z., Matter, M., and Habashy, B.M., "Bonding between Cu(II) Complexes with Celluloses and Related Carbohydrates", Journal of Polymer Science, 14, pp. 1773-1781, 1976.

5. Paskins-Hurlburt, A.J., Yanaka, Y., and Skoryna, S.C., "Carrageenan and the Binding of Lead", Botanica Marina, 19, pp. 59-60, 1976.

6. Volesky, B. and M. Isezos, "Recovery of Radioactive Elements from Wastewaters by Biosorption", Presented at the International ACS Conference, Honolulu, Hawaii, April, 1971.

7. Isezos, M. and Volesky, B., "Biosorption of Uranium and Thorium", Biotech. Bioeng., 23, pp. 583-604, March, 1981.

8. Polikarpov, G.G., "Radioecology of Acquatic Organisms", North Holland Publishing Co., Reinhold Book Division, N.Y., 1966.

9. Shumate II, S.E., Strandber, G.W., Parrott, J.R., Jr. and Locke, B.R., "Separation of Uranium from Process Waste Waters Using Microbial Cells as Sorbents", Paper presented at CIM Hydrometallurgy meeting held in Toronto, Canada, Nov. 1979.

10. Strandberg, G.W., Shumate II, S.E., Parrott, J.R., Jr., and McWhirter, D.A., "Microbial Uptake of Uranium, Cesium, and Radium", Second Chemical Congress of ACS, Las Vegas, Aug. 1980.

EFFECT OF TREATMENT PROCESSES ON THE REMOVAL OF SELECTED TOXIC CONTAMINANTS FROM POLLUTED WATERS & THEIR IMPACT ON RECEIVING WATERS

Roque A. Román-Seda* and Juan Puig**

ABSTRACT

A study was performed in order to ascertain the adequacy of primary treatment for removal of selected priority pollutants at Regional Wastewater Treatment Plants in Puerto Rico.

A continuous sampling program was performed for a period of sixty-six days collecting two-day composite samples at the Barceloneta Regional Wastewater Treatment Plant (WWTP). Barceloneta is a highly industrialized municipality located in the middle of the northern coast of the island.

The results, analysis and interpretation for the data at the Barceloneta WWTP are presented herein. Phenol, Copper, Chromium, Lead, Zinc, and total solids were analyzed for the influent, primary effluent, and secondary effluent from 48-hour composite samples. Time series, 14-day moving average series, removal efficiency and frequency analysis were performed on the data. The study showed that while for some heavy metals (Chromium and Copper) primary treatment may be sufficient at the observed influent levels, it is definitely not sufficient treatment for other heavy metals and/or toxic compounds (Phenol, Lead, and Zinc).

Therefore, in wastewaters where toxics amenable to adequate removal through primary treatment coexist with toxics not amenable to said treatment, secondary treatment is unavoidable if water quality standards are to be met. Even with secondary treatment, it was shown that for some toxics a heavy reliance will have to be made in the dilution effect of the mixing zones in order to comply with the Puerto Rico Environmental Quality Board's water quality standards. Frequency of occurrence for the 66-day period for the toxic pollutants studied was established, and could serve as a baseline for future reference and further studies.

INTRODUCTION

It has been proposed that **Primary Treatment** will bring pollutants to reasonable levels which will comply with water quality standards once the treated discharge undergoes dilution at the mixing zone of the plant outfall.

Existing short-term data tend to suggest that such treatment may not be sufficient to bring priority pollutants down to levels compatible with water quality standards. It is also a well known fact that primary treatment, as normally practiced in wastewater treatment, is not a particularly efficient process even from the standpoint of suspended solids, for which only 40-60 percent removal can be expected.

*Assistant Professor, Civil Engineering Department, University of Puerto Rico

**Civil Engineer, U.S. Geological Survey, Caribbean District

This project attempted to assess the environmental impact of omitting secondary treatment from Regional Wastewater Treatment Plants having significant industrial contributions through the systematic, rigorous, long-term observation of primary and secondary removal of selected priority pollutants and heavy metals.

OBJECTIVES

1. To trace the fate of selected priority/heavy metal pollutants present in domestic wastewaters due to industrial contributions in a Regional WWTP having both **primary** and **secondary** treatment.

2. To quantify the effect of both **primary** and **secondary** treatment on such pollutants.

3. To establish the frequency of occurrence of pollutant levels through a frequency analysis of the observed data.

4. To assess the probable impact of these contaminants, after treatment, on coastal receiving waters.

EXPERIMENTAL PROCEDURE

Automatic wastewater samplers (ISCO model 1580) fitted with teflon tubes and Nickel/Cadmium rechargeable batteries were placed at the plant influent (after degritting), primary effluent, and secondary effluent before chlorination. Thirty to fifty milliliter (ml) samples were collected every hour and composited for a period of 48 hours. Manpower and analytical capabilities did not permit shorter (ie., 24 hour) sampling periods. Ice was utilized inside the samplers to keep the samples cold for as long as was practically possible. Samples bottles were taken, moved and analyzed/preserved during the same work day. Thin layer chromatography and atomic absorption spectrophotometry was performed on the samples following Standard Methods for the examination of water and wastewater.

EXPERIMENTAL RESULTS

Phenol -

1. Figure 1 shows the influent, primary, and secondary effluent concentrations as well as the 14-day moving averages for phenol at Barceloneta Regional WWTP. Significant average concentrations (865, 569, and 107 microg/l with corresponding sample standard deviations of 223, 201, and 65 microg/l were observed in all three cases.

2. Greater removal was obtained with secondary treatment than with primary treatment.

Copper -

1. Figure 2 shows the influent/secondary effluent and primary effluent concentrations as well as the 14-day moving averages for the copper time series. Average concentrations of 0.13, 0.07, & 0.03 with corresponding sample standard deviations of 0.06, 0.07 & 0.02 ppm were observed for influent, primary and secondary effluents. The later is below the water quality standards while the former is slightly above.

2. Significant removal was obtained through both types of treatment. Primary treatment exhibited twice the removal than secondary treatment.

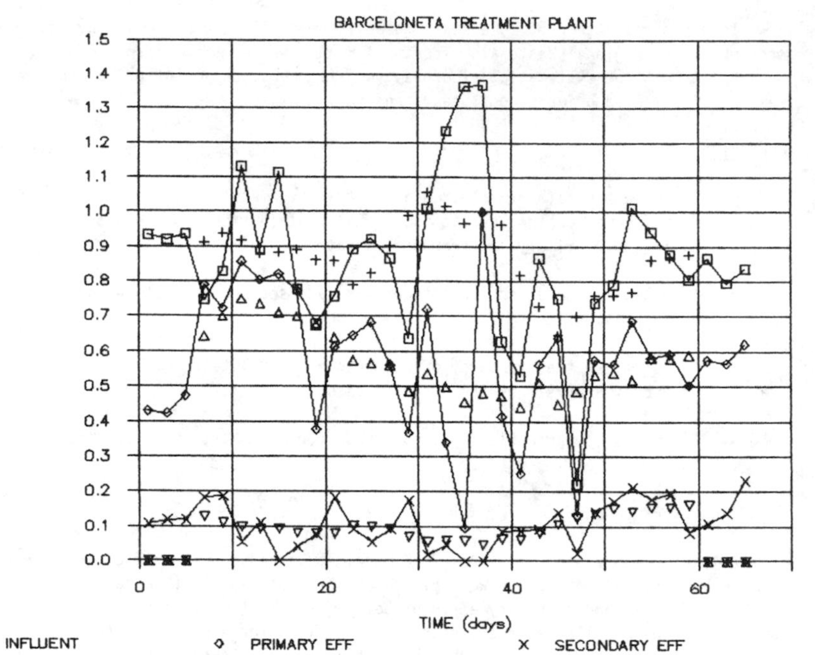

Fig. 1: Time Series for Phenol

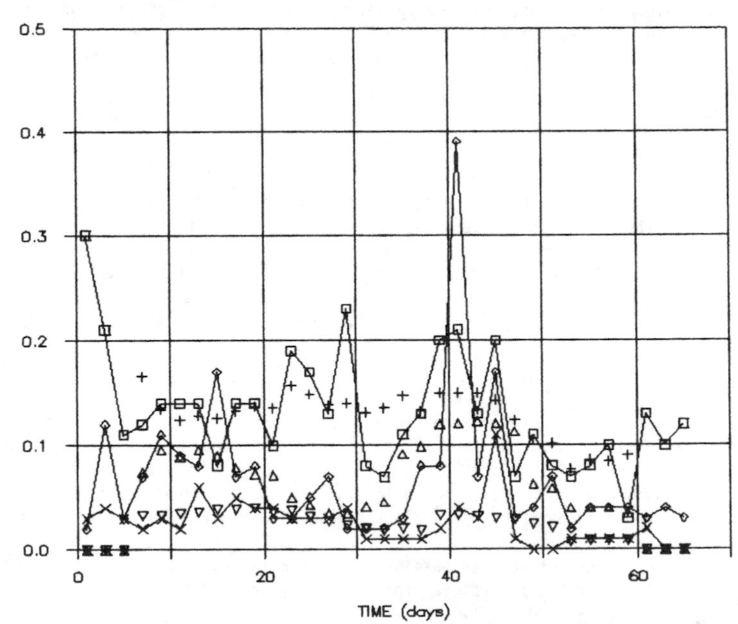

Fig. 2: Time Series for Copper

Chromium -

1. Figure 3 shows the results for the time series for Chromium. Average concentrations of 0.19, 0.1 & 0.04 ppm were observed along with respective sample standard deviations of 0.08, 0.07, & 0.03 ppm for influent, primary and secondary effluent. If chromium had been all in hexavalent form (tests were for total chromium only), then all effluent concentrations would have been in violation of water quality standards. If in trivalent form, none would have been in violation.

Lead -

1. Figure 4 shows the time series as well as the 14-day moving averages for lead. Average concentrations of 0.08, 0.06, & 0.04 ppm along with sample standard deviations of 0.04, 0.04, & 0.03 ppm were observed.

2. The secondary effluent level is approximately 2.67 times the water quality standard, and the primary effluent level is 4 times the standard.

Zinc -

1. Results for Zinc are shown in figure 4. Observed average concentrations were 1.83, 0.8, and 0.48 ppm and sample standard deviations were 1.2, 0.67, & 0.25 ppm for influent, primary and secondary effluents.

2. Average secondary effluent concentration is 10 times the water quality standard and the primary effluent concentration is 16 times the standard.

A summary of all results is given by Table 1.

CONCLUSIONS AND RECOMMENDATIONS

The major conclusions of this study are:

1. Secondary treatment is the principal remover of phenol (54% vs. 33% for primary treatment). The combined effect of primary and secondary treatment is 86.4%, which is quite respectable. **Nevertheless, at these levels the effluent phenol concentration requires the mixing zone to provide an average dilution of 10 for 50% of the time and a maximum dilution between 11 and 21 for 50% of the time. Primary treatment alone requires a mixing zone dilution of 82 times!**

   The net conclusion from this result is that for toxic organic compounds similar in physico-chemical behavior and biodegradability as phenol, secondary treatment is absolutely necessary and crucial.

2. For influent copper concentrations between 0.1 & 0.15 mg/l primary treatment can provide adequate treatment 50% of the time. Maximum influent copper concentrations of around 0.3 mg/l need to be guaranteed a mixing zone dilution factor of 3. Fifty-one percent of the overall 81% removal was contributed by primary treatment and 29% by secondary treatment. Primary treatment requires a mixing zone dilution of 3.1 to 4.4 50% of the tiem.

3. For influent hexavalent chromium concentrations of around 0.2 mg/l, primary treatment seems to provide adequate removal. For maximum hexavalent chromium of around 0.4 mg/l, a mixing zone dilution of around 2 is necessary for secondary treatment and of 3.8 for primary treatment. Primary treatment removed 47% and secondary

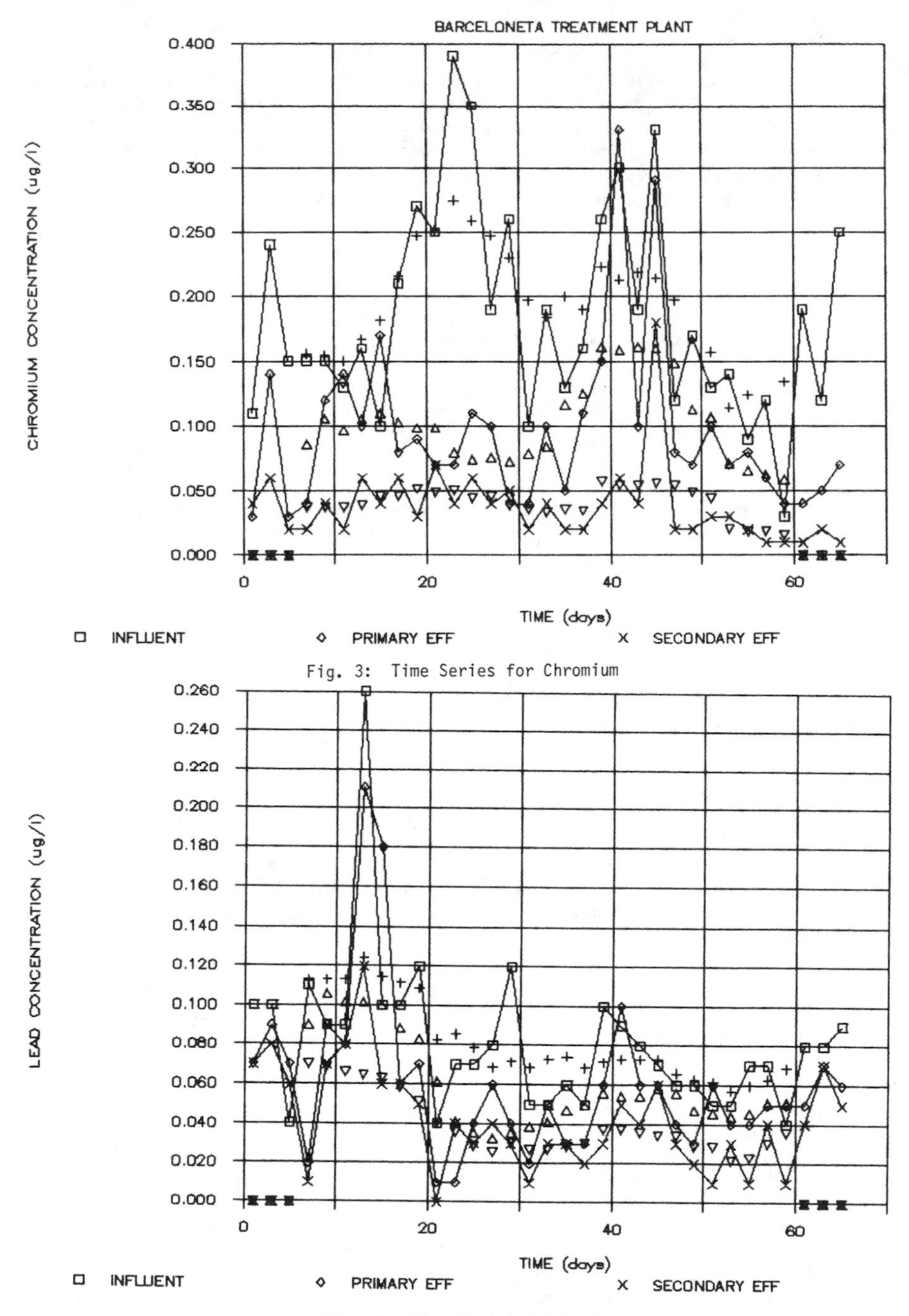

Fig. 3: Time Series for Chromium

Fig. 4: Time Series for Lead

Table 1: Summary of Results for Barceloneta Regional Wastewater Treatment Plant

Plant	Observed Avg. Concentration			Concentration 90 percentile			Probability of Exceedance		Concentration 90 percentile			Estimated/Required Dilution		% Removal		
Parameter	Plant Inf.	Prim. Eff.	Sec. Eff.	Plant Ing.	Prim. Eff.	Sec. Eff.	Avg. Conc.	Water quality Std.	Plant Inf.	Prim. Eff.	Sec. Eff.	Average	Maximum	Prim.	Sec.	Tot.
Barceloneta																
Phenol(mg/1)	865	568	107	640	270	30	0.53	0.95	1250	820	210	10	21	33	53.5	865
Max:	1369	1000	231													
Min:	220	94	0													
Lead(mg/1)	0.08	0.06	0.04	0.04	0.015	0.01	0.60	0.85	0.245	0.21	0.12	2.7	8	31	19	50
Max:	0.26	0.21	0.12													
Min:	0.04	0.01	0.00													
Chromium(mg/1)	0.19	0.10	0.04	0.09	0.04	0.01	0.56	6.45* -1.0+	0.46	0.19	0.10	None	2* none+	47	32	79
Max:	0.39	0.33	0.18													
Min:	0.03	0.03	0.01													
Copper(mg/1)	0.13	0.07	0.03	0.05	0.024	0.008	0.61	0.36	0.26	0.22	0.12	None	2.4	52	29	81
Max:	0.30	0.39	0.11													
Min:	0.03	0.02	0.00													
Zinc(mg/1)	1.83	0.80	0.48	0.60	0.32	0.18	0.46	0.98	5.6	2.80	1.40	9.6	28	51	12	63
Max:	5.06	3.32	1.62													
Min:	0.40	0.30	0.19													

* Hexavalent Chromium
+ Total Chromium

treatment 32% of the 79% total removal, both being significant fractions of the whole.

4. For influent lead concentrations of around 0.08 mg/1, the secondary effluent requires an average dilution of approximately 3 times to comply with the water quality standard 50% of the time. Dilutions between 4 to 8 times would be required 40% of the time for secondary treatment. Maximum dilutions of up to 14 times would be required 50% of the time for primary treatment alone.

5. Zinc concentrations are highest for all metals studied. Both primary and secondary effluents exceeded the water quality standard. Maximum dilutions for zinc would require between 10 and 28 times for secondary treatment and between 17 and 56 times for primary treatment for 40% of the time.

6. Copper and chromium seem to comply with the standard 50% of the time. Nevertheless, secondary treatment provides a significant amount of additional removal which serves as a safety factor against metal spills.

7. Phenol, lead and zinc need substantial dilution in order to comply with the water quality standard even with secondary treatment.

8. This study has shown that while for some toxic pollutants primary treatment may be adequate, for others it is definitely not.

9. Since there does not exist an effective/practical way to control the kinds and quantities of toxic pollutants that find their way into municipal wastewaters, one has to conclude that secondary treatment is necessary, since at least some of the toxic pollutants present in the heterogenous water matrix will require it.

This study has shown that unless a mixing zone can consistently dilute to 20 times the original concentration, there will be violations of the water quality standards for toxic pollutants a significant portion of the time (at least 10% of the time) even when secondary treatment is provided. If only primary treatment is present, a dilution factor of around 80 willhave to be provided for at least 10% of the time, to avoid violation of the water quality standards.

A Sanitation System Selection Process for Small Community Housing Projects

Gregory J. Newman*

Introduction

This paper presents a preliminary method for choosing among alternatives to conventional sewerage for small community housing projects. The paper is intended to assist community planners in formulating and comparing sanitation alternatives for on-site and off-site disposal of wastewater. It presents a step-by-step methodology for determining technically feasible sanitation alternatives.

The methodology accounts for technical factors that affect wastewater disposal. These factors include lot size, population density, project layout, physical characteristics of the site, engineering constraints for on-lot disposal and sewerage and requirements for off-site wastewater treatment and disposal. The impacts of general assumptions about development project layout, water use, waste generation, the physical environment and the cultural setting are discussed.

The sanitation alternatives considered include the following:

- pit latrines,
- septic tanks,
- soakaways,
- leaching fields,
- truck cartage systems,
- small bore sewers, and
- conventional sewerage.

Design criteria for each of the sanitation alternatives are presented. The reader is referred to the sources appended to this paper for detailed descriptions and discussions of the proposed alternative sanitation technologies(Baumann,1980; Feachem,1978; Kalbermatten,1982).

Site Development Consideratons

The establisment of lot size, population density, overall development layout and other factors affect the feasibility of providing the alternative sanitation technologies. Prior to establishing these parameters for a housing project, a planner should be familiar with how these parameters will limit the planner's options for sanitation.

Lot Size. The lot size may determine the feasibility of on-site disposal systems. The total lot area required for an on-lot disposal system can be calculated by estimating the house area, a "reserve" area, and the area required for wastewater disposal (size of a leaching

*Environmental Engineer, Camp Dresser & McKee Inc., One Center Plaza, Boston, MA 02108.

field). The reserve area includes space for trees, walkways, carriage ways, septic tanks, seperation between soil absorption units and other structures. A preliminary estimate of the reserve area may be calculated as 1.5 times the design house area. The area required for soil absorption will vary with site specific absorption rates and the waste load generated on each lot, as discussed below. In general, small lots may not be capable of on-site treatment, while large lots would allow a margin of safety and thus more reliable service for on-site disposal systems.

Project Layout. The physical spacing and location of water closets, houses on the lots, roadways, and other structures within the housing project will determine the feasibility of using a truck cartage system of waste disposal. The truck cartage system requires that trucks are able to easily approach the vaults storing the waste from the water closets at each house.

Area allotted as "open space" within the housing project could potentially be utilized as a communal on-site disposal facility (i.e. a communal leaching field). Again, allowance should be made for distribution boxes, septic tanks, trees, and other structures. It is recommended that only half of the open space be considered available for use as a leaching field.

It may be possible to utilize the open space in a staged sequence as a temporary on-site wastewater disposal facility prior to 100 percent development of the lots on a housing site. If a site develops slowly, a communal septic tank and leaching field may be appropriate for a period of years while many lots are undeveloped. After some years the total waste flows may become too great and the system can then be upgraded to a sewered system. This has the advantage of delaying the high costs of a sewered system.

Design Population. The design population density will influence the estimated waste load generated per lot and for the housing project as a whole and thus affect the area requirements for on-site disposal. A greater design population density will generally result in a greater volume of waste and require a greater land area devoted to disposal.

The rate of population growth in housing projects plays a significant role in the feasibility of staged sequences of sanitation alternatives. The housing sites may have a limited capacity for on-site disposal of wastewater when fully developed due to the area requirements of soil absorption. Prior to maximum development of the sites, however, the population density and corresponding wastewater flows may be much less, thereby decreasing the land requirements for soil absorption. If population growth in the housing sites is slow, temporary on-site disposal systems may be feasible. The temporary on-site disposal system can be upgraded to a sewered system, if necessary, at a later date.

The rate at which housing projects are developed also affects the feasibility of sewered systems. Sewers require minimum flow velocities to prevent clogging. If the rate of population growth and utilization of the site is slow, then minimum velocities may not be achieved on a regular basis and the sewers system may fail during the initial years of a housing project.

Waste Loads. In general, wastewater flows are proportional to the water supplied to housing projects. Total wastewater flow would be the sum of grey water, flush water and excreta, however, for the purposes of preliminary design it can be be estimated as 80 percent of the water supplied.

If water is supplied to each house, an estimated water consumption of 80 lcd, or greater may be assumed. However, actual consumption may be as low as 25 lcd if communal water service is provided. The result would be a drastic decrease in the expected wastewater flows which affects the feasibility of various sanitation alternatives.

The actual amount of grey water produced affects the feasibility of on-site disposal such as soakaways and leaching fields. Soakaways are generally feasible when grey water flows are less than 25 lcd. The area required for a leaching field is directly proportional to the wastewater flow to be absorbed by the soil. A higher grey water flow requires more leaching field area and may limit the use of a leaching field for disposal of grey water.

Small bore sewers and conventional sewers must achieve minimum flow velocities (0.3 m/s and 0.6 m/s, respectively) on a regular basis in order to resuspend solids and prevent clogging. If water use is less than 25 lcd then there exists a high probability of small bore sewer system failure, unless a sewer flushing program is instituted. If water use is less than 75 lcd there exists a high probability of conventional sewer failure. In final design, the evaluation of minimum pipe velocity should account for mean flow, peak flow, pipe slope and pipe size. These factors have been considered in the above preliminary criteria for water use.

Excreta production is generally within the range of 0.6 to 0.9 m^3/person-year. Based upon the excreta production rate and the design period for vault cleaning, the size of the vault for a truck cartage system can be estimated. The pit latrine can be sized for yearly cleaning, based upon the expected rate of excreta accumulation within the pit (approximately 0.1 m^3/person-year). Excreta production also affects sludge production in septic tanks, and consequently plays a similar role in designing septic tank size and period of pumping. A representative sludge accumulation rate within septic tanks of 0.04 m^3/person-year can be assumed.

Site Topography. Flat sites will increase the cost of conventional sewer systems. Gently sloping sites are optimum for most sanitation alternatives. Steep sites increase the likelyhood of failure for on-site disposal systems. Steep slopes may have steep groundwater gradients and raising of the groundwater locally may decrease slope stability over the long term as well as create an opportunity for polluted springs to well up.

Soils. Rocky and hardpan soils generally pose construction problems for each sanitation alternative. In rocky soils, preference will go to alternatives which require less excavation. The permeability of the soils may determine the feasibility of on-site disposal systems such as soakaways and leaching fields. Soil infiltration of wastewater is dependent upon soil characteristics and is affected by the development of a biological layer within the drainage pit or trench. Poor permeability or percolation rates may render infeasible the use of on-site grey water disposal via leaching fields or soakaways. The long term infiltration rate (hydraulic loading rate) of a leaching field is dependent upon the ability of wastewater to infiltrate through the biological layer which develops in the surrounding soils.

On-site disposal, via soakaways and leaching fields, requires soil permeabilities greater than 10 l/m^2-day. Permeability may be estimated by using Table 1 and knowledge of the soil types. This type of preliminary estimate must be confirmed by a percolation test. Percolation tests are empirical estimates of the soil suitability for subsurface disposal of wastewater. Such tests, however, do not give reliable estimates of long term soil permeability to wastewater. The design infiltration rate for sizing on-site disposal facilities is limited at 10 l/m^2-day, due to biological mat formation, unless more reliable data on long term infiltration rates is available (Griffin,1984; Laak,1980).

Table 1: Typical permeabilities for common soil types.

Soil Type	Typical Permeability for Wastewater
Sands	50 l/m^2-day
Silt-Loams and Silty Clay Loams	50 l/m^2-day
Fine to Medium Sands	34 l/m^2-day
Sand-Loams, Loams	30 l/m^2-day
Clay-Loams	14 l/m^2-day
Clays, some Clay-Loams	6 l/m^2-day

Groundwater. Groundwater elevation plays an important role in determining the feasibility of on-site disposal systems. The peak groundwater elevation should be at least a meter below the bottom of a drainage facility. High groundwater may result in flooding of leaching fields, soakaways and septic tanks and, consequently, may create a health hazard. High groundwater should also be taken into account when designing conventional sewerage systems. Sewer pipes should be sized to carry the higher levels of infiltration that result from high groundwater.

High groundwater conditions necessitate the use of conventional sewerage. The depth to high groundwater should be at least 1.0 m for pit latrines, septic tanks, vaults and leaching fields. The depth to high groundwater should be at least 1.0 m below the design depth of a soakaway (typically, depth to high groundwater is greater than 3.0 m).

The design peak groundwater elevation can be determined by placing three monitoring wells across the housing project site. More than three wells may be necessary for larger sites or to check areas where a shallow depth to groundwater is suspected. Monitoring wells are small diameter perforated pipes placed in the soil at least 3 meters deep. Water levels in the wells are monitored throughout the wet season. The minimum depth to groundwater is noted.

Health. Sanitation systems which permit greater water consumption (e.g. sewers) will help promote better sanitation practices (e.g. bathing and handwashing) by using more water, which may in turn improve health. Pour-flush fixtures help prevent the spread of disease by creating a water seal which aids in preventing flies and mosquitoes from breeding in excreta and entering the house. Proper use and maintenance of the systems discussed will promote a healthy environment. The relative reliability and the impact of a system failure, however, should be considered. An individual on-site disposal system may be more likely to fail due to poor maintenance, but such a failure will have fewer health implications than the failure of a shared facility.

Maintenance and Reliability. The reliability of each sanitation systems is a function of design, construction and maintenance. It is assumed that the quality of design and construction will be consistent regardless of the alternative selected. It should be noted, however, that conventional sewers do require a higher degree of skill to design and construct properly than the other systems. Reliability of each system may be most easily related to the degree of user maintenance required. The vacuum truck cartage system and on-site disposal systems may be considered less reliable due to the high degree of maintenance required.

System Selection

Selection Algorithm. Figure 1 presents a selection algorithm applicable to many small community development projects. This algorithm does not consider costs. It considers only the technical feasibility of the various sanitation alternatives and incorporates the assumptions stated previously in this paper.

The following set of questions require answers in order to proceed through the selection algorithm:

1. Will the lots be accessible by truck?
2. What is the high annual groundwater elevation?
3. Is there area available for on-site disposal?
4. Does soil permeability permit on-site disposal?
5. What is the per capita water consumption rate?

In order to answer these questions, the following data must be gathered and interpreted for each question.

QUESTION NO. 1: Will the lots be accessible by truck?
A site plan should be available showing roads, carriage ways, housing units and the approximate location of the water closets within the housing units. A positive response to this question means that a truck is easily able to approach each house near its water-closet.

QUESTION NO. 2: What is the high annual groundwater elevation?
The groundwater monitoring program described above should provide the answer to this question.

QUESTION NO. 3: Is area available for on-site disposal?
This question should be analyzed in two parts; is there area available on the housing lots for on-site disposal, and is there area available in the communal open space for on-site disposal?

Housing lot on-site disposal can be evaluated with the following data:

(a) Housing lot size (m^2)
(b) Maximum house size (m^2)
(c) Reserve area (m^2)
(d) Maximum number of persons per lot
(e) Per capita water consumption rate (lcd)

Calculate the wastewater flow per lot (= d x e x 0.8). Calculate the on-site disposal area (below). If the lot size is greater than the sum of the house size, reserve area (= 1.5 x b) and on-site disposal area, then Question No. 3 may be answered affirmatively.

The area required for on-site disposal may be estimated as follows:

$$A = (s.f.) \times Q_w/I$$

Where:
A = area required (m^2)
(s.f) = safety factor (use 1.5)
Q_w = wastewater flow (l/day)
I = limiting infiltration rate (10 l/m^2-day)

Communal on-site disposal can be evaluated with the following data:

(a) Site area (m^2)
(b) Number of lots per hectare
(c) Maximum number of persons per lot
(d) Per capita water consumption rate (lcd)
(e) Site communal open space (m^2)

Calculate the site wastewater flow (= a x b x c x d x 0.8). Calculate the area requred for on-site disposal of the site wastewater flow (above). If the on-site disposal area required is less than half of the site communal open space, then Question No. 3 may be answered affirmatively.

QUESTION NO. 4: Does soil permeability permit on-site disposal?
Knowledge of the soil types and the results from percolation tests should be available. This information can be compared to the criteria presented Table 1, in order to answer Question No. 4.

QUESTION NO. 5: What is the per capita water consumption rate?
An estimate of water consumption rates should be available based upon similar sites. It is assumed that consumption is 80 lcd for in house water service and 25 lcd for communal water service.

The alternatives selected using this algorithm are determined by the assumptions made. It is important to note that some of the assumptions are open to modification by planners (e.g. number of lots per hectare) and will affect the selection of technically feasible sanitation alternatives.

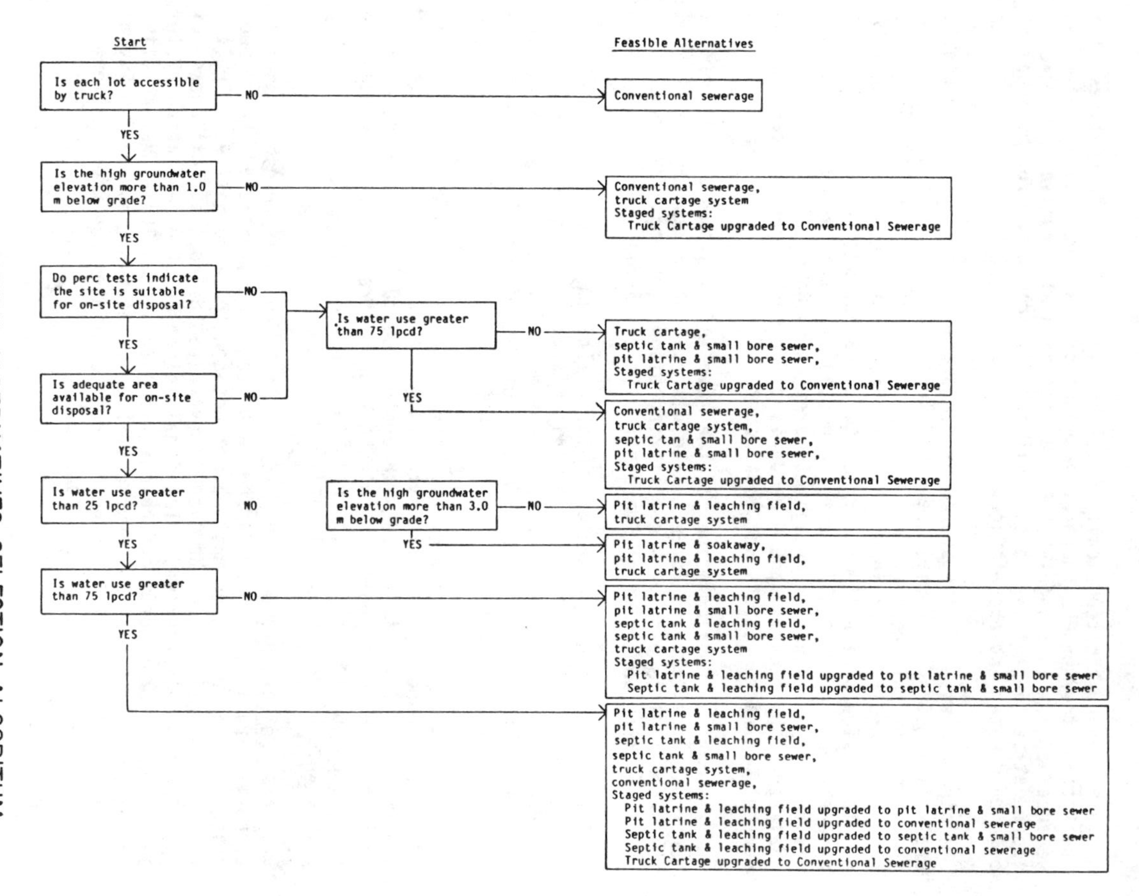

FIGURE 1: SANITATION ALTERNATIVES SELECTION ALGORITHM

Cost. Once the technically feasible sanitation alternatives are identified, they can be compared based on capital, operation, and maintenance costs. It should be noted that each alternative in Figure 1 represents a complete waste disposal scheme, accounting for disposal of both grey water and excreta are.

Actual costs of the systems and a presentation of the economic analyses required to derive present values from future costs are beyond the scope of this report. All operation and maintenace (O & M) costs can be expected to be incurred in the future at regular periods. A single design period (typically 20 to 30 years) should be used for all alternatives evaluated. Costs expected to be incurred within the design period should be brought to present value by standard economic analyses. Unit costs should be derived for each component based upon current contractor prices and local costs.

Treatment. Consideration of the level of treatment and various treatment technologies for ultimate disposal of wastewaters is beyond the scope of this paper. A simplifying assumption is to use wastewater stabilization ponds for ultimate off-site disposal.

The ponds should be designed with a minimum 30 day retention time and a depth of 2.0 meters. The minimum area (m^2) requirements can be calculated by multiplying the total wastewater flow (m^3/day) by 30 days and dividing the result by 2 m. Allowance must be added to include area for dikes, access road and a buffer zone.

The cost of the treatment facility should be added to the capital costs developed for each off-site wastewater disposal scheme. If the treatment facility is shared with another community, then the capital cost to the housing project should be proportional to the percentage of flow the housing project contributes. O & M costs should include maintenance of the treatment facility.

Final Selection and Design. Once the technically feasible sanitation alternatives and costs for each feasible alternative have been determined, the planner must select between the alternatives. Preliminary cost analyses can be used to evaluate the cost effectiveness of the feasible alternatives. The final selection of a sanitation system should not be based upon cost alone, but should evaluate the relative social acceptability, constructability, maintenance and reliabilty of the alternatives. Final design and cost estimates can then be prepared for the selected sanitation technology.

REFERENCES

The following is the list of principal referneces used in preparing this paper and those deemed most useful for further detailed evaluation of alternatives to conventional sewers.

Baumann, W. and Karpe, H.J. Wastewater Treatment and Excreta Disposal in Developing Countries German Appropriate Technology Exchange, University of Dortmund, Federal Republic of Germany, 1980

Feachem, R. and Cairncross, S. Small Excreta Disposal Systems. Ross Institute Information and Advisory Service. London, 1978

Griffin, R. Wastewater Management Alternatives for Rural Lakefront Communities. Departmant of Civil Engineering, University of Massachusetts, USA, 1984.

Kalbermatten, J.M. et al. Appropriate Sanitation Alternatives: A Planning and Design Manual. World Bank Studies in Water Supply and Sanitation Volume 2. John Hopkins University Press, 1982.

Laak, R. Wastewater Engineering Design for Unsewered Areas. Ann Arbor Science Publishers, Inc. Ann Arbor, Michigan, 1980.

IMPACT OF NON-POINT SOURCES ON PUERTO RICO'S WATER SUPPLIES

(PRESENT CONTROL STRATEGIES)

Santos Rohena Bentancourt*

Abstract

Puerto Rico's water bodies are being seriously affected by non-point sources of pollution. These comprises activities related to agriculture, cattle raising, chicken farms, construction activities, sand mining, and other activities related to the extraction of materials from the earth's crust; as well as activities pertaining to land disposal of wastes, urban runoff, and storage tank failures.

In our small and densely populated Island, most human activities occur on or near the watersheds. Rivers originate in the central mountain region, which extends from east to west.

Inadequate disposal of wastes by livestock enterprises is considered a major non-point source of water pollution. The Environmental Quality Board (EQB) created the Non-Point Sources Division to encourage individual farmers to install adequate animal waste management systems. Farmers are required to install these systems, and certification is renewed annually to verify that the systems are operated and maintained properly.

Improper utilization of pesticides and fertilizers also leads to non-point pollution problems. Excessive amounts are spread on crops and carried down the watersheds during rainy seasons. The EQB has developed a demonstration project designed to identify and to inventory the pesticides and fertilizers used in La Plata Basin.

Pollution from construction activities is being controlled by the Erosion and Sedimentation Control Plan (in Spanish our CEST Plan). Construction projects involving more than 900 square meters and a proximity to a water body are required to implement a CEST Management Program for its operations in order to prevent or minimize erosion and the consequent sedimentation of water bodies.

Urban runoff is a serious problem in Puerto Rico metropolitan areas. The problem is compounded in some areas of metropolitan San Juan where illegal connections of storm sewers to sanitary sewers have been detected. The EQB has establish a program to detect and correct this anomally.

Some small communities also contribute to non-point pollution because they lack adequate wastewater disposal systems. An interagency committee organized by EQB is addressing this problem and shall present recommendations for its solution.

Puerto Rico has 65 municipal landfills, 24 of which are located near water bodies. They represent potential pollution sources and must be kept under surveillance.

*President, Environmental Quality Board, Commonwealth of Puerto Rico, Office of the Governor, Santruce, Puerto Rico.

Organic pollution constitutes a serious problem for groundwater resources, specially those of the north coast because of the high porosity of its calcareous soil. Various agencies have started a 5 year groundwater study program to obtain the data needed to develop a management system for groundwater utilization and protection.

Short and long range plans are being developed to identify, quantify, and control non-point sources of pollution in Puerto Rico.

Water and Sanitation - Technology Transfer to Developing Countries

T. Viraraghavan[1], F.ASCE and T. Damodara Rao[2]

Abstract

"Appropriate technology" should not be a euphemism for inferior science or technology. Some of the options for the development of appropriate technology are as follows: 1) adapting or improving existing indigenous technology; 2) adapting evolving modern technology; 3) reviving old technology; and 4) developing new technology and transferring technology within developing regions and countries. The transfer of technology from developed to developing countries in the context of adapting evolving modern technology or developing new technology should lead to a technology which is locally appropriate from technical, economical, social and environmental perspectives.

The areas in which technology transfer would be desirable and realistic are discussed. The technology transfer in respect of community water supply may relate to improved designs of slow sand filters, manufacture of varieties of pipes and pumps, prospecting for groundwater, assessment of the need and development of systems for removal of trihalomethanes, use of alternate disinfectants, techniques for water conservation, leak detection and control, water pricing and disposal of water plant sludges. In the area of community sanitation measures, the technology transfer may involve transfer of knowledge and information on design of aerated lagoon systems and manufacture of aerators locally, the use of natural or synthetic liners to minimize seepage from lagoons, design and adoption of well engineered land treatment systems for wastewater and sludge, sludge dewatering devices and evaluation of new treatment techniques for use in developing countries.

---

[1]Professor and Coordinator, Regional Systems Engineering, Faculty of Engineering, University of Regina, Regina, Saskatchewan, Canada.

[2]Superintending Engineer, Tamilnadu Water and Drainage Board, Madras, India.

In the area of technology transfer, there should be international cooperation. There is a need to set up an international research centre for environmental engineering for developing countries in a developing country and to establish local research and development centres in each country. The international research centre is expected to act as a focal point in technology transfer to developing countries and may achieve an orderly and an appropriate transfer instead of the present fragmented and frequently careless and ineffective technology transfer.

Introduction

The 1981-1990 decade has been declared as the International Drinking Water Supply and Sanitation Decade. The task of the Decade is to provide the global population two of the most basic human needs - clean water for drinking and sanitary disposal of human wastes - a stupendous task when we consider that an estimated population of 1.1 billion in the developing world lack adequate sanitation and almost the same number lack access to safe water.

Widespread unemployment and underemployment in many developig countries have been major causes of concern in recent years. It is generally accepted that productive employment opportunities should be made available to large segments of population to foster economic and social development. In this context the present dominance of highly capital-intensive technologies in capital-scarce, labour-abundant economies has brought into question the appropriateness of the technology being transferred from the developed to the developing countries.

The following paragraph extracted from THE WORK OF WHO IN 1975 (World Health Organization, 1975) outlines the basic philosophy that should underline any technology transfer from developed to developing contries in the general area of environmental engineering: "Developing countries need labour-intensive techniques that are geared to existing local skills, equipment requiring minimum maintenance, a high degree of standardization, and simplified designs that can be manufactured locally. It is necessary to provide incentives to engineers, scientists and manufacturers in both developed and developing countries to deal realistically with these problems". In short, any transfer of technology should be symbiotic with locally appropriate technology as well as socially and environmentally acceptable. The following quotation from A HANDBOOK ON APPROPRIATE TECHNOLOGY (Canadian Hunger Foundation, 1976), summarizes this concept: "The only situation in which tools, technologies or knowhow from the industrialized parts of the world should enter a developing area is when they assist in activating and promoting local initiative, local understanding and local know-how, so that the responsibility for the technology can be assumed by the local people. This responsibility in part means the capability to reproduce the technology in a form adapted by the indigenous populations to suit their conditions and needs".

Water Supply

A majority of rural water systems in developing countries use ground water sources; smaller communities employ handpumps for obtaining the supply while the larger rural communities use mechanical pumping units to draw the supply and distribute it through a pipe system. Treatment usually involves only chlorination. Surface water sources are generally adopted in the case of urban areas - towns and cities; treatment generally involves sedimentation, slow sand filtration and chlorination in the case of medium-sized urban areas, while in the case of large urban areas, rapid sand filtration is used instead of slow sand filtration.

In developed countries, the techniques in respect of sanitation are standardized to an extent. Excreta collection is by means of the flush water closet; the wastewater is conveyed through sewers sized on the basis of self-cleansing velocities; the treatment leans heavily on techniques involving mechanical sophistication, and automation.

Though there has been no development of a completely new technique in this area for general use in developing countries, significant advances have been made in adapting and modifying existing techniques and processes to solve problems peculiar to local conditions. The low priority assigned to wastewater problems by most of the developing countries has been detrimental in achieving quick progress in innovative techniques through research and development.

Essentially, the schemes involve one of the following alternatives or some appropriate combination of them:

1) Water-carriage system
2) Air-carriage system (vacuum system)
3) Conservancy system
4) On-site disposal system

Though the water-carriage system is probably ideal from a technical viewpoint, its universal application is neither desired nor practicable, due mostly to economical and other constraints. Innovations like hand-flush latrines used in India that significantly reduce the quantity of water used, reduce transport and treatment costs to a great extent, in addition to conserving quality water. Vacuum systems would not be appropriate for developing countries generally, unless indicated for special locations especially when skilled operators could be provided. Though many developing countries use bucket systems and manual transfers in unsewered areas to disposal trucks, these systems present significant health hazards. Health hazards are minimized in some systems used in India where the excreta is washed down to a tank through a small pipe; the tank contents are emptied and hauled to a disposal site through a vacuum tanker.

On-site disposal systems like pit latrine, aqua privy, septic tank privy or regular septic tank soil-absorption system are used in many developing countries.

Technology Transfer from Developed Countries

A significant level of expertise is available in the developed countries and in some of the more advanced developing countries in the following areas of community water supply:

1) Improved designs of slow sand filters;
2) Manufacture of varieties of pipes and pumps;
3) Prospecting for groundwater;
4) Assessment and development of systems for trihalomethanes;
5) Use of alternate disinfectants;
6) Techniques for water conservation, including leak detection and control;
7) Water pricing; and
8) Disposal of water plant sludges.

The technology transfer may relate to any one of the above areas, depending upon the country and the project.

In many rural areas of the world especially in developing countries, sewerless sanitation would alone appear to be the feasible and appropriate method of providing sanitation. Most of the homes do not have individual house connections for water supply. Use of simple on-site excreta disposal systems built with locally available materials is probably the only answer to the problem. Innovations in developing suitable local systems and protecting public health should be encouraged. Much of the research and development work done in developed countries on systems such as compost toilets, septic systems and others, should be made available to local centres for adaptation and modification to suit local conditions. It would probably be necessary to evolve a scheme of inspection of these systems to ensure their proper functioning.

If on-site disposal systems are not feasible in some locations, use of tanker trucks to collect the wastewater from each household and haul it to disposal sites should be considered. Though less desirable than a sewer system, economic conditions would normally dictate the use of these systems for many years in such countries. Technology transfer from the west would involve, in this context, the manufacture of suitable tanker trucks locally (within the country) with vacuum draw-off arrangements if feasible, and opportunities to train people to operate these units.

When sewer systems are feasible on economic and other considerations it would be realistic to examine the possible use of a partial on-site system (septic tank or aqua privy) with effluent from each house conveyed to a single treatment system (tile field, lagoon, etc.) through a sewer system, with smaller-sized pipes, allowing less than conventional velocities. These types of systems may be feasible in suburban areas, city extensions and affluent villages. Technology transfer from developed countries in this scheme should aim at transferring information on experiences in their own countries as in U.S.A. (California) or in Australia. Experiences from within developing countries as the Zambian experience should be documented and made freely available to designers and planners in developing countries.

Many of the large or medium-sized cities in developing countries are partially or fully sewered. These cities would probably continue sewer extensions conventionally designed as needed. Developed countries could transfer technology to build plants for the manufacture of pipes suitable to local conditions, based on the availability of the pipe material within the country. Proper planning and market analysis would be needed before venturing on such a project.

The technology of wastewater treatment has been developed in industrialized countries. Their capital-intensive economy is reflected in a treatment technology that employs extensive mechanization, instrumentation and automation. It is quite possible that when engineers from industrialized countries serve as consultants in developing countries, they may recommend solutions, based on their own experience, which are not in harmony with the economic, technological or manpower resources of the developing country. Generally speaking, a treatment system that is simple but capable of satisfactory treatment of wastewater at a minimum cost should be considered for adoption in developing countries. This implies that waste stabilization ponds, aerated lagoons, oxidation ditches and similar low-cost alternatives would have to be preferred. Since lagoon systems require more land area, land use constraints would have to be investigated. It would be realistic in many of the tropical developing countries to use the effluent from these treatment plants for irrigation, especially to grow grass for cattle feed. Technology transfer for these treatment systems could involve the following:

1) Help establish local criteria for the design of these systems through local research, making use of information developed in the West and in other developing countries as a starting point.
2) Establish local plants to manufacture floating aerators (for aerated lagoons) and cage rotors (for oxidation ditches), instead of importing these units.
3) Use of natural or synthetic lines to reduce seepage from lagoon systems.

Lagoon systems would normally not involve sludge disposal problems. In the case of oxidation ditches, simple sludge drying beds could be used and the dewatered sludge hauled for land use.

This leaves one question unanswered. Should sophisticated technologies be ever transferred to a developing country and if it should, what should be the constraints? Factors such as appropriateness, the existence of an industrial infrastructure in the area, and the availability of skilled personnel to operate the plant would have to be considered. Appropriate technology need not be "low-level" or "inferior" technology. There are a range of options for the development of appropriate technology and one of the options is adapting evolving modern technology. The applicability of some of the promising recent developments in wastewater treatment for use in India and other developing countries was examined recently (Viraraghavan and Damodara Rao, 1985). Some of the new developments considered were oxygen activated sludge, rotating biological contactor (RBC), deep shaft process, anaerobic fermenters/filters and sludge dewatering devices.

Many of the medium-sized and probably all large cities in developing countries would require conventional sewer systems and wastewater treatment facilities; extension of onsite concepts for many of them would be impractical. Provision of "appropriate" treatment facilities would be essential. It is quite possible that some of the new developments could be used in the developing countries for certain situations with advantage.

Summary Comments

In the area of technology transfer, there should be international cooperation. It would be suicidal for a developing country especially in its initial stages of development, to seek transfer of technology from many countries in the solution of one facet of an environmental problem like water and sanitation. There are many competing technologies available in the West, which for obvious business reasons, could find their way into an unwary developing country and create chaos, if the developing country is not careful. There is a need to set up an International Research Centre for Environmental Engineering for Developing Countries, located in a developing country, and to establish local research and development centres in each country to evolve techniques suitable for that country. Some countries like India have already established research centres for environmental engineering research (National Environmental Engineering Research Institute, Nagpur, India) and such centres could serve as local contact centres for the proposed International Research Centre. Such an international centre could serve as a focal point in the transfer of technology to developing countries and possibly achieve an orderly and an appropriate transfer instead of the present fragmented and frequently careless and ineffective technology transfer.

Appendix - References

A Handbook on Appropriate Technology, Canadian Hunger Foundation, Ottawa, Canada, 1976.

The Work of WHO in 1975, World Health Organization, Geneva, Switzerland, 1975.

Viraraghavan, T., and Damodara Rao, T., "Recent Developments in Wastewater Treatment and Their Applicability to India and Other Developing Countries", Advances in Water Engineering, T.H.Y. Tebbutt, ed., Elsevier Applied Science Publishers, Barking, Essex, England, 1985.

Social & Political Aspects in Sewerage Planning

Anastasio López Zavala, Fellow ASCE*

Abstract

This paper deals with the sanitary sewerage planning and executive project and its related social & political aspects in a portion of the southwest part of Mexico City. Since the portion under study has at the present time a partial sanitary sewerage and many of the outfalls discharge either to a deep ravine or just to a crack above ground, it becomes apparent that a pollution problem arises in the downstream area because of flooding during rain season or in the groundwater reservoir due to the sewage infiltration.

It was intended to solve both problems by providing the planning and project of sanitary sewer interceptors to collect all discharges and transport them to 3 - 60 cm. diameter combined sewers in the lower portion of the area. However, at the time of the surveying of all existing facilities, it was learned by the project manager and authority supervisor of a strong opposition by the neighbors and land owners to carry out the project.

Introduction

The area under consideration is located in the southwest part of Mexico City within a very variable altitude beginning with elevation at 2 300 m s n m (meters above sea level) up to more than 3 000 m s n m. Economic status of the people is poor. The soil constitution is made out of basalt rock with its very high rate of infiltration due to numerous crackings. The whole present population of the area is near 25,000 inhabitants and provisions were taken to design for a future population of almost 60,000 inhabitants. The current sanitary conditions are very poor. Just 60% of streets have sewerage system. Houses without sanitary services fulfill partially their needs into the open air or through letrines or pit privies without a proper sanitary engineering design. The area is drained by two deep ravines connected downs tream to a main ravine. In

*Centro de Ciencias de la Atmósfera,
Universidad Nacional Autónoma de México.
Ciudad Universitaria, México, D. F., 04510.

the rain season these ravines convey a considerable 15 m^3/seg storm flow which discharges in the plain low area about 15-20 km from the higher elevation area, causing serious flood problems where a first class residential subdivision exists.

As may be understand, people living in the area under study, with low or not income at all, and as a consequence, without actually paying no federal taxes on property, are requiring considerable money expenditure from the federal district government to provide essential services as water and sanitary sewer, which they are not prepared to pay for.

This kind of situation is faced by almost all the big cities in the country. People migrating from rural areas to the city, looking for jobs for which they are not quite ready for, settle their poor houses down wherever they want or can, usually places without facilities, causing the problems already described in the above paragraphs.

Sanitary Sewerage Planning.

The planning of the Sanitary Sewerage System, contemplated a design horizon to the year 2000 for eight communities and took into consideration the existing discharges and a total future population of 60,000 inhabitants assuming a per capita water consumption of 150 liter per day and peak factor of 1.9 for covering both daily and hourly peaks. From the resulting value a factor of 0.8 was considered for maximum sanitary flow. No provision was taken for infiltration flow since the water table elevation is very low.

The planning was decided after the study of 2 or 3 different alternates. The final and acceptable alternate was that of a lower cost one but was not possible to locate the interceptor lines in government property land since all land belongs to the so called "ejido" areas and their original owners have sold part of their properties to these poor people.

Through a very careful inspection and surveying, all the present sanitary discharges either to the ravines or to the ground were recorded, including amount of flow, pipe diameter, elevation and location. If was not a matter of this planning to implement the design of the streets having no sanitary sewer but the control of all these discharges for their conveying by means of the interceptor off the higher elevations to a combined sewerage system located down in the low plain area of Mexico City.

The planning of a sewage treatment plant in this low area and the use of its effluent as an irrigation water was not accepted for its high cost involved and the lack of land.

Due to a recent construction of storm sewer facilities by some local authority, a problem was created since this new storm sewers were connected to the existing sanitary sewers. The sanitary sewerage planning took this into consideration and no attempt was made to disconnect the storm sewer because of the cost involved. Fortunately, the storm drain area is of no importance.

It was decided to connect the total maximum sanitary flow of about 200 lps (liters per second) (4.5 MGD) into 3 - 60 cm (24 inches) reinforced concrete combined sewers to alleviate the hydraulic load in only one sewer. The federal district authority did the approval of this decision.

By the planning and project of the interceptor sewers it was decided, after the approval, to commence construction works based on the cost estimate of the study.

Social and Political Aspects.

During the preliminary inspection phase of the construction program and throughout the site visit it became apparent the evidence of strong opposition for the construction from the part of the neighbors and land owners. Besides this situation it was also apparent the reluctance of the district authority to allow the project sanitary interceptor to pass through some social benefit building complex. This fact was instrumental to stop the whole project.

Conclusions and Recomendations.

The migration of people living in rural areas to the industrialized cities due to the lack of work creates as a consequence a human burden on the big cities. Generally speaking, these people are not ready to cope with the requirements of a fair job and very soon they find themselves as a jobless or underpaid employees, with the economic, social & health consequences that this brings along at their homes. The net result is the formation of the so called "misery belts" in the periphery or outskirts of the cities, where hardly they can account for services as electricity, water, sewers, garbage collection, gas, etc. The government is doing great efforts to afford as much as possible these services in these areas, although it is a well known fact that users many times can not pay their corresponding bills, causing an economic problem to the city authority. With this situation, the government is compelled to use the taxes from the middle class people in the city, not only to fulfill their needs but also those from these poor areas. If this situation goes on and on for years a dramatic economic crises results in these cities.

If is recommended then, in order to avoid this problem or at least to alleviate it, the procuring of the following measures:

1.- Federal District authority most not permit the migration of people living in the provinces unless they prove to have a god job in the city. This has to be done by means of an agreement with the local authority where the people have their permanent address.

2.- Sanitation of the areas already built shall be accomplished by properly taxing the properties. If people can not pay this taxing program, government may confiscate their property.

3.- Open space areas shall not be permitted to be occupied by neither people from provinces nor by unemployed ones.

DON'T EMPTY THE PIT

Bjorn Brandberg*

Outline to an Analytical Method for Dimensioning of Semi-Permanent Pit Latrines

---

INTRODUCTION

Can a conventional pit latrine be dimensioned to last a lifetime?

The author is of the opinion that even with reasonable pit volumes, convensional pit-latrines can be designed to last for 50 years or more, hence outlasting more expensive, complex and less reliable systems for sanitation.

The improved pit latrine has gained the reputation of being the affordable technology for appropriate sanitation in developing countries. Technical improvements, including ventilation of the pit, have also made them hygienic and free from smell and flies (VIP-latrines). Limited lifetime of pits, however, has often justified more expensive methods for sanitation, especially in urban low income high density areas.

Conventional water-borne sanitation system has proven to be expensive and often complicated to maintain. Pilot experiences from mechanical emptying (which is cheaper in operation than water-borne systems) has not proven less complicated, the emptying cost is about USD8.00 per M3 emptied volume, while manual digging in normal soil is only about USD2.00 per M3. The nature of the pit contents (high percentage of organic material which is subject to biological decomposition) and many identified cases of extremely low accumulation rates, (say 10 l/p/y) has led the author to analyse the factors influencing sludge accumulation in pit latrines, with the objective of calculating necessary pit sizes for long-lasting (semi permanent) pit latrines.

* Sanitation Adviser to the Government of Malawi, UNDP/World Bank, P.O. Box 30135, LILONGWE 3. Malawi.

SLUDGE ACCUMULATION

Traditionally, filling up rates have been assumed to be constant, say 60 l/p/y or 40 or 30 or any other constant factor. (The formula x l/p/y = no of litres accumulating (x l) per person (p) per year (y).)

Considering the nature of the pit contents, such an assumption must be false, as we both eat and defecate organic biologically degradable material (and not sand and stones that would accumulate constantly in the pit). A time factor is obviously of importance for the rate of accumulation. We can suspect that large pits, with long lifetimes, will allow for better decomposition of the organic material, and hence give lower accumulation rates.

Another factor of importance is the liquids. Saturation of the soil walls of the pit might be a factor of importance, where water is used for anal cleaning and/or where latrines are used for personal hygiene.

Accumulation in the pit will therefore depend on three components:

- biologically degradable material (BDM)
- non-biodegradable material (NBDM) and
- liquids

The total accumulation rate will consequently be the sum of the three components. Under certain conditions which shall be discussed later, the liquid component will be zero and the NBDM component will be insignificant. The accumulation rate would then depend on the BDM. We would then find that the BDM factor alone would give us pits with a very long lifetime. How long?

Only future research can give a reliable answer on that question, but existing experience indicates that large volume pits are a better alternative than both waterbone sanitation and pit emptying systems.

Nothing lasts for ever. A person can in very exceptional cases live 100 years. "Permanent" houses use to have a mortgage time of 40 years for the longest loans, but they normally stand longer, say 100 years.

For the purpose of this paper let us call 100 years permanent and 50 years semi permanent.

BIOLOGICALLY DEGRADABLE MATERIAL (BDM)

The BDM (biologically degradable material) entering into the pit is principally fecal matter, anal cleaning material (say 5%) plus some non fecal related BDM (say 10%).

We defecate around 75 kg per year (0.2 kg/day). Fecal matter has a moisture content around 70% (Urine is liquid, and shall be dealt with later.)

The dry weight of the fecal matter is 30% of 75 kg (22 kg). Add 11 kg for non fecal material (15% of 75 kg) which makes (11 + 22) 33 kg dry weight.

The rate of biological decomposition in latrines is not well known. Experiences from septic tank emptying indicate possible sludge accumulation rate of 20 kg/p/year. An estimated dry content of 30% gives us (30% of 20 kg) 6 kg. Reduction factor (33/6) 5.5.

Experiences from Zimbabwe (Stangle, 1985) indicate an average sludge accumulation rate over a nine year period of around 18 l/p/year. This figure includes everything (fresh and old BDM + NBDM + liquids). Moisture content 50% (estimated) and spec. weight 1.5 (estimated) gives us (18 x 0.5 x 1.5) 14 kg/p/year. Deducting say 5 kg for NBDM makes BDM 9 kg/p/year, 27% of the original 33 kg. Reduction factor 3.7 over a 9 year period.

From these two examples we can draw some conclusions:-

1. The dry content of our BDM latrine deposits is around 33 kg/p/y.

2. It is possible to achieve a 5.5 times reduction of the dry biodegradable material (33/5 = 6.6 kg/p/y).

3. The rate of biological decomposition depends on the environment. (Faster in wet than in dry environment).

4. Considering a time of 9 years, a reduction of approximately 3.5 times is likely.

5. For another period of 10 years, further reduction of the BDM of 50% is not unlikely (50% of 9 = 4.5 kg/p/y).

6. Further reduction is not likely (and not critical).

NON BIODEGRADABLE MATERIAL (NBDM)

As earlier stated, we eat biologically degradable material (like cereals, vegetables, meat, etc) and not NBDMs (sand, stone, glass, metal, etc). The normal anal cleaning materials (if any) are also biodegradable (paper, leaves, grass, etc). Water, which is frequently used in some regions, is liquid and shall be dealt with later.

The amount of NBDM that enters (floor sweepings, glass, metal etc) is normally a matter of misuse rather than a real need. The floor can for example be designed with a small step, or even a pedestal seat, not to encourage sweepings from the latrine floor to go down into the pit, except for what is very close to the hole (fecal matter).

Ashes are sometimes introduced into the pit to control smell and flies. The method is only effective where large volumes of ashes are introduced. Installation of a screened ventpipe (VIP-Latrine) is a very much better alternative, as ashes accumulate in the pit as NBDM.

LIQUIDS

Given that the area of absorption in the pit is big enough, liquids do not accumulate in the pit. They are soaked up by the surrounding soil. Water also contributes positively to accelerated biological decomposition. (Urine might have the opposite effect). A conservative infiltration rate is 20 l/m2/day (clayey soil).

If the toilet is used as a bathroom, you can expect a loading rate of 20 l/p/day. That is 200 l/family/day (10 people). Area required: 10 m2.

(Note that the pit should be lined if the side walls are used as infiltration area, due to the risk of erosion and possible pit collapse.)

If not used as a bathroom a maximum of 5 liters per family per day can be expected from daily cleaning. Infiltration area required 0.25 m2.

Large volume pits normally have more than 10 m2 infiltration area. The presence of water will also speed up the decomposition rate. If not used as a bathroom, the question of liquids can in most cases be ignored for large pits.

DIAGRAMMES

The accumulation of BDM and NBDM have been computed together with moderate assumptions on moisture contents and specific weights for different values of NBDM (see annexed diagrams).

DESIGN EXERCISE

Let us choose the most serious case of NBDM accumulation (10 l/p/y).

From diagram No. 4 (NBDM = 10 l/p/y) we find that for a 50 years lifetime of the latrine we should need a 990 l pit volume per person (say 1.0 m3).

Assume 10 people in the family.

DIMENSIONS

Pit volume (10 x 1.0) 10 m3 + 1 m3 free space = 11 m3

Assume a pit depth of 5 m
Horizontal pit area (11 /m5) 2.2 m2

Diameter 1.68 m, say 1.70 m

Using a pit collar or a lined pit, a slab of 1.80 m diameter would do.

CONTROL OF INFILTRATION AREA

Assume that the latrine is used as a bathroom for 10 people each using 20 l/day for personal hygiene. Volume to be infiltrated 200 l/day. Infiltration capacity 20 l/m2. Required area 10 m2. Bottom area 2.2 m2. Rest for the side walls: 7.8 m2. Circumference 5.6m (Diameter 1.8 m). Required free side height: (7,8-5.6) 1.39m. (The sides should be lined with open honey comb brickwork to allow for infiltration at sides without erosion.)

EVALUATION

The cost for digging and eventual lining of pits is approximately USD 5/m3 (Malawi figure). Pumping cost, including transport and operation, maintenance and capital costs of vehicle has been estimated to USD 8 per m3 (or more depending on transport distance).

It is therefore cheaper to dig than to pump the latrines. No foreign capital is required, and no interruptions will occur due to breakdowns and lack of spare parts. No sanitation alleys are needed to give access to suction vehicles. And no searching for safe sites for depositing the pumped pit contents.

As we can see from the curves of the diagrams (See appendix), latrines become more economical the longer they have been in use. A latrine with an expected lifetime of 50 years or even 100 (!) is not unrealistic at all.

With ventilated large volume pits the annual cost will be kept low enough to compete not only with conventional pit-latrines and conventional sewer systems but also with small diameter sewer systems.

For the sites where large volume pits can not be built, ventilated septic tank based latrines ("septic latrines") could be a viable alternative. The "septic latrine" is the integration of the septic tank and the ventilated improved pit-latrine (VIP). It is hygienic and odorless as the VIP and permanent as the septic tank. It does not require any water for flushing but provides safe disposal also for sullage water. A closer description of the septic latrine is, however, the subject for another paper.

PROPOSED RESEARCH

Several factors in the calculation of the diagrams (see appendix) have been estimated. As a rule, conservative estimates have been used.

A research project could investigate, through laboratory analysis, factors such as the ratio of biodegradable and nonbiodegradable material, moisture contents, specific weights, etc from samples taken from different latrines at different depths. The age of the samples could probably be analysed using the C12 method.

In order to verify the accuracy of the theoretical method, a number of reference latrines should be monitored on a monthly basis to assess actual as well as average accumulation rates.

It is very possible that we will find that the graphs in the appendix are too conservative, that required volumes can be reduced or that expected latrine lifetime will become still longer.

CONCLUSION

With the experience we already have, we know, that semi permanent latrines can be designed today. 50 or even 100 years of expected lifetime is possible.

One cubic meter of pit volume per person should be enough !

May be even less would do, but that future research has to show.

ACKNOWLEDGMENTS

Many people with long experience of low cost sanitation have contributed with ideas and critical thinking. Special thanks to Mike Stangle for sharing his experience and findings from Zimbabwe. I should also like to thank Sandy Cairncross, Peter Hawkins, John Briscoe, Peter Morgan, Geoffrey Read, Bill Boehm, Mexon Nyirongo and many others, who all have been helpful in different ways. The responsibility for conclusions and eventual mistakes is, however, my own.

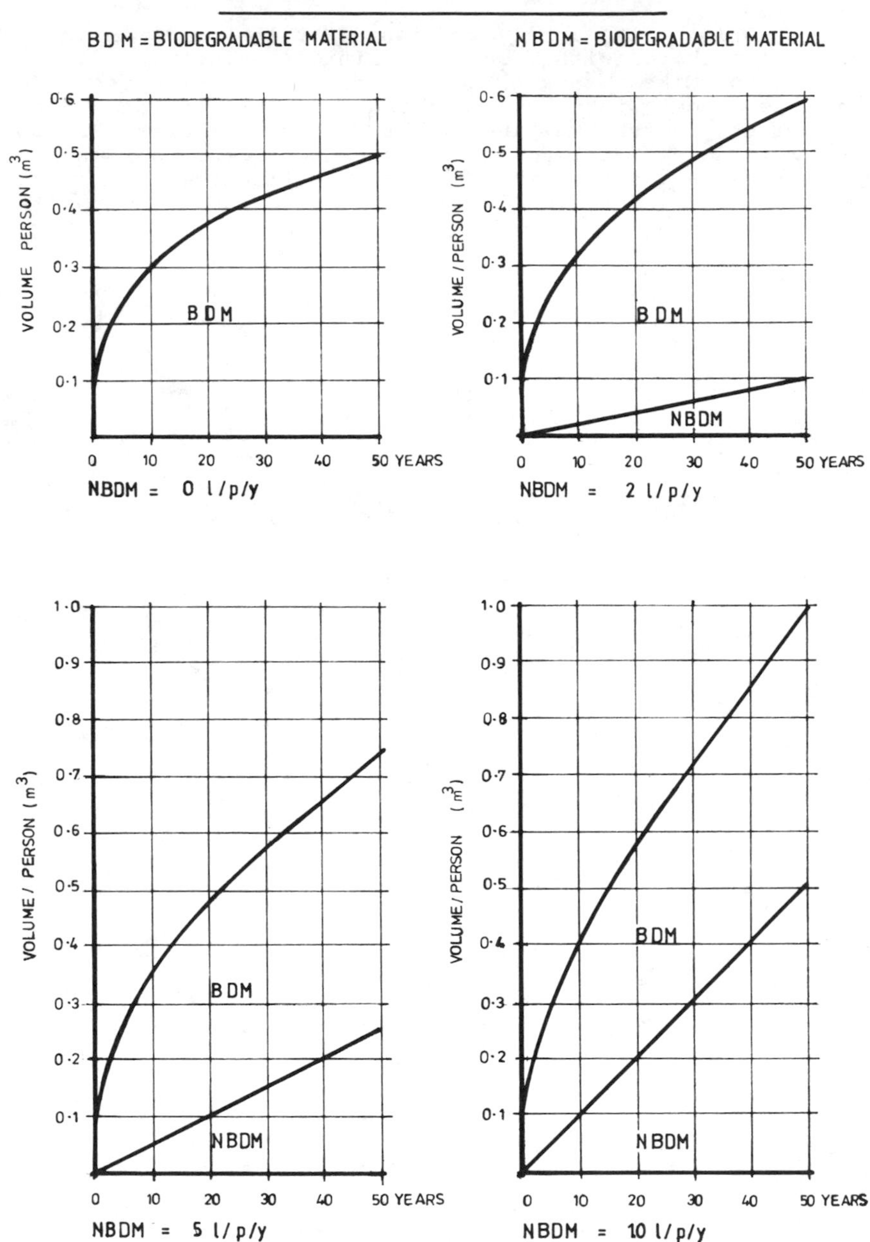
REQUIRED PIT VOLUME PER PERSON
B D M = BIODEGRADABLE MATERIAL
N B D M = BIODEGRADABLE MATERIAL
VOLUME PERSON (m^3)
0·6
0·5
0·4
0·3
0·2
0·1
B D M
0
10
20
30
40
50 YEARS
NBDM = 0 l/p/y
VOLUME / PERSON (m^3)
B D M
NBDM
NBDM = 2 l/p/y
VOLUME / PERSON (m^3)
1·0
0·9
0·8
0·7
B DM
NBDM
NBDM = 5 l/p/y
VOLUME / PERSON (m^3)
BDM
NBDM
NBDM = 10 l/p/y

Intensive Drogue Studies To Assess Outfall Impacts

Jonathan A. French, M.ASCE, and Carl R. Johnson, M.ASCE*

Introduction

For many years, drogues were the only dependable means of measuring ocean currents and nearshore circulation. Within the last quarter-century, rugged and practical in-situ recording current meters have been developed and are in extensive use. Over the past two decades, numerical models have been developed to simulate coastal and estuarine transport and dispersion, and are now in wide use.

Despite the advent of current meters and numerical models, drogues continue to be a useful and popular means of examining currents. Among their advantages are that they are inexpensive, easily built, and may be used by any crew with a safe boat and commonly available navigational skills and instruments. For questions of siting an ocean outfall to discharge wastewater or thermal effluent, drogues provide direct empirical answers to the question of where the effluent will go once it is discharged.

Drogue studies are therefore a part of most outfall siting studies. For a well-funded siting study for a major metropolitan outfall conducted in several stages, the preliminary reconnoitring and scoping surveys may include drogue deployments to supplement extant current information before embarking on a full-scale program of drogues, meters, and models. Modest-budget siting studies for small outfalls often rely entirely on drogues for information on local circulation.

A drogue for circulation surveys is an object designed to float at a specified depth beneath the water surface, and be carried by the currents, with its position determined at frequent regular intervals. Of many useful drogue designs, one that has proved very suitable in coastal circulation consists of a submerged sail, in this case two rectangular 2.5m x 0.6m surfaces intersecting perpendicularly. A fine wire tethers the sail to a small 0.1 sq m surface float, carrying a brightly visible flag. The sail is slightly weighted to keep the tether taut and vertical so that the sail will remain at the specified depth to which the tether has been cut. The necessary surface float and flag introduce the possibility of error due to wind drift, but this is minimized by keeping the float and flag as small as possible, and has been found to be negligible for practical purposes.

*Both Principal Engineers, Camp Dresser and McKee Inc.,
One Center Plaza, Boston, MA 02108

Commonly, a survey crew will bring their boat in early morning to a designated deployment site of interest, usually a site being considered for location of a wastewater outfall discharge. The crew will then deploy

about six drogues, and note the time and exact location of each deployment. The six drogues may have varying tether lengths, hence varying survey depths. At roughly half-hourly intervals, the crew will bring their survey vessel alongside each drogue's surface float, note the time, and determine the position, being careful not to disturb the float. At the end of the survey day, the drogues are recovered.

Figure 1 is an example of the resulting plot, showing the results of two days of drogue trackings from site "D5." Those tracks tending to move northwesterly were obtained one day on a flood tide during the hours of deployment. Those moving southeasterly were for a day with an ebb tide. Tabulations of time and position for each drogue at each of the 9 to 12 fixes are routinely provided with the plots.

Position-fixing may be by any means that gives desired accuracy. Electronic systems such as Autotape or Miniranger are usually the most desirable, but horizontal sextant, sighting compass, or LORAN may in some cases be acceptable or even preferable.

Strategies for Drogue Deployment

For outfall siting studies, drogues offer simple, inexpensive, and direct answers to the question of effluent movement and dispersal after discharge. Yet an obvious disadvantage is that a single drogue deployment provides only a very small sample of the long-term current pattern at a given site. The outfall will be discharging continuously, and the sea is continuously changing. How can a few drogues adequately describe the full circulation pattern?

For an adequate siting study, more than one drogue deployment is needed. Hundreds of drogue deployments would be ideal. Constraints of time and money usually limit the number of deployments to the order of five to ten. It is essential that the most information possible be inferrable from a few deployments. In planning a drogue study, one first gathers what information is already known about local circulation. Are tides dominant? What are the neap, mean and spring ranges? What are the daily and monthly tide patterns? Are winds important? Are there predominant daily patterns, or seasonal monsoon patterns?

It is useful to plot tidal information as in Figure 2, which is a continuous trace of tide elevations over a month. In Figure 2, daily predictions of high and low water were plotted, and the continuous curve interpolated. The figures show clearly that for the location studied the tide is principally semidiurnal, but on some days one tide is much greater than the other. It is clear that there is a great difference between spring and neap tides at that location.

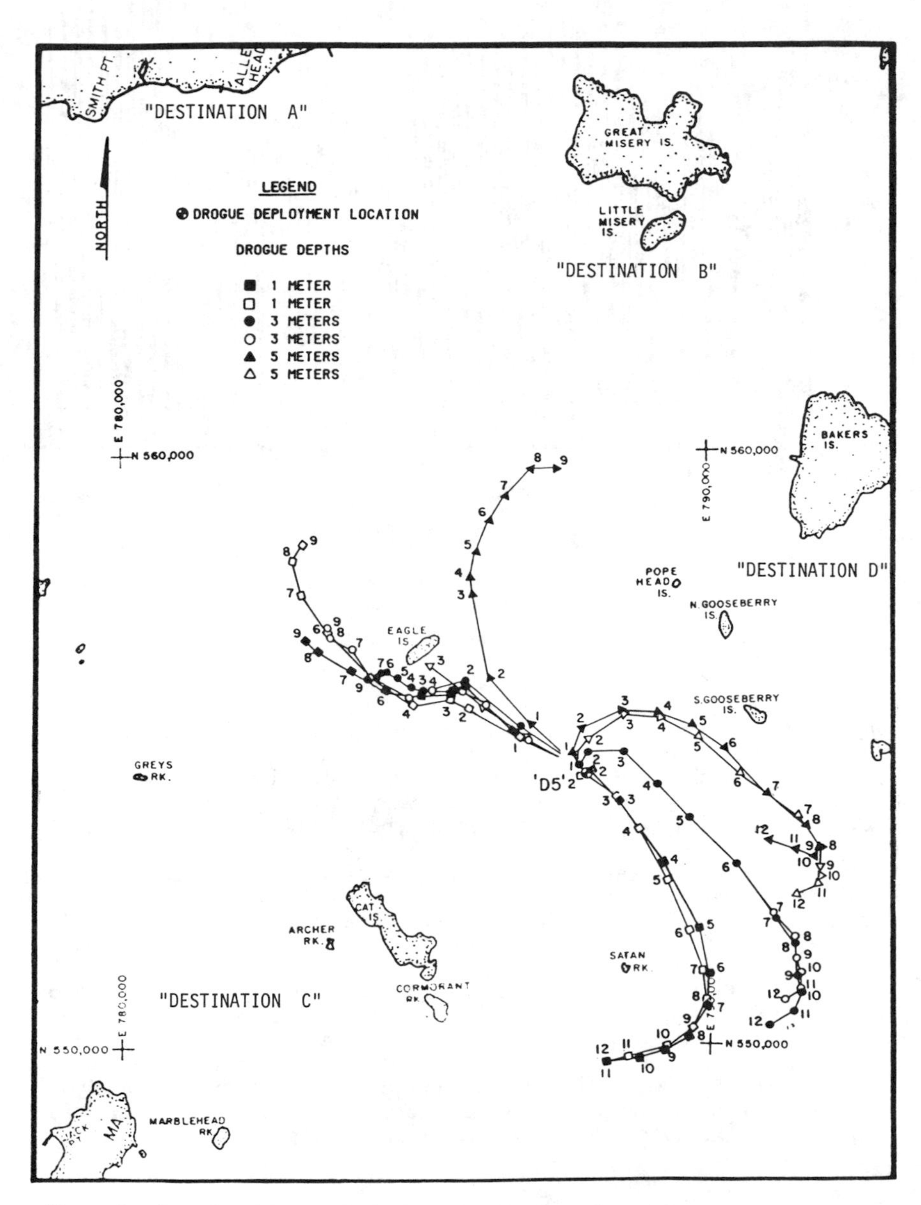

Figure 1. Example of Drogue Path Plots for Two Days from a Common Deployment Site ("D5")

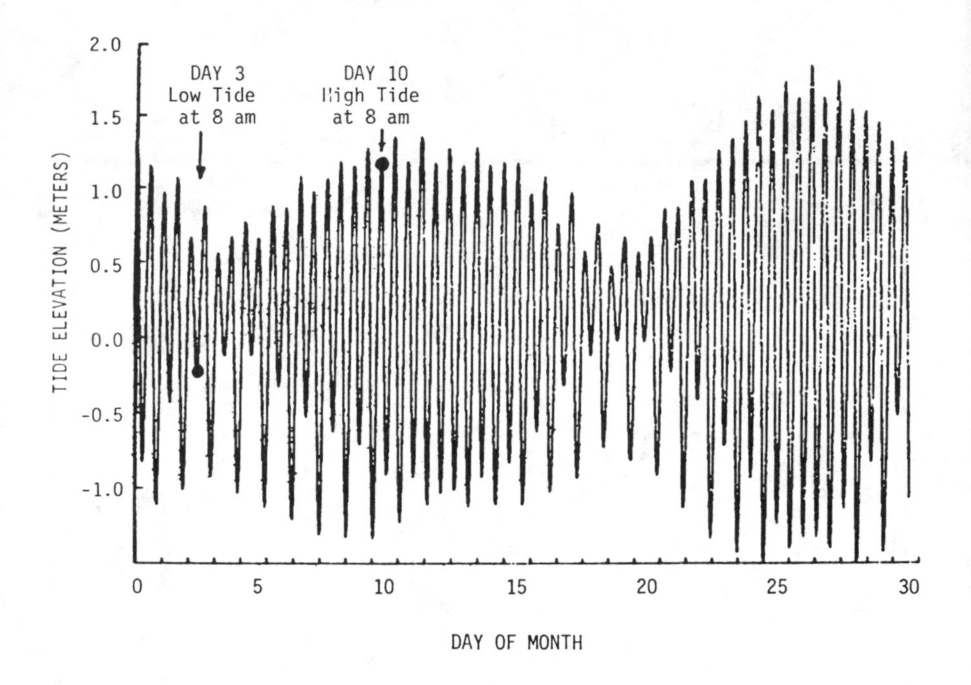

Figure 2. Tide Elevation Over the Course of a Month

Winds affect surface currents of the sea. In a monsoon climate, it clearly would be well to deploy drogues during each of the two seasons, to sample, say, both the northeast winds and the southwest winds. In temperate climates, records may show that wind patterns do not vary so strongly to warrant deployments in more than one season.

Figure 2 shows that there are two roughly similar tide patterns during a Ideally, one would wish to sample over a whole month, but if time and money are serious constraints, one of the two "lunar fortnight" cycles in the month may be sampled.

It is a practical fact that drogue studies are most easily conducted during daylight hours, say 8 a.m. to 5 p.m. The tide phase found at 8 a.m. on a given day depends on the lunar phase that day. For example, in Figure 2, it is low tide ("slack; flood begins) at 8 a.m. on Day 3; but by Day 10, it is high tide ("slack; ebb begins") at 8 a.m. By choosing the date carefully, one may focus on a flood (incoming) tide during daylight hours (e.g., Day 3) or on an ebb (outgoing) tide (e.g., Day 10).

Of the 14 days in a lunar fortnight, which should be chosen for drogue deployments if one has a budget for only 5 days of study? First, one may wish to get a well-distributed sampling of all tide phases, and sample on days 1, 4, 7, 10 and 13. Possibly, however, one is less concerned with ebb tides, which can be assumed to carry effluent shoreward towards important bathing beaches. In that case, sampling may be concentrated on flood tides (days 1, 3, 4, and 5 with a token ebb sample on day 10). On the other hand, one usually is most interested in sampling during "spring tide" conditions, when the tidal range is greatest, i.e., about Day 10, in Figure 2. Similarly, attempts should be made to thoughtfully sample known diurnal wind patterns e.g., morning and evening calm, or afternoon breeze. The wind speed and direction should be recorded hourly during deployment.

From this planning process a surveyor may conclude that it is worth a special effort to start very early in the morning or well into the evening on some days; or even to equip drogues with lights for nighttime deployments.

When the survey field work is complete, the data for each day may, as a first step, be presented as shown in Figure 1. However, while this format is informative, it does not by itself yield all the information that can be inferred from such a study.

"Probability of Destination" Diagram

As an example, we describe here a procedure found to be useful when the study discharge site is in a confined estuary or among a group of islands. To use this method, one must be able to apply geographic labels to one or more areas visited by each drogue during its voyage, e.g., "Deployed at Study Site D5, and travelled to or near Destinations A, B, and C." The time of initial drogue deployment, the average wind velocity during deployment, and the times of high and low tide on the day of deployment must also be noted.

Figure 3 is an example of what now may be done with these data. The vertical axis of the graph represents the tide phase, in degrees from 0 to 360. The horizontal axis of the graph represents a measure of the onshore component of local wind velocity.

Figure 3 is in fact a joint probability plot for ambient tide and wind conditions. In this case, the hypothesis is that if effluent were to be discharged at a given study site, tide and onshore wind effects will most likely determine the movement of the plume as it is carried away from the discharge point and dispersed. Any conceivable combination of tide and wind condition is represented by some point in Figure 3. A few of the points in Figure 3 represent the tide and wind conditions for the drogue studies undertaken. For those conditions, the drogue studies have shown where the plume of effluent will go, so we may write the destination directly onto the graph. After that is done, we may infer the likely destination of the effluent plume for any wind-tide combination, by noting the observed destinations for the most similar conditions tested in the drogue studies.

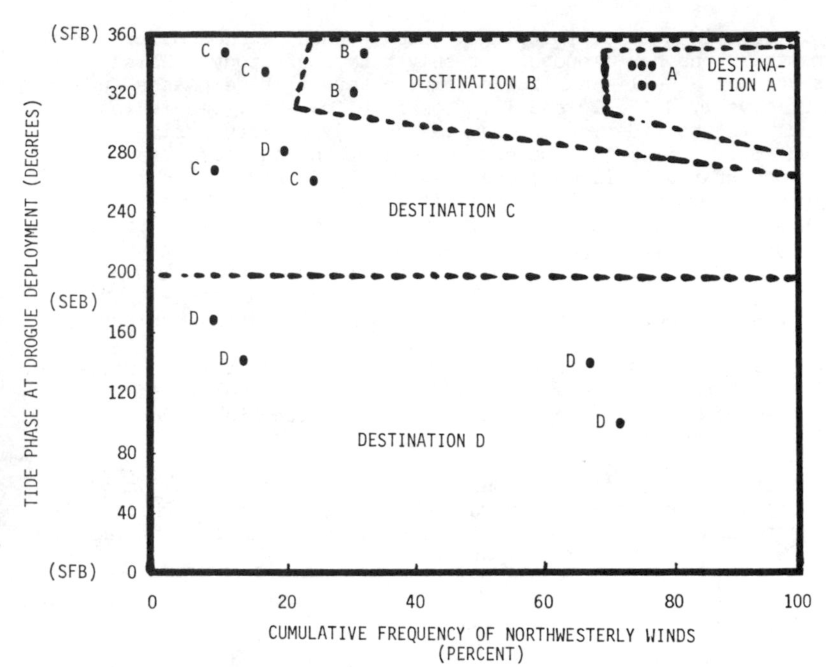

Notes: SFB--"Slack;Flood Begins" (Low Tide)
SEB--"Slack; Ebb Begins" (High Tide)
X ● -- Plotted wind and tide condition under which a drogue travelled to Destination X

Figure 3. "Probability of Destination" Diagram

In Figure 3, the scales for both axes have been selected to provide uniform probability density at all points on each scale. On the horizontal axis, notice that the onshore windspeed component scale is expressed as a percentile. To calculate percentile values, one needs a long-term wind record for the study area, best obtained from a local permanent meteorological station. One computes the onshore windspeed components (an offshore component is included as a negative onshore wind) for each wind velocity in the record. The results are ranked and plotted in a cumulative frequency distribution, with percentile values indicated.

In this example, the vertical scale is not distorted, as all tide phases have equal probability of occurrence at any randomly selected moment in time.

The result is that if Figure 3 were to be divided into a grid of small squares of uniform size, all squares would have equal probability of representing the tide-wind combination seen at a given moment.

After the drogue destinations have been written into the figure, the space may be divided into destination zones, using the drogue results as guidance, as indicated by the dashed lines of Figure 3. The destination zones may then be planimetered to find out what the probability of each destination is. In the example of Figure 3, there is about a 7 percent probability of effluent travelling to Destination A, while to B it is 17 percent and to Destinations C and D it is 33 and 43 percent, respectively. The probabilities sum to 100 percent.

Figure 3 therefore provides a simple, direct assessment of impact of a planned wastewater discharge. When such a figure is developed for each of several alternative sites, the site whose plume imposes least frequent impact on critical beaches or other critical resource areas may, by this measure, be considered functionally superior.

Conclusion

Despite the advent of recording current meters and numerical models in recent decades, drogues retain a role of importance in coastal oceanographic studies. By intelligently scheduling a limited number of drogue deployments to sample currents at a predicted variety of tide and wind conditions and appropriately graphing the results, one may infer sewage plume movement over a full range of possible tide and wind combinations.

Acknowledgements

The drogue design described herein and the drogue position plots of Figure 1 were provided by Ocean Surveys, Inc., Old Saybrook, CT, USA.

Notice

While the techniques presented herein were developed in the course of site-specific studies, the example presented is hypothetical and general, bearing no information about any actual site.

Dracunculiasis Eradication and the Water Decade

Ernesto Ruiz-Tiben, Ph.D.*, and Donald R. Hopkins, M.D., M.P.H.*

Efforts initiated in 1981 to take maximum advantage of the opportunity created by the Drinking Water and Sanitation Decade for global elimination of dracunculiasis (Guinea-worm disease) have steadily gained momentum. Currently, 19 of 21 endemic countries have anti-dracunculiasis activities underway. The public health and socio-economic importance of dracunculiasis are discussed, and current eradication efforts are described.

Introduction

The United Nations designation of the 1980-1990 period as the International Drinking Water Supply and Sanitation Decade (IDWSSD) recognized the absence of safe drinking water and sanitation as impediments to good health. The principal goal of the IDWSSD is to promote and facilitate the provision of both safe drinking water and sanitation to all populations by end of the Decade. Since dracunculiasis (Guinea-worm disease) is the only communicable disease which can be eliminated if affected populations are provided with, and use, safe drinking water, the Steering Committee of the IDWSSD in 1981, endorsed the concept of elimination of dracunculiasis from all endemic countries as a subgoal of the Decade. A resolution by the World Health Assembly in 1981 (WHA 34.25) on the IDWSSD recognized the unique opportunity of utilizing a decline in prevalence of dracunculiasis as a practical measure of the success of the Decade. In 1986 the thirty-ninth World Health Assembly passed resolution 39.21 which (1) endorsed efforts to eliminate dracunculiasis country by country in association with the IDWSSD, (2) called on affected member countries to establish national plans of action to eliminate Guinea-worm disease and to strengthen national surveillance, (3) invited bilateral and international development agencies, foundations and private voluntary organizations to assist affected countries, and (4) urged the Director-General of the World Health Organization to assist with the above, and to submit a report on the status of these activities in the regions concerned to the Forty-first World Health Assembly in May 1988.

* World Health Organization Collaborating Center for Research, Training, and Control of Dracunculiasis. Centers for Disease Control, 1600 Clifton Road, N.E., Atlanta, Georgia 30333.

This paper describes the disease, its impact on the health and economy of affected populations, and the current status of the global eradication initiative.

The life-cycle of the parasite and the disease:

Human dracunculiasis is a parasitic infection resulting from ingestion of Dracunculus medinensis larvae contained within the body of fresh water microcrustacean of the genus Cyclops which act as intermediate hosts. It is the only disease transmitted solely by drinking water. Cyclops spp. are typically found in shallow pools, ponds, and stepwells used as sources of drinking water. When humans ingest infected Cyclops the larvae are released in the stomach. D. medinensis larvae penetrate the intestinal wall and migrate to subcutaneous tissues. Mating occurs about 90 days post-infection, when the parasites become sexually mature. In about 90% of cases the adult female worm migrates to the lower extremities about nine months to a year post infection. During this time gravid female worms, which have grown to a length of 70-100 cm.(2-3 feet), become ready to emerge and release larvae (Muller, 1971).

Local itching, urticaria, and burning pain at the site of a small blister (caused by the worm) are usually the first clear symptoms of dracunculiasis. Soon thereafter the blister ruptures and becomes an ulcer containing a protruding worm. No other infection is likely to be confused with the picture presented by an adult D. medinensis emerging through an ulcer. In endemic areas villagers are able to accurately diagnose, and recall, Guinea worm infections. Because of its length, the emerging worm must be slowly and painfully extracted; usually, by the ancient method of rolling a few centimeters each day on a stick or twig. The open ulceration and physical extraction of the parasite usually lasts from 5-8 weeks (Kale, 1977). Pain associated with worm emergence or with secondary bacterial infection often prevents the patient from standing or walking. In some cases, deep abscesses, septic arthritis, or tetanus may ensue due to the chronicity of the open wound. In Nigeria, for example, the period of incapacitation ranges from 3 weeks to 9 months, which often coincides with the farmer's planting season (Kale, 1982). Fatalities are rare, but permanent disablity is estimated to occur in about one percent of cases.

The disease cycle is perpetuated through the release of millions of larvae by the female worm when the affected limb is immersed into water; often done to relieve the pain and burning sensation. The released larvae are then ingested by Cyclops spp., and reach the infective stage in about two to three weeks. Thereafter, whenever infected Cyclops are ingested via drinking water, the released third stage larvae infect humans and develop into a new generation of worms one year later .

Transmission of infection from man to Cyclops and from Cyclops to man is markedly seasonal. The transmission season is limited by the presence of shallow pools of stagnant water which serve as drinking water sources. These pools become transmission sites during the rainy season in arid areas, and during the dry season in regions with higher rainfall. Dracunculiasis is generally considered an infection of dry regions, where water sources are few and shared by many.

The seasonal occurrance of lesions and concomitant disability among rural populations causes dramatic impact on agricultural productivity. Incidence may reach its maximum (often grater than 50% of the 15-45 age group) during the labor intensive sowing or harvesting season, adversly affecting agricultural output to the detriment of nutritional status of the community. Impact on school attendance is also substantial in endemic areas (Nwosu, et al. 1982).

There is no evidence of acquired immunity to dracunculiasis, and the endemic community usually suffers annual attacks. However, the infection is self-limiting because the worm dies after the one year incubation period. A person must be reinfected annually for the disease to recur. Since nonhuman hosts are not believed to be involved in transmission of the disease to humans, total interruption of transmission for just one year can theoretically eradicate the disease from a community.

Distribution

Dracunculiasis occurs in 19 African countries, parts of India, and Pakistan. A few small foci of infection may still exist in Yemen and adjacent parts of Saudi Arabia. Approximately 140 million people live in Guinea-worm endemic areas, and 5-15 million cases occur each year. Almost without exception, this is an endemic affliction of poor rural populations. Affected countries may be grouped into 4 broad categories according to the known status of the problem (no implied ranking within each category):

Category			
<u>SEVERELY AFFECTED</u> (high prevalence rates, extensive geographic area involved)	Benin Ghana Togo	Burkina Faso Nigeria	Ethiopia Sudan
<u>MODERATELY AFFECTED</u> (moderate prevalence, or lesser known endemic geographic area	Mali Uganda	Mauritania India	Senegal Pakistan
<u>SMALL NATIONAL PROBLEM</u>	Cameroon Kenya	Ivory Coast	Guinea
<u>STATUS UNKNOWN</u>	C.A.R.	Chad	(Somalia?)

Table 1. Status of Dracunculiasis in Affected Countries

In general, dracunculiasis tends to be more prevalent in endemic African communities (prevalence rates of over 40% are common), than in India (prevalence rates above 20% are unusual). Whereas in Asia, step wells are common sources of transmission, ponds are the usual foci of infection in Africa.

Control Measures

Elimination of dracunculiasis is possible if affected populations alter water usage habits, receive and use new or improved water supplies, or if existing water sources are treated (Hopkins, 1984). Currently there is no effective therapy for dracunculiasis. Prevention is the preferred means of controlling this disease, though the life cycle can be attacked at several points.

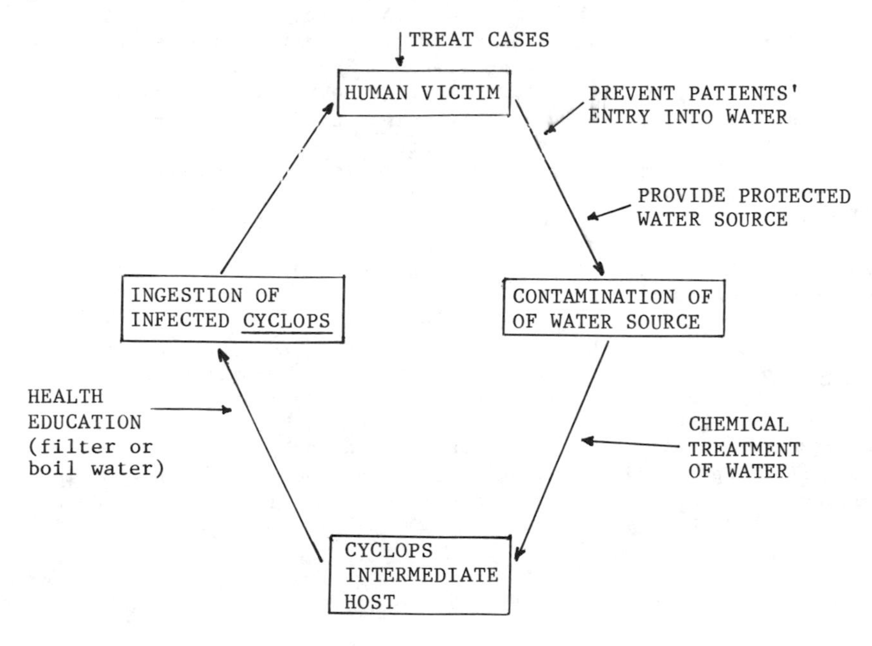

Figure 1. Control Measures Against Dracunculiasis at Different Points of Intervention.

The ancient practice of slowly winding the emerging worm around a stick is still practiced widely. It is important not to break the worm, since this may cause a severe systemic reaction. Aspirin can help relieve the local pain. Use of antibiotics can help prevent secondary bacterial infections, and dressing of wounds may help prevent transmission by keeping those with emergent worms from immersing their limbs in water. Where available, tetanus toxoid should be given to prevent a lethal secondary infection. When the worm is visible beneath the skin before it emerges, it may be removed through a superficial incision. Once the worm begins to emerge, an immunologic reaction apparently creates much more resistance to its removal.

Providing affected villages with reliable sources of drinking water is the most suitable long-term measure. While it is relatively expensive, it also provides other major benefits besides reducing the incidence of dracunculiasis, such as reduction in the time required to gather water and reduction in transmission of diarrheal diseases. In the 1960s piped water provided to the Nigerian village of Igbo-Ora with a population of about 30,000 reduced the incidence of dracunculiasis from 60% to 0% within 2 years. Other less well documented examples abound.

Behavior modification is another important effective means of attack. A key change in behavior that is desired in villages in which transmission occurs is for persons with blisters or emerging worms not to enter a pond or stepwell used as a drinking water source. Another important alteration of behavior is to convince villagers to filter or boil their drinking water. Fuel for boiling water is scarce in many areas, but infected *Cyclops* may be removed by filtering the water through a piece of cloth. In Burkina Faso an inexpensive, easily cleaned, and durable filter made from monofilament nylon material suitable for rapid filtering has been developed recently. Akpovi and his colleagues in Nigeria demostrated the ability of health education to drastically reduce transmission in endemic villages.

Temephos (Abate ™) is the chemical of choice for control of *Cyclops* spp. in ponds or wells used for drawing drinking water. Applied at the recommended concentration of 1 mg/l at 4-6 week intervals during the transmission season, this insecticide is safe for human consumption, and inocuous to fish and plants in the water. A group of Indian investigators achieved a 97% reduction in incidence of dracunculiasis within one year of using this method in a village of 3700 persons.

The Guinea-worm eradication initiative

With the advent of the Decade increasing attention and action have been directed against dracunculiasis. Key milestones in this regard include the following:

1. The convening of an International Workshop on Opportunities to Control Dracunculiasis in Washington, D.C., in 1982.

2. The designation of the Centers for Disease Control as the WHO Collaborating Center for Research, Training and Control of Dracunculiasis in 1984.

3. The convening of the first National Conference on Dracunculiasis in Nigeria in 1985.

4. The adoption of a resolution for Dracunculiasis Elimination by the thirty-ninth World Health Assembly in May 1986.

5. The convening of the the first African Regional Workshop on Drancunculiasis in Niamey, Niger in May 1986.

6. The agreements reached in late 1986 between the Governments of Pakistan and Ghana and Global 2000 Inc. of the Carter Presidential Center, Atlanta, Georgia, in which the latter agreed to help those two countries develop and implement national eradication campaigns.

The Strategy for Eradication

The strategy is based on the development of country-specific national plans of action to:

(1) Improve surveillance in each country to map the distribution of dracunculiasis, and monitor its occurrance over time.

(2) Provide safe (at least Cyclops-free) drinking water sources to endemic communities through construction of new or improvement of existing wells, and chemical control of Cyclops.

(3) Provide community education regarding the life cycle of the parasite and personal preventive measures such as the use of filters to remove Cyclops from drinking water.

(4) Promote intersectoral collaboration between private and public agencies involved with public health, agriculture, and water supplies and sanitation.

Status of anti-dracunculiasis activities in endemic countries:

Asia:

Prior to the begining of the IDWSSD, dracunculiasis had been deliberately eliminated from the Southern USSR in the 1930s, from Iran in the 1970s, and had disappeared in the wake of improved living conditions in many middle-eastern countries.

India began a National Guinea Worm Eradication Program in 1980 using active surveillance, water supply, health education, and vector control. By 1984 these efforts had reduced the total number of cases occurring by 10%, from 44,189 to 40,443. Between 1984 and 1985 the total number of cases had declined another 25% to 30,134. Over the same two years, the number of affected villages in India was reduced from 11,332 to 7,600. In 1985, Tamil Nadu State in southern India declared dracunculiasis eradicated.

Pakistan initiated a national eradication plan in 1987. A national survey is currently underway and a national meeting is anticipated in July. Global 2000, Inc. of the Carter Presidential Center is providing assistance.

Africa:

Benin, Burkina Faso, Cameroon, Ghana, Niger, and Togo have developed national plans of action and Guinea-worm control activities are underway. Although Ethiopia, Mali, Mauritania, Nigeria, Senegal, Sudan and Uganda do not yet have national plans, these countries are conducting surveys to assess the problem or are carrying out control projects in conjunction with existing water supply programs. In Nigeria for example, an aggressive elimination program was inaugurated in December, 1986 in Anambra State. This program is being funded by the Federal Government, UNICEF, and Japan. In Guinea the disease may have been eliminated as a result of provision of improved water supplies to populations in the endemic area. However, this outcome has not yet been confirmed.

A small foci of infection exists in Kenya near its border with Uganda and Sudan. Control in this remote area may have to be carried out in conjunction with the activation of programs in the surrounding countries. The extent of the Guinea-worm problem in Chad, Central African Republic and Gambia is uncertain. Ascertainment of the situation is being pursued. Lastly, an aggressive rural water supply program in Ivory Coast over the past two decades has dramatically reduced Guinea-worm infection. In 1966, the annual total of reported cases was 67,123; in 1976 the total was 4,971; and by 1985 it was 592.

As illustrated above, the principal thrust of control efforts is the provision of new or improved sources of safe (at least Cyclops-free) drinking water to populations affected by Guinea-worm. Therefore, in endemic countries collaboration between national and international agencies administering water supply programs, and Health authorities is essential to the process of giving endemic populations priority for safe water supplies. Through the years the dracunculiasis eradication effort has assumed steadily growing importance as potentially the single greatest legacy of the Water Decade. Receiving originally little or no recognition outside affected countries, the eradication effort has become a central part of the Decade.

REFERENCES

Akpovi, S.U., Johnson, D. C., and Brieger, W. R., (1981). Guinea worm control: testing the efficacy of health education in primary care. International Journal of Health Education, 24: 229-37.

Hopins, D. R. (1984). Eradication of Dracunculiasis. In: Water and Sanitation: Economic and Sociological Perspectives, Peter G. Bourne (Editor). Academic Press, New York, New York. pp. 93-114.

Kale, O. O., (1977). The clinico-epidemiological profile of guinea-worm in the Ibadan District of Nigeria. Am. J. Trop. Med. Hyg., 26 (2): 208-214.

Kale, O. O., (1983). Epidemiology of dracunculiasis in Nigeria. Opportunities for the control of dracunculiasis: Report of a Workshop, June 16-19, 1982. Washington, D. C.: National Academy Press. pp. 33-48.

Muller, R., (1971). Dracunculus and Dracunculiasis. Advances in Parasitology, 9: 73-151.

Nwosu, A.B.C., Infezulike, E. O., and Anya, A. O., (1982). Endemic dracunculiasis in Anambra State of Nigeria: geographical distribution, clinical features, epidemiology and socioeconomic impact of the disease. Annals of Trop. Med. and Parasit., 76: 187-200.

AQUALIFE - An innovative biological wastewater treatment technology for developing nations.

by N. Tufts and J. Stog

The paper describes the design and operation of an innovative biological wastewater treatment system, which is a self-contained, transportable, floating unit. With one moving part. It has the capability of treating the effluent from a typical N. African village of pop. 10,000. The system will be evaluated in 1987 in three Egyptian rural communities.

Aqualife is a biological wastewater treatment system develc by the late Prof. F. Evens, who was director of the Departr of Ecology at the University of Antwerp, Belgium. Initially successfully tested in 1974 in a brook-pond system in the munici park of Boom, an industrial town of 3500 located South of Antwe rights were later acquired by Stog in Waltrop, W. Germany, and the system engineered for worldwide marketing.

Since then, Aqualife has been successfully used to revive a "dead" lake, Bodenlos in Budapest, to treat effluent form a Sauerkraut factory and a sugar beet process plant in Germany before discharge to surface water, and to treat the biological wastewater from a village discharging into a canal leading to the River Schelde in Belgium

In 1987 four American manufactured Aqualife systems will installed to treat sewage collected from three villages, distance from Cairo, Egypt and in the Nile Delta. This techno is being evaluated for rural sewage treatment application Egypt by the Ministry of Local Government along with several o systems and processes under a US AID-financed project.

Aqualife may be described in sanitary engineering terms as an aeration system connected to a fixed-film biological reactor, fostering extended aeration and biological treatment in a natural recirculation zone. This introductory paper will emphasize how Aqualife differs in five major respects from many other systems which also rely on aeration to assist the biological purification process.

* Nathan Tufts is President of Agresource,Inc 101 River Rd. Merrimac, Ma. USA 01860 and a partner of Synergy International. Jochen Stog is President of Stog-Tec; 13 Dorfmuller Str.; 4355 Waltrop, W. Germany.

Aqualife is one of the few waste water purification systems conceived and designed by a biologist, one in which conditions for micro - and macro - biological growth and "in situ" treatment are carefully maximized. It may be particularly appropriate and cost effective to consider its use where a very high value is placed on agriculturally productive lands.

Designers have kept in mind simplicity, reliability, and low installation and operating costs. There is only one moving part, the electric motor.

Five unique features to the system:

- Aqualife operates "in situ", floating or fixed 90% submerged in the wastewater or polluted water body. The rigid pipe structure acts as flotation, and as the lower half is flooded as ballast, to maintain its position relative to the surface. In a shallow ditch, basin or stream, it may also sit on the stream or lagoon bottom, supported by its own structure.

- Aqualife pumps the polluted water, through a venturi to entrain and thoroughly mix oxygen with the water volume entering the system from down stream.

- The aerated water is forced through the system, containing all its biological contact media, and counter to the flow of polluted water. Good diffusion is achieved, and a very high efficiency transfer of oxygen into water per unit of electrical energy expended.

- Surface or biological contact area can be adjusted by adding or removing plates and plastic media.

- After a build up of biological growth to a certain thickness on the media plates, the organisms release themselves from the bio-contact area, and are ejected out of the unit, where they are carried downstream by the flow; thus, a zone of recovery is created between the unit and the intake pump site, a stretch of or more feet. It is the intent to establish a viable and permanent aerobic ecosystem in teh lagoon, canal, or sewage pond which the Aqualife maintains.

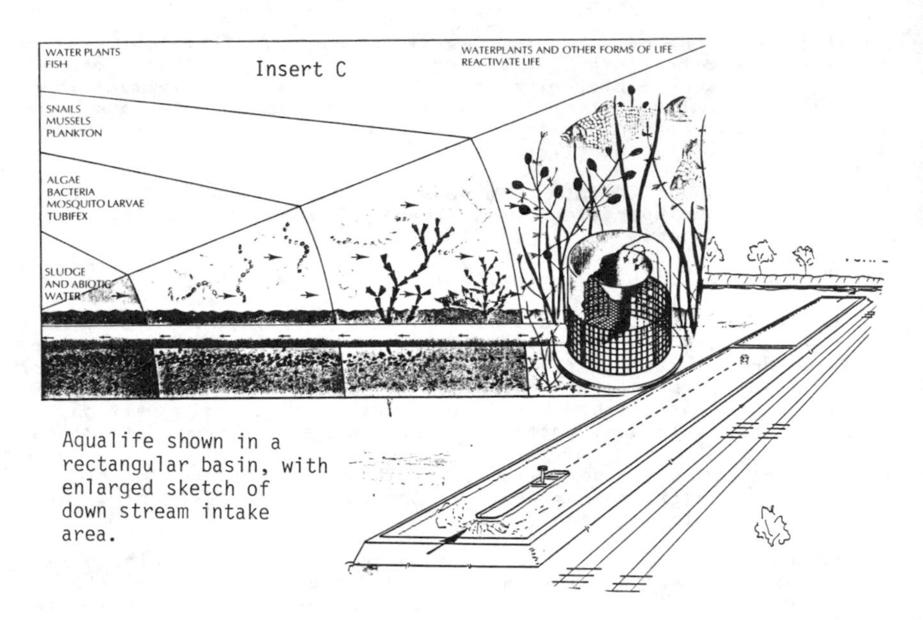

Aqualife shown in a rectangular basin, with enlarged sketch of down stream intake area.

High efficiency in the oxygen transfer aspect is achieved by this design, special contact or plates, very small initial bubble size and inherent efficiency of the venturi.

The whole system is designed to act as a primary growth and breeding area for a vast variety of micro-organisms living in the water. After growth on the interior surface is built up to a maximum, excess micro-organisms in great numbers are gently flushed out of the up-stream end of the unit, and carried downstream toward the suction head where partially treated polluted water is recirculated. On average, this recirculation factor is 10-11 times before discharge in high strength waste treatment applications. In a polluted lake, or larger water body, it may be much less, while achieving good results over longer time periods.

It is important to note that the region between the Aqualife unit, and the downstream in-take site is an area for active secondary treatment. It was the designers intent and belief that this exterior "natural" area between suction and primary treatment would be zone of regeneration. Low over-all costs depend partly on maximizing treatment conditions in the artificial lagoon, as in Egypt's case, where water depth and shallow basin perimeters are chosen to increase treatment.

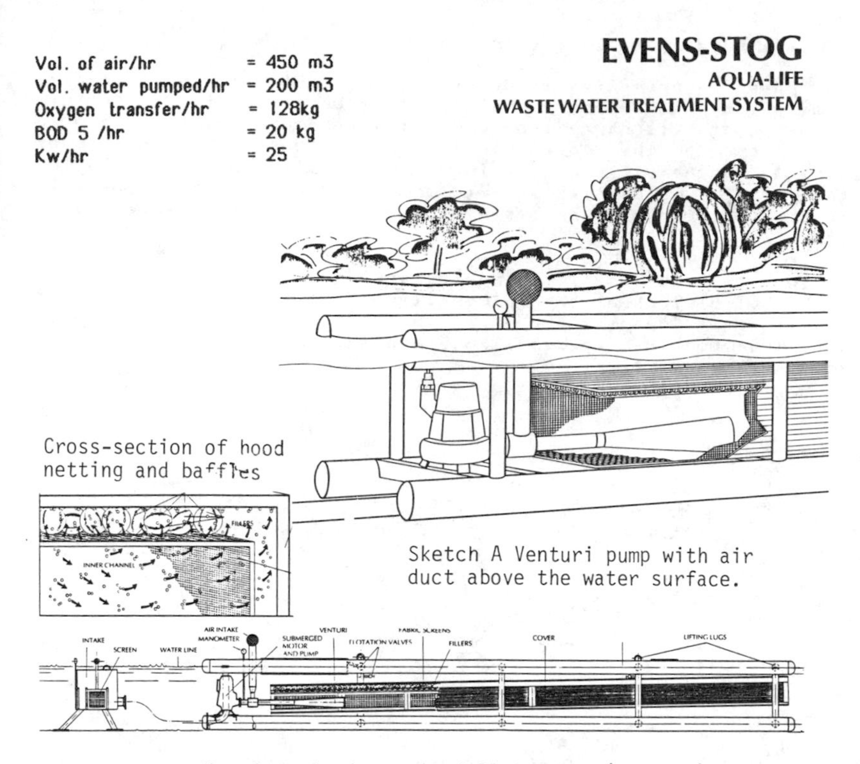

Sketch A Venturi pump with air duct above the water surface.

Sketch B showing unit full length -(42 feet)

As shown by the <u>sketch A</u> the first third of the Aqualife system comprises a venturi air entrainment section supported in a rigid tubular frame which also provides buoyancy. The intense mix of oxygen and water from the jetstream is driven forward into the fixed film biological reactor section some 10 meters in length. Contact time is maximized by the combined inner and impermeable outer channels, or hood over the unit, which lies just below the water surface, inclined slightly upward. The high point is about 200 mm below the water surface at the upstream end of the system. A combination of special nylon netting, vertical plates, and expanded-surface plastic media in voids, give the desired amount of surface for microbial growth. Fine bubbles are prevented from rising to the surface. <u>SKETCH B</u>.

Data from a Belgian Installation

In this paper we present some details on an installation in Bornem, Belgium which has been operating since December, 1985. In this case, the sewage discharge from 500 families found its way into a canal system on the edge of the River Schelde. A number of restaurants in the village contributed seasonally to the pollution of the canal, which was sufficiently impacted by the sewage to create a considerable proble of odors and unacceptable water quality.

Monitoring of the systems performance was undertaken by the Department of Biology at the University of Antwerp, and some of the early data results are shown in Table 1. The period of data shown is November 1895 to June 1986, and the analysis is continuing through the present time.

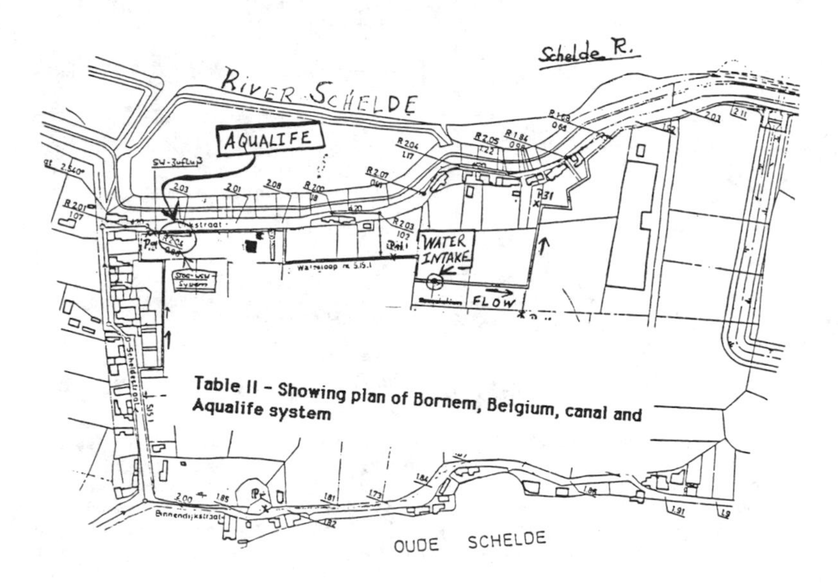

Table II - Showing plan of Bornem, Belgium, canal and Aqualife system

It may be said in general that operation of the Aqualife system has resulted in dramatic improvements in all enviromental/water quality parameters, and elimination of objectionable odors withing a short period of weeks.

Sewage treatment by the system, which was equipped with a smaller, electric motor (11 kw) due to the low initial BOD loading, was so satisfactory, that the village government plans to divert additional wastewater flows from new development and other houses into the canal following the successful experimental period. This is now an operational system meeting or exceeding its design objectives.

Installed in polluted canal receiving the biological waste from approx. 500 families.
Discharge to the River Schelde - Belgium

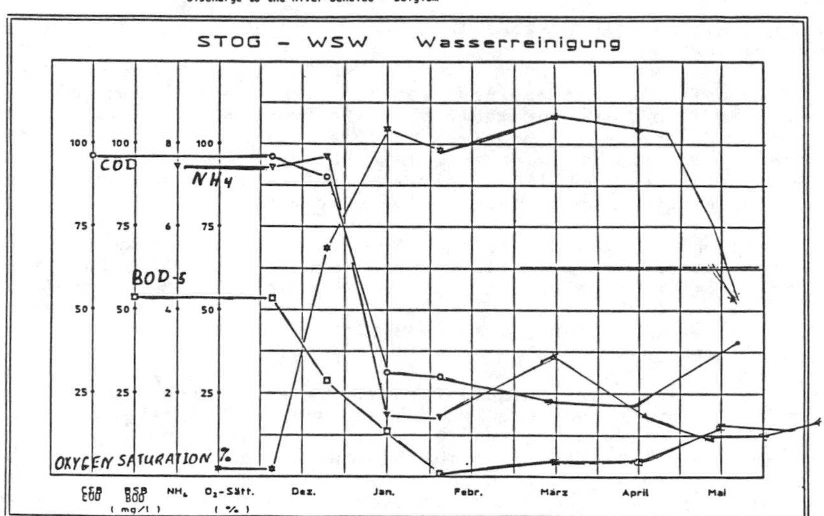

Monitoring of the systems performance was undertaken by the Department of Biology at the University of Antwerp, and some of the early data results are shown in Table 1. The period of data shown is November 1895 to June 1986, and the analysis is continuing through the present time.

BOD5, (at 5 locations), COD ammonia, and percent oxygen saturation were selected for periodic measurement at a number of points in the canal system.
TABLE II shows the village, River location and canal plan, with the location of the unit and its intake.

It may be said in general that operation of the Aqualife system has resulted in dramatic improvements in all enviromental/water quality parameters, and elimination of objectionable odors within a short period of weeks.

Sewage treatment by the system, which was equipped with a 50%-rated electric motor (11 kw) due to the low initial BOD loading, was so satisfactory, that the village govenment plans to divert additional wastewater flows from new development and other houses into the canal following the successful experimental period. This is now an operational system meeting or exceeding its design objectives.

Consideration for the developing world.

Some advantages to considering such a system in developing countries are as follows:

- The unit is standardized, easily assembled with a minumum of skilled labor, transportable over the road, and installed in existing orslightly modified polluted water bodies without the need for expensive or complex shore-based infrastructure. A deisel generator will supply alternative or emergency power to the electric motor which drives the venturi.

- It combines the efficient and continuous aeration of large water masses with a biological treatment area which protects and promotes the rapid growth of organisims. A yield of 3.5 to 5.0 Kg O2/Kwh is more than conventional aerators alone can provide, being made possible by the geometry of the underwater unit and good retention time.

- Low in capital cost, simple in design, and energy efficient, AQUALIFE needs no highly trained staff to operate. Required maintenance is very low as well. Site preparation costs vary from low to moderate where a simple retention basin must be dug out or enlarged. If down time is a key factor because of power failures, an electric generator set is provided, as in the case of the Egyptian village installations.

- The recirculated waste to be treated are pumped from a few hundred meters away, being downstream or at a predetermined depth away from the unit, which will promote circulation of large oxygen-rich water masses. The system will operate in winter, where turbulence of the water in the aerated ecosystem prevents or reduces the formation of ice.

- Sludge build-up is minimized by this system, which promotes aerobic biodegradation. Ideally, macro-organisma and higher life forms consume much of the material between the unit and its water intake.

- A typical treatment facility would involve 1/2 acre of land, a receiving basin or lagoon of 2800 to 3000 cubic meters volume, (occupying a space of about 40 x 50 meters) and a BOD 5 removal requirement of 500 Kg or 1100 lbs/day. In the Egyptian facilities, 95% removal will be achieved to the Public Law 48 standard of 60 mg/litre (BOD 5). The volumes are a typical in that basin or lagoon volumes were minimized, giving a 4-6 day average retention time.

Four installations for Damietta and Menofiya Governorates, Egypt

In the installations planned for Egypt, four systems are being specified for villages in Menufiya and Damietta. Two units will operate in a specially constructed lagoon for villages of 10,000 to 12,000 pop. respectively, and two units will operate for a third larger village of approximately 20,000 pop. * See illustration D. (Damietta)

The systems will be installed in cooperation with Chemonics International Inc, Cairo, and an Egyptian engineering firm which will do the site preparation work and be involved in monitoring performance following the anticipated operation in the Fall of 1987. Total cost of the four systems, including extensive site work and an access road, is $725,000. The project is being financed by US AID.

One factor in choosing what may be called intermediate technology, as opposed to natural lagoons or pond systems, is the very high value placed on agricultural land. Approximately 8-10 times as much land would be needed to provide comparable levels of treatment without any mechanical/biological treatment enhancement.

Conclusion

Aqualife has shown itself to be a promising and reliable biological water treatment technology in several situations in Europe, from treatment of food wastes to biological sewage from a European village.

In the USA one sees numerous applications in food waste effluent treatment, polluted ponds in urban settings, supplemental treatment in sewage lagoons or WWTP's not meeting their discharge standards. The unit is manufactured in Buffalo, New York.

The project in Egypt has the potential of proving cost effective operation in a developing rural village, with implications for treatment of gross pollution of urban canals, lakes and streams throughout the developing world. If good results are achieved, they will be a tribute to the vision of the late Prof. F. Evens of the Univ. of Antwerp, Belgium, who conceived of a simple and low cost system, Aqualife.

***

Illustration D (Damietta)

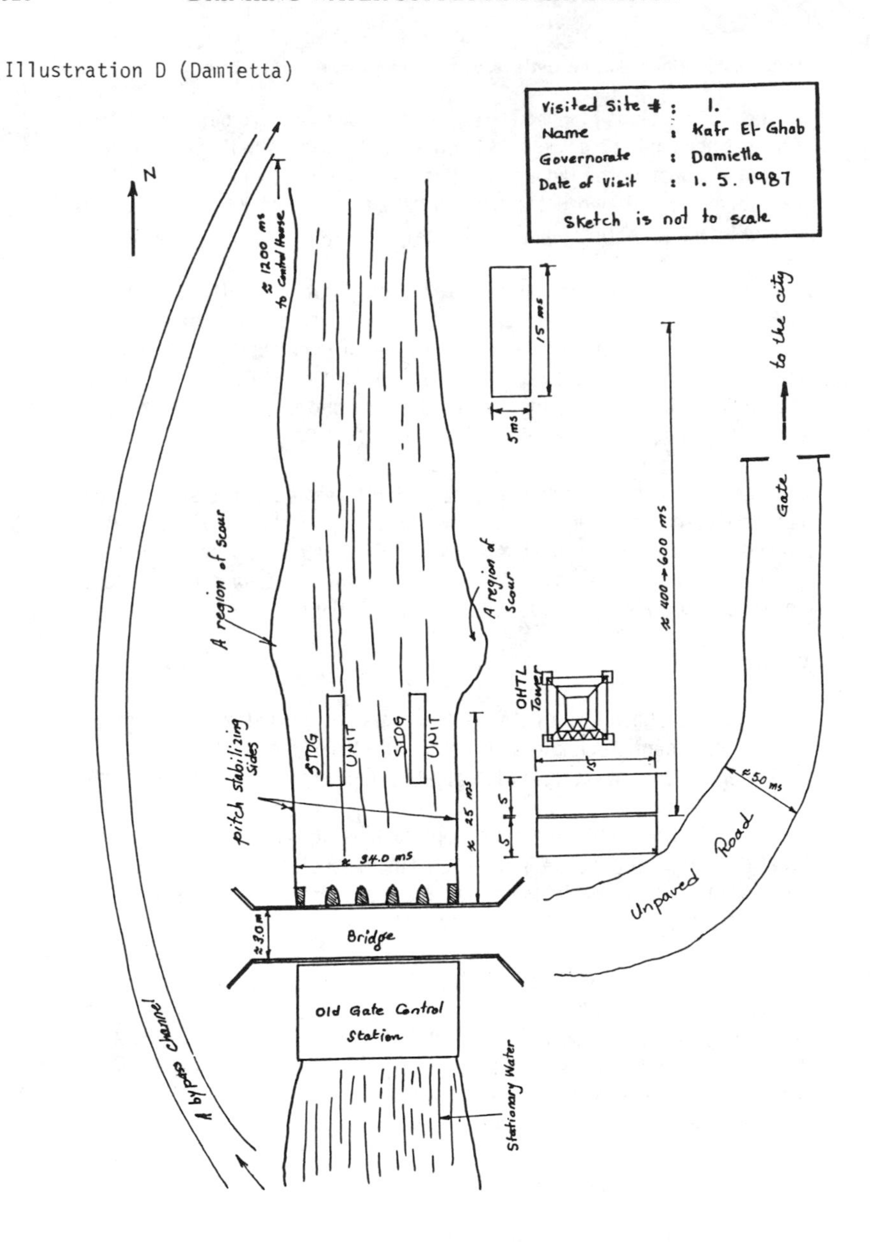

30,000 IMPROVED LATRINES IN MOZAMBIQUE
or
Should a latrine look like a house?

Bjorn Brandberg*

A COMMERCIAL SUCCESS

Local latrine building traditions, and difficulties in supply of building materials led to the development of a new improved latrine in Mozambique, which has become a commercial success.

During colonial times, little attention was paid as to how the "native" population lived. Construction of latrines in the urban areas developed in respect to a felt need of privacy, and not through interventions by authorities to secure public health.

* Sanitation Adviser, UNDP/World Bank, P.O. Box 30135, LILONGWE 3. Malawi (ex National Coordinator of the Mozambique Latrine Programme, INPF, Avenida Acordos de Lusaka 2115, Maputo, Mozambique).

A FENCED OFF CORNER OF THE PLOT

The latrines being built were genuine "privies"; i.e. a fenced off corner on the plot with a pit covered with poles, scrap material and soil. The conventional house type latrine designed and promoted by public health officers had difficulties to make its way in the spontaneously developing latrine building tradition.

THE FIRST NATIONAL LATRINE BUILDING CAMPAIGN

After independence the new government asked itself how they could improve living conditions in the country. This resulted in the First National Latrine Building Campaign. The whole country was stopped during one day (November 1, 1976) when everybody was told to build latrines. Those who already had a latrine or a water borne toilet should assist at a school, or a clinic or wherever a latrine was needed. Everybody should have a latrine.

Over one day the latrine was institutionalized in Mozambique. The vast majority of latrines built, however, were of the genuine privy type; the fenced off corner on the plot, with a pit covered with poles, scrap material and soil.

An example of a self-built urban latrine in Mozambique. It is a simple pit constructed by the user in a fenced of corner of the plot.

Though the latrine building campaign was a tremendous achievement, the Mozambican Government was of the opinion that improvements were required.

"PROJECTO LATRINAS"

In 1979 a multi-institutional team coordinated from the National Direcorate of Housing (today INPF, the National Institute for Physical Planning) dedicated itself to the task of defining appropriate technology for improved pit latrines for poor urban areas: "Projecto Latrinas" (the Latrine Project).

Compost latrines failed; The people did not like the idea of emptying them. Ventilated (VIP) latrines were found inappropriate; there was no material for the ventpipe, nor for the roof, and people did not use houses for defecation.

RETHINKING

We had to start from the beginning. The problem was hygiene, child safety and structural stability. Smell was not a problem in the unroofed latrines and fly breeding was not serious (at least not in Maputo).

What we, in the first place, needed was a hygienic and safe latrine slab suitable for mass production. After having looked into the possibilities and constrains of various materials and designs we opted for a locally produced round dome shaped latrine slab of nonreinforced concrete, with sloping smooth surfaces, elevated footrests, a key-hole shaped drophole and a tight fitting lid. Hygienic and safe.

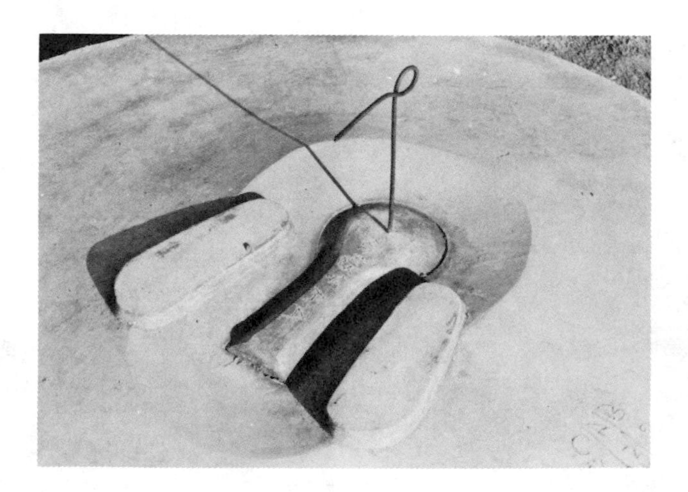

The drop hole has a lid that is cast in situ. It gives a fly-proof fit between lid and slab.

Material requirements were less than a bag of cement, some sand and broken stone for the concrete plus a piece of reinforcement bar for the handle of the lid. This was enough to make a large slab (diameter 1.50 m), .

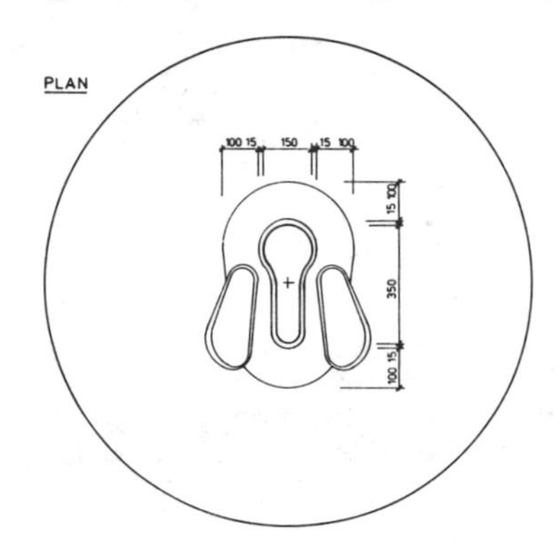

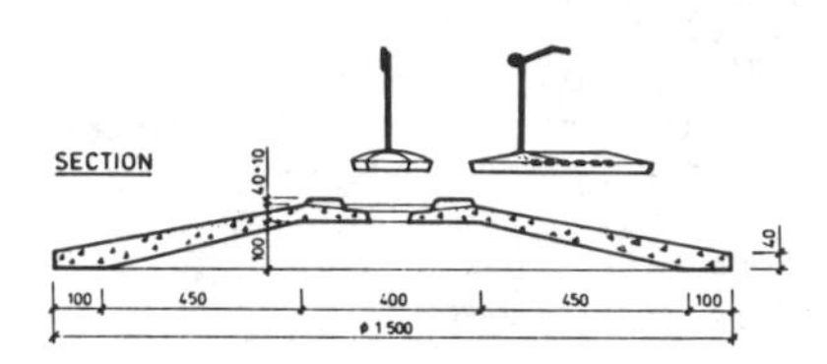

Engineering drawing of the Mozambican non-reinforced latrine slab. Dimensions in millimetres.

For unstable soil conditions the round pit was lined with sand-cement blocks, and where it was difficult to dig due to high ground water or rocky soil the pit lining was extended above the ground level in fully mortared masonry over the ground level.

SELF HELP BUILDING

In the majority of cases a family could simply buy a slab from one of the cooperatives (for the equivalent of USD 10.00, which included full recovery of costs, overheads and profits). They simply dug a pit and put the slab on. The fence was built of any material, normally reed, and so the latrine problem was solved.

The demand for slabs exceeds the production. We have people queuing when we start selling each week's production

INCREASING DEMAND

Having gone through a phase of initial difficulties, the new latrines became a commercial success. The project workshop in Maxaquene where the new latrines were produced and test sold, had difficulties to cope with the increasing demand. Private latrine producing building-cooperatives were formed with the assistance of local political leaders. A planned promotional campaign had to be stopped as the cooperatives could not meet the demand.

30,000 latrine slabs have been sold for self-help construction of improved latrines in Mozambique.

The constraint in the continued expansion of the programme was material and transport, and to some extent qualified field coordinators.

BUILDING-COOPERATIVES

The casting of the improved slab, construction of eventual lined pits and elevated pits as well as production of the sand-cement blocks for lining was normally the task of the specialized labour of the building cooperatives.

Material distribution, transport and supervision of bookkeeping is today handled by a cooperative union (Intercooperativa) owned and managed by the prosperous latrine building cooperatives.

EXPANSION

With the assistance of several international organisations, an increasing number of families had their improved latrines built.

The programme which started in the capital, Maputo, a city of over one million inhabitants, has now expanded to six of the ten major urban centres (with another two cities in the pipeline) and is successively moving out to smaller urban centres and the rural areas.

The slabs are produced in the areas where they are being installed.

In spite of war, food shortages, material supply problems, transport problems etc the Mozambican people are building improved latrines. 30,000 families have today built their improved latrines. Many more, however, are on the waiting list.

REASONS FOR SUCCESS

Mozambican latrines have no roofs and no ventpipes. But surprisingly few flies and no smell. They are happily accepted by the people, the are built in large numbers and they do protect peoples health.

"Projecto Latrinas" (the Latrine Project), once a small research project grew to a successful national programme. What was the reason for the success? Can it be repeated in other countries? These questions are difficult to answer, but:-

(1) The Mozambiquan Government (FRELIMO) had since the liberation war promoted the construction of latrines, and wanted to see a simple but effective latrine building programme being implemented.

(2) The appropriate technology is actually low cost, and can successfully compete with traditional technologies.

(3) The new latrines did not differ very much from the type of latrine people were used to see. Improvements were understandable to the people; the durability of the concrete slab and the lined pits, and the child safety of the key-hole shaped drophole. (What people often did not understand was the hygienic improvements, but they got the health benefits built in and installed together with the safety and durability of the slab).

(4) We had managed to solve a felt need. The low cost approach, with many beneficiaries and the simplicity of the technology, was attractive not only to the people but also to the government and to donor institutions.

(5) Finally, purchase power in Mozambique is fairly strong as there is a serious lack of consumer goods on the market. There is obviously a will and some capacity to invest.

TECHNICAL DATA ON THE MOZAMBIQUE LATRINE SLAB

Diameter:	1.2-1.5m
Thickness:	40mm
Total height:	100-150mm*
Weight:	100-200Kg*
Concrete mixture:	1+2+1.5 (volume parts of cement, clean river sand and 1/4 inch coarse stone.
Reinforcement:	None
Surface:	Concrete smoothed with a steel float
Strength:	Each slab is test-loaded with the weight of 6 persons after 7 days minimum of curing time.
Labour:	0.8 mandays/slab at continuous production
Price:	1.5m diameter 2,000 Metical (US $ 10.00) 1.2m diameter 1,500 Metical (us $ 7.50)

* depend on diameter

WOULD YOU LIKE TO KNOW MORE?

A complete Project Manual (in Portugues) including drawings and photos, training material and a slide programme (in English or Portugues) are available from INPF (the National Institute for Physical Planning), Avenida Acordos de Lusaka 2115, Maputo, Mozambique for the price of USD 15.00 and 250.00 respectively. By mailing a cheque chargeable to one of the major banks or an international money order, the material will arrive by air mail.

ACKNOWLEDGMENTS

Many people and organisations have contributed to the success of the Mozambique Latrine Programme. The Minister of Health, Mr Helder Martis was the strong man behind the First National Latrine Building Campaign. Jose Forjaz, the Secretary of State for Phyiscal Planning (ex National Director of Housing) and the Director Joao Nhamposse gave us all the institutional support we needed. Sandy Caincross's knowledge, analytical and critical thinking made us question the solutions found in the textbooks. Don Kossik and John Piper developed and implemented the cooperative model. Carlos Muge, Paula Brandberg and Alexandre Junior made the Slide Programme and assisted in the preparation and printing of the Manual, not to mention Paulo Oscar Monteiro, Mike Muller and Carlos Macave, the people at the Ministry of Health and all the field staff who actually implemented the programme.

I also want to thank Swedish SIDA, UNDP/HABITAT and the Mozambican Government who funded my 6 years assignment in Mozambique, IDRC who funded the research projects, UNDP/WHO, CUSU/SUCO - Canada, SAH-Switzerland, the Dutch Government and private people in Trollhattan Sweden, who all have contributed to the programme in different ways.

Potential of Biogas in Developing Countries

Timothy G. Ellis, AM. ASCE*

ABSTRACT

The potential of biogas in developing countries depends on research and development and further proof of its profitability. Of all the renewable energy technologies, biogas is the most established, with China leading the field, having built seven million digesters. Biogas digestion has the potential not only to recycle organic fertilizers and provide sanitation, but also to provide a renewable and clean energy source. As the organic matter in the digester is fermented to the state of a soil topping, it gives off methane gas, which is collected and used for cooking and lighting. Digesters retain the excreta and other wastes long enough to ensure that most of the pathogens are destroyed, thus breaking the cycle of infection. Despite social benefits from improved health and a new fuel source, biogas technology has to be financially profitable before it is widely used in developing countries. The cost of a family size unit ranges from $100 to $400, which in some countries exceeds annual family income. With the help of economic incentives and biogas education from the governments of developing countries, economic and sociological hurdles to biogas implementation can be overcome. Once the fundamentals of biogas technology are better understood it has the potential for improving the living conditions for a large segment of the world's population.

INTRODUCTION

The process of digesting organic matter in an anaerobic environment produces a flammable gas which is commonly referred to as "biogas" in developing countries. The potential of biogas as a renewable energy source has been recognized as early as the end of the nineteenth century in England where the gas was used for street lighting (NAS, 1977). In recent years, biogas technology has received renewed interest, especially from developing countries, most of which depend on imported fossil fuels and domestic firewood, and consequently, have severe deforestation and financial problems.

The breakdown of organic compounds in anaerobic digesters produces, in addition to biogas, an effluent slurry with significant fertilizer value. The application of digester effluent to overcultivated farmland can enrich deficient soils with nutrients. Furthermore, anaerobic digestion can destroy most of the pathogenic organisms which are the major cause of enteric diseases in developing countries.

* Engineer in Operations, Back River Wastewater Treatment Plant, 8201 Eastern Blvd., Baltimore, Maryland 21224, U.S.A.

In spite of the advantages outlined above, a number of economic and social drawbacks associated with biogas technology have hampered its dissemination in developing countries. Additionally, a large number of digesters installed in various countries are now inoperative or operate inefficiently due to faulty design and construction. This paper presents a brief review of the technical aspects of digester design and operation, a discussion on the economic and socio-political aspects of biogas technology, and a look at overcoming the hurdles of implementing this promising technology in developing countries.

TECHNICAL CONSIDERATIONS

Climate, type of feedstock, economic conditions, gas utilization, and cultural implications are all important considerations in choosing a proper biogas design for a family or community. The simplest and least expensive design is a batch digester. This unit, usually constructed with brick and mortar, is filled with organic matter and water, covered, and allowed to sit for three to twelve months. Gas production usually commences in about three weeks. In order to maintain a continuous gas supply while the digester is being emptied and filled, several units should be operated simultaneously. The relatively inexpensive cost of the batch digester is offset by the additional labor required for emptying and filling the unit. Semi-continuous digesters, however, are either connected directly to latrines and animal pens, or fed periodically with human excreta (nightsoil) and animal dung.

Chinese designed, fixed dome units are the most common digesters in developing countries. Agricultural residues are usually loaded twice a year, and dung and nightsoil are fed every day. The gas produced is contained within the fixed dome as it displaces liquid in the digester. The pressure that develops in this type of digester (up to 1.4 psi) has been blamed for the high percentage of inoperative digesters due to gas leaks (Stuckey, 1983). This has led to the development of other designs and gas storage facilities.

Most of the biogas digesters in India are the floating-cover design. These units store the gas in a cylindrical metal cover that traps the gas and forms a water seal with the digester slurry. The corrosion of the metal covers by hydrogen sulphide has led to the development of fiberglass and plastic covers.

The Taiwanese bag digester sits half-below ground and is made of PVC, nylon, or red mud plastic (a material made from aluminum refinery wastes). Production rates in these units can be quite high due to their plug-flow characteristics and their ability to trap incident solar radiation which can increase the digester temperature by as much as 3°C over temperatures observed in other designs (Stuckey, 1983).

Floating-cover and fixed-dome digesters can provide enough gas pressure to use the gas directly for cooking and lighting purposes. Digesters that use balloons or plastic bags for storage cannot be used for these purposes without repressurizing the gas. They are practical, however, for use with internal combustion engines which draw in the gas by the action of the pistons.

One design (see Figure 1) that is currently being developed uses the advantages of the bag digester's plug flow characteristics and flexible membrane cover and also has internal baffles. The baffles create sludge blankets that help to retain the methane producing bacteria in the digester for longer periods of time, thus the volume and cost are reduced. This design is modeled after McCarty and Bachmann's anaerobic baffled reactor, ABR, and is especially promising for the Third World due to its anticipated capture of pathogens in the sludge blanket layer. One such unit is being planned for a demonstration project at the Heifer Project International's Learning and Livestock Center in Perryville, Arkansas.

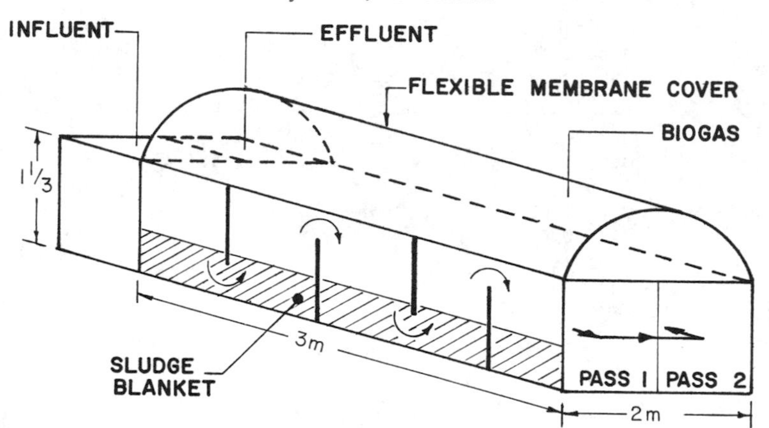

Figure 1. Plug Flow Baffled Reactor with Flexible Membrane Cover. Dimensions shown are for an $8m^3$ family size unit.

ENERGY ASPECTS

In most developing countries there is a need for non-commercial renewable energy sources. Traditionally, firewood, charcoal, and cow patties are used for cooking. Approximately one-third of the world's population depends on wood for fuel. This dependence puts a strain on the world's remaining forests, 40 percent of which are estimated by one source to be completely destroyed by the year 2000 (Polsgrove, 1981). Furthermore, alternative energy sources must be investigated and adopted to alleviate the financial burden of importing fossil fuels.

Biogas, which is 50 to 75 percent methane, has the potential for providing a renewable and clean energy source for developing countries. The calorific value of biogas is about 23,000 KJ/m^3 or $1m^3$ of biogas is equal to about 0.65L of gasoline or 0.57L of diesel fuel. In most applications in developing countries, however, the gas is used for cooking and lighting.

The biogas production in a digester is dependant upon such factors as temperature, retention time, mixing, and characteristics of the influent material. Table I shows the approximate gas yields from various waste material for certain temperatures and retention times (NAS, 1977). Other environmental conditions such as pH, alkalinity, availability of macro and micro nutrients, and absence of toxicity are recognized as the most important process parameters which affect the methane producing bacteria, but are difficult to monitor in a Third World situation. In general, high efficiency of digestion and biogas production can be acheived by maintaining the temperature near 35°C. For domestic units, insulation and construction of a greenhouse over the biogas unit are two simple methods that can be adopted for temperature control. To maintain high efficiency at lower temperatures, the retention time must be increased from the usual 30 days to 50 days or longer.

Table I. Yield of Biogas From Various Waste Materials (NAS, 1977).

Raw Material	m^3 of biogas per kg of total dry solids	Temperature °C	CH_4 content %	Fermentation time (days)
Cattle manure	0.20-0.50	11-31	-	-
Beef manure	0.86	35	58	10
Poultry manure	0.26-0.30	33	58	10-15
Swine manure	0.49-1.02	33-35	58-68	10-20
Sheep manure	0.37-0.61	-	64	20
Nightsoil	0.38	20-26	-	21
Forage leaves	0.5	-	-	29
Sugar beet leaves	0.5	-	-	14
Algae	0.32	45-50	-	11-20

AGRICULTURAL ASPECTS

The loss of topsoil and over-cultivation of agricultural land is a severe problem in many of the developing countries. As a country becomes overpopulated, the available farmland is worked harder to meet economic and nutritional needs. In Haiti, for instance, the ancient practice of burning brush in an area to be planted is common among the rural population. The cleared area is planted for several seasons, and then the farmer moves to another patch of land and repeats the process. This causes severe loss of nutrients and topsoil since the humic matter is burned and nothing is left to replenish the soil. This practice of shifting cultivation is practiced on an estimated 30% of the world's tillable soils (Polsgrove, 1981).

The slurry discharged from a digester can be applied to agricultural land to restore deficient soils with nutrients. The slurry contains inorganic nutrients (nitrogen, phosphorus, potassium, and various trace minerals) and degradeable and refractory organic matter. It stimulates the growth of microorganisms which contribute to the solubility and thus, availability to higher plants, of essential nutrients contained in soil minerals (NAS, 1977).

During the breakdown of organic compounds in digesters, nitrogen is conserved in the slurry (except for the minute fraction that exits in the digester gas). The concentration of ammonia nitrogen, however, increases by up to 40% over undigested material (McGarry, 1978). Therefore, in applying the biogas slurry to farmland, care should be taken to insure that the slurry is worked into the soil to minimize the loss of ammonia nitrogen through volatilization.

A commune in Sindu County in the Sichuan Province experienced increased productions by 25, 16, and 15 percent in vegetable, wheat, and rice crops, respectively, after using biogas slurry as a soil conditioner and fertilizer. In other countries 28, 10, 25, and 13 percent increases were reported for crop yields of maize, rice, cotton, and wheat, respectively (van Buren, 1979). These increases in productivity are attributed to the slurry's soil conditioning properties as well as the approximately ½ - 1½ percent by dry weight of each of the three major nutrients: nitrogen, phosphorus, and potassium.

Biogas slurry has also been used as a food supplement for pigs and fish in 5 to 20 percent mixtures (Stuckey, 1982). At Maya Farms in the Philippines the slurry is mixed with regular feed at a one to ten ratio. The digested material from the biogas tanks is reportedly high in vitamin B_{12} which is an important nutrient for young animals and is expensive if purchased as a food supplement (Maramba et al.).

PUBLIC HEALTH ASPECTS

Approximately 80 percent of the rural population in developing countries lack access to clean drinking water and adequate sanitation facilities (Dale, 1979). As a result, enteric diseases are the leading cause of death and disability in most developing countries. Diarrhea alone kills six million children every year (UNESCO, 1981). Helminths, or worms, infect nearly half the Third World population and can kill or cause malnourishment. The Helminth Ascaris, commonly known as the roundworm, infects 60 to 90 percent of young children in developing countries. A single gram of feces from an infected person can contain up to 300,000 eggs.

The removal of pathogens in biogas digesters breaks the cycle of infection by eliminating the transmission of disease through water pollution. The major factors which influence the rate of die-off of pathogens during digestion are the temperature in the digester, the residence time, the pH, the ammonia concentration, and most importantly, the lack of oxygen. Table II shows the die-off of certain pathogens in biogas digesters at specific temperatures and retention times.

Table II. Die-off of Enteric Microorganisms of Public Health Significance During Anaerobic Digestion (NAS, 1977).

Organism	Temp. (°C)	Residence Time (days)	Die-off %
Poliovirus	35	2	98.5
Salmonella ssp.	22-37	6-20	82-96
Salmonella typhosa	22-37	6	99
Mysobacterium tuberculosis	30	not reported	100
Ascaris	29	15	90
Parasite cysts (exc. Ascaris)	30	10	100

In biogas digesters, Ascaris eggs are typically removed by 90 percent. Chinese data suggest 93 to 98 percent removal for general parasite eggs (McGarry, 1978, Feachem et al., 1983). Many of these eggs settle to the bottom and remain viable in the sludge layer for long periods of time. The settling characteristics are so critical that the addition of a drop board at the outlet of a biogas unit in China, for instance, increased the removal of parasite eggs from 80 to 98 percent (IDRC, 1976). When the digester is periodically cleaned (usually annually), the sludge must be stored for a period of six months to one year (Feachem et al., 1983). The die-off of Ascaris will continue during storage, as well as during the application of the sludge to farmland. Once applied to the field, the eggs may live from a few hours to several months depending on environmental and soil conditions.

Aside from settling, effective removal of Ascaris eggs can only be achieved through heating. The temperatures in biogas digesters are usually 34°C or lower and are insufficient to heat-kill the eggs. Thermophillic aerobic composting, however, often achieves temperatures up to 55°C, which if maintained for several hours is effective in complete pathogen elimination (Feachem et al., 1983). One hundred percent removal of parasite eggs was achieved when night soil was heated to 60°C using waste heat from engines fueled with biogas at a power station in Foshan, China (Ru-Chen et al., 1978).

The success of biogas implementation at improving health conditions has been recorded where digesters were used to treat human and animal wastes. Two years after the introduction of biogas digesters to five communes in China, investigations by the Sichuan Institute of Parasite Disease Prevention showed that the percentage of people infected with hookworm disease decreased from 64 to 5 percent (Chen). The average number of hookworn eggs per gram of human feces decreased from 500 to 50. In another commune in China the incidence of dysentery decreased from 15 cases per year to none after biogas technology was implemented (Stuckey, 1982). There were also fewer flies and mosquitoes due to the enclosed treatment and safe disposal of the dung.

The potential of biogas technology for providing sanitation alone merits its use in developing countries. As stated in a report by the National Academy of Sciences in Washington, D.C. (1977), "There is no other practical method of treating human excreta - whether for disposal or to return nutrients to the land as fertilizer that will reduce the burden of pathogenic organisms as much as anaerobic digestion."

APPLICATIONS

The primary domestic use of biogas, which can be obtained from either individual family size units (about 8m^3) or community size plants (about 40m^3), is for cooking and lighting. For a family of five or six people, it has been estimated that an 8m^3 digester producing 2m^3 biogas can be met from the dung (about 50kg) of four or five medium sized penned animals. The addition of human excreta into the digester will add approximately 0.025m^3 of biogas per person per day (Lichtman, 1983). Increased gas yields can be achieved by adding supplemental material, such as kitchen scraps, water hyacinths, and sugar beet leaves. Water hyacinths, for example, can increase gas production by 0.014m^3 biogas for every kg (wet weight) added (Wolverton et al., 1975).

The use of biogas for lighting requires approximately 0.085 -

$0.17m^3$/hour, boiling one gallon of water requires about $0.28m^3$ (improving biogas stove efficiency, however, would reduce this quantity substantially), and a biogas refrigerator requires about $0.07m^3$/hour (Lichtman, 1983). These applications for the end use of biogas are especially applicable in areas lacking electricity. In India, for example, the cost of electrical hookups is prohibitive: only 14 % of the households overall and 5 % in rural areas have electricity. Biogas offers an affordable, decentralized, and renewable source of energy.

SOCIO - POLITICAL ASPECTS

The major advances in the implementation of biogas technology in developing countries have been due to strong encouragement and support from the governments. In China a reported seven million digesters were built under Mao in the 1970's (Ru-Chen et al., 1978). There are twenty-five biogas extension offices set up within the various provinces. There are 5000 people within these extensions who train biogas technicians. The 30,000 trained technicians in the country are split into 6,000 teams to provide experienced help in the construction and maintenance of biogas units (Stuckey, 1983).

India has experienced a similar push for biogas advancement. Low interest credit and subsidies have been made available and have resulted in the installation of between 6000 to 18,000 digesters per year since 1974 for a total of 108,000 biogas digesters (KVIC, 1984). Other developing countries have also made efforts to utilize biogas. South Korea has built approximately 29,000 units, most of which were installed by the government between 1969 and 1975 (Stuckey, 1983). There are about 340 units throughout the Philippines, the most impressive of which is located at Maya Farms where the waste from 46,000 hogs is treated in biogas units.

Biogas systems offer a decentralized energy source to rural households. The economic advantages for the individual household, however, can not be realized as readily as for an industry. An $8m^3$, family-sized unit can cost from 100 to $400, which exceeds annual family income in some parts of the Third World (Stuckey, 1983). Furthermore, a minimum of 3-5 pigs or cattle are needed for each family-size digester. In India, only 10-15% of the rural population meet these requirements. Large water requirements for the digesters, i.e., 80-90% water content (van Buren, 1979, Lichtman, 1983) may limit their use in areas where water is scarce.

SUMMARY

Biogas has significant potential in developing countries for producing useable energy, destruction of putrescible organic material in excreta with a concommitant destruction in disease-causing organisms and an increase in the quality of public health, and production of an end-product slurry that is suitable as a fertilizer for increasing agricultural production. The main reasons for the lack of more widespread use of biogas are lack of understanding of the process fundamentals as they relate to process design and operation, and social, political, and economic considerations specific to each situation. With proper attention to the requirements of bacterial growth such as temperature, pH, and nutritional requirements, and to design fundamentals such as proper mixing, detention time, and

and reactor integrity (e.g., prevention of leaks), biogas technology will provide the benefits listed above. Governmental support will be required to implement biogas technology, and education of the local population will be required to overcome sociological barriers.

Biogas digestion is a rare type of technology; perhaps no other single technology offers such a wide spectrum of benefits for developing countries. It is affordable, useable, and adaptable to the variety of conditions found in developing countries. It is an especially appropriate technology that deserves more widespread support.

ACKNOWLEDGEMENTS

The author wishes to acknowledge Drs. R.E. Speece, Gene F. Parkin and Mirat D. Gurol for their assistance in collecting reference material, editing of the initial text, and commenting on its content at Drexel University, Philadelphia.

References

Chen, R.C., "Up-to-Date Status of Anaerobic Digestion Technology in China," Macau Water Supply Company, Ltd., Macau.

Dale, J.T., (1979). "World Bank Shifts Focus on Third World Sanitation Projects," Journal Water Pollution Control Federation, Vol.51, No. 4, Washington, D.C., Pg. 662.

Feachem, R.G., Bradley, D.J., Garelick, H., Mara, D.D., (1983). Sanitation and Disease - Health Aspects of Excreta and Wastewater Management, J. Wiley and Sons, Chichester, U.K.

Indian Farmer's Fertilizer Cooperative Ltd., (1982). "Biogas Plant-A Supervisor's Manual," New Delhi.

International Development and Research Centre, (1976). "Excreta Removal From Middle Layer of a Fully Enclosed Type Biogas Plant," Revolution Committee, District of Mien Chu, IDRC, Ottawa, Canada.

Khadi and Villiage Industries Commission, (1984). "Gobar Gas Plant. Why and How?" Directorate of Gobar Gas Scheme, Bombay.

Lichtman, R. J., (1983). Biogas Systems in India, Volunteers in Technical Assistance, Arlington, VA.

Maramba, S., Judan, A., Obias, E., "Fuel, Feed, and Fertilizer From Farm Wastes - The Philippine Experience," Liberty Flour Mills, Inc. Metro Manila, Philippines.

McGarry, M., (1978). "Compost, Fertilizer, and Biogas Production From Farm Wastes in the People's Republic of China," International Development and Research Centre, Ottawa, Canada.

Moulik, T.K., (1981). "Biogas: The Indian Experience," UNESCO Courier, Vol. 34, July.

National Academy of Sciences, (1977). Methane Generation From Human, Animal, and Agricultural Wastes, Washington, D.C.

Polsgrove, C., (1981). "The Vanishing Forests of the Third World," Progressive, Vol. 45, Aug.

Ru-Chen, C., Cong, H., Zhi-ping, X., (1978). "A Biogas Power Station in Foshan, China," The Guanzhou Institute of Energy Sources, Chinese Academy of Sciences

Stuckey, D.C., (1982). "Biogas in China: A Back to the Office Report on a Study Tour of China (Nov. 4-18, 1982)," International Reference Center for Wastes Disposal, Dubendorf, Switzerland.

Stuckey, D.C., (1983). "Biogas in Developing Countries: A Critical Appraisal," International Reference Center for Wastes Disposal for World Bank UNDP Global Project - Research and Development in Integrated Resource Recovery, Dubendorf, Switzerland.

UNESCO, (1981). "UNESCO and the International Decade," UNESCO Courier, Vol. 34, Feb., pg. 16.

van Buren, A., (1979) A Chinese Biogas Manual, Intermediate Technology Publications Ltd., London.

Wolverton, B.C., McDonald, R.C., Gordon, J., (1975). "Bioconversion of Water Hyacinths into Methane Gas, Part 1," Report No. TM-X-72725, U.S. National Aeronautics and Space Administration - National Space Technology Laboratories, Bay St. Louis.

Solid Waste Management in Developing Countries

Sandra J. Cointreau, ASCE Member*

Abstract

The priority of solid waste management services in lesser developed countries (LDCs) is to get the refuse out from under foot -- and at the lowest possible cost. Most urban areas in LDCs spend 20% to 60% of available municipal revenues on solid waste collection and yet service only 50% to 70% of the population. This paper, based on over 50 assignments in LDC's performed by the author, presents information on public health risks of inadequate solid waste management and discusses the key parameters for planning systems of appropriate technology.

Introduction

Every developing country has its own unique blend of social, political, economic and physical conditions that profoundly affect the choice of viable options from among solid waste management techniques. Nevertheless, as a group, LDCs do have some common ground of which planners should be aware. A sampling of just a few conditions is provided below:

- Dwellings in neighborhoods of low-income level are comparatively small, often only one to three stories, and located close together. Commonly, there are only unpaved footpaths or narrow lanes for access within areas housing thousands of people. As a result, there is little or no place for waste storage and trucks can not directly obtain access.
- Climates are typically warm, some with heavy rainy seasons. Therefore, wastes decompose quickly and insects which spread disease (e.g., flies) propagate readily.
- Residents are often so poor that purchase of a refuse container for each dwelling would be prohibitively expensive, and theft would be prevalent. Many residents prefer to carry their small parcel of refuse to the collection truck when it comes at a scheduled time and rings a bell,

* Owner/President, Solid Waste Management Consulting Services, Ltd., 11 Tamarack Lane, P.O. Box 308, Woodbury, Ct., 06798, USA.

or to leave it in a communal container on their walk to public transport to work.

- Traffic often moves very slowly and is congested, because of the shortage of available road network relative to the number of cars in some cities, or because of the predominance of bicycles and animal drawn carts in others. This significantly affects the desirable size and speed of collection vehicles.
- Cities in LDCs are commonly so sprawled out (e.g., some have as much as half their populace located in shanty-type dwellings) that lands for disposal purposes may not be available within a reasonable transport distance. And yet, few cities have considered or implemented transfer station systems, as it would be difficult to obtain the financing for such a major capital investment.
- Labor costs are relatively low, whereas costs for equipment and fuel (which most often are imported items) are high. Therefore, for cost-effectiveness to be achieved, optimization of vehicle productivity needs to be emphasized more than worker productivity.
- Workers are very poorly paid and supervision is practically nil. Therefore, it is difficult to curtail traditional practices that lead to low worker productivity (e.g., collecting tips from residents and salvaging recoverable materials while on their routes).
- Recycling can achieve important national benefits from limiting use of foreign exchange for imported feedstock and finished product, therefore ways of maintaining or rechanneling informal sector recycling practices (e.g., scavengers at dump sites) need to be carefully considered.
- With the decline of agriculture in many countries and the resulting huge influx of migrants from rural areas into squatter settlements and slums in urban areas, the tax base of cities has not been able to keep up with the need for urban services. New methods of financing services are needed, but often avoided for political reasons.

From the above examples of conditions found in LDCs, it is apparent that direct application of standards and techniques from industrialized countries (ICs) would be prohibitively expensive, would be inappropriate given the myriad of other urban priorities and needs, and in many cases would not achieve desired improvements.

Relation of Disease to Solid Waste

An important reason to strive for good solid waste management is public health, particularly of children as they are most vulnerable. Over 50 diseases in LDCs come from the inadequate control of human excreta. Diseases derived from feacal matter are the primary cause of death for children under the age of five, responsible for one quarter to one half of their deaths.

In LDCs excreta arrives in refuse from disposal of dirtied toilet tissues being placed in baskets instead of within the toilets, from use of small buckets for bowel movements, and from direct use of the gutters and curbsides and refuse heaps as "public toilets".

Diseases commonly found in LDCs and transmitted by feacal matter include: diarrhea, cholera, hepatitis A, bacillary dysentery, colonic ulceration, hookworm and typhoid fever. While urine is generally sterile some pathogens are passed by urine, namely: urinary schistosomiasis, thyphoid, and leptospirosis. These pathogens are not free-living outside of excreta, and therefore proper sanitation and hygiene would practically eliminate the diseases they cause.

Key Parameters in Planning Solid Waste Systems in LDCs

In LDCs, the weight of solid waste generated at each dwelling is only 20 to 30% of the weight generated per home in ICs. Furthermore, because there is less plastic, glass and metal packaging, and paper material, it is largely organic food wastes.

The result of the above differences leads to higher density wastes in LDCs, and the volume of solid waste per dwelling in LDCs is only 2 to 10% of that in ICs.

The organic nature of the waste, coupled with the warmer climate and smaller living spaces which prevail in LDCs, necessitates more frequent collection -- generally every one to two days versus the predominant collection frequency in ICs of every 3 to 7 days.

General characteristics of municipal solid waste, as sampled and reported from various cities around the globe, appear to vary according to the level of economic development. Ranges of waste generation and character common to low-income, middle-income and industrialized countries are shown in Table 1.

Table 1
PATTERNS OF MUNICIPAL REFUSE QUANTITIES AND CHARACTERISTICS FOR LOW, MIDDLE AND UPPER INCOME COUNTRIES

	Low-Income Countries[a]	Middle-Income Countries[b]	Industrialized Countries
Waste Generation (kg/cap/day)	0.4 to 0.6	0.5 to 0.9	0.7 to 1.8
Waste Densities (wet weight basis- kg/cubic meter)	250 to 500	170 to 330	100 to 200
Moisture Content (% wet weight at point of generation)	40 to 80	40 to 60	20 to 40
Composition (% by wet weight)			
Paper	1 to 10	15 to 40	15 to 50
Glass, Ceramics	1 to 10	1 to 10	4 to 12
Metals	1 to 5	1 to 5	3 to 13
Plastics	1 to 5	2 to 6	2 to 10
Leather, Rubber	1 to 5	-	-
Wood, Bones, Straw	1 to 5	-	-
Textiles	1 to 5	2 to 10	2 to 10
Vegetable	40 to 85	20 to 65	20 to 50[c]
Misc. inerts	1 to 40	1 to 30	1 to 20
Particle Size, % greater than 50 mm	5 to 35	-	10 to 85

a Includes countries having an annual per capita income of less than US$360 in 1978.
b Includes countries having an annual per capita income of more than US$360 and less than US$3,500 in 1978.
c This may be reduced in areas with household or commercial garbage grinders which discharge to sewers.

Because solid waste in LDCs is characterized by high organic contents, high levels of moisture, and high densities, stabilization of refuse in landfills occurs very quickly (often within 2 years in tropical climates), incineration would not be self-sustaining (calorific values of wastes in LDCs are about half those of ICs), compaction devices would achieve only modest volume reductions (usually only 1½:1 versus the 4:1 achieved in ICs) and mechanized processes for materials recovery would not be economical.

A necessary step in analyzing which collection techniques would be most cost-effective is to collect time-and-motion data on the stop-to-stop travel times, loading times, and haul times of various service areas within the city. Using this data to estimate the amount of waste which could be collected by various types and capacities of equipment and crew sizes, collection options are cost analyzed. Whether compaction vehicles are economically justifiable for an LDC would depend more upon the need for their relatively low loading height and large capacity body than their volume reduction potential.

It is not uncommon in cities of LDCs for collection vehicles to spend 30% to 50% of their time in the workshop. Most of the delays are attributed to lack of spare parts, which in reality stems largely from poor inventory and budget planning, restrictions on foreign exchange to import parts, and time-consuming procurement arrangements.

The argument is often made that it is cost-effective to "standardize" the fleet city-wide, as a means of optimizing maintenance, repairs and supervision. Cost efficiencies in maintenance are highly unlikely to override cost efficiencies of having the optimum collection method for each type of neighborhood. Maintenance is usually only 10 to 20% of the total annual cost of owning and operating a collection vehicle. Furthermore, it is possible to standardize major components of the fleet without standardizing the fleet.

In planning a collection system, traditional cultural practices need to be considered. In Bangkok, Thailand, city refuse workers are known to spend time up to 40% of their available time on collection routes in sorting out recoverable materials from solid waste before loading the trucks. Their earnings from recycling are about equal to their salaries. While political leaders complain strongly about these workers, it is privately known that the workers typically pay a substantial fee (equal to about seven months salary) to obtain their relatively lucrative employment. Efforts to improve efficiency in collection in Bangkok would somehow need to assure that workers' earnings are adequate. In Cairo, Egypt, refuse collection to middle-income and high-income level households is provided by private individuals in return for the rights to recover organics as pig feed and materials for industrial reuse. Cairo's local governments have experienced difficulties in servicing low-income level households which do not have wastes that are sufficiently valuable in recyclables for the private sector to collect. Rather than work on overlapping collection routes in competition with the private sector, the government has been working to help the private sector improve its efficiency and extend its service.

Landfill is still the lowest cost disposal option for cities of LDCs. Because the money is simply not available to implement sanitary landfill according to western standards, modified landfill designs need to address the basic principles of good landfill practice (e.g., siting to limit degradation of water supplies, gradients and drainage to minimize infiltration, soil cover to limit vector access, good roadways to limit damage to collection vehicles).

Small cities often can not afford landfill equipment, and compromises need to be found. For example, it may be affordable to rent a bulldozer only a few days a month and use the machine to cover the previous month's trench-filled refuse and dig a new trench for the next month's refuse, as in San Nicolas, Mexico. Similarly, in Thailand, studies determined that it would cost a small community of 10,000 residents the equivalent of 10$US/tonne of refuse handled to own and operate an 80 hp bulldozer at the town landfill, whereas 7 communities of 10,000 residents could own and operate a 140 hp bulldozer and a low loader to haul it from site to site for a cost of 1$US/tonne handled. In a large city, whether a landfill compactor's higher cost is justifiable over a bulldozer's cost would depend on the size of the fill (remembering that little compaction is actually achievable in LDCs) since a landfill compactor is more maneuverable and can therefore handle more material per shift.

Consider finding a way to integrate scavenging into disposal operations, as in Ecatepec, Mexico. The site has two levels; trucks dump their loads on the upper level; the scavengers sort through the loads on the upper level; when completely finished with sorting, the scavengers manually push the remainder of the load over the edge to the lower level; landfill equipment operate at the lower level to spread, grade, compact and cover the refuse daily. The site is fenced and only adult scavengers registered with the municipality are allowed to enter. Trucks are checked upon entry and those with potentially infectious or hazardous wastes are diverted to a section of the landfill where no scavenging is permitted.

Composting, while technically viable, is typically not economically justifiable because farmers operate on a subsistence level and can not afford the purchase and transport of compost product. Similarly, landfill biogas recovery, while technically viable, is not economic in many places because of the lack of a gas distribution network or nearby electric generating plant.

Planning too often tries to result in direct answers. Often it would be better to propose a process of pilot testing and demonstration. In this way, local people obtain experience in comparing options and operating alternative systems. Optimum designs and specifications for procurement can be developed before major expenditures are made. Residents can be educated and surveyed for their response to the technology, and adjustments made to the technology. Implementation problems, time frames, and costs can be well assessed and the results can be packaged into replicable modules for easy extension, as is presently be done in a national program of pilot projects in Mexico.

Public Cooperation with the Solid Waste Service

Many cities in LDCs are literally throwing away their money on inefficient and ineffective solid waste service. In commercial and high-income level residential zones, service is usually daily, and on some streets (as seen in Peru, Tunisia, and Thailand) trucks might provide collection 2 to 3 times a day. This is because residents behave with complete abandon, discharging garbage to the streets at all hours of the day and night, and placing it curbside without adequate containment. For example, in most Thai communities, sweeping comprises a significant 35 to 55% of recurrent expenditures, simply to clean up after littering and inadequate storage practices. In Manila, the Philippines, in 1982 there were 14,000 sweepers on the roles.

As a prerequisite to cost-effective city cleanliness, there needs to be a clear commitment at the highest political levels -- this was most obvious in a comparison of Indonesia's reputed "cleanest" city, Padang, versus a number of others visited. Only in the cleanest city had the Mayor made a long-term commitment to change the city's appearance. His commitment included visiting schools monthly to talk to the children about sanitation, and implementing strong vigilance and enforcement of regulations about littering and household waste storage and discharge.

In Thailand, political commitment would need to be at the provincial level to be effective. Although Thai cities have regulations to control resident behavior, the public blatantly disregards these. Local mayors usually have no police support for enforcing sanitation regulations, because police work for the Ministry of Interior. Only provincial Governors, who are Ministry of Interior employees of a status equivalent to the Department of Police Director would have the clout to obtain the police support.

Laws and Regulations

Equitable fair laws are essential to environmental improvement. We have experienced in ICs that industries are not likely to spend time and money on waste management and pollution control unless their competitors are required to do the same, and unless there is the threat of liability. Similarly, in LDCs homeowners are not likely to stop dumping refuse on open lands or clean refuse from their open drains if their neighbors are not required to do the same.

In LDCs, there are seldom national or provincial environmental legislation to provide clear standards for service. At best there might be an Environmental Act setting out goals and objectives. Indonesia has a new national act, about 3 years old, with broad goals and objectives. Thailand goes a step further and has a requirement for an EIS to be submitted for environmental approval on any new major development. Mexico has formed an agency at the central government level for setting standards of service and providing technical assistance.

Nigeria has gone further still with comprehensive, detailed and stringent acts developed by some states, coupled with semi-autonomous authorities to provide service, enforce the law and levy user fees. As an example of the Nigerian commitment, there is a national Environmental Sanitation Day monthly requiring every citizen to devote the day to cleaning property, streets and drains -- all unofficial transport and all commercial trading are suspending during the clean-up period, and people not productively engaged in cleaning are arrested and fined heavily. Mobile courts are used to act immediately on enforcement.

Local Government Institutional Arrangements

Planning too often results in recommendations for equipment, infrastructure and staffing. These are short-term remedies only, unless institutions are strengthened, personnel are trained, and the public is educated to enable regularly optimizing operations.

Because solid waste techniques rely on mobile equipment and people, rather than large civil works and stationary equipment (e.g., water supply and sewerage), the overall system must involve continuous planning, and cost recovery so that it is self-sustaining.

Bibliography

1. Cointreau, S., Environmental Management of Urban Solid Wastes in Developing Countries: A Project Guide, The World Bank, Washington, D.C., Urban Technical Series Paper Number 5, June 1982.

2. Cointreau, S., Armstrong, W., with contributions from EMENA Urban Division Staff and Water Urban Department Advisors, EMENA Urban Solid Waste Management, Development of Strategies, Confidential Report No. 4875, The World Bank, Washington, D.C., June 29, 1984.

3. Cointreau, S., Gunnerson, C. G., et. al., Recycling of Municipal Refuse: State of the Art Review, United Nations Development Programme Global Project on Resource Recovery, World Bank Technical Paper Number 30, Washington, D.C., January 1985. January 1984.

4. Cointreau, S., Chapter 10: Solid Waste Collection Practice and Planning in Developing Countries, in Managing Solid Wastes in Developing Countries, edited by John R. Holmes, John Wiley & Sons, Ltd., 1984.

5. Feachem, R., Bradley, David., Garelick, H., Duncan Mara, D., Sanitation and Disease: Health Aspects of Excreta and Wastewater Management, World Bank Studies in Water Supply and Sanitation 3, John Wiley & Sons, 1983.

Appropriate Sanitation Applications

Richard J. Scholze Jr. *
Ed D. Smith *
Stephen P. Shelton **

ABSTRACT

This paper presents a brief review of sanitation technology and research by a Corps of Engineers laboratory with possible application to developing nations. Sanitation methods applicable for both military field use and base development similarly to rural and urban development must be examined to determine the most appropriate alternative for any given scenario in regards to topography, climate, duration of use, numbers of users, etc. Results are also presented on the findings of an investigation of remote site waste treatment technologies.

INTRODUCTION

Sanitation is defined by the World Health Organization (WHO) as "the control of all those factors in man's physical environment which exercise or may exercise a deleterious effect on his physical development, health, and survival." The activity relating to disposal of human waste in developing countries and similarly to military field situations is important for a variety of reasons, primarily protection of individual and group public health. The collection, transport, treatment and disposal of human excreta are of the utmost importance in the protection of the health of any community. Large groups, communities, and individuals all must maintain high standards of hygiene to avoid debilitating chronic or acute disease.

* U.S. Army Construction Engineering Research Laboratory (USA-CERL), Environmental Division
PO Box 4005
Champaign, IL 61820-1305
1-800-USA-CERL, x 743

** University of New Mexico
Department of Civil Engineering
Albuquerque, NM

Human excreta (feces, urine and associated waste) are the principal vehicle for transmitting and spreading a wide range of communicable diseases, some of which are leading causes of sickness and death in developing nations. Prompt and thorough removal and disposal of human wastes is vital. Otherwise a slum, military encampment, or similar densely populated environment may become an ideal breeding and dissemination area for flies, mosquitos and other organisms which cause disease or are vectors in addition to the microorganisms traveling the fecal-oral route. Additionally, filth-borne diseases could become prevalent resulting in ineffectual performance by the afflicted population.

Over 50 infections, including cholera, typhoid and paratyphoid fevers, dysenteries, hookworm, and other infections or infestations can be transferred between people through various direct or indirect routes involving excreta. Developing countries are often recipients of tremendous human losses from these diseases. Excreta-related infections are caused by pathogens normally transmitted in human excreta. Control comes about through maintenance of high standards of personal cleanliness and hygiene concurrent with proper waste management.

Field expedient excreta disposal has historically been a vital concern for the military. Hygiene-related illness has been a major cause for reduced effectiveness among combat troops. The vast majority of hospital admissions during major recent wars was due to disease and noncombat casualties, i.e. World War II 85 percent. Because absence of controlled mechanisms and procedures for human excreta disposal have resulted in tremendous health problems, military field operations have always required procedures for human waste management. The selection of inappropriate waste management techniques is a common problem in field encampments, primarily due to a carryover mentality (i.e using what was used previously). However, other solutions to the human waste disposal problem are available which may be more appropriate for a particular site-specific application.

Rural sanitation in many situations is not primarily a technical problem. Except in densely populated areas, variations of pit latrines or composting latrines will be appropriate and inexpensive. A variety of designs are available which are able to cope with most adverse circumstances except high population densities. The problem with rural sanitation is to encourage the population to use and maintain the facilities. Programs have been in operation throughout the world to establish pit latrine facilities. Generally, these have been unsuccessful not because of construction problems, but following completion of the units they were either not used or not maintained. (Cairncross and Feachem 1983)

Simple solutions do not exist and this challenge of appropriate disposal of excreta is one of the greatest challenges which public health workers and sanitary engineers face.

Previous experience has produced a number of guiding principles which are summarized by Cairncross and Feachem (1983):

1) Excreta disposal is a sensitive matter about which people have strong cultural preferences. Therfore, local populations must be involved in the process to upgrade sanitation in an area and often modifification to existing technology will be more likely to succeed than implementation of an entirely different technology.

2. People require a reason or motivation for using a new kind of latrine. Generally, health improvement is not perceived as such a motivation. Possible motivations may include economic value from excreta reuse or a desire for privacy.

3) Any latrine needs good maintenance and will become fouled and offensive without it. If this is permitted to happen the latrine will either not be used or will become a major health hazard in its own right.

Urban sanitation supplies the major challenge in developing countries to provide appropriate systems suitable for high-density low-income communities. Conventional water-borne sewerage systems are by far the most convenient to the user. However, several reasons exist why they are inappropriate for most communities in developing countries according to Cairncross and Feachem 1983): their high cost, consumption of water (a valuable resource which should be conserved), technical requirements of construction and operation and maintenance, installation of sewers which destroys substantial numbers of houses, and misuse for disposal of inappropriate wastes. However, during activities in which military construction engineers are promoting good relations in host nations, construction of hospitals or waterborne sewage treatment facilities may be desirable from both the host nation's viewpoint and that of the engineers who are familiar with the requisite construction procedures. The technology which is utilized should be of the level which the local population can operate and maintain with expertise and materials existant in the local community, in addition to economic considerations.

Field environments encountered by military operations and their civilian assistance programs cover the complete range of conditions on the Earth's land masses, polar environments through tropical; desert through high humidity and rainfall conditions; swamps; forests; mountains; in short, anywhere man has ever traveled. Personnel present also vary from individuals and small groups through large populations which may require different options for their human waste management problem. Many options exist, however guidance is necessary to use the most appropriate technology in a given situation thereby maintaining health and morale at a high level.

Military concerns for proper sanitation are similar to urban situations in that large population densities may be involved. Also both field aspects and base or city development concerns can be addressed.

Extensive examination of field sanitation methods was required to determine the most appropriate alternative for any given scenario, such as topography, climate, duration of use, number of users, logistics, societal mores, etc.)

The scenario for waste treatment alternatives applicable to the military is variable in terms of numbers of users, duration of stay, area geography, and purpose.

Critical information needed for design and selection of sanitation systems are presented in Table 1.

One primary concern is the geography of the environment when selecting alternatives for human waste disposal. Knowledge of soil conditions and depth to groundwater are indispensable. For example, fractured rock will conduct pathogen-laden water downstream where it may be pumped out for use from the groundwater if a pit latrine comes in to the groundwater table. The climate also is important. Does flooding occur frequently or is it arid? Flooding can spread filth as in the above example or even wash the cover off buried waste if burial was too shallow exposing the whole area to hazardous condiitons. Another aspect of importance is the population to be serviced and for how long will they be present in the environment. There is no universal solution to the excreta-disposal problem. However, appropriate options are available.

The importance of conserving a safe, potable water supply can not be exaggerated. Surface waters can be hazardous for body contact and/or ingestion from some pathogens. Contamination of ground water by pathogens leaching through excreta in direct contact will spread additional diseases through ingestion.

Alternatives used for disposal of human wastes in the field include:

1. Cat-holes
2. Straddle trench latrines
3. Deep pit latrines
4. Burn-out latrines
5. Mound latrines
6. Bored hole latrines
7. Pail latrines
8. Urine soakage pits
9. Composting latrines
10. Chemical treatment

Construction of the minimum essential sanitary latrine facilities is one of the highest priorities for base development. Sewerage may be a desirable option with adequate water for flush toilets and other amenities if city development or an extended stay is planned. If that decision is made, the technology which is ultimately used for waste treatment should be appropriate for the using population in regards to operation and maintenance. In most locations, a variant of lagoon will be the best solution at providing high levels of pathogen reduction at an economical cost using local technicicians.

Table 1

Information Needed for Selection/Design/Location of Sanitation Systems

Climatic Conditions

Temperature ranges
Precipitation, including drought or flood periods

Site Conditions

Topography
Geology, including soil stability
Hydrology, including seasonal water table fluctuations
Vulnerability to flooding
Base layout

Population

Number, present and projected
Density, including growth patterns
Available skills
Available materials and components
Municipal services available, including roads, power

Environmental Sanitation

Existing utilities and status quo
Existing excreta disposal, washwater removal and storm drainage facilities
Applicable regulations

REMOTE SITE WASTE MANAGEMENT

Research on remote site waste management investigated applications at Army remote sites such as training ranges, guard stations, bivouac and recreation areas. Two technologies were thought appropriate to meet the needs of simplicity, ease of operation and maintenance, and appropriate carrying capacity while providing an environmentally safe, odorless, user-acceptable environment. The technologies were composting toilets and aerated vault latrines. Other technology options were also evaluated, but these technologies were the most appropriate to the Army's needs in that situation. Fundamental research at several universities supported the investigation by addressing topics such as does the process really work and how, where is it appropriate, what are the economics, what are the health concerns and more. These technologies have been demonstrated and proven on installations and are also appropriate for other countries in the proper environment.

Remote site waste management has received substantial evaluation with a widespread research program. Remote sites include a variety of locations such as training ranges, recreation areas, and ceremonial grounds. They commonly do not have piped water available, however, electric power is usually present. Investigation was directed toward solving sanitation problems at sites which would be used daily by approximately 200 people. Preliminary evaluation indicated that two technologies would be appropriate for further evaluation and investigation aerated vault latrines and composting latrines. They meet the goals of being able to meet the needs of a large number of personnel in a timely, environmentally safe, hygienic manner with low levels of operation and maintenance. Conventional technology at such sites included pit latrines, vault latrines and chemical latrines; each with their own problems. The investigation has been summarized in Scholze et al. 1986.

Both technologies provide high user acceptance, effective process operation, acceptable health considerations and avoidance of environmental contamination.

Aeration of vault latrines by means of bubble aeration units is applicable for both new construction and retrofit conditions. The technology (See Figure) consists of a motor/blower unit which is connected to a perforated pipe which rests slightly above the the vault floor in a longitudinal direction. Air continuously supplied to the waste supports the growth of aerobic microorganisms, which break down the wastes into carbon dioxide and water. Aerobic decomposition is generally about four times faster than the anaerobic decomposition which occurs in vaults functioning as holding tanks. Foul odors are eliminated or greatly reduced. Some minimal mechanical maintenance is required such as cleaning or replacing the air filter and checking the drive belt.

The composting latrines which were investigated and are discussed in this paper were the large continuous composter (multrum) variety as compared with the vault-type composting latrines which are common in tropical countries. This was done since the continuous composter varieties are less labor-intensive which is a driving consideration for military application. Vault composters will produce end-product appropriate for agricultural application and bin composters will also produce stable, pathogen-free endproduct. However, they are relatively labor intensive. Primary consideration for military application was volume reduction rather than producing a pathogen-free product. Continuous composting latrines were found to reduce volume, however, it was primarily a physical process due to evaporation rather than biological decomposition. Recommendations for finished end-product from a continuous composting toilet are that the material be bagged and landfilled to ensure no health problems. Alternatively, developing nations may elect storage for several months to insure pathogen destruction, similar to what currently occurs in double-vault

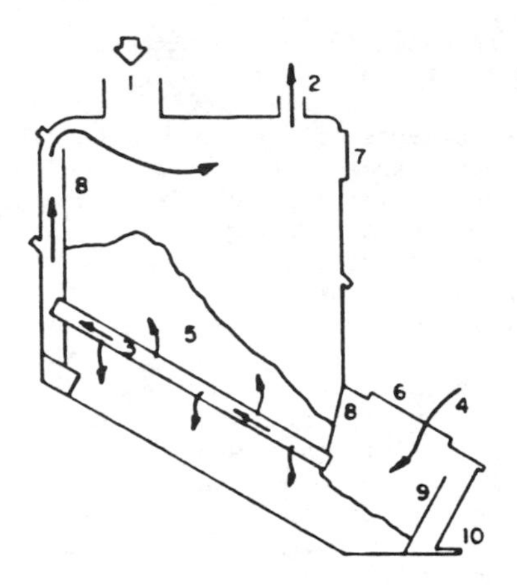

The Arrows indicate Air Flow Through the Toilet

Key :
1. Waste Chute
2. Vent Pipe with Fan
3. Air Duct
4. Air Inlet
5. Composting Mass
6. Emptying Hatch
7. Inspection Hatch
8. Waste Baffles
9. Liquid Baffles
10. Liquid Drain

Typical composting latrine.

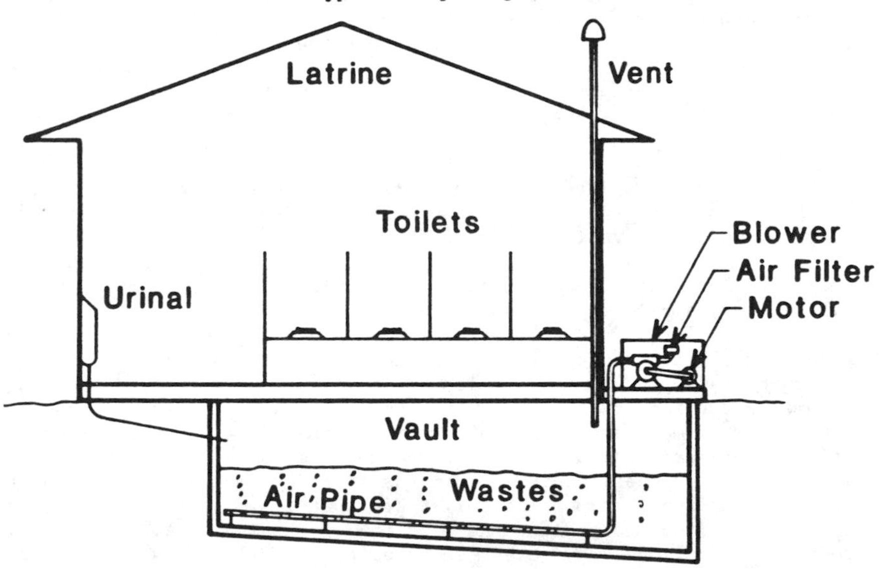

Aerated vault latrine.

composting latrines, or burial. Removal, handling, transporting and disposing of endproduct should be done carefully avoiding direct contact. Color and odor of the processed material do not indicate reliably whether the product is composted. There may be pathogens even in a black, odorless endproduct.

Composting latrines (See Figure) are large chambers in which wastes and organic bulking agents are placed for biological and physical breakdown into humus-like material by aerobic decomposition. Breakdown of this waste occurs naturally through the course of one to two years, without the addition of water or chemicals, by aeration using a series of air channels and baffles, and a continuously operating fan. In the absence of oxygen, anaerobic composting occurs slowly, producing offensive odors. Regular addition of bulking agent, monthly pile raking, periodic inspection, and semiannual removal of the final product are the only additional maintenance requirements besides routine latrine maintenance. Liquid buildup is a major concern in public situations and proper drainage must be available and operating.

SUMMARY

In summary, appropriate technology used for military field operations can parallel that used in developing nations both in rural and urban situations. Rural applications or waste treatment for small numbers of individuals need not be a complex technical problem with many simple, economical alternatives available. Urban populations and similarly large numbers of military personnel in a field situation need appropriate hygine and sanitation methods to avoid massive public health problems which can be easily transmitted in a lax environment. Remote site waste treatment options such as aerated vault latrines and large continuous composting latrines, while appropriate for training ranges may represent a larger capital investment than a developing country is willing to contribute. Alternatively, modifications to these technologies may mitigate these problems. For example, composting latrines using double-vault or similar technology have been effective for years in many countries. The emphasis with the recipient users must be on proper training and an understanding of why to use the facilities to avoid the facilities falling into nonuse and disrepair.

REFERENCES

1. Cairncross, S. and Feachem, R.G., Environmental Health Engineering in the Tropics: An Introductory Text, John Wiley and Sons, Chichester, 1983, 281 p.

2. Scholze, R.J., Alleman, J.E., Struss, S.R., and Smith, E.D., Technology for Waste Treatment at Remote Army Sites, USA-CERL Technical Report N-86/20, 1986, 177 p.

Introduction to Operation and Maintenance

Carl R. Bartone and James Jordan

Sessions on Strategical Planning for O&M in urban water and sewerage systems examine O&M issues for developing countries. Particular attention is given to the critical area of reduction and control of unaccounted-for water, and examples are given of successful programs in Latin America for water loss control.

A number of technical papers are devoted to Sewage Treatment Plant O&M, including means of upgrading plant performance through improved O&M.

The issues of O&M of Small Community Water Supply systems are examined, focusing attention on program planning. Methods for evaluating O&M components of water supply projects are analyzed, along with the concept of reliability engineering.

The critical interrelationship between operational considerations and design or rural water supply and sanitation systems are examined. New design concepts are presented, aimed at reducing operations problems. Also, attention is focused on the importance of using feedback from previous O&M experience in order to improve designs for small systems.

Attention is also give to monitoring and improving drinking water quality in rural systems in developing countries.

OVERVIEW OF OPERATIONS AND MAINTENANCE ISSUES

David B. Cook and Jens Lorentzen

The views presented herein are those of the authors, and they should not be interpreted as reflecting those of the World Bank.

Abstract

This paper provides an overview of key issues in operations and maintenance of water supply, sewerage and sanitation facilities, and related urban service delivery systems in cities in developing countries; the issues were identified through analyses of urban development projects which have been financially assisted by multilateral and bilateral donor agencies over the last decade.

The issues are presented and discussed in the broader context of urban management, and emphasis is placed on interlinkages between different functional areas of urban service delivery and related urban management functions. The rationale for increased future emphasis on operations and maintenance and general approaches to addressing operations and maintenance problems are discussed and supported by examples from World Bank-assisted projects.

Introduction

Operations and maintenance (O&M) of urban infrastructure and service delivery systems is a major issue in cities world wide. For a number of reasons, O&M has often had a low priority on the agenda of politicians and chief executives at the different levels of government.

Particularly maintenance has often received less than adequate attention, and deteriorating water treatment plants unable to sustain adequate levels of treatment, water distribution systems with unaccounted for water as high as 50% in some cases, etc. bear witness to the O&M problem.

Mr. Cook is Engineering Advisor, Water Supply and Urban Development Department, The World Bank, 1818 H Street, N.W., Washington, D.C. 20433, U.S.A.; Mr. Lorentzen is a Municipal Engineer seconded to the Department by the Danish International Development Agency (DANIDA).

Recent analyses of several urban projects assisted by multi and bilateral donors over the last decade, have led to the identification of a number of issues which should be considered when addressing operations and maintenance problems in LDC-cities.

An Overview of Key Issues in Operations and Maintenance

- Maintenance often has a low priority on the agenda of politicians and chief executives because it is not perceived as being as politically "visible" and "glamorous" as new capital works; the field of maintenance is a professional backwater in many countries and does not attract a high caliber of staff;

- Poorly maintained and unreliable urban infrastructure and service delivery systems hamper public and private sector productivity. They thereby affect cities' capacity to act as driving forces in their regional economies and, in turn, in the national economy of the countries concerned;

- The O&M problem has an important dimension of equity -- the urban poor are the most vulnerable to deficiencies in O&M. They often live in urban fringe areas or marginal lands with poor access and with infrastructure and utility services often provided in a haphazard manner which adversely affect systematic O&M efforts; also, because of their low-income, they have no alternative than to rely on public services;

- Lack of maintenance has direct and sizeable effects on the balance-of-payments. Inefficient operations and maintenance result in waste of scarce imports, e.g., energy and chemicals, premature needs for replacement of imported spares, and in rehabilitation works, which often have substantial foreign currency components.

- There is often a lack of clearly delegated responsibilities and accountability for operations and, particularly, for maintenance activities;

- Among many authorities responsible for O&M, a "management by crisis" situation exists. Basic information on assets and their condition is not available, and systematic plans and strategies for management of operations and maintenance rarely exist. More often than not, lack of management is the fundamental cause of deficiencies in O&M;

- Local resource mobilization is highly inadequate; cost recovery through taxes, fees and user charges does not sustain adequate levels of expenditure, particularly in the area of maintenance. Available resources are often fully committed to payroll and essential operating expenses like power supply;

- The fact that maintenance is frequently under-funded, may in part be the result of a lack of understanding of the costly downstream consequences of neglecting needed maintenance;

- Central-local government relations play important roles in two areas: (i) ad-hoc mechanisms for grant allocation hamper systematic planning and budgeting of O&M at the local level; and (ii) national or regional government decision makers are often isolated from the operational realities of local authorities, and because of this, maintenance at the local levels gets little consideration in the planning and budget cycle;

- Institutional arrangements whereby one authority is responsible for planning, design and construction of assets and another (often local), for O&M is unsatisfactory. Problems arise because the local authority is often insufficiently involved in the design process, and as a result, it is unprepared to assume the O&M responsibility, because these aspects were not adequately considered in design and specifications;

- International development agencies may in some cases, have biased borrowers against maintenance, because of their policy to lend for capital investments with little support for recurrent costs;

- Low pay scales and lack of incentives make it difficult for local governments and even parastatals to retain qualified and motivated staff, and very low productivity results. There is often overstaffing in the lowest echelons, while vacancies exist in critical managerial and supervisory posts. There are pronounced needs for training of local staff at all levels, particularly those engaged in O&M activities, because this is an area where educational institutions have very often failed to provide adequate pre-service training;

- Inefficiency and low productivity in operations and maintenance work carried out by force account, suggest there may be substantial scope and potential for contracting out activities to the private sector;

- Community participation in the area of O&M works well in a number of cases and has a further potential in some countries, but it requires planning and monitoring efforts by the responsible public authority, and will only succeed where there is political will and follow-up;

- Equipment for operations and maintenance is frequently inappropriate for developing countries, and decision makers often have a low level of understanding of the importance of appropriate technology in O&M. Specified equipment is not suitable for use in environments characterized by labor-intensive work methods; in some cases, uncoordinated development efforts by different donors have resulted in the proliferation of a wide range of unstandardized equipment systems. Spare parts are frequently unobtainable; costly vehicles and equipment are cannibalized to obtain parts for other units;

- Materials for maintenance are often inappropriate because standard specifications, laboratories and testing facilities are not available. Foreign exchange to procure even simple imported materials is not available;

- Urban services are strongly interdependent at technical or functional levels, and require coherent, coordinated O&M efforts across several service sectors, e.g., road maintenance and traffic mangement are severely affected by uncoordinated diggings for water mains and pipe trenches which are not properly reinstated; efforts to improve public health may be jeopardized by stagnant water caused by failure to properly maintain drainage systems; drains do not function properly without a proper street sweeping and garbage collection system, etc. There are many such interlinkages which underscores the need to address O&M problems in a broader urban management context; many interventions in the past to improve O&M have not succeeded because they were too narrowly focused on individual sectors;

- There is rather limited expertise on O&M available in the consultant profession; twinning arrangements (i.e., a professional relationship between a local authority in a developing country and a similar but more experienced organization in a different part of the world), sometimes assisted by consultants, have in a number of cases proven to be a successful mechanism for technical assistance delivery in the area of O&M.

Approaches to Improved Operations and Maintenance

1. Convince Decision Makers:

Not least of the above issues is the apparent psychological bias against O&M which exists in LDC's. Politicians prefer to be seen opening or commissioning new assets and see little political mileage in O&M activities. Indeed, one of the most difficult features of the O&M problem is its attitudinal dimension. Attentiveness to O&M needs is, in the industrialized countries, to a large extent a result of a certain professional pride among operatives. In the developing countries, the many resource constraints and the difficulties in compensating individual performance in public authorities, makes it a difficult challenge to develop the necessary "esprit de corps" and the individual commitment towards better O&M. Technical assistance and training to increase decision makers' understanding of O&M issues and to close skill gaps among operatives, are obviously crucial elements of efforts to improve O&M.

It is clearly important to justify increased budgets for O&M on the basis of direct costs and benefits, or perhaps even more importantly, the consequences of neglect. On a present value basis, the direct costs to the responsible authority of premature reconstruction or replacement resulting from neglect of maintenance needs can be several times the cost of a timely maintenance effort.

But operations and maintenance of urban services should also be seen in a macroeconomic perspective. In most countries, more than half of the Gross Domestic Product (GPD), is produced in urban areas and this nonagricultural share of GPD is increasing. Cities, thus, serve as increasingly important driving forces in the national economies of developing countries, but their ability to sustain efficient public and private sector activities is contingent upon well maintained and reliable urban infrastructure and service delivery systems.

In many cities in West Africa, for example, the situation is so bad that manufacturing firms have to provide their own backup facilities for unreliable public water supply and other services and thus do not enjoy the benefits of the economy of scale inherent in collective service delivery. The Bank's 1983 urban sector review in Lagos, Nigeria estimates that as a consequence of industry being forced to provide its own facilities, production costs for goods and services have been increased by as much as 30 percent, making Lagos one of the most expensive cities in the world. Many small, new firms and informal sector enterprises are unable to afford their own infrastructure facilities and may be forced to idle capacity during public service "downtime" or to forego production and leave the market altogether.

2. **Make Choices Between Strategic Options and Establish Policies**:

Within the broader framework of urban management, the strategic options available to improve local performance in operations and maintenance of specific services must be identified. Political decision-makers and urban managers must take decisions in principle and develop related policies in areas of particular importance for O&M such as: (i) definition of institutional responsibilities where these are unclear; (ii) development of accountability centers in the organizations concerned to institute clearly defined accountability for specific O&M functions; (iii) staffing and training; (iv) achievement of cost-effective balances between execution of O&M works by direct labor and by contract, this may include the option of complete privatization of certain service functions; (v) enhancement of community participation; (vi) strategic choices between maintenance, rehabilitation and reconstruction/replacement of specific assets; (including trade-offs between preventive vs. corrective, and routine vs. periodic maintenance options), and (vii) policies on cost recovery, local resource mobilization, e.g., through property taxes, and rationalization of intergovernmental financial transfer systems.

3. **Develop Specific Action Programs**:

The strategic choices and policy decisions should then be translated into programs of action which specify how to undertake the necessary improvements. A number of immediate and obvious priorities for improvements can often quickly be specified in a short term action program. Actions which require legislative changes such as increases of tariffs and taxes, institutional reorganizations and privatization, etc. or surveys and condition assessments to improve asset registers, will require longer lead times.

This "action planning approach" has been successful in several World Bank assisted efforts and has led to early identification of priorities for improvements in O&M and other aspects of urban management. The broader view of services in an urban management framework provides a good basis for sector-balanced identification of investment needs and a more efficient achievement of economic and social development goals.

Examples From Recent Work Assisted By The World Bank:

In 1982-83, two World Bank Missions visited Lagos, Nigeria to undertake an Urban Sector Review. The city in 1982, had a population of 4.5 million which represented a ten-fold population increase over a 20-year period. Lagos is on almost flat, sea-level land, that is laced with streams and lagoons and has a high water table. Poor soils and heavy seasonal rains, coupled with inadequate and clogged storm drains have resulted in frequent, and often severe flooding.

Using a recently completed UNDP-Supported Master Plan as a data base, the Bank teams and the Lagos State Government staff commenced the sector review. As this broad review of the issues facing Metropolitan Lagos moved from planning, efficiency, equity and resource issues to the day-to-day urban management problems facing the parastatals and local authorities, it became increasingly clear that the city was in crisis. All the classic issues facing metropolitan centers in developing countries were manifested, i.e.: The city was growing annually at almost half-a-million people. It was the private sector, mainly the informal private sector, which was building the city; municipal service deficiencies proliferated; development on the flat swampy land without adequate services was giving rise to extensive slum communities. Despite substantial efforts on the part of the federal government, which provides major transportation infrastructure, and the State Government which supports the parastatals and local authorities, the pace of growth and change was overwhelming available financial resources. The local fiscal base, so essential to under-pin urban management was weak; large numbers of properties were "illegal" and not included in rate rolls, and the billing and collection process was not keeping up with demand. Staff resources available for urban management tasks were poorly paid and unequipped for the job; the local authorities, in particular, were weak. The situation with regard to O&M was critical. In this situation, it was clear that early action was required, and with the cooperation of the Nigerian authorities, the sector mission converted itself into a project identification mission.

These and subsequent efforts led to the appraisal of a major solid waste management and drainage project before the completion of the sector report. In the opinion of the authors, the fact that this project was identified within the framework of the broader urban management issues led to a more meaningful and sector-balanced project. It also led to the inclusion of property valuation, institutional support and technical assistance subcomponents with metropolitan wide significance. The broad assessment of O&M leading to the project took several important intersectoral linkages into account, e.g., programs to improve cleaning of existing drains and construction of new drainage, coupled with programs to improve solid waste removal, curtail flooding, assist road maintenance (nothing destroys roads more quickly than poor drainage), enables drains to act as a surrogate for sewerage systems, and improve environment and public health.

The sector work also ranked other issues facing the city. Second to solid waste management and drainage came water supply, followed by slum upgrading and transport issues. A water supply project has been appraised, and a recently completed follow-up sector mission is recommending a further urban project aimed at addressing growth, urban efficiency improvements through investments in the transportation sector and productivity increases by area upgrading and related measures. The water supply project intends to address the issue of distribution system management. The present level of 50% unaccounted for water could possibly go higher when production and pressures are increased, and the project allows for the appointment of consultants, whose first task will be to hold the unaccounted for water percentage to current levels, and hopefully, slowly bring it down. The program put forward at appraisal aims to keep politicians and financiers interested by targetting the larger users first, in the hope of proving how cost effective good O&M can be. In this way, it is hoped to divert some resources from costly water production works to the difficult task of improving distribution systems in dense residential or mixed development areas.

Recent work in Ghana gave a Bank mission team an opportunity to apply urban action planning techniques in a group of secondary cities. Among the mission's objectives was to carry out broad assessments of infrastructure needs with a view to identifying urban management and infrastructure projects in these cities. It was found useful to structure the discussions with local authorities according to issues in the following three main categories: (i) O&M of existing assets; (ii) Residual deficiencies, i.e., the need to extend services to residents not yet served; and (iii) Urban growth, i.e., the need to cope with additional future service requirements caused by rapid urban population growth.

In almost all cases, the attention and enthusiasm of the key officials was obtained; the need to consider interlinkages was appreciated, and in many instances, the approach led to identification of issues which had not been previously considered. In several cases, staff of the authorities concerned, demonstrated their ability to modify the approach to their own particular needs and objectives and develop action programs which they considered realistic and achievable.

Reduction and Control of Unaccounted
for Water in Developing Countries

Harvey Garn and Arumukham Saravanapavan[1/]

During the past 20 years of World Bank lending, an extensive review of data from 54 water supply expansion projects in developing countries shows, unaccounted-for water has remained at an average of about 34% of production, which increased from 73 million to 93 million cubic meters per year in the average project. The rate of unaccounted -for water has been surprisingly resistant to reduction. This paper analyses the data, draws conclusions, and explores ways for tackling the problem.

Introduction

Water supply authorities all over the world face the continuing challenge of ensuring that the maximum amount of water produced reaches its intended users. This challenge is especially acute for water supply authorities in developing countries, since they are often faced with the further challenge of inadequate funds for capital expansion and ongoing operation and maintenance.

The amount and percentage of water that is produced but not accounted for by sales are measures of how well this challenge is being met by a particular authority. This paper is intended to provide summary information regarding the magnitudes of unaccounted-for water typically experienced in water supply institutions in developing countries and to present practical suggestions on how to reduce it in cost-effective ways.

Summary Data

There are many reasons for concern about the problem of unaccounted-for water. It tends to represent a substantial share of total production. It has proven to be difficult to reduce. It causes the supply institution to incur costs that yield no offsetting revenue. If extensive and persistent enough, unaccounted-for water can result in the necessity to expand production facilities earlier than would have been necessary if rates of unaccounted for water could have been reduced.

1/Harvey Garn, long term Consultant, and Arumukham Saravanapavan, Senior Sanitary Engineer in the Water Supply and Urban Development Department of the World Bank. The views presented herein are those of the authors, and they should not be interpreted as reflecting those of the Bank.

This general pattern was observed in projects in all of the countries in the six regions[2] borrowing from the World Bank throughout the twenty-year period during which projects have been reviewed. These data are summarized in Tables 1 and 2 on page 8. Table 1 reveals that the average percent of water production unaccounted for at the time of project appraisal ranged from 22% in supported projects in East Africa to 43% in South Asia. By the sixth year it was expected that the percent unaccounted for would be reduced by at least two percentage points in all regions and that there would be reductions of 14.4 and 10.8 percentage points in the two regions with the largest initial levels--East Asia and Pacific and South Asia (Table 2).

Although the average initial percent unaccounted for was reduced in three regions--East Asia and Pacific; Europe, Middle East, and North Africa; and West Africa--in no region was the percentage point reduction as large as expected, and in half of the regions the percentage of water unaccounted for actually increased. A similar picture emerges with regard to the volumes of unaccounted-for water. As Table 2 shows, the average incremental production and sales volumes achieved tended to be lower than had been expected. Even so, the actual volumes of unaccounted-for water were larger than expected in all regions except the Latin America and Caribbean region, where the shortfall in production and sales was the largest. In over 85% of the individual projects reviewed, the actual share of production unaccounted for exceeded expected levels.

Financial Implications and Benefits from Reduction in Unaccounted-for Water

Substantial volumes and percentage shares of unaccounted-for water impose serious financial penalties on water supply institutions. If there is sufficient demand for the water produced, unaccounted-for water results in unnecessarily reduced revenues to the supply institution. The lost revenues, in turn, inhibit cost recovery on existing assets and limit the contribution that the institution can make to future investments. Even if additional output cannot be sold, given local demand conditions, unaccounted-for water imposes unnecessarily large operational costs to achieve those sales. Put slightly differently, the reduction of unaccounted-for water provides financial benefits to the supply institution in the form of either increased revenue or reduced costs.

It is worth noting that the revenue response to a reduction in unaccounted-for water provides the water supply institution with an important indicator of the current relationship between demand and supply. If, for example, leak reduction leads to the additional water being quickly sold, it is clear that there is additional demand at the current price of water. If the additional water is not quickly sold, it may be possible to reduce output and achieve savings in operating costs. Such responses can, therefore, provide important clues about when it would be desirable to invest in additional supply capacity.

2/The Bank is administraively divided into six regions

TABLE 1

Region	Initial Production		Initial Sales		Initial UFW (Volume)		Initial UFW (Percent)	
	Expected	Actual	Expected	Actual	Expected	Actual	Expected	Actual
ALL	77.8	72.7	54.7	47.4	23.1	25.2	29.7	34.7
EA	12.2	13.3	9.2	10.3	3.0	3.0	24.7	22.1
EAP	189.6	162.9	118.9	97.8	70.7	65.1	37.3	39.9
EMENA	99.8	95.8	76.1	64.8	23.7	31.0	23.7	32.3
LAC	66.6	57.9	50.1	38.0	16.5	20.0	24.7	34.2
SA	28.6	27.9	15.7	15.8	12.8	12.1	44.9	43.3
WA	30.9	33.3	20.2	20.9	10.7	12.4	34.7	37.3

TABLE 2

Region	Incremental Production		Incremental Sales		Incremental UFW (Volume)		Incremental UFW (Percentage Point Change)	
	Expected	Actual	Expected	Actual	Expected	Actual	Expected	Actual
ALL	40.0	20.6	35.3	13.0	4.7	7.6	-6.1	0.50
EA	8.1	5.6	7.0	1.6	1.1	4.0	-4.3	14.7
EAP	55.2	32.8	61.1	30.9	-5.9	1.9	-10.8	-5.7
EMENA	35.7	40.8	29.9	29.1	5.8	11.7	-2.0	-1.1
LAC	91.9	24.1	78.3	15.5	13.6	8.6	-5.7	0.3
SA	13.4	21.2	13.4	11.9	0.0	9.1	-14.4	0.1
WA	11.3	14.5	9.3	9.3	2.0	5.2	-4.3	-0.4

Note: EA - East Africa; EAP - East Asia and Pacific; EMENA - Europe, Middle East, North Africa
LAC - Latin America and Carribean; SA - South Asia; WA - West Africa

The potential magnitudes of these benefits in the typical situation of Bank-supported projects can be derived from the review data referred to earlier. These magnitudes can be expressed in terms of the revenue enhancement or cost reduction elasticity of reducing unaccounted-for water. The elasticities are defined as the percentage increase in revenue or percentage reduction in operating costs associated with a given percentage reduction in unaccounted-for water. For the 54 reviewed projects, a 10% reduction in unaccounted-for water would produce about a 6% increase in revenue or reduction in operating costs. Using the average values from the projects reviewed, a reduction of unaccounted-for water from 34% to 30%,for example, would result in an annual increase in revenue of about $700,000 or a cost reduction of about $590,000 (1980 U.S. dollars) in a typical project. The potential revenue gains (if there is excess demand) are larger than the operation cost savings (if there is excess supply) for this group of projects because the average tariff is slightly higher than the operation costs per cubic meter.

The Present Situation

In many cities of the world, ranging from Turkey in the Middle East to South Korea in the Far East, concern about the high levels of unaccounted-for water is increasing. The degree of interest in taking action varies from country-to-country, although it is far greater than it was a decade ago, when funds for the construction of large new water production facilities were more easily obtained. Long-term commitment is difficult to predict. However, existing resource constraints are forcing many utilities to take a harder look at rehabilitation, repair, improved maintenance, and reduction of unaccounted-for water as "interim" steps for meeting increasing demand, until major capital water production works are affordable. It is therefore important that water supply professionals take this opportunity to do everything possible now to upgrade Operation and Maintenance (O&M) standards in developing countries, to demonstrate the cost-effectiveness of reducing unaccounted-for water, and institutionalize the appropriation of adequate budget allocations. When funds for major capital works are once again available, it is important not only that adequate budget levels for O&M be maintained but also that distribution networks continue to be carefully designed so as to ensure minimum leakage during their working lives. At the same time, it should be stressed that investment in production facilities should not be made in cases where rehabilitation works and reduction of unaccounted-for water would be more cost-effective.

Taking action

The following is a list of places where recent concerns about high levels of unaccounted-for water is resulting in action:

Turkey: Six towns in the Cukorova Region do not measure water production, but based upon pumping volumes unaccounted-for water is estimated at 60%. Though future increases in water production are anticipated, a program has also been proposed for the reduction and control of unaccounted-for water. Consultants are preparing the program.

Nigeria: Lagos, the capital, is reported to have unaccounted-for water levels of about 50%. New facilities have been proposed and consultants are to be appointed to work toward maintaining, and then gradually lowering this figure.

South Korea: This rapidly industrializing nation has begun a national leak detection program with assistance from consultants. The results are not yet available.

Philippines: Manila, the capital city, had unaccounted-for water levels of about 40% in 1976. Following a doubling of production capacity, unaccounted-for water increased to 60% because of higher system pressures. The city is presently engaged in a major integrated program to reduce unaccounted-for water to more reasonable levels.

Indonesia: The two largest cities, Jakarta and Surabaya, have levels of unaccounted-for water of about 50% and 60% respectively. Indeed, a small production increase of 1 cumec in Surabaya "vanished" without noticeable impact on revenues. Both cities are taking serious steps to ensure that future planned production increases do not meet with the same fate. Consultants are assisting these cities with integrated programs to reduce unaccounted-for water. Some smaller towns are reported to have unaccounted-for water levels of about 20%. Their challenge is to maintain this level through sound operation and maintenance practices. Any extensions of the distribution system must would be soundly engineered and carefully built to obtain low levels of leakage. Corrosion protection would be an important design feature.

The widespread prevalence of unaccounted-for water - a symptom of poor maintenance - is a serious matter. What is the reason for these past failures? We have failed because cities have given low priority to the reduction and control of unaccounted for water in expansion projects. The complementary actions required for sound engineering and construction and for improving overall operation and maintenance have run into manpower and institutional problems. Leakages have been caused by corrosion, failure to carry out specified pressure tests on newly laid pipes, or because of poorly made service connections. Meter selection, calibration, and routine replacement programs have not been given the needed support. So called "Action Plans" to reduce unaccounted-for water in projects have floundered because of operational problems and lack of management commitment. For example, very recently in Colombo, Sri Lanka, consultants trained crews for 12 specially equipped vehicles to carry out leak detection surveys. Only seven months after the consultants left the country, three of the vehicles were diverted for other use, and two were not operational. The slippery downward slide had started, and without timely intervention, another seven months would probably have seen the demise of the program. Whenever consultants are called in, there should, therefore, be a commitment from senior management to continue the program, and to the concept of controlling unaccounted for water. It requires persistent effort over several years (five to ten) for long-term benefits to be achieved. For instance, according to one report Tokyo needed about 20 years to reduce unaccounted-for water from 50% to 20%. There is no quick fix and, of course, this does not make such a program attractive to decision makers who are often in office for less than 5 years.

Faced with the need to take action on unaccounted-for water, some city authorities are daunted by the complex nature of the problem and give up almost from the beginning; they do not know where to start. When consultants are brought in, they usually have narrow terms of reference limited, perhaps to carrying out mapping and pilot leakage surveys. Their success could be marginal because important financial, institutional, and manpower development aspects are often overlooked.

Tackling the Problem

The World Bank is making a renewed effort to assist committed cities and countries in their efforts to tackle the problem. As a first step, a set of comprehensive guidelines is in the making; their approach is to demonstrate, from the beginning, that investment in reducing unaccounted-for water can be financially attractive. Since corrosion is a problem in some systems, another set of guidelines is also under preparation for protecting pipes against corrosion in water supply and wastewater systems.

The long-term goal is to achieve competent management of the water supply network. The first step to efficient management of a distribution system, assuming that it has been soundly engineered and well constructed, is to develop good maintenance and operational practices. Many developing countries are unwilling or unable to provide the required resources; furthermore, decisionmakers, much as they may agree with the need for sound operation and maintenance, tend to cut budgets in this very area when there is a resource constraint. However, if it could be demonstrated that a meaningful impact on revenues can be achieved through improved maintenance by starting with the reduction and control of unaccounted-for water, and that it is cost effective to invest in maintenance, there is a good possibility that the needed resources would be provided. This is the basic thrust of the guidelines. The pressure for large investments in new water production facilities, which often increase unaccounted-for water, may thereby be decreased.

There is nothing new in regard to technology in the guidelines. What they attempt to do is organize the work required for reduction and control of unaccounted-for water in such a way that meaningful benefits accrue within a few months of the program's implementation. The result should be to provide encouragement for decisionmakers to continuously evaluate the work done and gradually provide more resources so that sound operation and maintenance practices are initiated.

In almost every town there are large users of water. These are usually government institutions, large commercial undertakings and, in bigger cities, light and heavy industry. These users may represent about one-quarter of total consumption, and because of higher tariffs, many contribute almost half the revenue of some water supply agencies. But because of incorrect sizing of the large meters used to measure this consumption, meters that do not work, or that underrecord flows because they are not maintained, the water authority may be losing far more revenue from these large volume users than through illegal connections or even malfunctioning domestic meters. For example, in the town of Lowell, Massachusetts, one large meter is calculated to be equivalent to about 400 domestic meters. In developing countries the proportion could be greater, making it more cost-effective to start the unaccounted-for water reduction program with the large water consumers. A large meter rehabilitation and repair program could, therefore, be the first step in the long road toward reducing and controlling unaccounted-for water.

The Task Force Approach

The guidelines recommend that a water authority that is committed to the reduction of unaccounted-for water establish a task force headed by an engineer of adequate stature to work in collaboration with finance, O&M, production and distribution, and training managers. One of the first steps to be taken by the task force should be to check on the large users of water. The finance manager should ensure that the billing and collection of revenues for these large consumers is correctly organized so that within a six-month period a substantial increase of revenues would be achieved, justifying further expenditures on the program. Though not always necessary, it may be wise to hire consultants to provide the specialized expertise.

Basic Data Checks

Typically, a consultant hired to address the overall problem faces an unresponsive environment in which the management is not yet convinced of the need for implementing sound operation and maintenance practices. Such managers are usually awaiting the end of financial constraints in order to continue the traditional policy of improving the supply of water either by developing new sources or expanding treatment works. Data relating to the existing supply may be inadequate for a realistic assessment of the supply situation; available figures on the quantity of water produced and consumed may be so inaccurate as to call for a good deal of interpretation by a seasoned professional. In many developing countries production meters, if provided, do not work, and those that do work may not have been calibrated in years. Local expertise is usually non-existent as are spare parts for repair and equipment for calibration. Varied techniques such as using test meters, pumping into a reservoir and measuring volume change to arrive at pump outputs, diversion to waste through a weir etc., would yield a reasonable estimate of total production. For similar reasons, estimation of water consumption may be even more difficult. Pilot studies may have to be carried out in different kinds of social areas in towns where there is no domestic metering or where most domestic meters are inaccurate or do not work. Consumption estimates will also have to be made of large users such as industry, government, or institutions through spot checks, and reference to historical data, followed, if necessary, by extensive checks on large meters. Next, tests should be made to discover any major hidden leaks on trunk mains, which are rare, and for leaks from reservoirs or losses by overflow caused by uncontrolled pumping, which are more common. The guidelines describe in detail these and other areas in which basic data must be obtained.

Strategy

If the basic data checks described above reveal that the level of unaccounted-for water is over 25% of net production (in some water-short countries 15% may be considered to be excessive), a preliminary financial analysis is required to justify budget allocations for a more detailed study, including leakage surveys, and to obtain a commitment from management for developing and implementing a 2-3 year action plan to improve O&M as part of a broader 5-year program. The task force could be phased out at this point, having, with the help of consultants, obtained the commitment of management, which by that point should be aware of the cost effectiveness of the program and the good possibility of delaying major investments in production facilities. By the end of the 3rd year, solutions to metering and leakage problems, which are usually the most evident, would have been developed. At this point a cost benefit analysis should be undertaken to justify further investment in improved mapping and meter- reading procedures, rehabilitation, training of the work force, and ultimately to work toward a distribution system management that is organized to be proactive rather than reactive. This could take another 5-10 years. In this brief paper it is not possible to describe the detailed procedures described in the guidelines, which are expected to be published later this year.

Conclusion

The experience indicates that on a project-by-project basis, unaccounted-for water has decreased very slightly or has increased as a percentage of production. The overall trend is for the level of unaccounted-for water to remain high. Consequently it would be advantageous to the supply institution to undertake additional efforts to reduce unaccounted-for water if there is a reasonable chance that the loss reduction program would cost less than the potential revenue gains (if there is excess demand) or the potential operating cost savings (if there is excess supply). While the figures cited here for potential benefits should be treated as indicative only, they suggest that many supply institutions in developing countries would find it in their interest to seriously investigate improved approaches to the reduction of unaccounted-for water. From the social point of view, water that is produced and then lost through leakage (which is often a major cause of the problem) represents a nonrecoverable loss in potential benefits to water users or a nonrecoverable financial resource cost. Those interested in both the viability of water supply institutions and the social benefits of improved water supply have a stake in seeking ways to reduce unaccounted-for water.

THE OPERATIONAL DEVELOPMENT PROGRAM OF THE WATER SERVICE COMPANIES IN BRAZIL

Author: Paulo Roberto Carraro Bastos *

The purpose of the paper is to offer a view of the Program launched in Brazil in 1982 for the Operational Development of the water supply companies.

Emfhasis is given to the role played by the company at the operational and commercial areas, bearing in mind that these areas gather the bulk of the company personnel. Attention is drawn to the fact that the goals of the water company are closely related to the performance of these areas.

BACKGROUND

In the seventies and eighties, the sanitation sector in Brazil went through an accelerated expansion, in order to provide brazilian cities with water supply systems and, therefore, cover the existing deficit.

In effect, if, in 1973, the state companies operated aproximately 500 water systems, in 1976 this number increased to 1200, reaching today more than 4.000 systems.

This accelerated expansion process led to an increase of the scope of activities of the sanitation companies, with an expressive growth of the number of systems in operation. As all efforts were directed to the construction of new systems, a great activity took place in the development of projects and construction, while operation and maintenance were left in a secon place.

Around 1981, we were concerned with one aspect of the problem: while we were living through continuous process of water production systems growth, a project with 10 year scope would be saturated in 4 or 5 years.

At the same time, the companies would develop financial difficulties, since they needed large investments to maintain their levels of services.

BNH - Avenida Chile, nº 230 - 17º andar - Rio de Janeiro
CEP: 20.039 - Brazil

As a consequence, the elaboration of a research and diagnosis project about water leakage was promoted this research led to the solution of the problem, in terms of the so called State Program of Leakage Control, developed by all the Companies.

After consolidation, in the national level, of the data obtained, the average level of leakage was found to be around 50%, as well as that the system operation control was below the expectations.

CAUSE ANALYSIS

Starting with the above mentioned data, an analysis of the causes of the elevated water leakage was conducted, as well as on its consequences on the eficiency and productivity of the Company.

In the brazilian case we could mention, in general, some of the aspects that led to the detected problems:

- A process of accelerated expansion, linked to the lack of resources for the operation, maintenance and commercial sectors;
- Analysis based on project indicators, without attention to the levels of operational control in under wich the systems would be operated;
- Privileged treatment to the production sectors in regardless of the distribution sectors;
- Increasing investments in the ampliation of production systems, due to the increase in the distance between production and consumption places, which causes the elevation of flow and pressure in the existing networks;
- Lack of control in the operation of water systems, usually characterized with only one indicator: "LACK OF WATER".

A CYCLIC PROCESS

We found out that we were living through a cyclic process, where defficient control (the only indicator used was "LACK OF WATER") led to solutions that implied in an unilateral decision in the project and construction sector, in the sense of increasing water production and/or distribution, without consideration to the fact that the volume of water treated and lost could represent a supply at a practically zero distance and at a low cost.

STATE OF THE OPERATIONAL SECTORS

The operational sectors of the sanitation companies have shown, in general, the following deficiencies:

- Insufficient and inadequate material resources;
- Lack of sistematic training of human resources;
- Lack of institutional resources, specially in terms of the planning and control of operational development actions;
- Lack of financial resources.

BASIC ASSUMPTIONS

Those facts given, the creation of a program that involves changes in the cultural, technological and institutional aspects of the companies is needed. It starts in the operation maintenance and commercial sectors and develops through all the company, eventually coming to the sanitation sector, in terms of actions related to the services of water and sewage.

OPERATIONAL DEVELOPMENT PROGRAM

The operational development program has as an objective to supply the sanitation companies with resources to structure and develop its final activities - operation and commercialization-through the execution of the following activities:

a) Pitometry and macro-measuring;

b) Technical Cadaster;

c) Consumer Cadaster;

d) Integrated system of service provision and public relations;

e) Micro-measuring;

f) Maintenance and reabilitation of operational Units;

g) Improvement of materials and equipment quality;

h) Review of project and construction criteria;

i) Operation development;

j) Bill preparation and collecting;

l) Commercialization.

We should now make some considerations about some fundamental aspects of the strategy established for the development of the Program:

- Priority, within PLANASA, for the operational control Program in the definitions of the companies investment Program;

- Execution of the Program in stages, guarantering a regular flow of resources for the development of all activities;

- Creation, within the Companies, of a planning and operational control Unit, with the following responsabilities:

a) Planning the development of operational areas integrated to the Company global planning;

b) Implementation of the operational control proceedings;

c) Management of the operational control Program.

- Definition of performance indicators that would allow the monitoring of the Program's results during it's evolution;

- Development of a technology interchange process amongst the Companies.

EXPECTED RESULTS

With the objectives defined and the strategy for Program Development established, the following results were expected:

. Direct Results:

- Reduction of operational expenses;

- Postponing of investments in the ampliation of production systems; and

- Increase in the income collected, with a consequent elevation of operational income without any tariff increase;

. Indirect Results

- Increase in sector productivity;

- Increase in covering minor costs;
- Improvement of the company image in the community;
- Consumer satisfaction;
- Increase in the availability of resources for investments in sewage systems; and
- Reduction in tariffs in a medium and long range.

USING THE CCP APPROACH AT HERKIMER, N.Y.

John K. Esler, P.E.,* and Timothy J. Miller**

Using a two step approach to improving plant performance, an operator-engineer team from New York State Department of Environmental Conservations' Operations Assistance Section describes their program for this 1.7 mgd activated sludge facility. Following an initial comprehensive performance evaluation which revealed numerous causes for the plants poor performance, a program of technical assistance was implemented to address all the major performance-limiting factors.

Introduction

The Herkimer Wastewater Treatment Plant was constructed in 1970 to serve an upstate New York village with a population of 8,200. It was designed to treat 1.7 mgd using the conventional activated sludge process for lower flows and the contact stabilization mode for flows approaching the design rate. Due to seasonal high flows from infiltration and inflow, the plant was always operated in the contact stabilization mode.

This plant was designed with two parallel flow trains: two complete-mix stabilization basins and two complete-mix contact basins, followed by 2 rectangular clarifiers. The major treatment units are:

- hand-cleaned bar screen
- comminutor
- 3 influent pumps rated at 1250 gpm at 30' TDH
- aerated grit chamber
- 12" Parshall flume
- 4 aeration basins: each 35' x 35' x 12' SWD
- 2 rectangular clarifiers: each 66' x 16' x 8' SWD
- 2 aerobic digesters: each 35' x 70' x 12' SWD
- 2 digested sludge decant tanks: each 45,000 gallons capacity
- 1 one meter sludge dewatering belt press

In the Fall of 1984, Herkimer was ordered by the Court to take several actions to improve its effluent:

- o Commence an Infiltration/Inflow reduction program.
- o Replace one constant speed influent pump drive unit with a variable speed driver.
- o Construct a bypass for relief during high flows.
- o Install a polymer addition system to improve solids separation in the clarifier and, if the polymer system was not sufficient, construct a third clarifier.

While the court order was being developed, the Village committed

* Chief, External Development, NYSDEC, Albany, NY 12233-3506
** Wastewater Course Instructor, NYSDEC, Albany, NY 12233-3506

themselves to an annual budget appropriation for the reduction of I/I. They also field-tested a polymer addition system ahead of the final clarifiers; this was found to have an excessive operating cost for chemicals.

At about the time of the court order, the Village became aware that our Department had a technical assistance program that could benefit their plant. They had also heard of our recent work upgrading final clarifier performance. Faced with a court-mandated deadline for improving their operation, the Village requested our assistance to work on their final clarifiers. We agreed to take on this project, but only if we first conducted a thorough evaluation of the plant and their operation. This evaluation was to follow the format recently introduced by the USEPA in their handbook called "Improving POTW Performance Using the Composite Correction Program Approach."

The "CCP" Approach

The CCP Approach was developed by EPA as the result of hundreds of site visits to "troubled" plants...plants that were unable to meet their designed treatment expectations...plants that didn't have major I/I problems or design deficiencies, or major problems with industrial wastes. These site visits eventually led to detailed evaluations of over a hundred of these plants to identify the most predominant reasons for their failures. This resulted in listing of common problems in all four major areas that affect plant performance: design, administration, operation, and maintenance. The main conclusion of this effort was that each plant usually has a number of performance-limiting factors that are unique to that facility. Over 70 of the most common factors are listed in this CCP handbook as a part of a new operations auditing technique called the "Comprehensive Performance Evaluation"...the "CPE".

The CPE

The objectives of a CPE are two-fold: first, to identify all the major factors limiting performance, and second, to determine whether either a major facility upgrade is necessary, or, if a "Composite Correction Program"...a "CCP"...is capable of producing the desired effluent quality.

There are five main activities in a CPE:

1. an evaluation of the major unit processes
2. an identification of the performance-limiting factors
3. a prioritization of the performance-limiting factors
4. assessing the ability of a CCP to improve performance
5. writing the CPE report

Although these are different activities, some are conducted concurrently with others.

Conducting the CPE at Herkimer: Preliminary Office Work

In order to make the on-site activities more productive, one of our first steps was to search our files for the recent compliance monitoring reports. In addition to these reports, we also reviewed the quarterly inspection reports from our Regional Office. This helped us identify the key people involved in this plant and gave us a good feel for the problems that had previously been identified. As a final office activity, we looked over the contract drawings and calculated the volumes of the various treatment units.

Field Activities

Our initial on-site activities were devoted largely to the collection of data required for the evaluation of the major unit processes. We also observed the total operation so that we could make the subsequent identification of performance-limiting factors. Following our routine "kick-off" meeting, we then toured the plant, looking at the equipment and asking related questions. This is time when experience in plant operations.....and tact.....are the most important ingredients.

The attitudes of the plant staff about their operations and their relationships with the various regulatory people and consultants who had been active at the plant in the past was another important facet of this evaluation. The total time allotted to this field activity was three days in the field by a two-man evaluation team. This team consisted of an engineer and former wastewater treatment plant operator, both of whom were familiar with the operation of activated sludge plants.

It should be noted at this point the importance of using a team with field experience in performing this on-site evaluation. The teamwork is important both in dividing the work and in maximizing the use of the different talents of the operator and the engineer. It is also axiomatic in this type of work that "two heads are better than one" in analyzing situations and especially in formulating possible solutions.

Back to the Office

After the field work was done, the team returned to the office to assemble its observations in a logical order and fill in the gaps in the data needed to complete the evaluation. This step is simplified by using the forms provided in Appendix D of the CCP Handbook. This leads to the next step which is the evaluation of the major unit processes.

There are two ways to do an evaluation of the major unit processes: use the forms and the technical guidance presented in the Handbook, or use a simpler software package developed by the EPA for use on the IBM PC. This evaluation examines the three major unit processes at a treatment plant: the "aerator" (meaning the biological reactor--aeration basin, trickling filter, etc.); the secondary clarifier; and the sludge handling capability. Key loading and process parameters are calculated and the results for each parameter assigned a score by comparison with standard tables. Then, each of the three major unit processes receives a total score by adding together the points assigned to the loading and process parameters. At the conclusion of this step, the facility is judged to be either:

a Type 1--all major unit processes are adequate; or,
a Type 2--major unit processes are marginal; or,
a Type 3--major unit processes are inadequate.

At the Herkimer plant, the "aerator" was a Type 2. This marginal rating resulted from the high organic loading. The "clarifier" was a solid Type 3 on the basis of its configuration, shallow side water depth, sludge removal method, and surface overflow rate. The "sludge handling capability" was a Type 1, although it was difficult to evaluate because of its dependence on a single decant tank in the winter, and a single mechanical method of sludge dewatering. Based on the standard CPE criteria, this plant was ranked as a Type 3, using the "worst case" determination. We then proceeded with the identification of the performance-limiting factors.

The Performance-Limiting Factors

The previous major unit process evaluation is used to broadly categorize performance potential by assessing only the physical processes. In contrast to this, the evaluation of PLF's focuses intensely on the individual facility and the factors unique to that facility. This part of the evaluation identifies the capabilities and weaknesses of existing operation and maintenance practices and administrative policies. In evaluating the design aspects, the focus is on the operability and the flexibility of the various unit processes. At Herkimer, we identified an unusually large number of PLF's as requiring correction, 19 in all. These are shown in the following listing with a notation of the particular reference code from the Handbook. The most serious factors were:

- staff motivation (A2b1)
- poor process controllability (C2c2)
- deficient clarifiers (C2g)
- sludge treatment limits; sludge wasting (C2g)
- ultimate sludge disposal limited (C2h)
- cold weather inoperability (C3l)
- limited sewage treatment understanding (D1c)
- insufficient process control testing D2b
- insufficient application of concepts & testing to process control (D3a)

Also of importance for their effect on performance were:

- excessive I/I (C1f)
- lack of unit bypass for aeration (C3c)
- inappropriate level of certification for operator (D1b1)
- non-attendance at available training courses (D1b2)
- delays in procuring critical parts (B3b)
- problems due to hydraulic surges (C1b)
- inadequate preliminary treatment (C2a)
- insufficient alarm systems (C3e)
- equipment inaccessible for maintenance (C3k)
- inadequate O&M manual (D4a)

Presentation of CPE to Administrators and Staff

The presentation of the evaluator's findings is an important aspect of the CPE. Even if the owner elects not to proceed with a composite correction program, the CPE can form the basis for planning future plant improvements. In most cases, there are some factors identified that relate either to the operator's lack of sewage treatment understanding or a deficiency in the application of his knowledge or test results to control the process. At Herkimer, both of these factors were identified as severely limiting the performance of the plant. The challenge for the evaluator is to present this type of finding in such a way that the owner and his operator recognize this as a problem that needs attention, but do not feel personally affronted by what can be perceived as a challenge to their worthiness as operators.

It is not essential at the CPE presentation to go into detail concerning how the plant can address each deficiency. Many of them are self-explanatory. It is beneficial, however, to go over the factors that the program will address and give some indication as to which deficiencies are correctable without major expenditures of money. Furthermore, assuming the CPE gives the evaluator enough hope that the plant can be brought into compliance through a composite correction program, the report should give an estimate of the time and level of effort that would be involved in this program.

At Herkimer, the owner was encouraged by the findings of this CPE that they might be able to avoid a major capital expenditure for a clarifier and still improve their effluent if they were diligent enough in undertaking the program suggested.

The Composite Correction Program: Activated Sludge

The clarifiers had previously been the focus of the blame for the plant's poor performance. We determined, however, that the first element of our program had to address the weaknesses in the activated sludge control program, for indeed, there was no existing strategy for controlling the process.

Based on previous experience, we selected a program using a target level for the pounds of solids to be maintained in the system. This was preceeded by a slow, informal presentation of general information about how the activated sludge process works, as well as a discussion of what kind of things can happen to upset the process. The major controls that were established included a trial rate for return sludge and a method for calculating daily the proper amount of sludge to be wasted. This required major changes by the plant staff since sludge had not previously been returned continuously, and wasting was done only when the wash-out of solids seemed excessive. The high flows due to the I/I problem had often taken care of much of the need for wasting. The new need for daily wasting involved a considerable commitment of additional time. It required sampling and analysis seven days a week for the actual wasting routine, and additional time and money for processing and disposing of the waste sludge.

The results, however, were well worth the effort. Within weeks the process had stabilized and the task of finding the optimum solids to maintain for the activated sludge inventory could begin.

There was still a problem in controlling the process during periods of high flows. One detail that we then discovered was that there were three submerged ports connecting the stabilization basin to the contact tank. These were in addition to the two wall openings that were visible at the surface. This seemed to explain the fact that the mixed liquor concentration in the stabilization basin was only slightly above that in the contact. By blocking off the submerged ports, we hoped to be able to maintain a higher inventory of solids in "stabilization" and protect them from washouts. We soon discovered, however, that the backmixing that resulted from just the two surface openings was enough to prevent the desired build-up of "stabilization" solids. The solution that we finally settled on was to completely block off one of the surface ports, and to partially block the remaining port so that it acted as an overflow weir. This prevented the contents of the contact basin from entering the stabilization basin. This modification resulted in a doubling of the concentration in "stabilization", which provided the necessary reserve of activated sludge solids that could then be protected during high flows.

Clarifier Improvements

Having established working controls for the activated sludge process, it was then possible to try to improve the clarifier performance. Our first step was operational. The blanket levels, which before this time had not been subject to control, were now monitored twice daily. This was done in order to maintain as close to a zero blanket level as possible without seriously diluting the return sludge concentration. This helped limit the amount of solids that could be washed out at high flows.

During the previous year, we had shown in similar clarifiers at another plant how a slotted mid-tank baffle could be installed that would markedly improve suspended solids capture. This same baffle was then installed at Herkimer with a similar degree of success; the effluent suspended solids was improved by approximately 30 per cent. We then experimented with several other baffle configurations and locations, some of which were an improvement and some of which didn't work as well. The configurations that seemed to work the best in these clarifiers were: (a) a solid baffle at mid-tank, whose top was about three feet under the water surface, and (b) a set of slotted baffles, one at mid-tank and one at approximately the two-thirds point. With these clarifier baffles and with a stable activated sludge process, Herkimer was now able at times to produce an effluent suspended solids level of less than 20 mg/l. The daily average suspended solids had been reduced from a level of 44 mg/l during the year preceeding this project to 28 mg/l during the first eight months following the project.

Sludge Wasting Procedures

One of the most important aspects of the new process control program was the sludge wasting operation. Each morning, in order to calculate the solids inventory, the various mixed liquors and return sludge concentrations were determined. Then, by comparing the current inventory with the target set-point for the solids inventory, the correct amount to waste could be determined. Since the waste sludge flow meter was inoperative, a system was devised using the waste sludge concentration and a target depth of sludge to be wasted into the decant tank in order to know when the correct volume of sludge was wasted. After the wasting operation was completed, a sample from the mixed contents of the decant tank was then analyzed to determine the actual pounds of sludge that was wasted. This system allowed the waste sludge volume to be controlled to the nearest 200 pounds.

Dampening Flow Surges

The constant speed influent pumps each had a capacity of 1.7 mgd at the design condition. Since each pump discharged independently to a distribution box, the combined discharge from two pumps was 3.4. mgd. During periods of normal flow, one pump would cycle ON and OFF, causing the flow to surge through the system and through the clarifiers at a rate of 1.7 mgd. When the flow exceeded this rate, the second pump would then commence pumping with an additional 1.7 mgd rate of flow. Both pumps would continue at this high rate until the wet well was drawn down; then, both would shut off at the same time.

Our first modification was to adjust the low level "cut-offs" so that the "lag" pump shut off at a higher level, thus keeping the "lead" pump in nearly continuous operation. This eliminated a major source of ON-OFF cycling. Then, in order to dampen the remaining surges, the 35 foot long effluent weir from each aeration contact tank was raised 6 inches along most of its entire length, leaving a slotted opening at one end that was approximately 30 inches wide and 4 inches high. This opening acted as a control orifice which permitted normal flow to pass and allowed the surge due to the second pump to be stored in the contact tank until the water surface rose 6 inches. It would then overflow the new weir. Although this did dampen the intensity of the surge, its effect on the plant operation could only be evaluated subjectively since there were always several other factors affecting the plant effluent at the same time.

Improving Flow Measurement

As the project progressed, we observed that the flow was approaching the Parshall flume with a considerable amount of energy...so much energy that there was often a hydraulic jump just ahead of the flume. This created extremely turbulent conditions in the flume and was also an indication of a high velocity of approach to the flume. Using a portable magnetic current meter, we found that the velocity often exceeded 3 feet per second. With the 12 inch Parshall flume, this caused the flow to be underestimated by approximately 25%. At other times, the turbulence in the approach channel would cause a wave under the ball float, resulting in a higher flow indication. These conditions were eliminated by placing a series of restraining baffles in the channel leading to the flume in such a manner that the velocity of approached was reduced to about 2 feet per second, which is within the acceptable range for this flume. As a part of our training effort, the operator was also taught how to use the Parshall flume tables and a measuring rule to check the calibration of flow indicator.

Other Improvements in Operation

In addition to these changes, there were numerous other activities undertaken during the CCP:

- The operating staff revised piping and equipment covers to prevent freezing during the long winter.
- The chief operator attended the training courses needed for certification.
- We assisted the staff in locating and sizing a plant bypass for use during seasonal high flows.
- Changes were made in grit and rag removal to improve efficiency.
- The operating staff was shown how to make use of pressure guages and pump curves to monitor the operation of their centrifugal pumps.

Summary

This plant was typical of many that are unable to meet their effluent limits. A thorough evaluation of all aspects of the plant operation showed that there were a number of reasons for the poor performance; reasons that were related to the design, the operation and the maintenance of the plant. By addressing these deficiencies in a systematic manner, the process was brought under control, design deficiencies were corrected, and the plant was able to meet its effluent limits without a major expenditure of money. Even more importantly, the plant staff acquired new skills that enabled them to continue to improve the plant and function as a team with justifiable pride in their work.

OPTIMIZING FINAL CLARIFIER PERFORMANCE

John K. Esler, P.E.*

The performance of a final clarifier is a function of many different factors, including the design, maintenance and operation. In addition to the clarifier itself, there are a number of external factors that influence its performance. This paper discusses the designer's and operator's role in optimizing clarifier operations, as well as the techniques that can be used to evaluate and improve clarifier efficiency.

INTRODUCTION

The performance of a final clarifier is a function of a number of different factors. These factors are related to the overall design of the treatment plant as well as the specific design of the clarifier. They are a function of the overall operation of plant as well as the operation of the clarifier and its appurtenances. It follows, then, that if we are going to optimize the performance of this unit, we should carefully assess not only the clarifier, but also its relationships to the rest of the plant...for, indeed, there are as many opportunities for improvements outside the clarifier as there are within it.

CLARIFIER DESIGN OPTIONS

a) Delivering the Flow

Assuming the flows have been accurately predicted, the initial concern is to deliver the flow smoothly to the clarifier. The surges due to ON-OFF pump cycling are only slightly dampened by primary and secondary treatment units. These surges reach the final effluent weirs in minutes, with peak flows corresponding to the initial pump output. Variable speed pump controls can avoid this situation.

The accurate division of flows among clarifiers is the next major concern, with tapered channels and symmetrical layouts the normal engineering solution. In most projects, however, the results are difficult to measure. In the studies reported by Crosby and Bender[1], and in several plants measured by the author, these flows have been found to vary by as much as 30 percent among parallel or symmetrical units during normal operating conditions. An even more serious problem routinely occurs when one or more clarifiers is taken out of service for maintenance or repairs, or, when the units "planned for the future," but not yet built, are needed to achieve symmetry. One solution that accomplishes the design objective is the use of an up-flow distribution chamber. With adjustable overflow weirs, flow

Chief, External Development, NYSDEC, Albany, NY 12233-3506

can be evenly proportioned to all units, or varied according to weir settings.

b) Choosing a Clarifier Configuration:

There are so many clarifier shapes to select from, each with a number of different weir and launder configurations, with almost no real-life performance comparisons. This decision is often simplified by resorting to the "standard" of using the same design that was used on the last project. However, there is such a tremendous difference in the hydraulic retention times between the many options available that the designer should consider optimizing this particular parameter when making his selection.

Two general types of clarifiers that I have found to yield the poorest actual retention times are the circular clarifier with the single perimeter weir, and the rectangular clarifier with a cluster of launders at the far end. Considering their dismal hydraulic efficiency, in the range of 20 percent of their theoretical values, these types should not be used for secondary treatment without the use of some additional devices to improve their performance.

On the other hand, the types of units that have consistently yielded the highest retention times, in the range of 45% to 75%, are the circular units with a peripheral spiral feed system and circular units with one or more inboard launders. Also in the same class as these is the circular unit with full-surface radial launders, either on the surface or submerged.

There's one shape that has proven to be an operational "headache" in virtually every installation in New York State. We call this the "squircle"--a square clarifier with an adaptation of the circular clarifier's scraper mechanism. Apparently, designers are attracted by the opportunity for economical, common-wall construction. However, this design has two major flaws: first, it promotes an uneven flow pattern favoring the corner weirs; and second, the articulated corner sweeps are notoriously prone to mechanical failure.

c) Inlet Design

With rectangular clarifiers, the designer should provide multiple, adjustable inlet gates that are visible to the operator. Gates that function as overflow weirs have the advantage of providing even flow distribution while permitting floatable materials to pass on to the clarifier.

With circular clarifiers, the designer may select either a peripheral or center-feed design. The peripheral feed systems rely on some form of tapered channel or raceway to evenly distribute the flow, although it is questionable that this objective is actually accomplished. One type of inlet imparts a spiral motion to the clarifier contents, which may account for its high hydraulic retention time. The main operational objection to peripheral feed is the problem of scum accumulation in the inlet channel. Although rotating skimmers can be provided, it would probably be helpful if the inlet skirt were lowered to a position just above the water surface, thereby preventing the normal build-up of floatables.

The center-feed designs have a similar problem with the floatables that do not seem to be released by the usual openings in the centerwell. Again, a lowering of the centerwell will minimize the build-up. As an alternative, a simple slide-gate can be used to decant the scum to the return sludge well in clarifiers with hydraulic sludge removal devices.

One main advantage of center-feed units is that they can be designed to enhance flocculation of the biological solids, which should produce better settling and clarification.

d) Sludge Withdrawal

The vast majority of rectangular clarifiers are provided with counter-current sludge scrapers and sludge hoppers at the inlet. The major concern with this arrangement is the possible scouring of return sludge near the inlet baffle. In larger units, say over 150 feet in length, the sludge hopper is often moved to a location close to the mid-length. In some installations the sludge scapers move with the current to hoppers at the far end of the clarifier. Although this has performed well in some cases, there is an inherently higher hydraulic loading throughout the clarifier since the total flow from inlet to outlet includes not only the normal through-flow, Q, but also the recycled sludge flow, R. This produces an additional velocity component, due to R, that may promote the transport of solids to the effluent weirs.

With circular units, there are numerous options. The suction type withdrawal system, using an "organ pipe" configuration, is favored if rapid sludge is a design objective. These systems are more difficult to regulate if the tube openings are submerged in the centerwell than when they are provided with a telescopic valve overflow arrangement. The plugging problem that often plagues the operation can be minimized by closing off the alternate suction tubes on each scraper arm to increase the veolicty in the remaining tubes. These centerwells also have a periodic maintenance concern with constant wear of the rubber or neoprene seal that is designed to prevent the mixed liquor from diluting the contents of the return sludge well.

Another type of hydraulic sludge removal uses a submerged single or double suction arm with varying port openings along the length of the arm. If the preliminary treatment permits rags to reach the clarifier, the smaller of these ports can plug without the operator being aware of the condition.

Probably the most trouble-free of sludge collection systems is the rotating plow which scrapes the sludge to a center hopper. These systems are virtually free of operating problems and have proven to be adequate with all types of sludges.

e) Scum Removal

There are two important considerations in circular clarifiers for good scum removal: first, the skimmer should be full-radius and have some backward pitch from the centerwell in order to encourage the scum to move to the perimeter baffle, and, second, the scum hoppers should be located on the down-wind side of the clarifier in order to take advantage of the effect of the wind pushing the scum. Although continuous scum removal is required by most design standards, it should probably also be considered for the chlorine contact tank, since scum will always collect there.

f) Weir Placement

In rectangular clarifiers, weirs should be kept away from the end wall and should probably cover at least 30 percent of surface. Weir launders placed parallel to the flow line will minimize interference with the flow pattern while providing adequate weir length to minimize upward velocities in the clarifier.

With circular clarifiers, the designer should provide multiple launders placed away from the side walls to minimize carry-over of solids caused by the bottom currents. Although a system of radial surface launders would provide for the optimum hydraulic condition, they would preclude any scum collection.

THE OPERATOR'S ROLE

A well-designed clarifier is like any other well-designed machine. Given the proper maintenance and operated in accord with the designer's intentions, it will perform without fail. It is the role of the wastewater treatment plant operator to see that these conditions are met. In doing so, he must always keep in mind the demands of the biological processes in order to meet the total treatment objectives.

a) The Biological Process

In order to achieve the best clarifier performance, the biological process must be operated to avoid a number of pitfalls that can drastically effect suspended solids removal. Some of these problems are:

- filamentous bacteria growth that prevents compaction
- excessive Nocardia surface scum
- denitrification
- rising sludge due to anaerobic conditions
- dispersed growth leading to poor solids capture
- excessive solids inventory

Although a good clarifier can sometimes compensate for these problem conditions, they often lead to an excessive loss of solids in the effluent.

b) Sludge Blanket Control

Controlling the sludge blanket is often the operator's most productive activity for maintaining good performance. Although some blanket may be necessary to achieve the desired return sludge concentration, the sludge blanket should generally be kept to a minimum, particularly with rectangular clarifiers.

c) RAS Flow Control

The operator has two concerns with return sludge flow control: meeting process control demands, and withdrawing sludge uniformly from all clarifiers. Although the latter objective is readily met with a separate RAS pump dedicated to each clarifier, in a surprising number of installations the operator is required to constantly monitor and adjust RAS rates because of poorly-designed manifold systems.

d) Scum Removal

The condition of the surface scum provides a useful indication of the status of the biological process. Its removal is routine with surface skimmers, sometimes aided by surface spray water. However, during times when Nocardia is excessive, or when denitrification is occurring, most scum removal systems prove to be inadequate. As a short-term solution the operator can apply a hypochlorite spray to reduce Nocardia scum, or increase RAS rate to avoid denitrification. However, a useful corrective action for Nocardia control is often to reduce the mixed liquor concentration. If nitrification is not a treatment requirement, adjustments to the biological process can be made to reduce nitrification.

ANALYZING CLARIFIER PERFORMANCE

The performance of a final clarifier has been shown to be a function of numerous factors, both internal and external to the clarifier. Any analysis of a clarifier must seek to identify <u>all</u> those conditions that might limit its performance. Prior to conducting field activities, a review of contract drawings and plant data will help to evaluate the design factors. Then, there are a number of field activities that will help identify the problem areas:

a) Sludge Blankets

Sludge blankets should be examined using a transparent core sampler. By releasing measured portions of the core sample, a solids profile can be developed. The entire core sample is also useful as a true composite sample of the clarifier contents for calculating sludge quantities.

b) Diurnal Sludge Blanket Patterns

The actual shifting of solids to the clarifier can be quickly determined by the use of cross-sectioning techniques with a core sampler. Sample amounts, say, at one or two foot intervals, are obtained from the sampler, and sludge density contours can then plotted on a vertical profile of the clarifier.

c) Weir Levels

Weir levels should be checked visually at low flows to find the high and low points. An unlevel weir will induce an imbalanced current in that direction.

d) Diurnal Effluent Samples

Using a 24-bottle composite sampler, the daily change in effluent suspended solids will indicate any time pattern for loss of solids.

e) Biological Process Performance

In addition to the normal process control observations, the examination of the settlometer supernatant for turbidity and suspended solids will establish the baseline for the clarifier effluent. The use of a good phase contrast microscope will also aid in identifying settling problems due to filamentous bacteria.

f) Flow Measurement

A major cause of poor clarifier performance is the imbalance of influent flows among clarifiers and within the individual unit. In order to measure flows to or from a clarifier, or battery of clarifiers, the investigator usually must install some type of job-built weir or flume. These flow measurements need not be precise, but must indicate the relative differences in flows between units. Uneven RAS withdrawal will also create differences in flows through a unit, as well as unexpected sludge blanket variations. If RAS flow meters exist, their measurements _must_ be verified by at least one other independent flow reading.

g) Hydraulic Patterns

There are three levels of sophistication that can be used to identify the hydraulic conditions that are unique to each clarifier:

1) An intense dye slug can be used to visually pin-point short-circuiting currents and the existence of reverse currents.

2) A controlled dye slug introduced at the influent will yield effluent samples to develop a dye dispersion curve. This curve will indicate the degree of short-circuiting as well as the actual hydraulic retention time in the clarifier.

3) A dye tracer, continuously injected in the influent, can be sampled at several stations along the clarifier, at set time intervals, to produce a contour plot of the actual short- circuiting current as it progresses toward the effluent weirs.

Each of these tests will give a certain insight to the nature of the hydraulic currents. The dye tracer test, however, will precisely identify the cause of many clarifier failures.

SIMPLE MODIFICATIONS TO IMPROVE PERFORMANCE

As we've shown, the failure of a clarifier can be due to a number of design and operational causes. In times of plenty, the remedy of choice is often to increase hydraulic capacity by adding one or more new clarifiers. The approach that is more cost-effective, however, is to first make the indicated adjustments in operations and design in order to optimize the existing facility. Once we're satisfied that these factors have been addressed, there are usually short-circuiting currents remaining that will still severely limit the performance of the clarifier, particularly during times of hydraulic stress. The next step, then, is to improve the hydraulic condition of the clarifier by adding one or more in-tank baffles.

In-tank baffles were first publicized by Crosby and Bender[1] based on their work on circular clarifiers. In one project, a center-feed unit with a single perimeter weir was retro-fitted with a continuous perimeter baffle. In the other project a continuous circular baffle was installed at mid-radius. This latter baffle revolved with the collector mechanism. In both cases, the in-tank currents were dampened and the resultant effluent suspended solids were reduced by over 25 percent.

In recent years in New York State, we have applied a similar approach to improve rectangular clarifiers. After first locating the bottom short-circuiting currents (with velocities as high as 12 feet per minute), we were able to interrupt these currents with slotted mid-tank baffles. In each of these projects, the effluent suspended solids was reduced by almost 30 percent. In our second project, at Herkimer, we found several other in-tank baffle configurations that produced similar improvements.

Since then, we are aware of at least four other projects in New York where the plant staff has installed baffles in rectangular clarifiers. In each case, the operators have reported a dramatic improvement in effluent suspended solids capture.

SUMMARY

In order to optimize clarifier performance, the designer has the initial responsibility to provide an operable clarifier system with even flow distribution and with good hydraulic characteristics. The plant operator's role then is to optimize the operating conditions in order to produce the best effluent quality. Finally, since most clarifiers have short-circuiting currents that reduce their efficiency, in-tank baffles should be added to produce an even better effluent.

1. Crosby, R.M. and Bender, J.H., "Hydraulic Considerations that Affect Secondary Clarifier Performance", USEPA Technology Transfer, March 1980.

MATHEMATICAL MODELING OF PHYSICAL-CHEMICAL WWTP

By Edmundo D. Torruellas[1] and Dale D. Meredith[2], Members ASCE

ABSTRACT

Various mathematical models are developed which by computerized means simulates the operation of a complete physical-chemical treatment plant. It is the primary objective to optimize the treatment scheme selected and it is aimed to obtain the required treatment efficiency at a minimum economic cost. A method for determining the minimum cost design or operation of a physical-chemical treatment plant is presented. The method utilizes an existing search technique (Box complex algorithm) to formulate the design as an optimization problem with nonlinear cost functions and realistic mathematical models to represent each of its constituent unit processes. The level of cost resolution adopted derives from a series of economic models developed which basically make a Present Worth Analysis of each unit within the system. The economic data are mostly based on updated cost information obtained from Environmental Protection Agency Technology Transfer Manuals. Among the benefits directly derived from this study protrudes the development of a relatively new design concept which underlines the need of the simultaneous consideration of all the components of a physical-chemical treatment plant for the most economic design or operation.

INTRODUCTION

The present status of technology in the field of municipal and industrial wastewater treatment offers two general approaches. These can be most easily grouped as biological conversion and nonbiological or physical-chemical (P/C) removal processes. Physical-chemical system have shown the ability to produce an excellent quality of effluent in terms of the primary pollution indicators, suspended solids, organics and phosphorous. Unfortunately, P/C treatment requires large quantities of expensive chemical additives and energy resources. Also, the operation and maintenance costs are of considerable magnitude. So, its major obstacle is therefore economic. It is this same obstacle that have just recently make some investigators direct their efforts toward

---

Note: This paper is part of the author's Ph.D. Thesis submitted to the Graduate School of the State University of New York at Buffalo, Buffalo, N.Y.

[1] Chief, Region II, Sewer Proy. Design Dept., Puerto Rico Aqueduct and Sewer Authority, Santurce, Puerto Rico

[2] Professor, Department of Civil Engineering, SUNYAB

mathematical simulation and conclude that the optimum design of a treatment system consisting of several interrelated units should consider all elements of the treatment plant as a unified whole. This can be obtained by representing each unit with a realistic model and using this resource to find out how downstream units are affected by the fluctuations of the variables and parameters of linked preceding units.

METHOD OF SOLUTION

A series of mathematical models were combined and computers programs were developed to simulate the operation of a complete P/C wastewater treatment plant. The scheme of treatment modeled (See Figure (1)) included processes such as coagulation, flocculation, surge storage, sedimentation, sand filtration, and adsorption on granular activated carbon. Also, sludge treatment processes such as gravity thickening, vacuum filter dewatering, and hauling and disposal were modeled. Plant influent flowrate and its chemical and physical characteristics were some of the input parameters. Flowrate (Q), suspended solids (SS), and total organic carbon (TOC) were adopted as main quantitative describers of flow. The inflow was properly adjusted to account for incremental solids, hydraulic, and organic loads imposed by recycled flows and considered as influent to the first unit process. Design parameters were fixed and variables values selected from a preset range for each particular application for this and every subsequent unit. Then, utilizing this data the particular unit was designed and mathematically simulated. This provided effluent data information (Q, SS, TOC) and amounts of sludge produced (when applied) which were used as influent data to the next unit process linked in the sequence of processes. The same procedure was successively applied to each unit process each time utilizing previous produced information from its preceding unit as source of incoming flow description.

MODELS DEVELOPMENT

The criteria used for the selection and development of the mathematical models were: 1) generality in use and acceptance, 2) moderation in complexity but never neglecting aspects of fundamental nature for the process, and 3) existence of experimental verification of the model. Because of space limitation in this publication, these models will be only briefly discussed. The author encourages all those persons interested to go more deeply into the details of this study to contact him or the Graduate School of SUNYAB for this purpose.

PRELIMINARY PROCESS - No modeling of this process was intended in this work. Future study is recommended to include at least the grit removal process within the treatment scheme because of its effect on the sedimentation and sludge thickeners units.

COAGULATION MODELING - The purpose of the coagulation process modeling was to determine the additional mass load contributed to the flocculation unit (next unit in the treatment scheme) as a result of the reactions of the destabilizing chemicals used. The model assumes that the optimum coagulant dose (OCD) have previously been determined from jar tests as per EPA Suspended Solids Removal Manual (21) and that it consisted in the application of 90 mg/l of Ferric Chloride and 0.5 mg/l of polymer. It was also assumed that these doses produced a 95% reduction in phosphorous. The process chemistry was represented by the stoichiometry of three predominant kinds of reactions for the destabilizing chemical. Namely, these were 1) reaction with alkalinity, 2) reaction with phosphate, and 3) effect of iron complexation in water. Before these equations were solved, the solubility constants were corrected for temperature effects. The Van't Hoff relationship (25) was utilized to express this dependence. The output obtained from this model consist of: a) required tank volume and power, b) stoichiometric quantities of Phosphorous and Ferric Chloride precipitation, and c) total mass concentration contributed to the flocculation chamber.

FLOCCULATION MODELING - Due that the size of particles colliding affects the agglomeration rate constants according to Kavanaugh (14), the effect of particle size distribution was considered in the flocculation process. The distribution of particles within the system was represented utilizing concepts of Cumulative Size and Power Law Distributions (16). The formulation used is based on the work of Friedlander (9) and Lawler (16) for an heterodisperse system which describe the change in concentration of particles of a particular size as they combine with particles of other sizes during the flocculation process. The significant transport mechanisms were assumed to be Brownian diffusion and fluid shear. Expressions for the collision frequency functions for these mechanisms were used as derived by Smoluchowski (24). The author's model made two variations to Lawler's model: First, the model was developed in a fashion that virtually any number of particles sizes may be considered. Second, instead of arbitrarily fixing the largest particle size, a maximum stable flocc size was assumed according to Parker's et al theory (20). The effect of a continuous flow complete mix stirred tank reactor was also incorporated into the model (27). N equations (one for each particle size interval) were then numerically integrated using an algorithm developed by Gear (11). The output obtained from the flocculation model includes: a) required tank volume and power, and b) mass conc. of each particle size at preselected values of time intervals.

SEDIMENTATION MODELING - The effects of heterogeneity in particle size and unequal vertical distribution of particles because of settling were described by dividing the particle size range into N equally spaced logarithmic size units and by segmenting the tank depth into M equal units. Each of the M boxes was assumed to be completely mixed so that within each box the particles remain uniformly distributed. Within any time interval, particles can flocculate within each box and can settle from one box to another. The formulation used is also based on the work of Friedlander (9) and Lawler (16). The significant particle transport mechanisms were assumed to be Brownian diffusion and differential settling. M x N equations were integrated numerically utilizing Gear's algorithm (11). Another modification was made to Lawler's model. This time was the inclusion of the hindered velocity of particles (which affect the rate of particles entering or leaving each sedimentation cell) as given by Richardson (22) into the model. The volume of sludge produced in the sedimentation tank was accounted for by considering the "loss" of particles from the bottom cell to be transferred to the first sludge processing unit (sludge thickener). Similarly, the effluent from the top cell was assumed to constitute the influent to the filtration unit. Both the effluent from the sedimentation unit (influent to filtration unit) and the underflow (influent to thickener) were completely characterized in terms of flowrate, mass concentration distribution of each particle size, and volumetric sludge rates during the simulation of the events of the interactions occurring within a front of particles moving across the tank under the assumption of plug flow. The hydraulic design of the sedimentation tank consisted in the determination of required volume, area, and depth of tank given the hydraulic retention time and surface loading rate, main variables of this unit.

FILTRATION MODELING - The filtration process model predicts the effluent SS conc. and the head loss development at any depth of a sand filter. This was accomplished modifying Mohanka's multilayer filtration model (19) to make it apply to continuous size graded filters as caused by backwash of the media. To this effect another modification was done which consisted in the addition of a term to provide for the surface headloss development characteristic of these filters. The formulation utilized combines the Iwasaki's equation (12) and the continuity equation for the conc. of SS with the expressions of Ives (13) as modified and presented by Mohanka (19). These equations were then formulated as finite difference expressions and solved incrementally in time and depth of filter. The modified Carmen-Kozeny equation (19) which apply to "dirty" filters was used to describe the variation of headloss through the depth of the filter as filtration proceeds and solids builds up. To account for the surface headloss, an additional term sometimes known as Boucher's Law (2) was considered. The media size distribution was assumed to be completely characterized by the effective size and the uniformity coefficient. The hydraulic design of the filtration model consisted in determining the required filter area, depth, and volume of sand and gravel. The filtration model permits the description of the existing conditions in the filter at predetermined select time intervals. These conditions include mass conc. of particles, headloss, and specific deposit (volume of SS deposit per unit bed volume) at any filter depth increment. The model simulates the operation of the sand filter up to the point where any of the established maximums in time, effluent conc., or headloss is equaled or exceeded. A "record" is kept of the filter effluent conc. at each time increment so that an average effluent conc. (influent to adsorption unit) may be obtained.

SAND FILTER BACKWASH MODELING - The Amirtharaja's model (1) was adopted for modeling the backwash process. The expanded height of a sand filter bed can be predicted within an accuracy of aprox. 1/2 in. using this method (1). The model permits the calculation of the velocity required and headloss that will develop through a fluidized bed at a prefixed operational expansion of media. These calculations permit the sizing of the pumps and motor required and the estimation of the O/M expenses.

GRANULAR ACTIVATED CARBON ADSORPTION MODELING - The Michigan Adsorption Design and Applications Model (MADAM) program as presented by Ying and Weber (28) was adopted for this simulation. Essentially, this sophisticated model predicts the dynamic organic effluent concentration for the combined process of adsorption and biological activity in an upflow activated carbon column. Finite-difference numerical methods used for the solution of the associated algorithms were validated by Ying and Weber (28) by confirming that model simulations and predictions converged under limiting conditions on those of correspondingly simplified models of other authors. A maximum operation time and Total Organic Carbon (TOC) allowable in effluent was imposed by the author to their models. This corresponded to the fact that real operation dictates column stops for backwash or when effluent guidelines are equaled or exceeded. The design of this unit of treatment consisted in determining the depth, surface area and volume of carbon required for a particular application given the TOC characteristics of an influent approaching at a specific velocity.

CARBON FILTER BACKWASH PROCESS MODELING - Amirtharaja's model (1) was also adopted for modeling the granular activated carbon (GAC) backwash process. Although Amirtharaja's predictions for graded coal systems were not as successful as for sand filters (1), no other model was found on the literature for this process.

CHLORINATION PROCESS - The chlorination process was not modeled under this study. Because actual design practice is mostly based on the provision of a minimum detention time for this basin prior to plant flow discharge, it was felt that the economic trades inferred from the interrelation of this with other P/C treatment units was not relevant.

SURGE STORAGE PROCESS MODELING - The surge storage unit could be modeled as a flocculation basin. However, because of the uncertainty in the characterization of the recycled flows, this process was not strictly modeled. Instead, a mass balance was performed on influent and effluent flow and mass loading rates. This gave at least a gross characterization of the recycled loads composed of the heavily mass loaded flows from the backwash of the sand filter and carbon adsorption process. The hydraulic design of the surge storage basin consisted simply in determining its required holding capacity and the mass concentration of outflow to be directed to the head of the plant. Also the power requirement for tank mixing was calculated.

SLUDGE THICKENING PROCESS MODELING - Sludges do not normally behave like Newtonian Fluids. This reason precluded the utilization of the sedimentation model for this process. On the other hand, although the methods deviced by Yoshioka (29) and Dick (8) have gained popularity in use, still most designers rely extensively upon experience acquired from studies at full size installations for the design of the thickener unit (5,18). Because of this, the parametric approach based on the hydraulic and solids loading was choosen to represent the thickener model. Also, its simplicity and generality in use contributed to this decision. The thickener was designed to provide an underflow concentration of 10% SS (23). The design was based on the traditional mass balance between feed, underflow, and overflow hydraulic and solids load. Hydraulically, it consisted in the calculation of the required tank dimensions as well as the flowrate and SS conc. of overflow and underflow which will be directed to the head of the plant and vacuum filter unit, respectively.

VACUUM FILTRATION PROCESS MODELING - The formulation used in this model correspond to the basic theory of the filtration process developed by Carmen (4) and extended by Coakley (6,17) for conditions of streamline flow by application of the Poisseville's and Darcy's laws. The recommendations given by Gale (10) for compressible sludges were also taken into account. The model estimates the expected yield of the vacuum filter for a previously characterized fed sludge and obtain the necessary filter area for that expected performance. Additionally, it performs a mass balance on flow and SS for the characterization of the filtrate and the produced cake in terms of their production rate and percentage SS. It also take into consideration the fraction of time the filter will be operated due that the filter yield will depend on this factor.

SLUDGE DISPOSAL BY HAULING AND DUMPING - The sludge disposal model was developed completely from the economic point of view. The details are given by Torruellas in his Ph. D. thesis (26).

TOTAL PHYSICAL-CHEMICAL TREATMENT SYSTEM MODELING - The complete modeling of the P/C treatment system was performed by converting the previously developed computer models into subroutines of a comprehensive program simulating the total system. The incremental loads imposed by the substantial recycled flows were accounted for by an iterative procedure to balance the incremented differences of recycled mass (SS), flow, and TOC obtained from successive simulations. It was not until the incremented values in these three parameters were equal to or less than some previously established preset limits that a single simulation for the whole P/C treatment plant was considered complete. The economic analysis phase was then commenced. A Present Worth Analysis (PWA) for each unit process modeled was performed making use of the previous design information obtained (size and capacity of tanks, equipment, etc.) and a Total Present Worth Sum (PWSUM) calculated by adding the PW cost of each individual unit. This constituted one single point of the Box's "complex" (set of points within the feasible region). The optimization of the referred treatment scheme was accomplished by minimizing this total amount but at the same time complying with preset quality standards and/or obtaining the treatment efficiency required.

BOX-COMPLEX OPTIMIZATION OF THE PHYSICAL-CHEMICAL WWTP - The Box-Complex Algorithm (3) is a multivariable constrained numerical search that can be used to find the maximum (or minimum) of a general nonlinear function $f(x_1,...,x_n)$ of **N** independent (explicit) variables $x_1,...,x_n$ subject to **M** constraints of the form

$$g_i \leq x_i \leq h_i \; ; \quad i = 1,...,M \quad \text{.......................(1)}$$

in which $x_{n+1},...,x_m$ are implicit variables and are functions of $x_1,...,x_n$; and lower and upper constraints g_i and h_i are either constants or functions of $x_1,...,x_n$. Constraints on the variables $x_1,...,x_n$ are called explicit constraints and the constraints on $x_{n+1},...,x_m$ are called implicit constraints. To find a maximum (to find a minimum, $-f(x_1,..., x_n)$ is maximized). The method has been programed for computer use by Joel A. Richardson and given for popular use by Kuester and Mize (15). The details of the algorithm are given by Craig (7) and Torruellas (26).

Figure (1) shows the P/C treatment system considered in this study. The objective of the system optimization was to minimize the sum of the Capital Costs plus the PW of the annual O/M expenses. The system design variables considered are given in Table (3). Upper and lower constraints for the system variables were selected after a search of typical ranges from textbooks on wastewater treatment, EPA Technology Transfer Manuals, and American Waterworks Association Standards. The selected values for the constraints and their dimensional units are shown in Table (1). Mathematically, the objective function can be expressed as

$$\text{Minimize} \quad f[\ \theta_c,\ \theta_f,\ G_f,\ \theta_a,\ L_a,\ V_f,\ H_f,\ d_f,\ U_f,\ V_a,\ H_a,\ P_v,\ R_v,\ C_v,\ f_v\ ] \quad \ldots\ldots\ldots\ldots\ldots\ldots (2)$$

in which f[ ] = PW of the complete treatment system including Capital and O/M Costs.

RESULTS AND DISCUSSION

The computations for the P/C system simulation and optimization were carried out on a CDC CYBER 174 computer with a FTN compiler. The program required about 40 CP secs for compilation and an average of 500 CP secs execution time for each simulation. The results obtained from the optimization phase are summarized in Tables (1) and (2). The baseline problem was ran two times, each one beginning with a different initial starting point within the feasible region as required by Box's method. The initial point for the first run was choosen as the rounded mean value in the feasible region , while an extreme point (upper constraint) was selected for the second. Columns (7) and (11) of Table (1) show the corresponding final values of the variables for these two points. It may be noted that the variables with indexes 5, 9, and 11 have converged perfectly, those with indexes 3, 7, 8, 10, and 13 have fairly converged, while the others still need extra iterations for full convergence. The savings implied by the use of the Box-Complex Optimization for the design of the P/C treatment system may be inferred from the progressive results obtained which are shown in Table (2).

CONCLUSIONS

The design of a P/C treatment system was optimized utilizing the Box-Complex algorithm which proved to be a valuable and efficient procedure in achieving, at least partially that objective. The implementation of this new design concept requires experimental data for the determination of the various parameters used by the models. This is necessary if input data is to be representative of the real conditions at plant site. Also, although the criteria used in the selection of the mathematical models which were to represent the P/C system in this study precluded those that have not been actually tested, some assumptions were deemed to be necessary for their improvement. These assumptions need validation through further experimentation with new pilot plant studies.

The model may be used also for the optimization of the operation of a P/C system. However some changes are necessary, principally in the variables defining the system which result less numerous because some of the unit process design variables previously considered become fixed parameters in the new problem. These are: 1) H_f, depth of the sand filter, 2) d_f, effective size of sand, 3) U_f, uniformity coefficient, and 4) H_a, carbon bed depth. Some new variables may be brought into the system. These are: 1) R_1, the fraction of incoming plant flow Q to be recycled, and 2) θ_s, the hydraulic retention time at the surge storage tank. Because of their direct dependence, both R_1 and θ_s may be expressed as functions of Q. Mathematically, the objective function for the operation of a P/C plant becomes:

$$\text{Minimize} \quad f[\theta_c, \theta_f, G_f, \theta_s, L_s, V_f, P_v, R_v, C_v, f_v, R_1, \theta_s] \quad \ldots\ldots (3)$$

in which f[ ] = PW of the operation of the P/C WWTP.

TABLES AND FIGURES

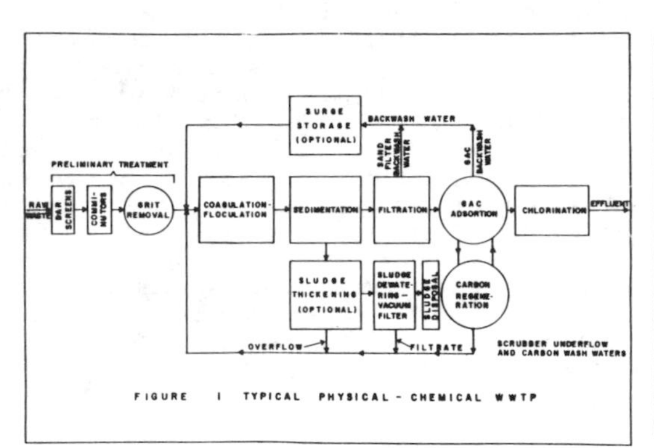

FIGURE 1 TYPICAL PHYSICAL - CHEMICAL WWTP

TABLE (3)- P/C SYSTEM DESIGN VARIABLES

VARIABLE	PROCESS	DEFINITION
1 θc *	COAGULACION	HYDRAULIC RETENTION TIME
2 θf	FLOCCULATION	" " "
3 Gf	"	VELOCITY GRADIENT
4 θs	SEDIMENTATION	HYDRAULIC RETENTION TIME
5 Ls	"	SURFACE LOADING RATE
6 Vf	FILTRATION	FILTRATION RATE (APPROACH VEL)
7 Hf	"	DEPTH OF SAND FILTER MEDIA
8 df	"	EFFECTIVE SIZE OF SAND
9 Uf	"	UNIFORMITY COEFFICIENT
10 Va	ADSORPTION	APPROACH VELOCITY
11 Ha	"	CARBON BED DEPTH
12 Pv	VACUUM FILTER	APPLIED VACUUM
13 Rv	"	RATIO OF FORM/CICLE TIME
14 Cv	"	CYCLE TIME (TIME FOR ONE DRUM REVOLUTION)
15 fv	"	FRACTION OF TIME VF IS OPERATED

* Note: The velocity gradient in the coagulation process Gc was considered as f(θc).

TABLE 1 - RESULTS OF BOX - COMPLEX COMPUTATION FOR TWO DIFFERENT STARTING POINTS SELECTED AS MEAN AND EXTREME (UPPER CONSTRAINT) VALUES

VARIABLES		RANGE		MEAN VALUES				EXTREME VALUES			
INDEX (1)		LOWER CONSTRAINT (2)	UPPER CONSTRAINT (3)	I P[a] (4)	I C[b] (5)	F C[c] (6)	F B[d] (7)	I P (8)	I C (9)	F C (10)	F B (11)
1 θc in	secs	10	80	40	43.2	78	77.4	80	56.5	14.2	13.8
2 θf	secs	600	3600	1800	2155	2212	2571	3600	2755	621	626
3 Gf	1/sec	10	100	50	74.4	100	100	100	91.1	75.5	77.7
4 θs	secs	3600	9000	7200	7529	8887	8856	9000	8129	5425	5169
5 Ls	m/hr	1	2.1	1.5	1.2	1	1	2.1	1.4	1	1
6 Vf	cm/sec	0.1	0.7	0.4	0.5	0.7	0.7	0.7	0.6	0.4	0.4
7 Hf	cm	30	150	90	91.4	52	49.4	150	111.4	40	36
8 df	cm	0.04	0.15	0.09	0.1	0.05	0.06	0.15	0.12	0.04	0.04
9 Uf	undim	1.1	1.7	1.5	1.3	1.1	1.1	1.7	1.4	1.1	1.1
10 Va	cm/min	8	41	25	32.9	41	41	41	38.2	33.3	34
11 Ha	cm	300	915	600	504	356	300	915	609	300	300
12 Pv	cm Hg	25	66	38.1	42	38.4	38.9	66	51.3	26	25.8
13 Rv	undim	0.1	0.4	0.25	0.33	0.4	0.4	0.4	0.4	0.3	0.3
14 Cv	secs	240	480	360	388	480	480	480	428	296	291
15 fv	undim	0.2	0.85	0.5	0.56	0.57	0.59	0.85	0.68	0.27	0.25

[a] I P = initial point
[b] I C = initial centroid (after correction of "worst" point)
[c] F C = final centroid
[d] F B = final "best" point

TABLE 2 - RESULTS OF BOX - COMPLEX COMPUTATION FOR THE VALUES OF THE OBJECTIVE FUNCTION FOR TWO DIFFERENT STARTING POINTS SELECTED AS MEAN AND EXTREME (UPPER CONSTRAINT) VALUES

	Values of the Objective Funtion ($ 10[6])			
	Mean Run Values		Extreme Run Values	
	Initial Complex	Final Complex	Initial Complex	Final Complex
F (1)	21.588	7.563	18.895	7.600
F (2)	13.227	7.778	13.277	8.144
F (3)	11.178	7.739	11.197	8.935
F (4)	45.741	7.738	45.829	7.686

REFERENCES

1. Amirtharajah, A., "Predicting Expansion of Filters During Backwashing", Journal AWWA, Vol. 64, Jan 1972, pp 52-59.
2. Boucher, P. L., "A New Measure of Filtrability of Fluids with Applications to Water Engineering", Institute of Civil Engineers, London V. 27, 1947, pp 415-447.
3. Box, M. J., "A New Method of Constrained Optimization and a Comparison with Others Methods", Computer Journal, Vol. 8, No. 1, 1965, pp 42-52.
4. Carman, P. C., "A Study of the Mechanism of Filtration, Parts I-III, J. Soc. Chem. Ind. Vols. 52-53, London, 1933, 1934.
5. Clark, J. W., Viesmann, W., and Hammer, M. J., Water Supply and Pollution Control, Third Edition, Crowel, Harper, and Row.
6. Coakley, P., and B. R. S. Jones, "Vacuum Sludge Filtration", Sewage and Industrial Wastes, Vol. 28, No. 6, 1956.
7. Craig, Meredith, and Middleton, "Algorithm for Optimal Activated Sludge Design", Journal ASCE, EE 6,December 1978, pp 1101-1117.
8. Dick, R. I., "Role of Activated Sludge Final Settling Tanks", J. San. Eng. Div. ASCE, 96, 423, (1970).
9. Friedlander, S. K., "Smoke, Dust, and Haze"; Wiley:New York, 1977.
10. Gale, R. S.,"Some Aspects of the Mechanical Dewatering of Sludges", Filtr. Separ., 5, 2, 133, 1968.
11. Gear, C. W., "Numerical Initial Value Problems in Ordinary Differential Equations"; Prentice Hall: Englewood Cliffs, New Jersey, 1971.
12. Iwasaki, T., "Some Notes on Sand Filtration", Journal AWWA ,29, 1591,(1937).
13. Ives, K. J., "The Physical and Mathematical Basis of Deep Bed Filtration", water (Netherlands), Vol. 51, No. 24, November, 1967 (English Text), pp 439-446.
14. Kavanaugh, M. C., Toregas, G., Chung, M., and Pearson, E. A., "Particulates and Trace Pollutant Removal by Depth Filtration", Progr.Water Technol. 1978, 10 (5,6),pp 197-215.
15. Kuester, J. L. and Mize, J. H., "Optimization Techniques with Fortran", McGraw-Hill Book Co., New York, N.Y.,1973.
16. Lawler, D. F., Omelia, C. R., and Tobiason, J. E., "Integral Water Treatment Plant Design", Chap. 16, Particulates in Water, Characterization, Fate, Effects, and Removal; Advances in Chemistry Series 189.
17. McCabe, B. J., and W. W. Eckenfelder, Jr., Biological Treatment of Sewage and Industrial Wastes, Vol. 2, Reinhold, New York, 1958.
18. Metcalf & Eddy, Wastewater Engineering,Treatment/Disposal /Reuse, 2nd Ed.,Mcgraw-Hill.
19. Mohanka, S. S., "Theory of Multilayer Filtration", J. San. Eng. Div., Proc. Amer. Soc. Civ. Engrs., 95, SA6, 1079, 1969.
20. Parker, D. S., Kauffman, W. J., and Jenkins, D., "Flocc Breakup in Turbulent Flocculation Processes", J. San. Eng. Div. ASCE 98, 79 (1972).
21. Process Design Manual for Suspended Solids Removal, U. S. Environmental Protection Agency Technology Transfer, EPA 625/1-75-003a, January 1975.
22. Richardson, J. F., and Zaki, W. N. in Sedimentation and Fluidization: Part I,Trans. Instn. Chem. Engrs., Vol. 32, 1954.
23. Sludge Treatment and Disposal, USEPA Technology Transfer Design Manual, 1979.
24. Smoluchowski, M., "Versuch einer Mathematischen Theorie der Koagulations - Kinetik Kolloider Losungen", Z. Physic. Chem., 92, 129-168, (1917).
25. Snoeyink, V. I. and Jenkins, D., Water Chemistry, John Wiley and Sons, pp 72,264-268
26. Torruellas, E. D., "Mathematical Modeling of Physical-Chemical Wastewater Treatment Plant", Ph. D. Thesis submitted to the Graduate School of SUNYAB ,Buffalo, N.Y.
27. Weber, W. J., Physicochemical Processes for Water Quality Control, Wiley Interscience Series, 1972.
28. Ying, W., and Weber, W. J., "Bio-physicochemical Adsorption Model Systems for Wastewater Treatment, Journal WPCF, Vol 51, No. 11, pp 2661-2677, Nov. 1979.
29. Yoshioka, N. et al, "Continuous Thickening of Homogeneous Flocculated Slurries", Chemical Engineering (Tokyo) 21, 66 (1957).

THE IMPORTANCE OF OPERATION & MAINTENANCE EXPERIENCE FEEDBACK FOR IMPROVEMENTS IN PLANNING AND DESIGN

Fred M. Reiff*

INTRODUCTION

Over the past decade the term appropriate technology has been used to convey an increasingly wide range of concepts in the water supply and sanitation sector. It means something slightly different to the social scientist, the politician, the economist, the engineer, the planner and the people served by the technology in question. Nevertheless, there are three parameters of appropriateness, which are usually incorporated in each of the respective definitions regardless of their other differences. They are utilization, operation and maintenance. To a great extent utilization is dependent upon successful operation and maintenance so in this sense operation and maintenance become the fundamental parameters for the ultimate measure of success of water and sanitation facilities.

Nevertheless, the feedback of operation and maintenance experience into the planning and design of systems is largely neglected. In many of the developing countries the planners and designers of water and sanitation systems are isolated from the subsequent operation and maintenance of the facilities for which they have worked so diligently to bring into existence. Because of the high cost time of these professionals they may never visit even the construction of their works much less conduct follow up visits to evaluate the operation and maintenance. This is particularly prevalent in countries where planning and design are highly centralized and it is further exacerbated in those in which the well educated professional believes that working with the hands carries a negative social stigma. This is unfortunate not so much from a social standpoint but in that it seriously stunts the professional growth of the planners and engineers who fail to take advantage of the most critical and demanding educator of all, which is the harsh reality of witnessing just how well the end product of all their rationale, creativity, theory, training and effort has actually survived the real world for which it was ultimately intended.

There is no equal substitute for a few first hand inspections of the actual operation and maintenance but this is not possible on a frequent basis. For this reason several less costly parallel pathways of feedback which involve other qualified personnel should be established.

*Regional Adviser, Pan American Health Organization (PAHO), 525 23rd Street, N.W., Washington, D.C. 20037

The continuous and accurate feedback of operation and maintenance experience has far reaching economic, technical and even political consequences and it is folly to neglect this aspect. When properly implemented and utilized the information obtained can make the difference between mediocre and virtuoso engineering, between an economical and a debt ridden utility, between a durable, safe and dependable service and one that is totally unsatisfactory and between a well satisfied and a disgruntled user (or politician).

More definitively there are five principle reasons why this feedback is so important if not essential for improvement in the planning and design of water and sanitation systems.

1. Reduction and prevention of recurring problems
2. Support of technological development
3. Human resource development
4. Improved decision making
5. Cost reduction
6. Quality control

REDUCTION AND PREVENTION OF RECURRING PROBLEMS

The replication of mistakes is one of the most common problems produced by the lack of corrective feedback. It is difficult to prevent the making of mistakes in design and planning but, it is even more difficult to correct a faulty or erroneous design, an inferior material, a bad construction methodology, planning practice (and the like) after they have become a standard. There are a number of reasons why deficient standards and practices may continue to be utilized for many years, but the principal one is that it takes so long for them to come to the attention of the people who have the authority and the ability to correct them. This in turn is brought about by a number of factors.

Since the expected design life of many of the components of water and sewerage facilities might range between a number of years and a number of decades, a deficiency or problem may take a number of years to manifest itself. Another factor is that the operator usually tries and usually succeeds to a considerable extent to ameliorate problems thereby masking the cause of the problem. An additional contributing element is that frequent turnover in personnel will severely limit "institutional memory." When all of this is coupled with a situation in which there is little concerted effort to assure the flow of feedback, the personnel who have been responsible for a faulty decision will frequently be gone before the disclosure of the problem and the replacements will rely on the established standard assuming that it has withstood the test of time when just the opposite may have been true.

The misapplication of a standard material or device is another common cause of repetitive problems which could readily be identified and remedied by operation and maintenance feedback.

When feedback from O & M experience is only sporadic and not systematized it is usually viewed with skepticism or considered to be unreliable. Then, when the deficiency finally comes to the attention of designers, planners and other decision makers it is all to often blamed on problems in operation and maintenance rather than analyzing the reality of the situation and the conditions which made the application inappropriate.

When the feedback mechanism does not distinguish between preventive maintenance, corrective maintenance and repairs, the cause of the problem is somewhat clouded. This distinction should be made in any effort to correct repetitive mistakes.

SUPPORT OF TECHNOLOGICAL DEVELOPMENT

Many of the important advances in technology in water supply and sanitation have been at least in part due to observation and analysis of operation and maintenance. When this feedback is coupled with laboratory analysis and research it provides a powerful set of tools for the development of technology . A full scale system can provide almost all the oportunities for analysis that is obtained from pilot projects if there is adequate feedback.

The viewpoint of the operator is also an excellent source of ideas for areas which need improvement. They are not isolated from the reality of conditions which do not conform to the laboratory or the textbook. They frequently have excellent recommendations for improvements in planning and design based on years of witnessing recurrent problems, inefficient processes and existence of conditions not anticipated in the planning and design. The operators frequently have to make modifications in the facilities in order to improve the capacity to meet increased demands. The immediate pressure and necessity of making improvements faced by the operators foster innovative and creative solutions. When their modifications, changes and new applications work especially well they should be brought to light by the information system and considered for future incorporation in planning and design. Many advancements in equipment, materials, and design have had their roots in operational feedback.

HUMAN RESOURCE DEVELOPMENT

The recognition of a problem is usually a stimulation for its solution. This can be utilized to develop the creative aspects of human nature. Everyone, at all levels of an institution can learn by problem solving experience and this can be an important part of human resource development. In the water supply and sanitation sector of developing countries there is no lack of problems and many are problems for which the successive approximation method is far more practical than rigorous scientific analysis. This situation can lend itself well to the establishment of a positive atmosphere in which an individual can learn by mistakes made by himself or by others with little embarrassment or loss of face.

Usually the shorter the length of time between the commission of an error and its being brought to the attention of the person who made it and its rectification, the more likely the individual involved will not repeat it. From a human resource development perspective it is important that operation and maintenance experience feedback be fast, accurate and unbiased.

IMPROVED DECISION MAKING

Well planned and executed operation and maintenance feedback improves decision making in a number of ways. It informs management as well as the engineers and planners of what is functioning well, under what conditions it functions well, and what measures can be taken for further improvement. It will also facilitate the identification and dissemination of successful corrective measures which have been taken at the operational level to compensate for deficiencies. Additionally, it will provide an early warning of problems to be anticipated so that timely action can be taken to minimize or avoid the problem altogether. This is particularly important during natural disasters or other emergency conditions.

Feedback in combination with mini and micro-computers makes it possible to sort and analyze the enormous amount of information generated by the operation of water and sanitation facilities. This enables the prompt detection of important anomolies, trends, and symptoms which would otherwise have gone unnoticed.

COST REDUCTION

In developing countries the annual cost of operation and maintenance of water and sanitation systems ranges from 15% to 30% of the total capital cost of the facilities. This usually means that more than 2/3 of the monthly costs are due to operation and maintenance. Thus, from an economic standpoint it is probably more important to strive for reduced operation and maintenance costs rather than to devote effort and resources to reduction in capital cost. This in itself should be sufficient reason to develop a reliable accurate system for operation and maintenance experience feedback.

It is always important to look at costs from a standpoint of actual life rather than estimated design life. It is also important to determine the real cost of a component of a system which has not functioned as intended or has not functioned at all. When it is taken into consideration that in Latin America more than 30% of the total water storage capacity is not utilized because there isn't enough water to keep up with the demand or because of control problems, and that more than 50% of the disinfection units are inoperative because of lack of repairs or chemicals, and that there is more than 40% unaccounted for water in many of the water utilities and the majority of small package waste treatment plants in developing countries fail to provide the treatment expected and a very large percentage of them have failed altogether,the magnitude of the economic loss becomes apparent. When these facilities are financed through loans, payments must continue to be made regardless of whether they are functioning or not. A significant portion of the money being paid for water and sanitation is paying for nothing.

In the water and sanitation sector it is usually less expensive to avoid or prevent a problem than to correct it. It is also necessary to avoid premature failure of system components by better planning, design and operational efforts in order to minimize the cost of a system. One of the major purposes of the feedback mechanism is to disclose the causes of premature failure. This will make it possible to alter material or equipment specifications in a timely manner to avoid repetition of premature replacement. It will indicate what measures are feasible to extend the life of facilities as well as to possibly increase their capacity. It will also provide valuable information to estimate operation and maintenance capability of a community or agency when potential future projects are under consideration. One of the advantages of tapping feedback from the operation and maintenance personnel is that many of their solutions are practical and cost effective.

QUALITY CONTROL

Quality control is used herein in the broad sense to include not only the quality of drinking water or treated waste water but also quality of service and quality of design, materials and construction. Again the information generated during operation and maintenance provides parameters for estimating all of these aspects. With this information it is also possible to develop a strategy for resolving the problem(s) by correlating deficiencies in operation and maintenance with deteriorating quality.

The quality of drinking water is of critical public health importance. It is of such importance that surveillance is almost always conducted by the appropriate health authority in addition to whatever quality control measures are utilized by the water utility or agency. The availability of operation and maintenance data and records is useful for both of them. In an investigation of a suspected waterborne outbreak of disease the operation and maintenance records are always one of the first sources of information reviewed. With a system of feedback of operation and maintenance experience it is possible to reduce the number of episodes of waterborne disease by identifying trends and precursors of serious problems and resolving the causative factors before they give birth to an emergency situation.

CONCLUSIONS

A reliable and accurate system for the feedback of operation and maintenance experience is one of the most powerful management tools available for managers and decision makers in the water and sanitation sector. It provides information of immediate as well as long term usefulness, and the breadth of usefulness is considerable.

Such a system is especially important in developing countries in which conventional solutions to drinking water and sanitation have not always been satisfactory.

Collection of data is not enough. It must be sorted, analyzed and utilized by personnel in a position to make management and engineering decisions. The volume of data will be such that it will almost be essential to utilize a computer to accomplish this.

A reliable and accurate system for feedback is always worth the cost and the effort in terms of system reliability, quality of service, user satisfaction and savings in overall costs.

APPENDIX

1. "Assessment of the Operations and Maintenance Component of Water Supply Projects," WASH Technical Report No. 35, June 1986, USAID.

2. "Maintenance Anomalies" Fred M. Reiff, U.S. Public Health Service, April 1965.

3. "Manuales DTIAPA" Números, c 1 - 13, CEPIS, Pan American Health Organization, 1983-1985.

4. "The Potable Water Project in Rural Thailand," AID Project Impact Evaluation, Report No. 3.

5. "Effective Record System for Maintenance," Paul Trot, Operations Forum, Vol 2, No. 2, Feb. 85, WPCF.

6. "Appropriate Technology for Water and Sanitation, Sociocultural Aspects of Water Supply and Excreta Disposal," Mary Elemdorf and Patricia Buckles, Dec. 1980, World Bank.

7. "Preventive Maintenance of Rural Water Supplies," Nov. 1983, World Health Organization.

8. "Informe de Viaje a Colombia", J.A. Hueb, Set. 1983, CEPIS/PAHO.

9. "Informe Final de Misión", Ing Luis Sánchez, Nov. 16, 1983, San Salvador.

10. "Manual Modelo de Normas de Operación y Manteniniento de Equipo de Sistemas de Agua Potable", Proyecto Gua/84/007, Feb 1986, COPECAS, OPS/OMS/PNUD, Guatemala.

11. "Operation and Maintenance of Rural Drinking Water and Latrine Programs in Honduras," WASH Field Report No. 152, August 1985, USAID.

12. "Modelo de Gerencia de Operación Y Mantenimiento", Programa de Salud Ambiental, HPE, Organización Panamericana de la Salud, 1986.

A QUEST FOR APPROPRIATE SANITATION TECHNOLOGY FOR BANGLADESH

Ali Basaran *

ABSTRACT

Results of a pilot study covering several areas representing different socio-economic and geographical settings in Bangladesh are presented. The test areas were selected to test basically four different types of low-cost sanitary latrines (LCSL) for technical, social and economical appropriateness. A total of 180 LCSLs of four different types namely, International Voluntary Services (IVS), Ventilated Improved Pit (VIP), Vietnamese (Viet), and Chute-Pan (CP) were distributed free of cost for use by villagers and evaluated. In addition, the study also included several pour-flush water seal (WS) LCSLs as part of the on-going national programme in the study areas to allow evaluation of a total of five different types of LCSL technology. The later has emerged as the technology of choice.

INTRODUCTION

Bangladesh is basically a low-lying country made up of a flat alluvial plain and extensive network of rivers which contribute significantly to the socio-economic life of the nation. Diseases associated with poor hygiene and poor sanitary methods and habits for disposal of human wastes continues to be of national concern (1,2,3). The prevailing system of excreta disposal in the country can best be described as very primitive (1,4,5,7). Thus, enteric diseases like diarrhoeal attacks continue to prevail, resulting in a high rate of infant morbidity and mortality (1,2,3). Controlling these diseases requires provision of sanitary excreta disposal facilities for a 1985 population of over 100 million people (15 million households) of which 87% live in rural areas under the greatest poverty levels and underdeveloped state (1,3,4). A baffling problem of searching for an appropriate low-cost sanitary latrine (LCSL) technology for the country, which could be socially acceptable, financially affordable, technically sound and yet meet the minimum sanitation requirements, has been investigated in the field and from time to time modified (4,5,7).

The Village Sanitation Scheme (VSS) which started by the Department of Public Health Engineering (DPHE) in 1974 had produced a total of 135,000 direct pit water seal (WS) latrine sets as of June 1974, resulting in a coverage of about 2% of the 11 million rural and about 22% of the one million urban households (4,5,6,7). Therefore, there was an urgent need for a sanitation programme of big magnitude (4). It was felt that development and introduction of different types of LCSL sets

*World Health Organization, Dhaka, Bangladesh

and components meeting a minimum sanitary standard and being appropriate to different social, cultural, economical and physical conditions prevailing in the country, could increase the potential for wider coverage and improve the current health status (4,5,6).

METHOD OF APPROACH

Projects were planned and implemented in two villages to field-test different LCSL technologies over a two year period. The objective was to identify a suitable technology under various social, economic and geographic conditions of rural Bangladesh. Two communities (villages) comprising of typical socio-economic groupings were selected. One of these was Rahamaterpara (PA-1), located in a coastal area in the south eastern part of the country, approximately 18 miles north of Chittagong City and one mile east of sea shore. The other was Ramgovinda (PA-2), located in an inland water scarce area in the north western part of the country, approximately 3 miles east of Rangpur Town.

Types of LCSLs that were designed and constructed in the test project areas were as follows:

1 International Voluntary Service (IVS) Type: Bamboo squatting platform placed directly over an un-lined pit.
2. Ventilated Improved Pit (VIP) Type: Reinforced cement concrete (RCC) squatting slab with ventilation pipe placed directly over a pit lined with RCC rings.
3. Vietnamese Type: Twin, brick-in-cement, vaults with sludge withdrawal doors covered with RCC squatting slab.
4. Chute Pan (CP) Type: RCC squatting slab fixed with cement mortar pan having a short, lateral projecting chute fixed directly over a pit lined with RCC rings.

All of the test LCSLs, including the pits, were constructed up to the squatting platform level by DPHE through contractors and without any cost to the households. However, households were to construct the latrine superstructure at their own costs.

The test villages, being located within the DPHE's routine VSS area, also had WS type LCSLs with RCC squatting slabs and (FCC) WS pans placed directly over a pit lined with five RCC rings. However, these latrines were installed by households at their own cost after purchasing a set for Taka 100.- (1 US$ = Taka 24.-) from the DPHE Sanitation Centers (VSPC).

Both project villages were visited on several occasions by DPHE staff including Health Educators, UNICEF and WHO Staff. Such visits were made before, during and after construction of the latrines for promotion, and observation of use and maintenance.

Surveys were conducted to collect baseline information in each of the two project villages. The baseline survey information was thereafter used to select households for the test latrines. The selection was based on:

TABLE 1: SUMMARY OF BASELINE SURVEY INFORMATION

	Whole Vill. (Base Surv)		TEST LATRINES							
			IVS		VIP		Vietnam		Chute	
	PA-1	PA-2	PA-1	PA-2	PA-1	PA-2	PA-1	PA-2	PA-1	PA-2
By Occupation:										
Business	9	24	4	1	–	3	–	1	1	1
Farmer	44	66	7	11	12	11	2	2	5	9
Service	47	29	4	1	6	2	2	2	7	5
Labourer	6	80	1	6	1	1	–	–	–	3
Other	20	3	4	–	1	–	1	–	2	–
By Annual Income:										
Tk 6000	53	41	–	12	–	5	–	–	–	5
Tk 6000-18000	47	133	–	7	–	10	–	2	–	12
>Tk.18000	17	28	–	–	–	2	–	3	–	1
Not known	15	–	–	–	–	–	–	–	–	–
By land Ownership										
Not land	26	–	1	–	–	–	–	–	–	–
House only	17	112	5	12	3	3	1	–	2	11
<1/2 acre	57	19	9	1	14	4	2	–	9	1
1/2-1.0 Acre	12	15	5	4	1	3	1	1	3	1
>1.0 Acre	16	56	1	2	2	7	1	4	1	5
Not known	4	–	–	–	–	–	–	–	–	–
By literacy:										
Upto primary	158	140	–	7	–	6	–	2	–	3
Upto secondary	39	51	–	1	–	1	–	–	–	5
Above secondary	8	22	–	–	–	–	–	2	–	4
Illiterate	613	1203	–	11	–	10	–	1	–	6
Family size:										
Upto 2 persons	8	9	1	–	–	2	–	–	1	2
3-5 "	45	82	5	5	8	4	2	1	2	5
6-10 "	70	87	14	12	9	9	2	2	11	10
> 10 "	9	24	–	2	3	2	1	2	1	1
By Latrine type:										
Water Seal	1	7	–	1	–	–	–	2	–	1
Septic tank	1	2	–	–	–	–	–	–	–	–
Open/Surface	109	179	20	18	20	17	5	2	14	14
Pit latrine	–	14	–	–	–	–	–	1	–	3
No latrine	21	–	–	–	–	–	–	–	1	–
By Water Source:										
a)Drinking										
Tubewell	48	105	10	5	12	11	3	3	7	13
Pond	72	–	10	–	8	–	2	–	8	–
Ringwell	–	25	–	12	–	6	–	2	–	5
Other	14	2	–	2	–	–	–	–	–	–
b)Other purposes										
Tubewell	–	105	–	5	–	11	–	3	–	13
Pond	132	–	20	–	20	–	5	–	15	–
Ringwell	–	95	–	12	–	6	–	2	–	5
Other	–	2	–	2	–	–	–	–	–	

a) A few households representative of each different social and economic groups for each type of LCSL to be tested.
b) Households' willingness to construct the latrine superstructure at their own costs, for all types of test latrines.

The baseline survey results are summarised in Table 1.

EVALUATION RESULTS & DISCUSSION

The households who were provided with different types of test LCSLs were adequately representative of the socio-economic conditions within the test villages. The construction quality and finish (upto platform level) done by contractors had much room for improvement in both villages. Some households in both villages expected the contractor to also build the superstructure and therefore, did not build their own. The absence of superstructure was a common reason given by households for non-use of the latrine. There was dissatisfaction among some households who believed that "better" latrines were provided to others and therefore, they were indifferent to use and maintenance.

The evaluation is based on data collected through field surveys by a team of surveyors from DPHE, UNICEF and WHO. The results are summarised in Table 2, and discussed briefly below by type of latrine tested.

<u>IVS Type</u>: In total, 9 out of 39 units built were in use. All the unlined pits collapsed after a short period. The bamboo floor of some latrines in PA-1 also collapsed. Users of all 20 units in PA-1 complained that this type was inferior in quality in comparison to other types constructed in the same village.

<u>VIP Type</u>: In total, 29 out of 37 units built were in use. The large defecation hole was considered to be unsafe for children by some users in PA-1. Some users complained of water splashing during use, arising from high water table and the open hole. In all cases in PA-1 the vent pipes were constructed inside the superstructure and as such, were not exposed to sunlight. Furthermore, in 9 out of 17 vent pipes in PA-1 the fly screens were either not in place or were found damaged. There were few instances where the vent pipe opening was fully closed, preventing ventilation and light.

<u>Vietnam Type</u>: In total, 8 out of 10 units built were in use. Users in PA-1 considered the defecation hole to be too small. The cover for the hole was correctly used in only 3 out of 10 units. Ash/soil cover was not placed after use in both PAs. In some instances both chambers were in regular use simultaneously. In PA-1 excreta was exposed through defective sludge removal doors in the chambers.

<u>Chute Type</u>: In total, 29 out of 33 units built were in use. The chute was found broken in few instances in PA-2. There were few instances where excreta was exposed on the pan due to poor flushing.

TABLE 2: SUMMARY OF EVALUATION SURVEY INFORMATION

	TEST LATRINES								PRG.LATR	
	IVS		VIP		Vietm		Chute		W.S	
	PA-1	PA-2	PA-1	PA-2	PA-1	PA-2	PA-1	PA-2	PA-1	PA-2
Total constructed	20	19	20	17	5	5	15	18	20	13
Total surveyed	20	19	20	17	5	5	15	18	11	8
Total in use	8	1	17	12	5	3	14	15	9	7
Appr. cost(Tk) to (w/o superstr) :										
DPHE	-	-	520	520	400	400	350	350	350	350
Contractor	428	428	184	184	1191	1191	145	145	145	145
Household	-	-	-	-	-	-	-	-	100	100
Total	428	428	704	704	1591	1591	495	495	495	495
Use and maint.										
Total in use	8	1	17	12	5	3	14	15	9	7
Platform clean	0	1	16	11	2	3	13	13	8	5
Odour	5	INA	6	1	3	INA	4	INA	5	4
Flies present	5	INA	3	INA	2	INA	7	INA	4	0
User group										
Adults & child.	0	1	4	8	3	2	9	6	4	5
Adults only	5	0	13	4	2	1	5	9	5	2
Children only	3	0	0	0	0	0	0	0	0	0
Reasons for non-use										
No superstr.	10	-	2	2	-	-	1	2	-	-
Pit collapsed	-	18	-	-	-	-	-	-	-	1
Bad smell	-	-	-	1	-	1	-	-	-	-
Dislike latr.	-	-	-	1	-	-	-	-	-	-
Dislike site	-	-	-	1	-	-	-	-	-	-
Other	2	-	1	-	-	1	-	1	2	-
Total non-use	12	18	3	5	-	2	1	3	2	1
Superstructure										
with roof	0	1	11	4	2	2	9	7	2	3
without roof	2	0	6	10	3	3	5	8	7	4
No superstr.	18	19	3	3	0	0	1	3	2	1

INA : Information not available.
Tk : Taka, Bangladesh monetary unit (during the study 1 US $= 24 Tk)
PA-1: Study area/village 1 (i.e., Rahamaterpara village)
PA-2: Study area/village 2 (i.e., Ramgovinda village)

As mentioned earlier, both PAs also had households having WS latrines. These had been installed at owners' costs, including the purchase of WS pan and five RCC rings for pit lining at Taka 100.-, from VSPCs. Therefore, survey information on these LCSLs have also been included here.

Water Seal (WS) Type: Out of the 19 units surveyed 16 were in use. Of the 16 in use, 12 were being used with the WS broken and 4 with the WS intact.

Numerous other types of LCSL components, for squatting slabs and pit linings, have been under experimentation and study at VSPCs in collaboration with UNICEF and WHO. A comprehensive account of these will not be attempted here, but briefly speaking: 5" brickwork; tar coated bamboo mat; earthen clay rings; earthen jars and pots; oil barrels, etc. were tried as pit linings; and concrete slabs reinforced with bamboo strips; Malaysian type PVC-WS complete with PVC bend and pipe; jute-plastic WS pan of local manufacture were tried as slabs.

A number of non-governmental organisations (NGO) are also experimenting with WS, non-WS, direct pit, offset pit options. Some known LCSL components include pans of baked clay together with baked clay connector pipes; pit lining and pit covers of bamboo.

Another technology under field study in urban areas is the WS double offset pit LCSL

Many of the experimental units studied by DPHE were found to be not feasible for further studies or development due to cost and durability factors. However, of these presently under study by DPHE, rings made of cement sand mix of 1:3 reinforced with No. 10 steel wire and baked clay rings for pit lining seem to have potential for further development.

As a result of all these, 1.5" thick rectangular squatting slab of RCC fixed with FCC water seal pan; and five rings of 1.5" thick RCC, with 30" diameter and 12" hight, emerged as the standard latrine unit produced at DPHE's VSPCs during the VSS Phases I and II.

CONCLUSIONS AND RECOMMENDATIONS

The village test projects indicated that the chute-pan type LCSL had good acceptance and use, as a non-WS option. This type of latrine is suggested for further testing in one selected village.

The VIP type latrine did not function as designed due to defective location of the vent pipe in relation to superstructure. Further tests, in another selected village could provide more specific information.

The study indicated low potential (if any at all) for any further studies or development of the IVS type or Vietnamese type of latrine.

TABLE 3: PROBABLE COMBINATIONS OF LOW-COST SANITARY LATRINE (LCSL) UNITS

Lining / Floor	Cement C. Rings 30"dia.,12"high (set of five)	Cement-sand rings 30" dia.,9"high (set of seven)	Earthen rings 24" dia.,6" high (set of twelve)
RCC slab with FCC	232	205	120
Water seal pit	107	107	107
Total unit cost	239	312	227
FCC slab with	232	205	120
FCC water seal pan	90	90	90
Total unit cost	322	295	210
FCC slab, FCC pan	232	205	120
with chute	107	107	107
total unit cost	339	312	227

The test latrines should be under continued study. The responses and practices of households when pits are filled up should be studied and considered in improving the technology and design.

Further test project implementation could be improved by:

a) Full-time supervision, including monitoring the use of latrines;
b) Planning in consultation and collaboration with local Administration and Health officials, identifying their specific responsibilities in promotion and monitoring activities;
c) More specific efforts to explain test objectives to households, in addition to general promotion and education;
d) A degree of participation by households in the construction of the latrines (in addition to the construction of the superstructure).
e) Avoiding use of contractors.
f) Avoiding different types of latrines in the same village.

The provision of necessary moulds and trained staff to the DPHE VSPCs for the wider production of FCC squatting slabs should be expedited.

The 1" thick cement-sand ring for pit lining should be introduced for field studies at some selected location. The quality of production, handling in transport and installation should be monitored further

The use of earthen (baked clay) rings, as pit lining, should be introduced in some selected locations. The set of latrine units, sold to households, would then include the DPHE produced squatting slabs and (say) 12 earthen rings (each 6" high). An appropriate sale price should be fixed. (For the introductory phase, DPHE could purchase the earthen rings from local manufacturers. In such cases, the concrete or cement rings should not be produced by DPHE at those locations).

The evaluation, after an adequate period should include: damages in transport and installation, community acceptance, local production potentials, costs.

Probable combinations of LCSL units are shown in Table 3. The units specified are those used in routine DPHE programme and those experimentally identified (to date) as feasible for further development. All figures in Table 3 are production costs in Taka, except for earthen rings. The costs of earthen rings are estimated average local market prices. The number and costs of all pit lining rings are for a common pit volume of approximately 16.0 cu. ft. (below ground level).

ACKNOWLEDGEMENT

Author wishes to acknowledge the contribution made by the DPHE, UNICEF and WHO staff in carrying out the field work in this study.

REFERENCES

1) Anonymous, 1985, Area and Population, Household and Housing Charact.; and Educ., Health, Fam. Plann. Soc. Welfare and Sports, in "Stat. Pocket Book of Bangladesh 1984-85," Bangladesh Bureau of Statistics, Stat. Div., Min. of Plann., Dhaka, Aug. 18, 1985.

2) Anonymous, 1985, Country Report on Primary Health Care in "The Pyongyang Conference: Primary Health Care in Action," WHO , SEARO Regional Health Papers No. 6, New Delhi, 1985.

3) Anonymous, 1985, Phys. Plann. Housing and Water Supp. in "The 3rd 5-yr Plann, 1985-90," Plann. Comm. Min. of Planning, Dhaka, 1985

4) Anonymous, 1950-1987, DPHE Records and files in "Rural Water Supply and Sanitation Programme & Annual Development Programmes/Plans for 1950-1987", Dept. of Public Health Engineering, Min. of LGRDC, People's Rep. of Bangladesh, Dhaka, 1986.

5) Anonymous, 1986,Country Situation Presentation in "Draft Proc. of Inter-Country Workshop on Accel. of Nat'l Progr. on Sanitary Disposal of Human Excreta," WHO SEARO, New Delhi 27-31 Oct. 1986

6) Anonymous,1985, Rev. Report of the Working Group on Rural Water Supply and Sanitation in "Planning Towards the Third Five Year Plan (1985-90)," Dept. of Public Health Engr., Min. of LGRDC, Dhaka, 1985.

7) Basaran, A., 1985, WHO Experiences on Water Supp. and San. Progr. in Bangladesh, in "Proc. of Workshop on Water & Sanitation Interv. Related to Diarrhoeal Disease in Bangladesh, Dec. 17-19, 1985," A. Basaran and R. Islam ed., Int'l Centre for Diarrhoeal Disease Res., Bangladesh, Dhaka, 1987

PUTTING IT ALL TOGETHER: APPLIED USE OF MICROCOMPUTERS IN THE MANAGEMENT OF WATER AND WASTEWATER SYSTEMS

Gary Kovach*

Computers are solving the complex problems of time and distance in the management of rural and municipal water and wastewater systems in Puerto Rico. The routing of operational and maintenance information from remote utilities to a central administrative location in San Juan sometimes took as long as 186 days by conventional manual means. Many factors contribute to the problems in the communications network:

Distance - There are over 100 wastewater and 50 water treatment facilities distributed throughout 3,500 square miles of island.

Accessibility - Difficult roads, jungle environment, weather and mountain terrain pose particular travel difficulties that prevent timely movement between treatment plants.

Communications - Most rural facilities are not connected by phone. These facilities rely on inter-office mail delivered by manual courier. All regional offices have telephones.

Environmental Conditions - The combination of hot, moist weather creates a particularly difficult maintenance environment that drains valuable manpower time. One concern is the constant humidity and high operating temperatures for mechanical and electrical equipment. Also, heavy rainfall and resulting plant growth places heavy demands on grounds maintenance to beat back the jungle encroachment.

Put it all together. That was the request of the Puerto Rico Aqueduct and Sewer Authority (PRASA) of Metcalf & Eddy in the role of project management. PRASA is undertaking an ambitious repair and rehabilitation project that will affect nearly all of the 110 wastewater systems on the island. They require a reporting mechanism that will deliver operating records and discharge monitoring reports to management and regulatory agencies in a timely manner. Timely response to O&M needs and problems could result in equipment cost savings, improved services, and compliance with discharge permit requirements. One of the main concerns in this project was to find a better way to collect,

*Metcalf & Eddy de Puerto Rico, San Juan, Puerto Rico

manage, and utilize the huge volume of information generated from these facilities.

Due to problems in the communication system of PRASA, the flow of information was often disrupted which delayed submittals of discharge monitoring reports (DMR's) to the regulatory agencies. Manhours were better served in operations and maintenance than in preparing reports for the regulatory agencies. The data management scheme implemented for PRASA in the Bayamon Region has successfully freed operations from the reporting task and drew the far flung treatment plants into a closely linked network of data gathering and report generation. In the areas of water and wastewater treatment, challenges like those presented in Puerto Rico have frequenty been encountered but less frequently conquered.

The challenge set before the Operations and Maintenance group of Metcalf & Eddy was to:

1) create a system of data management using "state of the art" hardware and software;
2) provide an output of required data for discharge monitoring to regulatory agencies (NPDES reports);
3) provide a means to track critical budgetary items;
4) account for preventive and corrective maintenance;
5) free supervisory personnel of the reporting task;
6) draw the far flung treatment plants into a closely linked network of data gathering and report generation.

The challenge of "putting it all together" was met with a data management system consisting of:

- word processing capability
- spreadsheet design
- communications
- data storage and retrieval

The Operations and Maintenance group of Metcalf & Eddy has recognized the need for comprehensive data management in the areas of operations and maintenance and has developed two software packages that address those particular needs:

- RODA/M for Records and Operational DAta Management; and,
- COPE/M for COmputerized Parts and Equipment Maintenance management.

Additional software such as wordprocessing, spreadsheet design, communications, and self-designed data base management are essential to the total picture of data management like a complete set of tools in the managers "tool kit". All these programs are PC-based computer programs which provide convenient "management tools" to assist plant managers in their role as decision makers.

RODA/M

RODA/M is the means by which observations from the operation of a water or wastewater treatment plant are complied, stored, and made ready for output into customized reports. Observations made in the treatment facility are divided into several area of data gathering:

a. laboratory analysis,
b. operator gathered information,
c. budget and cost accounting,
d. maintenance and equipment status.

A typical setup for a RODA/M program would include the following elements from those areas of information gathering:

1. Equipment Status;
2. Chemical, Utility, and Manpower Unit Costs;
3. Chemical, Utility, and Manpower Usage;
4. Inventories of Chemical and supplies;
5. Environmental Factors;
6. Plant Observations of Wastewater Characteristics; and
7. Laboratory Analysis of Treatment Process.

From the data base formed by the collection of data and observations, the manager of the treatment process generates operational control strategy reports, budgetary reports on chemical, utility, and manpower use and costs, trend plots, and a means for reporting to regulatory agencies.

COPE/M

COPE/M consists of (5) basic parts:

1. Predictive preventive maintenance scheduler;
2. Repair/corrective maintenance tracker;
3. Spare parts inventory system;
4. Equipment library for report generation; and
5. Preventive and corrective maintenance history reporting and tracking.

It is sometimes difficult to picture the entire maintenance requirement of a facility when the need to maintain is secondary to the primary concern to correct or repair equipment. When maintenance problems become too great, discussions on the subject always refer to it as an entity. The problem is narrowing the perspective to specific components of the system and their individual maintenance needs. Once the maintenance requirement is broken into manageable individual parts with assigned PM tasks in a data management program such as COPE/M, the logic need not be sorted manually again. COPE/M provides reliable scheduling of PM according to the programmed frequency and the tracking of consumed parts and supplies.

Wordprocessing

WordPerfect was chosen by the client because it best suited the bilingual nature of correspondence with the federal government and in-house agencies. Supportive documentation for operations or maintenance can accompany permit monitoring reports and be retrieved from memory within minutes.

Spreadsheet

Lotus 1-2-3 is a very useful tool in creating rows and columns of numbers or data. Mathmatical as well as engineering relationships can be identified to these columns and rows to present predictive management of information such as for budget preparation.

Communications

The software for communications provide a dialing directory for storage of phone numbers. By selecting from a menu in the program, the system operator can be connected to another computer within the network for text, voice, or micro to micro communications. Data, reports, or correspondence can be exchanged across the island in the time it takes to make a phone call.

Data Base Management

RODA and COPE are considered data base managers but very specific in the nature of the data managed. It was then necessary to add another data manager for data fields or environments that develope in the future. dBase III plus was chosen for its fexibility and all around application. The system operator would simply design the input labels and how the output would be represented.

SETUP AND IMPLEMENTATION

The island of Puerto Rico has been divided into 7 regions by PRASA. Because the greatest population is centered around the San Juan area, the Bayamôn Region was selected as the first region to undergo computerized data management. The Bayamôn Region encompasses 12 wastewater and 8 water treatment plants in a westerly suburb of San Juan. The most distant facility is within a radius of 20 difficult road miles from the regional office and regional laboratory. The primary facility that serves the majority of the population is the Bayamón Regional facility. The regional facility has its own maintenance division while the laboratory serves the analysis needs of the entire region. A mobile maintenance crew provides preventive and corrective maintenance for all other facilities and all pumping stations.

Touring each facility, assessing the operational needs, and reviewing NPDES permits provided the basis for generating a listing of all parameters and observations needed to set up the data base within RODA/M. A separate RODA/M program was styled to the specific requirements of each facility.

An evaluation of existing practices and procedures dictated the type of output needed from the RODA/M program. Further review of PRASA's operational reports, checklists, and other documents provided the information for selecting the other supportive software. Once the software was chosen, the hardware was selected based on:

a. PRASA's request for 3 year resident data storage,
b. software compatibility, and a
c. convenient office "tool" at the immediate command of the regional director's management team.

An IBM AT personal computer with 640K RAM, 30 megabyte hard disk storage, and double diskette drives (one 360 kilobyte and one 1.2 megabyte) was chosen. A high speed printer, color monitor, and a 1000 watt, 60 minute battery backup were added to complement and support the system. A 1200 baud modem was added for communicating across an ordinary phone line.

DESCRIPTION OF THE BAYAMÓN REGION

The Bayamón Region has of 12 wastewater treatment plants:

1. Bayamón Regional Facility - a 40 MGD primary treatment plant with grit and rag removal, primary clarification, and chlorination for disinfection. Solids handling is accomplished by sludge thickening in circular thickeners and dewatering by belt presses. The sludge cake is sent to offsite multiple hearth incinerators for final disposal.

2. Bayamón Gardens WWTP - a 0.6 MGD trickling filter plant with a clarigester for solids removal and stabilization, and drying beds for dewatering sludge. Chlorination is provided for disinfection.

3. Dorado WWTP - a 1.2 MGD trickling filter plant with primary clarification, and two-stage anaerobic digestion for solids stabilization, and drying beds for dewatering sludge. Chlorination is provided for disinfection.

4. Las Teresas WWTP - a 0.375 MGD trickling filter plant with a clarigester for solids removal and stabilization, and drying beds for dewatering of sludge. Chlorination is provided for disinfection.

5. Royal Town WWTP - a 2.0 MGD trickling filter plant with primary clarification for solids removal, a one-stage anaerobic digester for solids stabilization, and drying beds for solids dewatering. Chlorination is provided for disinfection.

6. Santa Juanita WWTP - a 1.0 MGD trickling filter plant with two clarigesters for solids removal and stabilization, and drying beds provide sludge dewatering. Chlorination is provided for disinfection.

7. Corozal WWTP - a 0.5 MGD activated sludge plant with aerobic sludge digestion for solids stabilization and drying beds for sludge dewatering. There is no primary sludge removal or thickening. Chlorination is provided for disinfection.

8. Covadonga WWTP - a 1.0 MGD activated sludge plant with aerobic sludge digestion for solids stabilization and drying beds for dewatering. There is no primary sludge removal or thickening. Chlorination is provided for disinfection.

9. Naranjito WWTP - a 0.5 MGD activated sludge plant with aerobic sludge digestion for solids stabilization and drying beds for dewatering. There is no primary sludge removal or thickening. Chlorination is provided for disinfection.

10. Toa Alta Heights WWTP - a 0.5 MGD activated sludge plant with aerobic sludge digestion for solids stabilization and drying beds for solids dewatering. There is no primary sludge removal or thickening. Chlorination is provided for disinfection.

11. Toa Alta WWTP - a 0.1 MGD imhoff tank system with drying for solids handling. Chlorination is provided for disinfection.

12. Vega Alta WWTP - a 0.25 MGD imhoff tank system with drying beds for solids handling. Chlorination is provided for disinfection.

SUMMARY OF THE BAYAMÓN REGION

There are 4 types of treatment processes in use for the Bayamon Region:

1. Primary treatment - (1) Bayamon Regional Facility only;
2. Trickling Filter - (5) plants;
3. Activated Sludge - (4) plants; and,
4. Imhoff Tank - (2) plants.

It is evident that the data management system executed for the Bayamón region has it application for both large complex facilities as well as small package plants.

PUTTING IT ALL TOGETHER: O&M CONSIDERATIONS

The chore of putting it all together required a thorough examination of all operational and maintenance considerations. Excellent sources for plant information comes from the plant O&M manual (the "how to operate and maintain the whole plant" book) and the individual process equipment vendors O&M manual. There is no substitute for plant and equipment documentation in the expeditious setup of a dynamic computerized information system.

Some operational considerations noted in the development of RODA/M were:

- the frequency of data gathering and reporting (daily, monthly, or quarterly);
- laboratory capability to perform process control analysis as well as discharge permit monitoring; and,
- side stream flow measurements were not designed into the plants for calculating and balancing process loadings.

Typical maintenance considerations encountered were:

- availability of vendor's or manufacturer's O&M manuals;
- maintenance staffing and logistics;
- spare parts inventory and storage; and,
- preventive maintenance scheduling plan.

The plan to implement a data management mechanism to enhance the flow of information includes the six elements for controlling and regulating data or generating compiled information:

1. Metcalf & Eddy's RODA/M program;
2. Metcalf & Eddy's COPE/M program;
3. Word Processing with WordPerfect;
4. Spreadsheet Design with Lotus 1-2-3;
5. Communications with PCTalk; and,
6. Data Base Management with dBase III plus.

SUMMARY

Following the successful completion of the Bayamón Region, computerized data management will be extended throughout the island of Puerto Rico. The completed plan is to create an information network that will respond to regulatory agency requirements to expedite delivery of NPDES reports. In so doing, the network also provides a means to track valuable operational data that can be turned back into the facility for process, budgetary, and maintenance monitoring and control. The network provides a vital communications link for the exchange of information between regions and the central PRASA office through a normal phone line.

Before development of the computerized data management system, discharge monitoring reports (DMR's or NPDES permit reports) took as long as six months to process through regular channels and arrive in the hands of the regulatory agencies. It is now possible to generate any regional plant's DMR within several minutes. In addition, a compiled data dump of all the data used in the production of the DMR can be printed. The outputs can be transferred by phone to any location if required.

The RODA/M programming allows for the entry of text to the data base for each day to describe the state of operations, solids handling, and maintenance. A summary report of the monthly diary for each facility is then generated from the daily entries to provide supporting dcoumentation and explanations of permit violations.

For the most part, RODA/M would meet the needs of plant and process operation; and, COPE/M would provide individual equipment needs. COPE/M would take care of plant equipment by scheduling preventive maintenance care and lubrication on a planned frequency, track consumed supplies and spare parts, and deliver historical reports on maintenance performed. Researched information from the vendor manuals is systematically entered into the COPE/M data base. The beauty of the effort is that the correlation of maintenance information need only be sorted once. On a planned frequency, COPE/M can be called upon to schedule preventive maintenance and generate equipment work orders according to trade or location.

The idea of rapid data delivery is extremely appealing to management of water and wastewater utilities. The need for this type of reporting to regulatory agencies is great. Yet, in addition to DMR's, the mechanism is in place to allow for better process control and monitoring. Plant managers have output formats that detail compiled and calculated information for improved process efficiency and regular monitoring of budget.

The six programs resident on the IBM AT provide a means to better control data and have it readily accessible to all appropriate parties.

1. RODA/M manages operational records;

2. COPE/M eliminates concern that process equipment is maintained as outlined by manfacturers;

3. Word processing saves valuable file documents in corespondence, for generating checklists, and other forms;

4. Spreadsheet design work gives a new perspective to aligning values with calculations and engineering relationships for output into plots or printed copy;

5. Communications enables the computer to deal with other computers in the exchange of information; and,

6. The independent data base management program allows for creation of specialized data management, styled entirely by the system operator to manage other data as needs arise in the future.

In conclusion, the difficulties of distance and communication pose incredible challenges to standard means of data handling and transmission. Manual sorting of information to achieve direction in the decision making process has evolved to computer manipulated "smart" outputs that places the right information in the hands of those with a need to know. The application of micro-computers in Puerto Rico is rapidly solving communication problems.

REFERENCES

Cesario, Lee; "Microcomputers for Water Utilities". American Water Works Associations, 6666 West Qunicy Ave., Denver, CO 80235,(1986).

Drinking Water Quality Surveillance and Control in Developing Countries

Terrence P. Thompson, P.E.*

Introduction

The International Drinking Water Supply and Sanitation Decade is a cornerstone in the World Health Organization's (WHO) strategy to achieve health for all by the year 2000. Multinational and bilateral development agencies and national governments participating in the Water Decade also view public health improvements as major benefits to be attained through investment in the water supply and sanitation sector. Only limited health benefits can be realized however if water supply projects result in systems that deliver sub-standard water quality unsafe for human consumption.

In order to protect investments in water supply projects, and to achieve the public health benefits for which they are intended, drinking water quality surveillance and control programs must be implemented by water suppliers and health authorities. These programs can assure the proper operation of systems and the continuing delivery of safe drinking water long after the systems are turned over by development agencies.

Drinking Water and Health

The link between microbiological water quality and human health was developed by early researchers such as John Snow in his famous investigation of a cholera epidemic attributed to the Broad Street pump, in London, in 1854. The link was firmly established when water chlorination practices introduced in the United States in the early 1900s led to dramatic declines in the rate of typhoid and other waterborne diseases. Microbiological contaminants in drinking water can cause a variey of diseases, the more common of which are summarized in Table 1.

Chemical contaminants are also of concern although their health effects are often chronic in nature, as oppossed to the acute effects associated with microbiological contaminants. Table 2 summarizes the health effects of chemical and other contaminants in drinking water.

Aesthetic quality of water is also important inasmuch as water with unpleasant taste, odor, or appearance may cause consumers to turn to other unregulated, and possibly unsafe, sources of water if the latter are more aesthetically appealing.

*Deputy Director of Public Health Engineering, New York City Department of Health, 65 Worth Street, New York, New York 10013 U.S.A.

TABLE 1 - WATERBORNE DISEASES DUE TO MICROBES

WATERBORNE DISEASE	ORGANISM	HEALTH EFFECT
Gastroenteritis	Various Pathogens	Acute Diarrhea and vomiting.
Typhoid	Salmonella typhosa (bacteria)	Inflamed intestine, enlarged spleen, high temperatures - Fatal
Bacillary dysentery	Shigella (bacteria)	Diarrhea, rarely fatal
Cholera	Vibrio Comma (bacteria)	Vomiting, severe diarrhea, rapid dehydration, mineral loss - high mortality.
Infectious Hepatitis	Virus	Yellow skin, enlarged liver, Abdominal pain - low mortality - lasts up to 4 months.
Amebic Dysentery	Entameoba histolytica (protozoa)	Mild diarrhea, chronic dysentery.
Giardiasis	Giardia Lamblia (protozoa)	Diarrhea, cramps, nausea and general weaknesses - lasts 1 week to 30 weeks - not fatal.

Source: AWWA, 1979

TABLE 2 - HEALTH EFFECTS OF CHEMICALS AND OTHER CONTAMINANTS IN DRINKING WATER

DISEASE OR SYNDROME	SOURCE	HEALTH EFFECT
Toxicoses	Metal	Intake of metals in drinking water, food and air from both natural sources and human activities. These include arsenic, cadmium, copper, chromium, lead, mercury, selenium vanadium, zinc, et al. Can be important on a local basis, e.g. arsenic in parts of Argentina.
Toxicoses Cancers Mutations Birth Defects	Organic Chemicals	Intake of certain chemicals, esp. certain synthetic organic chemicals, including some pesticides. Also some trihalomethane by-products of chlorination are suspect carcinogens Not now a high priority problem in LDCs.
Cancers	Radionuclides	Natural and man-made radioactivity. Not now a high priority in LDCs.
Cardiovascular Disease	"Hardness"	Some epidemiological evidence indicates an inverse correlation of cardiovascular diseases with hardness of drinking water.
Fluorosis	Fluoride	Damage to teeth and bones resulting from long-term ingesting of high concentrations of naturally occuring fluorides.
Methemoglobinemia	Nitrates	Serious, sometimes fatal poisoning of infants following ingestion of well waters containing nitrates (NO_3) at concentrations higher than 45 milligrams/liter.
Endemic Goiter	Iodine	Iodine-deficient water or water containing goitrogens.
Asbestosis and Mesothelioma	Asbestos	Asbestos in lungs known to cause cancer. Fate in gastrointestinal tract unknown.
Hypertension	Sodium	Sodium-restricted diets necessary for parts of population.

Adapted from McJunkin, 1982

Health effects of various contaminants have been discussed in detail by the National Academy of Sciences (1977).

It should be noted here that drinking water quality programs should be applied in concert with other environmental health programs. Particularly in developing countries, consumers are exposed to microbial and chemical contaminants through numerous other routes of exposure: airborne, flies and rodents, intrafamilial, auto-infection, etc. Hence, other environmental health measures are needed such as: sanitation, pest control, food protection, solid waste collection and disposal, health and hygiene education, etc.

Establishing Surveillance and Control Programs

In order to protect drinking water quality, formal programs within clearly established institutional frameworks are needed. WHO has published guidelines to assist countries in planning and implementing such programs.

WHO Guidelines

The World Health Organization published its second edition of the European Standard for drinking water in 1970, and its third edition of the International Standards in 1971. Each of these standards have recently been reviewed, revised, combined, and issued in three separate volumes under the title "Guidelines for Drinking Water Quality" (WHO, 1984).

Volume I of the Guidelines, "Recommendations," presents recommended guideline values, not standards, together with information necessary to understand the basis for the guidelines. WHO recommends that individual countries develop their own standards from these guidelines, taking into account economic cost-benefit as well as local needs and circumstances. This is a departure from WHO's former philosophy of dictating firm, universal standards. Rigid standards, in circumstances where they cannot be met, are useless and only breed contempt and apathy for drinking water quality programs. The health criteria used in developing WHO's guidelines are documented in Volume II, "Health Criteria and Other Supporting Information."

The third and most practical volume, entitled "Drinking Water Quality Control in Small Community Supplies," covers simple methods for water sampling and analysis, sanitary surveys, and other means of investigating and controlling drinking water quality in small systems.

WHO recommends two-stage development of surveillance programs. A comprehensive, fully estblished program would have many diverse activities as shown in Table 3. However, such a program would be an unrealistic goal in the near term for many developing countries. WHO recommends that surveillance programs be established with basic activities, again as shown in Table 3, and that more sophisticated activities be integrated into the program with time.

TABLE 3: SUMMARY OF PRINCIPAL ACTIVITIES FOR VARYING LEVELS OF SURVEILLANCE

Activity by the surveillance activity	Level of surveillance	
	Initial	Advanced
Laws, regulations and policies	Basic	Complete
Enforcement	Basic	Complete
Drinking water standards	Bacteriological and some physical/ chemical	Numerous parameters as defined by WHO or equivalent
Technical assistance	Limited	Active
Training of		
- Staff	On the job plus short courses	As for initial level plus technical institute
- Water works operators	Seminars plus short courses	As for initial level plus technical institute
Sanitary inspections	All urban and some small communities	All urban, many small communities
Approval of sources	All urban and some small communities	All urban, many small communities
Sampling and monitoring	Urban areas	Urban areas and special rural situations
Standard methods of analysis	Bacterial and residual chlorine	WHO guidelines or equivalent
Remedial action	As needed	As needed
Establishment of Laboratories	Existing health laboratories	As for initial plus reference laboratory
Design standards or criteria	Advisory	Those available
Control of cross-connections	Advisory	Active programme
Plumbing Code	Advisory	Codified and enforced
Laboratory support services	Media and reagents should be available	Assess laboratories at large plants
Material and additives standards	Advisory	Approved listing

TABLE 3: SUMMARY OF PRINCIPAL ACTIVITIES FOR VARYING LEVELS OF SURVEILLANCE (CONTINUED)

Activity by the surveillance agency	Level of surveillance	
	Initial	Advanced
Regulation of special water supplies		
- institutional	Hospitals, major rail and air terminals	As for initial plus other concentration areas
- temporary	None	As above plus large camps, market fairs etc.

Institutional Responsibilities

It is highly desirable that surveillance responsibilities be assigned to an independent agency distinct from the water supplier(s). While each water supplier is responsible for controlling the quality of drinking water, the surveillance activity provides an independent check on water quality and exercises regulatory authority over water suppliers in order to assure that drinking water is safe and in compliance with local standards. The role of the surveillance agency is usually assigned to the health authorities. A quote from Dr. Bernd Dietrich (1986), former Director of Environmental Health, World Health Organization, is appropriate here:

> "There is an obligation, in light of primary health care, for national health authorities to expand their involvement rather than pay the usual lip service on a matter of such paramount health importance. As part of primary health care they have a responsibility for health and hygiene education, strengthening community involvement, and training community workers to support community water supply and sanitation programmes. They must likewise assume their responsibility for the surveillance of water quality."

Legislation, Regulations and Standards

While basic surveillance programs can be implemented without related legislation and standards, a firm basis in law is essential to assure their viability in the face of opposition, and to assure the gradual full development of the programs. Basic legislation should specify:

(1) the scope of authority, including authority over all drinking water within the agency's jurisdiction;

(2) the agency delegated to administer the law; and

(3) the right of the agency to establish and enforce regulations governing the protection, production, and distribution of water.

With enabling legislation in place, water quality standards may be developed and water supply regulations may be established. Most developing countries use WHO's water quality guidelines without modification and without formal adoption as their national water quality standards (although this approach is contrary to WHO's philosophy of developing local standards). In many countries, the only water supply regulations are those for the protection of water sources, and these are often geared towards surface waters with virtually no provisions for protection of groundwater.

Other regulations needed for complete protection of drinking water quality include: treatment standards; codes for facilities design and construction; plumbing codes; cross connection control regulations; material standards; permitting of individual wells; and pollution control regulations for protecting water supplies from wastewater and solid waste disposal.

Surveillance Activities

The two most essential surveillance activities are sanitary surveys and water quality monitoring.

A sanitary survey is an on-site inspection and evaluation of all conditions and practices in the watershed/recharge area and water supply system that pose, or could pose, a danger to water quality. Some key items to be checked in a sanitary survey include: pollution sources in the watershed/recharge area; effectiveness and reliability of treatment; operations and maintenance practices; provision of standby equipment, spare parts, and adequate tools and supplies; and need for personnel training. Volume III of the WHO publication contains general guidelines for conducting sanitary surveys.

Water quality monitoring certifies the safety of drinking water or indicates the need for remediation. Parameters for monitoring should be selected based on: the severity of health effects associated with each parameter, the potential for its occurrence in the water system, and resources available for monitoring. Bacteriological monitoring is therefore essential for allpublic water systems, and chlorine residual determinations are important for all systems practicing chlorination. Most systems can perform quantitative measurements of turbidity, pH, and conductivity, and all can do qualitative measurements of taste, odor, and appearance. These parameters, which are simple and inexpensive to monitor, give useful indications of system upsets or contamination. In many developing countries it may also be important to monitor for industrial and agricultural chemicals (Fano et al, 1986). If it is not practical to maintain laboratory facilities for this purpose, tests might be performed on a contract basis by an outside laboratory.

The results of all sanitary surveys and water quality analyses should be reported in writing. When findings indicate a need for prompt remedial measures, oral communication should be made by the quickest means possible. In all cases the surveillance agency should follow-up with the water supplier to assure that corrections are made.

Water Quality Control Strategies

While health authorities have the responsibility to carry out surveillance activities, only the water suppliers themselves can take the necessary actions needed to actually achieve and maintain acceptable water quality. This requires the implementation of multiple barriers against disease transmission, that is, the implementation of water quality control measures in every component of water supply systems:

Protection of Water Sources

Water quality must be protected at its source through land use controls and control of pollution discharges. Watershed rules and regulations give water suppliers authority in this area.

Adequate Treatment

Ideally, filtration and disinfection should be provided for control of pathogens. Where disinfection is not practical due to costs or operations and maintenance constraints, significant reduction in pathogens can be realized through filtration alone. Slow sand filtration, without the aid of coagulants, is particularly well suited to developing country needs. Prolonged storage also helps in this regard.

Secure Storage

Water storage tanks should be secure and protected from vandalism, wind deposition, and intrusion by rodents, birds, and other animals. Treated water should be stored in covered tanks.

Safe Distribution

Distribution systems should operate under continuous positive pressure in order to prevent any inflow of contaminated water. For the same reason, there should be an active leak detection and repair programs. Backflow prevention devices should be placed on the service lines of hazardous users, such as hospitals, mortuaries, laboratories, industries, etc. The maintenance of a chlorine residual provides some degree of protection, as a prophylactic measure. It may be necessary to occassionally superchlorinate distribution lines to control algal growth, and flushing may be necessary to purge systems of rusty or otherwise objectionable water.

Other Water Quality Control Measures

In order to assure the proper performance of water supply facilities (treatment facilities, distribution facilities, etc.) a formal operations and maintenance program is necessary. This entails the establishment of well defined operations procedures; preventive maintenance at scheduled intervals; the provision of spare parts, supplies, equipment and tools; and personnel training.

Finally, water suppliers should conduct their own water quality monitoring programs and programs of self inspection. Health authorities cannot be expected to do the entire job alone. Water suppliers, like any producer of a consumable commodity, have a vested interest in providing a safe product for their consumers' use. Furthermore, while health authorities can make periodic visits to watersheds and water supply facilities, water suppliers have daily access to these places.

Conclusions

Water quality surveillance and control programs are essential for protecting investments in water supply projects and for the attainment of public health goals associated with projects. Implementing such programs requires a clear delineation of responsibilities and authority between health authorities and water suppliers. Ideally, a spirit of cooperation should be fostered between the two entities with the realization that they hold a common objective: the production and distribution of safe drinking water that will protect and improve public health.

References

National Academy of Sciences, 1977. Drinking Water and Health. ISBN 0-309-02619-9, Washington, D.C.

Dieterich, B.H., 1986. "Drinking Water Challenges and Perspectives for the 1980s - the WHO Perspective." Proceedings of the American Water Resources Association's International Symposium on Water-Related Health Issues, November 11, 1986, Atlanta.

World Health Organization, 1984. Guidelines for Drinking Water Quality. ISBN 92 4 154168 7, Geneva.

Fano, E., Brewster, M., and T.P. Thompson, 1986. "Managing Water Quality in Developing Countries." Natural Resources Forum, United Nations, New York.

American Water Works Association, 1979. Principals and Practices of Water Supply Operations, Volume 1: Introduction to Water Sources and Transmission.

McJunkin, F.E., 1982. Water and Human Health. Prepared by the National Water Demonstration Project for U.S. Agency for International Development, Washington, D.C.

THE DEVELOPMENT OF A WATER SURVEILLANCE PROGRAMME FOR PERU

By

Barry Lloyd, Mauricio Pardon & Jamie Bartram

DelAgua, Robens Institute

University of Surrey, Guildford, U.K.

0. ABSTRACT

A **national programme** of potable water surveillance and improvement in Peru is being implemented by the Ministry of Health. The Technical Directorate for the Environment (DITESA) is the national reference authority responsible for environmental surveillance activities and houses the central reference laboratory.

The surveillance programme was initiated in 1983 with a **preplan** study of 60 water supply systems in health region XIII which comprised 6 health areas located in the central highlands and high jungle. The preplan demonstrated that 82-89% of distributed water was fecally contaminated and none of the rural supplies were chlorinated. Sanitary inspection data were deficient and therefore a sample diagnostic study of a further 40 systems was undertaken using reporting procedures modelled on WHO guidelines. The common deficiencies and points of risks were identified and summarised. It was demonstrated that the communities at greatest risk were those with no option other than surface water treatment. None of the sedimentation and filtration systems significantly reduced contamination and they all require rehabilitation.

1. INTRODUCTION

Peru may be divided into 3 geographical zones which are defined as coastal, andean highlands and jungle. The Health Region No. XIII was selected for the preplan project as it occupies a central position in Peru (Figure 1). The region comprises both high Sierra and high jungle. By far the highest proportion of the population, greater than 75%, is located in and around a single fertile valley, the Mantaro valley, in the high Sierra at about 3,000 to 4,000m above sea level. The Mantaro river together with the rivers Ene, Perene and Palcazu form important head waters for the Amazon. The Mantaro is already included in the Global Environmental Monitoring programme of UNEP/WHO (GEMS/WATER) and it was logical to extend monitoring to include the area as a Health Exposure Assessment Location (HEAL).

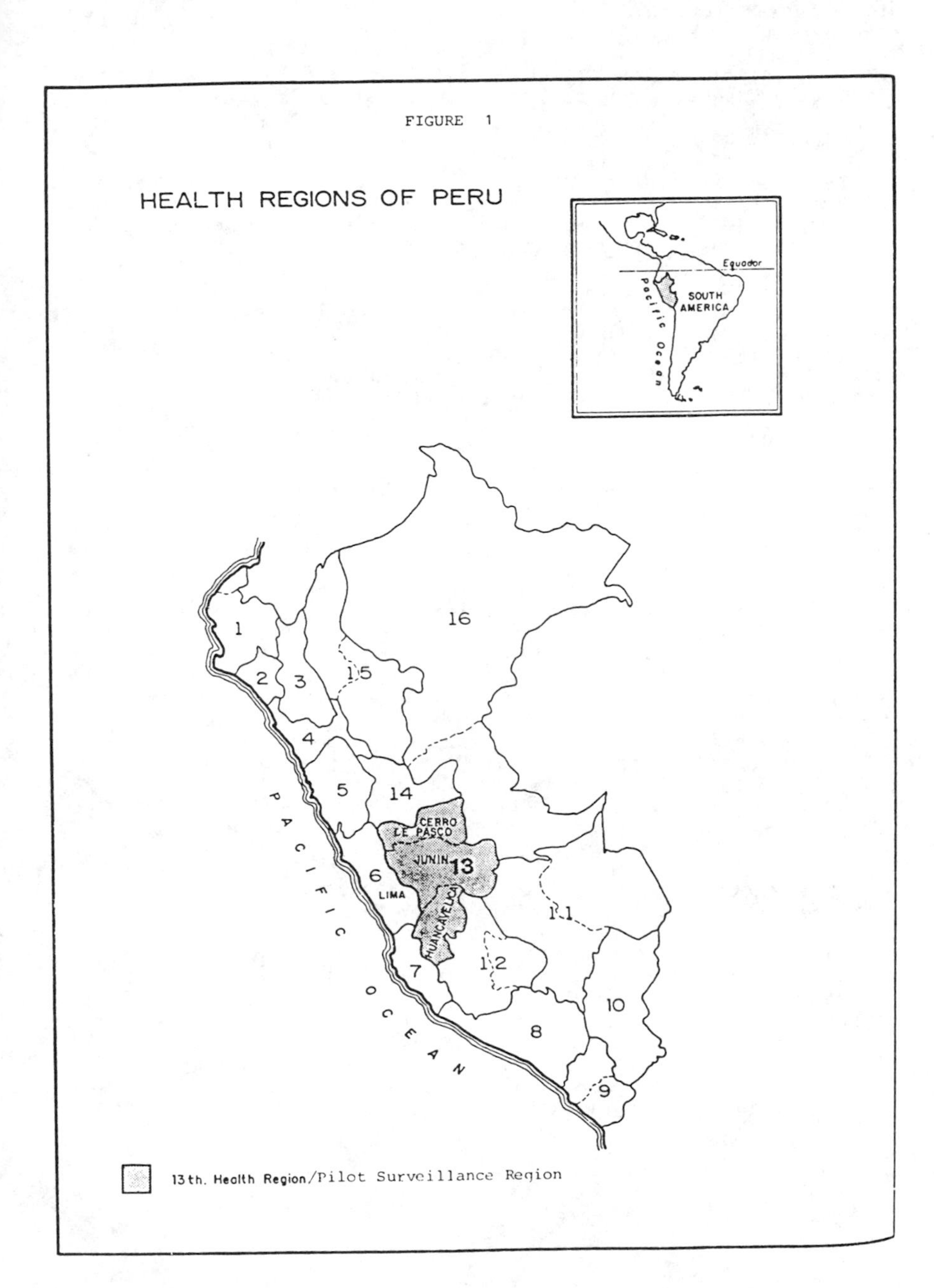
HEALTH REGIONS OF PERU
Equador
SOUTH AMERICA
Pacific Ocean
1
2
3
4
5
6
LIMA
7
8
9
10
11
12
13
CERRO DE PASCO
JUNIN
HUANCAVELICA
14
15
16
PACIFIC OCEAN
13th. Health Region/Pilot Surveillance Region

FIGURE 1

This was readily justifiable in view of the grossly contaminated nature of the Mantaro deriving from the complex of mines in the Oroya and Morococha areas as well as the sewage pollution from the town of Oroya.

The Health Region is subdivided into hospital areas and the population and water supply services are shown in Table 1.

2. OBJECTIVES

The objectives of the Surveillance Programme are to enable the Peruvian Ministry of Health, Technical Directorate of the Environment (DITESA) to undertake activities aimed at compliance with existing water quality legislation.

The legal, political and administrative basis of the Water Surveillance Programme derives from Peruvian WaterLegislation (Ley de Aguas, 13997, Codigo Sanitario D.L.17405 and Codigo Sanitario de Alimentos D.L. 102/03). These laws set out the needs for protecting drinking water sources, disinfection and treatment of supplies and water sampling.

3. STRATEGY

3.1 National Infrastructure

Peru has a mixed administrative authority for water and sanitation services. A national water authority (SENAPA) has responsibility for coordinating the activities of the urban centres whereas for example the capital, Lima, has an autonomous water authority (SEDAPAL). The rural populations are served by the Directorate of Basic Rural Sanitation (DISABAR) of the Ministry of Health. Surveillance should not be confused with quality control which, as the WHO Guidelines for Drinking Quality (1984) clearly recommend, are distinguished in the following way:

> "Monitoring for the routine control of drinking water is the activity of the water supplier; separate checking and testing should be carried out by a surveillance agency; it is highly desirable that the two agencies be separate bodies and independently controlled".

The Ministry of Health is responsible for surveillance and has a basic, centralised laboratory service within its Technical Directorate of the Environment (DITESA). It provides a unified Central Reference Laboratory fully equipped for microbiological, inorganic, organic and organoleptic analytical functions. Initially, as a decade objective, it has been recommended that DITESA should also develop health region laboratories and support surveillance using basic portable field test kits in the provincial towns and cities. A three tier surveillance organization is proposed to provide laboratory and infrastructural support at national, regional and provincial level.

Table 1
Relationship between health area populations, water supplies and trainees

Health area	Total population	Nº* Water supplies	Trainee surveillance staff
Huancavelica	181,600	34	3
Huancayo	549,062	>111	15
Jauja	189,805	82	2
Tarma	119,563	78	3
Junín	31,677	18	2
Selva Central			
- La Merced	280,688	21	2
- Satipo	22,000	18	2
- Oxapampa	52,000	8	2
Cerro de Pasco	160,000	>100	3
TOTALS	1,586,395	>470	34

* Nº of supplies include both rural and urban.

It was agreed to initiate a national surveillance strategy commencing with a preplan project in the central health region of Peru. Depending on the results of the evaluation of the preplan, the programme would be progressively expanded throughout adjacent departments.

3.2 Methods

Philosophy underpinning the Methods

The methods for implementing surveillance in provincial and rural areas are described in the WHO "Guidelines for Drinking Water Quality" Vol III. They include two key components, sanitary inspection and water quality analysis which have been adopted in this pilot project. Surveillance is an **investigative activity intended to evaluate all conditions that pose a danger to health.** We have therefore endeavoured to develop an inspection and assessment method which incorporates the **critical factors which control the quality of service and hence the risk to health.** The sanitary inspection report thus includes a resume of each water supply's

COVERAGE	(% of the total population served).
CONTINUITY	(hours/day and days/year water supplied).
QUANTITY	(total volume/capita/day supplied).
QUALITY	(classified primarily on fecal contamination).
COST	(domestic tarriffs paid/month).

This information is abstracted from the diagnostic report form which is a modified version of the WHO recommended sanitary report suitable for use in Peru. A complete report of the analysis of the quality of service in the pilot region will be available from DelAgua in August 1987. This paper deals primarily with the bacteriological quality of supplies.

Rationale for using Portable Test Kits

One of the major disadvantages of conventional water analysis is that, with the exception of real time telemetry (continuous on-site recording and electrical relay of data), the information produced is retrospective, following sample transfer to the laboratory. Although real time monitoring is, at present, impractical in the overwhelming majority of provincial and rural locations it is nonetheless desirable to produce, record, report and act on data in the minimum time. On-site analysis is therefore considered to be essential to ensure that corrective action can be initiated without unnecessary **delays.** This approach obviates the need to preserve samples and thus the risk of deterioration of samples during protracted transit times to the laboratory. Furthermore the sampler is unencumbered by the volume and weight of numerous sample bottles.

DelAgua has packaged five critical tests as a field kit (Figure2) which allows testing for fecal coliforms, chlorine

Figure 2

The DelAgua Portable Water Testing Kit Components.

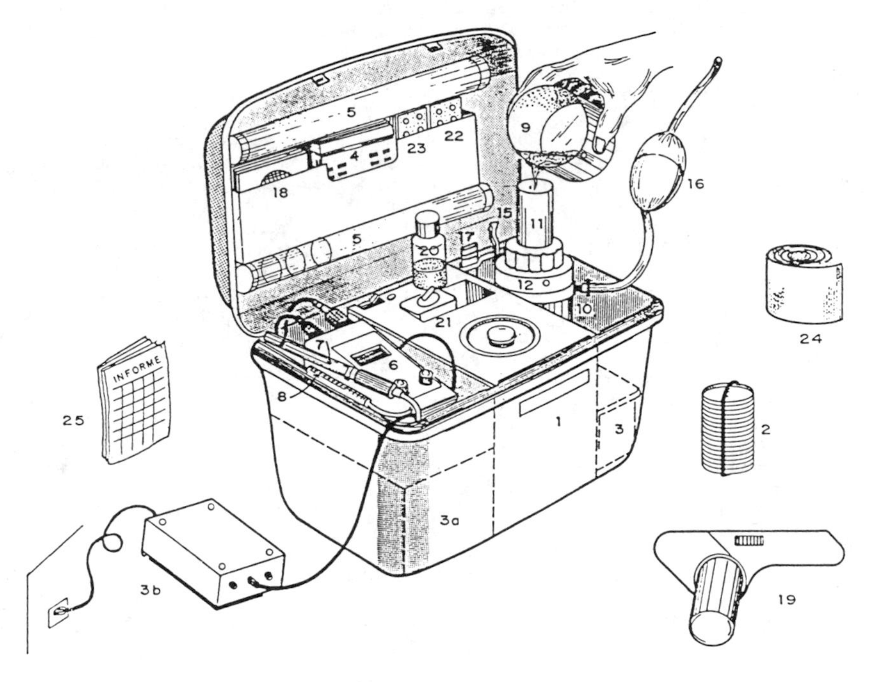

Daily checklist of equipment components and consumables.

ITEM

1 Carrying case with incubator
2 Aluminium petri dishes with carrier
3 Storage box with emergency 12 V power lead & spares (S)
3a Integral 12V battery
3b Charger unit (110V or 220V) to mains
4 Chlorine residual and pH comparator
5 Turbidity tubes, 5-10 TU and 20-2,000 TU
6 Conductivity and temperature meter
7 Conductivity probe
8 Temperature probe
9 Stainless steel sample cup
10 Stainless steel vacuum flask
11 Stainless steel filter funnel and locking collar
12 Aluminium filter assembly base

SPARES (S)

13 Upper and lower 'O' rings (S)
14 Bronze membrane support disc (S)
15 Stainless forceps
16 Suction pump
17 Gas lighter (S)

CONSUMABLES

18 Membrane filters
19 Pads & dispenser
20 Membrane broth FC
21 Methanol dispenser
22 DPD 1 & 3 tablets
23 Phenol red tablets
24 Higienic tissues
25 Daily report forms

residual, turbidity, pH and conductivity to be carried out on site. If required the integral incubator and battery permit 5-6 cycles of 14 hours each of incubation (at 44° C) before it is necessary to return to a laboratory base.

In surveillance programmes additional specific tests can be added for regional laboratories which require the conventional range of chemical analytical facilities, e.g. NH_3, NO_3, metals and toxics analysis; but the critical parameters remain the basis of routine surveillance and quality control in most developing countries because by far the commonest health risk is that deriving from microbial contamination.

3.3 Training of Personnel

It was proposed that surveillance should be carried out as a team activity and that the team should comprise as a minimum an area coordinator, a sanitary technician and a laboratory technician. The ideal arrangement might be one in which each health area dedicated a taskforce to water surveillance. The reality is that all sanitary and laboratory staff are multifunctional and it was necessary to train, where available, at least twice the number of staff per area, to allow for wastage and transfer to other duties. The regional capital, Huancayo required additional support and resources to cover the substantially larger urban population and disproportionate number of smaller water supply systems located within the area. It was concluded that the minimum number would be 21 (7 areas x 3 staff) but twice this number would be trained if available.

The selection of sanitary and laboratory technicians for surveillance training depended on the population served by water supplies, the number of water supply systems per area, and their location. This data is summarised in Table 1.

4. PRELIMINARY RESULTS OF THE DIAGNOSTIC STUDY and PREPLAN

4.1 Interpretation of Fecal coliform and chlorine residual results

The water quality results reported here have been reduced to a resume of fecal coliform and chlorine residual results in line with WHO Volume III Guidelines for rural supplies.

i. FECAL COLIFORM DATA

Of those systems where two sampling visits were possible during the preplan, only 7 of the 60 systems (11 per cent of the total)conformed to WHO guidelines acceptability for the more stringent fecal coliform count of zero per 100ml sample in all samples. All of these were spring water sources classified as gravity without treatment (GST). An additional 4 springwater supplies, visited only once, were also free from fecal coliform contamination. Thus 11 of 60, or less than **18 per cent of systems conform to WHO guidelines for bacteriological quality** based on the absence of fecal coliforms and were subsequently classed as class A. The preplan results compare favourably with the diagnostic of systems thus far analysed in the pilot region wherein (22 per cent) 69/307 are classified as class A.

Only bacteriological and chlorine residual data were under consideration in the preplan and at that stage no account was taken of service level, including quantity, continuity and coverage of the supply. However at the start of the decade an international assessment report summarised "the level of access to both safe water and supply services in the rural sector". Peru was categorised in the 21-40 per cent range for safe water and supply service level. However it has been reported that 54 per cent of the rural population of the health region enjoyed organised water supply services, and our preplan sample suggested that only 18 per cent of the 54 per cent i.e. less than 12 per cent of the rural population have regular access to fecally uncontaminated water. Thus if the WHO guidelines for bacteriological safety are accepted, using the more conservative fecal coliform test, and clearly the samples quoted are representative, then the 1980 categorisation for Peru was over optimistic. Had total coliforms been used as a criterion of water quality it is likely that almost all rural supplies would be shown to be contaminated given the total absence of disinfection control.

It is important to bear this concept in mind when we attempt to assess the significance of contamination in the remaining 82 per cent of preplan and 78 per cent of systems in the diagnostic study which have been demonstrated to be either sporadically or uniformly exposed to fecal contamination. It was therefore considered useful to categorise the bacteriological quality of water from each supply system. In order to do this, the range of contamination in a single reservoir and the corresponding distribution system was classified (A-D) for ready comparison with subsequent visits and other systems in study.

Fecal Classification:

Group	Range of fecal coliform contamination occurring in service reservoir and distributed supply.
A=	0/100ml in all samples on one sampling visit i.e. conforms to WHO bacteriological guidelines
B=	1-10/100ml in all or any samples on one sampling visit i.e. low level, often sporadic, contamination
C=	11-50/100ml in all or any samples on one sampling visit i.e. medium level with significant waterborne disease risk
D=	>50/100ml in all or any samples on one sampling visit i.e. grossly contaminated with high waterborne disease risk

It was gratifying to plot a frequency matrix (Table 2) for all 9 possible combinations of grades and find a majority of systems (28/53) exhibiting homogeneity in grading in consecutive quarters of the year, on successive sampling occasions.

The second most common category of water quality was at the second level of homogeneity (AB, BC, CD), containing 13/53 systems. Together the top two levels of homogeneity made up 78% of the classification.

It is noteworthy that the preplan survey was conducted prior to the main rainy season in the Sierra. Increased heterogeneity, and generally higher contamination, might have been expected had the survey overlapped with the rains.

ii) CHLORINE RESIDUAL DATA

Although villages sometimes claimed to be disinfecting their supplies at the reservoir prior to distribution, no chlorine, free or combined, was detected at any time in any of the rural systems sampled. The application of hypochlorite disinfection is recommended by the Ministry of Health but is rarely attempted and there is no strategy for assisting villages in implementation, operation or control.

4.2 Relationship between gross contamination and type of system

i. Sedimentation/Slow sand filtration systems (GCT) studied in the preplan.

The most worrying and obvious conclusion to be drawn from the preplan data sheets is **the total failure of treatment plants** constructed under the National Plan for Rural Water Supply. Not one of the systems in study operated effectively. This confirms the independent findings of DelAgua in similar systems in the department of Lima, evaluated in 1983/84 under a British Overseas Development Administration sponsored research scheme (R3760). Of the 10 Junin preplan supplies classified as uniformly grossly contaminated (DD), six were treatment systems. The remaining two treatment systems supplied water classed as BD and BC. Thus all the treatment systems in the preplan survey supply contaminated water to the distribution system, and 6 of the 8 systems in study supply grossly contaminated water (category DD) which carries a continuous high risk of water borne disease. Clearly the rehabilitation of these systems should take a high priority equal to that of the construction of new systems.

ii. Gravity systems without treatment (GST)

Only four out of 46 spring water supplies studies in the preplan sample fell into the most grossly contaminated category (DD) on both sampling occasions. One of these supplies was contaminated at the source. Another, had uncontaminated source water but contaminated reservoir water and the report failed to record an inspection of the conduction pipe between source and reservoir or the state of the reservoir. Thus the origin of the contamination could not be identified. The other two supplies

Table 2. Frequency matrix demonstrating the consistency of the fecal coliform system of classification of water supplies

Most homogeneous	AA 7	BB 7	CC 4	DD 10	=	Nº (%) 28 (53)
	AB 6	BC 5	CD 2		=	13 (25)
	AC 3	BD 4			=	7 (13)
Least homogeneous	AD 5				=	5 (9)
Nº of systems	21 +	16 +	6 +	10	=	53 (100)

Table 3. **Rural water supply systems in the pilot region classified according to fecal contamination in supply**

Health area	Fecal contamination group				Total systems (%)
	A (%)	B (%)	C (%)	D (%)	
Huancavelica	Incomplete report				
Huancayo	21 (19)	33(30)	19(17)	38(34)	111 (100)
Jauja	28 (34)	30(37)	11(13)	13(16)	82 (100)
Tarma	19(24)	32(41)	15(19)	12(15)	78 (100)
La Merced	Incomplete report				
Satipo	0 (0)	5(28)	3(16)	10(56)	18 (100)
Junín	1 (16)	11(61)	3(17)	3(17)	18 (100)
Totals	69(22)	111(36)	51(17)	76(25)	307 (100)

Gross fecal contamination |__25%_|

Medium & gross fecal contamination |____42%______|

All fecal contamination |_________78%___________|

Fecally uncontaminated |_22%_|

were deficient in source water data because the abstraction was sealed or locked. AD-represents maximum heterogeneity in contamination on separate sampling occasions and hence spasmodic gross contamination. It was not surprising to find that all 5 systems showing this characteristic were the gravity untreated type (GST). It was however disappointing to be unable to identify the source of contamination in 4/5 systems because the source spring inspection points were sealed and therefore not sampled. The point of contamination of the fifth system was localised within the distribution system.

The points mentioned above highlight both the importance of a detailed sanitary inspection report to complement the water quality report, and the need to provide advance notice and contact with the local Administrative Committee of an intended sampling visit in order to gain access to all components of a system. If both are available then it is usually a more straightforward task to organise the repair and improvement of the supply.

6. ACKNOWLEDGEMENTS

The investigations described in this report have been made possible by the generous support of the following organisations:

1. Overseas Development Administration, Government of the United Kingdom.

2. The Environmental Hazards Division of WHO, Geneva under an agreement with UNEP.

3. CEPIS-PAHO/WHO, Lima, Peru.

4. The Technical Directorate for the Environment, Ministry of Health, Peru.

The views expressed are the responsibility of the authors and not necessarily those of the supporting organisations.

REFERENCE NOTES

(1) WHO, Guidelines for Drinking Water Quality - Vol.1. Recommendations. 1984. p. 9.

(2) SOLSONA, F. Programa de Control de Calidad de Aqua de Consumo en la Republica del Peru.Consultancy Report 10th June - 10th July 1983. PAHO/WHO, Lima (unpublished).

(3) LLOYD,B; PARDON, M; WEDGEWOOD, K. & BARTRAM, J. Developing Regional Water Surveillance in Health Region XIII Peru, Phase I Report for The Government of Peru, March 1986. p. 14.

(4) SAENZ, R. *Estudio sobre el Sistema y Organizacion Requeridos para Optimizar el Funcionamiento de Motores y Bombas en los Acueductos Rurales del Peru. Informe Final.*
Investigacion No. 2. DTIAPA - CEPIS-PAHO. Sept. 1982.

(5) DELAGUA LTD. *The Rehabilitation of the Water Treatment System of the Rural Community of Cocharcas, Huancayo, Junin, Peru.* The Government of Peru. Technical Division of Environmental Sanitation of the Ministry of Health. Lima. July 1986.

ORGANOLEPTIC WATER QUALITY: HEALTH AND ECONOMIC IMPACTS

J.I. Daniels,[1] D.W. Layton,[1] M.A. Nelson,[2] A.W. Olivieri,[3]
R.C. Cooper,[3] R.E. Danielson,[3] W.H. Bruvold,[3] R. Scofield,[4]
D.P.H. Hsieh,[5] and S.A. Schaub[6]

ABSTRACT

Organoleptic properties of drinking water (i.e., characteristics perceptible to the senses) can affect the acceptance of water by the public. In this paper we present a risk-analysis methodology, along with supporting data, that can be used for assessing the relationship between the level of either (1) turbidity, color, and odor; or (2) total dissolved solids (TDS); or (3) metabolites of algae and associated bacteria in drinking water, and the fraction of an exposed population that could reject the water. We explain how this methodology can be used by public health authorities in developing nations as a rational approach for adopting pragmatic water-quality guidelines for these organoleptic constituents, and for accurately correlating concentrations of these organoleptic constituents with the need to commit manpower and resources to improve water quality in rural areas, small communities, and large cities.

INTRODUCTION

The organoleptic properties of drinking water (e.g., taste, odor, and appearance) constitute important parameters in defining its public acceptability. In developing nations of the world, organoleptically unacceptable sources of drinking water can have impacts on both public health and the economy. For example, the alternative to organoleptically unacceptable drinking water is obtaining water from another source. This could mean dissatisfied consumers choosing alternatives that are superior organoleptically, but inferior because of the presence of pathogens or toxic substances. In such situations there is an increased risk of disease. Illness that does develop in the population can reduce productivity, increase infant morbidity and mortality, and add to health care expenses for the government.

In this paper we present a data base and risk-analysis methodology that can be used to assess quantitatively the fraction of an exposed population that could refuse to drink water on the basis of the level of either (1) turbidity, color, and odor; or (2) total dissolved solids (TDS); or (3) metabolites of algae and associated

[1]Lawrence Livermore National Laboratory (LLNL), Livermore, CA. [2]LLNL. Present address: Washington State University. [3]University of California at Berkeley. [4]University of California at Davis (UCD). Present Address: ENVIRON, Washington, DC. [5]UCD. [6]U.S. Army Medical Bioengineering Research and Development Laboratory.

bacteria that it contains. We explain how this methodology can be used by public health authorities in developing nations as a rational approach for adopting pragmatic water-quality guidelines for these constituents, and for correlating concentrations of these organoleptic constituents with the need to commit manpower and resources to improve water quality in rural areas, small communities, and large cities.

RECOMMENDING GUIDELINES

Elevated levels of organoleptic constituents (e.g., turbidity, color, and odor, etc.) can make the taste, odor, or appearance of drinking water objectionable to consumers. Adverse health effects may also be associated with high concentrations of these constituents in drinking water, although causal relationships are not well defined. Consumers ceasing to drink water because the degree of objectionable organoleptic characteristics exceeds a tolerable limit for acceptability, will presumably seek an alternative supply. Alternative sources that are inferior because of the presence of pathogens or toxic substances could produce significant health and economic impacts. Identifying the locations where such impacts could occur in developing nations can be accomplished by recommending guidelines for concentrations of these organoleptic constituents of water and comparing existing concentrations in available water supplies to these guidelines.

Methods for recommending guidelines for the aforementioned organoleptic constituents of drinking water are described next. In adapting these methods for use in developing nations we assume that most of the population of a developing nation is confronted with a limited number of choices with respect to alternative sources of drinking water, should they dislike the water that is available or provided. For this reason, convenience in obtaining water and thirst are considered to increase the tolerance of such populations to any objectionable organoleptic properties of water. Therefore, we assume that consumers in developing nations will generally accept a taste, odor, or appearance in drinking water that might otherwise be found objectionable by consumers in developed nations.

TURBIDITY, COLOR, AND ODOR

Turbidity in water is caused by suspended material such as clay; silt, finely divided organic and inorganic matter; soluble, colored organic compounds; and plankton and other microorganisms (APHA, 1985). Measurements of turbidity derived by nephelometry (determining the intensity of reflected light) are expressed as nephelometric turbidity units (NTU).

Color in water may result from the presence of metallic ions (iron and manganese), humus and peat materials, plankton, weeds, and industrial wastes (APHA, 1985). Color is generally measured by the platinum-cobalt visual-comparison method (one color unit = 1 mg/L of platinum in water) and is expressed as units of color and not as mg/L.

Odor in water can be produced by a variety of different compounds, particularly organic substances (APHA, 1985). Because odors in water are usually complex and detectable at concentrations that are too low to permit isolating and determining the responsible agents, the test for odor in water is based on measuring its intensity. The threshold odor number (TON) is a frequently used measure of the intensity of odor in water.

Harris (1972) performed a study designed to systematically relate combinations of turbidity, color, and odor values in drinking water to the percentage (from 0 to 100) of the public accepting the water. He obtained acceptability ratings for 125 water samples containing different combinations of color, odor, and turbidity from three different consumer populations (i.e., consumers of bottled water, unfiltered tapwater, or filtered tapwater). In assessing a sample, the respondent observed and smelled the sample, then indicated the degree to which he or she could accept the water by selecting the appropriate rating on the action-tendency scale shown in Table 1.

According to Harris (1972), the borderline between acceptance and rejection for the population surveyed was located between statement 5 on the rating scale ("Maybe I could accept this water as my everyday drinking water") and statement 6 ("I don't think I could accept this water as my everyday drinking water").

From an analysis of Harris's data we derived the following multiple regression equation:

$$A = 86 - 0.5(C) - 1(T) - 0.1(S), \qquad (1)$$

where A = percentage of population rating water acceptable; C = color units; T = nephelometric turbidity units (NTU); and S = threshold odor number (TON). Approximately 80% ($R^2 = 0.794$) of the variation of the population's acceptance of drinking water is explained jointly by color, turbidity, and odor.

As stated previously, we assume the population of a developing nation to be less sensitive to the organoleptic characteristics of water than populations in developed nations, like the one Harris (1972) surveyed. Based on this assumption, we shift the threshold for acceptance and rejection for populations in developing nations from between rating statements 5 and 6 to between statements 6 and 7 (see Table 1). To determine the effect of this shift on Eq. 1, we examined how the shift would change the one frequency distribution that was presented by Harris (1972) as an example in his report. This frequency distribution reflected the responses of individuals from each consumer population to water containing 5 NTU, 15 color units, and a TON of 3; levels specified by drinking water standards of the U.S. Environmental Protection Agency (1975, 1984). Our examination revealed that modifying the acceptance-rejection borderline in the aforementioned manner increases the acceptability percentage by approximately 15%. Therefore, we assume that this increase is applicable to frequency distributions of responses to all other combinations of turbidity,

Table 1. Action-tendency scale for rating a water on basis of turbidity, color, and odor.

Rating	Statement
1	I would be very happy to accept this water as my everyday drinking water.
2	I would be happy to accept this water as my everyday drinking water.
3	I am sure that I could accept this water as my everyday drinking water.
4	I could accept this water as my everyday drinking water.
5	Maybe I could accept this water as my everyday drinking water.
Acceptance Threshold for Population Surveyed by Harris (1972)	
6	I don't think I could accept this water as my everyday drinking water.
Acceptance Threshold for Population of Developing Nation.	
7	I could not accept this water as my everyday drinking water.
8	I could never drink this water.

color, and odor. Consequently, the coefficients in Eq. 1 should be increased by a factor of 15% (i.e., multiplied by 1.15) for developing nations. A mathematical expression can then be derived for estimating the percentage of the population (P) in a developing nation that could refuse to drink water and need to seek another source, for any combination of color, turbidity, or odor in the water.

$$P = 1.1 + 0.575(C) + 1.15(T) + 0.115(S). \quad (2)$$

There is no apparent direct correlation between concentrations of these physical characteristics of water and adverse health effects. However, results from research like that conducted by Sproul _et al_. (1979) indicate that turbidity values above 5 NTU, caused by organic material, decrease disinfection efficiency, whereas inorganic sources of turbidity do not. Therefore, we recommend consideration be given to using the U.S. Environmental Protection Agency Standards (1975, 1984) for turbidity (5 NTU), color (15 color units), and odor (TON of 3) as guidelines for these organoleptic constituents of water in developing nations. However, even at these levels, about 16% of a population in a developing nation might still refuse to drink the water. Nevertheless, this is a reasonable estimate as color levels above the standard are considered to be objectionable to a large fraction of a population, and the standard for odor is considered to be a level that most consumers would find acceptable (U.S. Environmental Protection Agency, 1984). Additionally, detectability of color or odor by consumers may not necessarily be that undesirable.

TOTAL DISSOLVED SOLIDS

Mineral salts and small amounts of other inorganic and organic substances constitute the filterable residue content of water (i.e., the material that will pass through a standard glass-fiber filter disk). The concentration of filterable residue is expressed commonly as milligrams per liter (mg/L) of TDS (APHA, 1985). Typically, the ions of the mineral salts are the predominant constituents of the dissolved material; consequently, conductivity is a frequently used method for estimating the TDS content of a water sample.

The relationship between TDS concentration and the percentage of a population in a developing nation that could refuse to drink the water is shown in Figure 1. We developed this relationship based on the results of analyses performed by Bruvold and Ongerth (1969) of taste panel responses to California water supplies. The action-tendency rating scale upon which the responses were made was constructed according to a psychometric-scaling method and although eight of the statements are virtually the same as those used by Harris (1972) (see Table 1) the scale is not identical and consists of nine statements (see Fig. 1) with equal distances between scale values (Bruvold, 1968). However, the two rating scales do refer to the behavioral response of the individual concerning the actual consumption of water, and in both cases we assume that the threshold at which a population in a developing nation will reject the water will occur between the same two response statements. Finally, a normal distribution is assumed to exist around the lines best fitting the relationship between TDS concentrations and scale values derived by Bruvold and Ongerth (1969), and a constant standard error of estimation is assumed for the scale.

According to the information in Fig. 1, the guideline for TDS concentration should be below 2000 mg/L. This will limit the fraction of a population in a developing nation that could refuse to drink the water to a level less than or equal to about 15%. Additionally, TDS concentrations below 2000 mg/L should also limit the amount of magnesium and sulfate ions present, which are constituents of TDS and very effective laxative agents. In fact, magnesium and sulfate ions are administered clinically as saline laxatives and in fasted individuals single doses of 480 mg of magnesium or 1490 mg of sulfate are cathartic (Fingl, 1980).

METABOLITES OF ALGAE AND ASSOCIATED BACTERIA

Certain cyanobacteria (blue-green algae) and actinomycetes (gram-positive filamentous bacteria that grow in close association with cyanobacteria), which are aquatic microorganisms, have been identified as the source of taste and odor-producing biochemical compounds in surface waters (Amoore, 1986). Two of these compounds, geosmin and 2-methylisoborneol (MIB), are metabolites that have been identified as the two major causes of earthy-musty tastes and odors in water (Izaguirre _et al_. 1982). Threshold odor concentrations for these compounds are among the lowest detectable by the human sense of smell (Amoore, 1986)--for geosmin the lowest reported average threshold odor concentrations was 4 ng/L and for MIB the lowest reported average

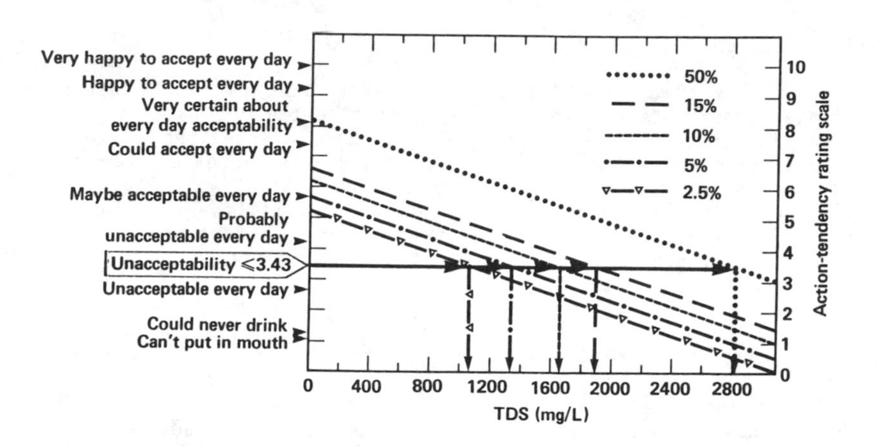

Figure 1. Relationship between TDS concentration and population of a developing nation rating water unacceptable on the action-tendency scale (i.e., ≤ 3.43, the boundary scale value defining unacceptability for populations in developing nations).

threshold odor concentration was 8.5 ng/L (Means and McGuire, 1986) and because these are average values many people can detect these compounds at lower levels.

Burlingame et al. (1986) studied an episode of unacceptable taste and odor attributed to concentrations of geosmin above 20 ng/L in a water supply in Philadelphia, PA. The highest levels of geosmin were present in this water when the floating masses of algae were most intense, and the maximum concentration of geosmin recorded exceeded 100 ng/L. Consumer complaints were greatest when the geosmin level was above 45 ng/L, and few complaints were associated with geosmin concentrations of 30 ng/L or less. On the basis of this data, Burlingame et al. (1986) concluded that 30 ng/L was an acceptable target level for acceptability. According to data presented by Krasner et al. (1983), MIB concentrations above 30 ng/L also are likely to produce moderate to strong organoleptic effects. Therefore, we conclude that MIB levels also should not exceed 30 ng/L for water to be acceptable.

For developing nations, a reasonable guideline for acceptable levels of geosmin and MIB is 45 ng/L in drinking water. We base this recommendation on the fact that the greatest number of complaints from Philadelphia consumers occurred at geosmin concentrations above 45 ng/L (i.e., 139/169 = 82%) and very few at levels below 30 ng/L (i.e., 6/169 = 3.5%) Furthermore, sensory fatigue is caused more readily by geosmin than by other odorants (Burlingame et al., 1986), which indicates that elevated levels can be tolerated and percentages of complaints are probably conservative estimates of the fraction of the population likely to refuse to drink the water. Additionally, there appears to be no adverse health effects associated with the presence of these metabolites in drinking water (Amoore, 1986).

OPTIMIZING RESOURCE COMMITMENTS FOR IMPROVING WATER QUALITY

Comparison of water-quality monitoring data with recommended guidelines for the aforementioned organoleptic constituents of water will identify drinking water sources that potentially need improvement. These waters can then be prioritized for the commitment of resources for their improvement by the following procedure. First, the population that could refuse to drink the present supply needs to be estimated using the algorithms just described. Next, the total cost for that population to obtain available supplies that are of better quality should be determined. Should no such supplies exist, then the cost of health care for the population that presumably would seek alternative supplies, which could be inferior because of the presence of pathogens or toxic substances, needs to be determined. Although such costs are hard to estimate precisely, they can be quantified. For example, White *et al*. (1972) estimated medical costs in East Africa attributed to water-related disease preventable by improving supplies. Finally, the medical costs or costs of obtaining better quality water from another source can be compared to the one for improving the quality of the existing water supply, then the magnitude of this difference can be used to indicate the water supplies needing attention first.

CONCLUSION

Uncertainties in our estimates of the fractions of populations at risk for refusing to consume water based on its organoleptic constituents can be reduced by performing psychometric testing with populations in developing nations and the water supplies they now use. Until this is done, we believe the methodology just described is an effective approach for recommending organoleptic water-quality guidelines and optimizing the commitment of resources.

ACKNOWLEDGMENTS

Work performed under the auspices of the U.S. Department of Energy by the Lawrence Livermore National Laboratory, Livermore, CA, USA, under Contract W-7405-Eng-48. Funding was from the U.S. Army Medical Research and Development Command.

REFERENCES

(APHA) American Public Health Association (1985). Standard Methods For the Examination of Water and Wastewater, Sixteenth Edition, American Public Health Association, American Water Works Association, Water Pollution Control Federation, Washington, DC, 1985.

Amoore, J.E. (1986). "The Chemistry and Physiology of Odor Sensitivity," J. Am. Water Works Assoc. 78, 70-76.

Bruvold, W.H. (1968). "Scales for Rating the Taste of Water," J. Appl. Psych. 52, 245-253.

Bruvold, W.H., and H.J. Ongerth (1969). "Taste Quality of Mineralized Water," J. Am. Water Works Assoc. 61, 170-174.

Burlingame, G.A., R.M. Dann, and G.L. Brock (1986). "A Case Study of Geosmin in Philadelphia's Water," J. Am. Water Works Assoc. 78, 56-61.

Fingl, E. (1980). "Laxatives and Cathartics," in Goodman and Gilman's The Pharmacological Basis of Therapeutics, A.G. Gilman, L. S. Goodman, and A. Gilman, Eds. MacMillan Publishing Co., Inc., New York, NY, 6th ed., pp. 1002-1012.

Harris, D.H. (1972). Assessment of Turbidity, Color, and Odor in Water, Anacapa Sciences, Inc., Santa Barbara, CA, Technical Report 128 prepared for U.S. Department of Interior, Washington, DC.

Izaguirre, C.J. Hwang, S.W. Krasner, and M.J. McGuire (1982). "Geosmin and 2-Methylisoborneol from Cyanobacteria in Three Water Supply Systems," Appl. Env. Microbiol. 43, 708-714.

Krasner, S.W., M.J. McGuire, and V.B. Ferguson (1983). Application of the Flavor Profile Method for Taste and Odor Problems in Drinking Water, Water Quality Technology Conference, American Water Works Association, Norfolk, Virginia, December 6, 1983.

Means, E.G., III, and M.J. McGuire (1986). "An Early Warning System for Taste and Odor Control," J. Am. Water Works Assoc. 78, 77-83.

Sproul, O.J., C.E. Buck, M.A. Emerson, O. Boyce, D. Walsh, and D. Howser (1979). Effect of Particulates on Ozone Disinfection of Bacteria and Viruses in Water, U.S. Environmental Protection Agency, Cincinnati, OH, EPA-600/2-79-089.

U.S. Environmental Protection Agency (1975). "National Interim Primary Drinking Water Regulations," Fed. Regist. 40, 59566-59588.

U.S. Environmental Protection Agency (1984). National Secondary Drinking Water Regulations, U.S. Environmental Protection Agency, Office of Drinking Water, Washington, DC, EPA-570/9-76-000.

White, G.F., D.J. Bradley, and A.U. White (1972). Drawers of Water, The University of Chicago Press, Chicago, IL, pp. 151-200.

STRATEGY TO IMPROVE DRINKING WATER QUALITY IN SMALL WATER SYSTEMS IN PUERTO RICO

Jorge Martinez, P.E.*

Introduction

We have confirmed that most private water supply systems in Puerto Rico supply coliform contaminated drinking water. Based on the Sanitary Surveys conducted, most systems were found to be vulnerable to significant contamination. The intent of the Safe Drinking Water Act is that all consumers served by public water systems be provided with safe drinking water. The nearly 90,000 persons relying on private water supply systems in Puerto Rico cannot be excluded. Therefore, timely and appropriate actions are to be initiated to reduce if not to eliminate the significant health threat that exists in these private water systems in Puerto Rico.

In order to address the Small Private Water Supply problems in a timely and efficient manner a detailed strategy has been developed which deals with the problem on four different fronts namely: technology, education, enforcement, and financial assistance.

Strategy

The Island of Puerto Rico with an estimated population of 3.2 million people and a territorial extension of 3,500 square miles, provides drinking water to more than 90 percent of its population by means of the government owned Puerto Rico Aqueduct and Sewer Authority.

PRASA, as frequently named by the Islanders, is one of the largest water supply utilities in the U.S.A. presently operating 240 Public Water Supplies (PWS), out of which over 80 are water filtration plants. The nearly 400 million gallons per day produced by PRASA comply with all the existing drinking water standards on a fairly continuous basis.

On the other hand, approximately 3% of the Island's population, about 90,000 people are being served drinking

* Environmental Engineer, U.S. Environmental Protection Agency, 1413 Fernandez Juncos Avenue, Santurce, P.R. 00909

water by some 300 non-PRASA Public Water Supplies. These non-PRASA systems are mostly located in the rural areas of Puerto Rico that are not served by PRASA, and are categorized as "very small" serving between 25 and 500 persons. These 300 systems historically have not been regulated by the Commonwealth. Some 200 of these systems are dependent upon significantly vulnerable surface sources, usually a creek, a spring, or a stream.

As these systems were previously unregulated and not operated by a qualified operator, they are either unaware or choose to ignore their responsibilities as a water purveyor.

Limited water quality data on these systems compiled by the Puerto Rico Department of Health indicated a significant water quality problem; over 80 percent were found to be positive for coliform bacteria.

In response to this significant problem, it was decided to implement a special initiative to address the non-PRASA systems. The Bayamon region of Puerto Rico was selected as the study area with the intent of characterizing the systems there in terms of treatment, deficiencies, water quality, and operation and maintenance practices.

All of the non-PRASA systems in the Bayamon region were visited and sampled at three locations: raw water intake, mid distribution and end-of- distribution system. Each sample was analyzed for total coliform, fecal coliform, and standard plate count. Eighty-five percent of the systems sampled were found positive for total as well as for fecal coliforms, in violation of the drinking water standard.

In addition the nine non-PRASA systems in the town of Comerio were selected for a more in-depth study, consisting of not only sample collection but detailed sanitary surveys. Based on the sanitary surveys the following findings of general nature were common to the nine systems:

1. None of the systems are presently providing formal treatment (disinfection, filtration) to their drinking water nor forsee doing so in the near future.

2. All sources are surface waters, (seven springs and two streams.)

3. There is a widespread lack of knowledge of the system's components, operation and maintenance

and the extent of the distribution system, among the community and persons in charge.

4. In general, communities show a lack of awareness in drinking water quality and judge present quality and service provided as good.

5. All of these nine non-PRASA communities are impoverished communities. Therefore, non-PRASA supplies are in poor financial condition to implement many, if not all, of the recommendations necessary to improve system conditions.

6. All the systems are categorized as "very small", not serving more than 300 persons.

7. All sources are fairly inaccesible and unprotected.

8. Most systems have possible inter-connections with other supplies.

9. All of the non-PRASA systems in Comerio are providing bacteriologically unsafe drinking water.

The following specific findings were also common to the nine systems:

1. On all the sources (7 springs and 2 streams) surface runoff gains access into the source, providing for contaminated water from higher adjacent lands to enter the source.

2. Most sources are unprotected from the entry of livestock or man.

3. For all of the springs, encasements do not prevent the entrance of surface drainage or debris.

4. In general raw water collection and distribution tanks do not have tight fitting and locking covers. Most overflows are not properly screened.

5. All systems have possible interconnections with independent rain collection facilities as well as with PRASA's water supplies in the area. Most houses are actually being fed from these three sources; rain, PRASA and the system itself. Interconnections are controlled by means of unreliable common valves.

Following is a list of the short and long range recommendations given to the nine NON-PRASA systems in Comerio as a result of the Sanitary Surveys:

- Short Range Recommendations:

  1. Surface water diversion ditches must be constructed uphill from springs. (On-earth ditches can easily be made by one or two persons.)

  2. Suitable spring encasements must be constructed. (This task although it appears complex, can be achieved fairly easy, as cheap labor is available and relatively small amounts of concrete will be needed.)

  3. Existing encasements as well as newly constructed ones must be disinfected utilizing readily available chlorine bleach.

  4. Clean and disinfect collection-distribution tanks. Provide tanks with tight-fitting covers and locking mechanisms and establish periodic tank cleaning schedules.

  5. A suitable sampling point, such as a protected unthreaded faucet, must be provided.

  6. Adequate positive pressure must be maintained at all times.

  7. Provide frequent water quality monitoring.

- Long Range Recommendations:

  1. Fences or other means of protection must be constructed surrounding sources to prevent entry of man or animals, which can bring contamination into the source. (Materials and labor could prove to be constraining in this task.)

  2. Investigate and eliminate possible cross-connections within the system. (These effort is extremely time consuming and will probably need to be performed by qualified personnel, not presently available at the system.)

  3. Provide disinfection or treatment by any suitable means, such as chlorination or filtration. (Any means decided here will be accompanied with technical, financial and operation and maintenance requirements.)

4. Provide stand-by power.

5. Prepare formal Operation and Maintenance Program.

As a result of the Bayamon experience, a long-term strategy has been developed to address the problems of the non-PRASA issue. The strategy deals with the problems from four different fronts, namely, technology, education enforcement and financial assistance.

A sanitary survey training course was presented in San Juan, Puerto Rico during February 1987 to train individuals, responsible for the inspections of public water systems, in the fundamentals of sanitary surveys.

In terms of technology, although many alternatives are available, slow sand filtration has demostrated excellent bacterial reductions with a minimum of operation and maintenance, which makes it the number one choice to be tried at non-PRASA supplies. To demonstrate the applicability of this treatment in a tropical environment, a conventional slow sand filtration facility will be constructed and operated at a selected community. Along these same lines, several package slow-sand filtration units will be installed at other smaller communities served by non-PRASA systems.

In order to raise community awareness to the public health aspects of drinking water quality, a public education program is being developed for use by the health educators of the Department of Health for presentation in the non-PRASA communities.

In many of the more remote communities, it is felt that the public health education program will only be marginally successful. Therefore, in order to address this problem on a long term basis, a formalized school teaching package on water supply at three separate levels of education through high school is being developed. It is hoped that this will be incorporated as a 1-year course in the school curriculum of the Commonwealth.

Enforcement will be undertaken by issuing Administrative Orders against selected non-PRASA systems to achieve compliance with the drinking water regulations.

And finally it is necessary to create awareness of the non-PRASA situation to other federal and Commonwealth agencies which can provide financial assistance through either grants or loans.

A directory of Federal and Commonwealth agencies having funding programs for water supply improvements, specific to P.R., will be developed. This directory should include; agency names, addresses and phone numbers, contact person, requirements to meet, application forms and any other technical information necessary to request or obtain financial assistance.

Monitoring for Indicator Bacteria in Small Water Systems

Edwin E. Geldreich, Eugene W. Rice and Eleanor J. Read*

The problem of monitoring for water quality in small or rural water supplies is a great concern because these are the public water supplies that experience most of the waterborne disease outbreaks. The reasons for the unsatisfactory state of public health may be found either in the use of poor quality raw source waters that are not adequately treated or in the ineffectual operation of existing plant processes.

Major waterborne agents contaminating water supplies include pathogenic bacteria, virus, protozoa and blood flukes, any of which cause a variety of gastroentestinal illnesses. These organisms originate in the feces discharged from infected humans, animal pets, farm animals and wild life and are transported by sewage and stormwater to the receiving waters that may be used as a raw source water in water supply. Engineered process barriers are designed to intervene in the passage of waterborne risks through public water supply. How effective these barriers are in prevention of pathogen passage in drinking water is the key purpose of the monitoring program.

While monitoring for pathogens in water is desirable, it is not practical for many reasons. First, the number of waterborne pathogens identified in water is estimated to be several hundred or more when consideration is given to the numerous species of Salmonella and types of viral agents. This list of pathogens has expanded in recent years with the discovery of new bacterial and viral agents that include Yersinia, Campylobacter, rotavirus, reovirus, parvovirus and the protozoans, Giardia and Cryptosporidium. Other pathogens will be discovered in the future as new breakthroughs in methodology unravel the mysteries of unidentified agents associated with waterborne outbreaks. In the United States alone (Table 1), unidentified etiologic agents accounted for 46.5% of all outbreaks during 1961-1983 and caused 86,740 individual illness cases (Craun, 1985; Lippy and Waltrip, 1984).

The problem of pathogen monitoring is further complicated by the lack of a single test that will detect all bacterial pathogens, viral

---

*EPA Senior Microbiologist; Research Microbiologist; Water Engineering Research Laboratory and Senior Statistician, Computer Sciences Corporation, Cincinnati, Ohio, respectively.

agents or pathogenic protozoans and the knowledge that a negative pathogen test is inconclusive, using current state-of-the-art techniques. Monitoring sewage for pathogens currently prevalent in the community is not completely adequate because there will be no input from pathogen shedders in the infected farm animal and wildlife population. While some methods are available for specific pathogens (*Salmonella*, *Yersinia*, *Campylobacter* and enterovirus) these tests are best done in the specilaized laboratory, not by the average technician using the limited resources available in the small laboratory. Cost becomes the other limitation on direct monitoring for pathogens. Estimates for bacterial pathogen examination in water samples can range from $20 to $50 per sample while a virus or *Giardia* examination can cost $100 or more. Obviously, these costs can not be reconciled to small water system operating budgets. Therefore, the only viable alternative is to select a surrogate organism or bacterial indicator of fecal contamination that can be detected in a simple, inexpensive laboratory test.

Table 1. Waterborne Outbreaks in the United States During 1961-1983*

Etiologic Agents	Outbreaks	Cases	Deaths
Bacterial			
Shigella	52	7,462	6
Salmonella	37	19,286	3
Campylobacter	5	4,773	0
Toxigenic *E. coli*	5	1,188	4
Vibrio	1	17	0
Yersinia	1	16	0
Viral			
Hepatitis A	51	1,626	1
Norwalk	16	3,973	0
Rotavirus	1	1,761	0
Protozoan			
Giardia	84	22,897	0
Entamoeba	3	39	2
Chemical			
Inorganic (metals, nitrate)	29	891	0
Organic (pesticides, herbicides)	21	2,725	7
Unidentified agents	266	86,740	0
TOTAL	572	153,394	23

*Data adapted from Craun (1985); Lippy and Waltrip (1984).

The perfect surrogate would be one that completely mimics the occurrence and survival pattern of all fecal pathogens that might be waterborne and traceable through inadequate treatment barriers. Unfortunately, such an organism or indicator system does not exist. What is available are several indicator systems (total coliform, fecal coliform and fecal streptococcus) that are found in all warm-blooded animal feces with densities ranging from 10^6 to 10^8 organisms per gram (Kowal, 1982; Geldreich, 1978). Historical studies of these candidates has placed general emphasis on the wide spectrum of intestinal organisms that comprise the total coliform population. Their detection in water supply would suggest fecal contamination has occurred and this contamination very likely includes some intestinal pathogen risk to water consumers. Furthermore, adequate treatment will remove all coliforms detectable in 100 mL, including those that are of fecal origin as well as the more ubiquitous environmental strains. While this assumption has generally been accurate, the concept is occasionally flawed by virus or protozoan occurrences not associated with detection of coliforms in 100 mL samples. However, an overwhelming data base from systems world-wide has supported the general assumption that coliform absence in public water supply indicates there is reasonable assurance that the water is safe and carries minimal public health hazard from pathogens (World Health Organization, 1984).

Basic Principals for Coliform Detection

Total coliform detection in water supply is not complicated nor does it require a large amount of specialized equipment or a professional microbiologist to perform the test. There are two approaches: measurement of coliform density in a 50 or 100 mL test portion or the simple determination of the presence or absence of coliforms in 100 mL sample volume (Figure 1). For quantitative measurements, many laboratories use a membrane filter (MF) technique which involves the cultivation of differentiated coliform colonies on the filter surface using a lactose type medium and incubation at 35°C for 24 hours before colony counts are established (American Public Health Association, 1985). The multiple tube fermentation test (FT) has been in use by some laboratories for many years as the alternative method for detecting coliforms (American Public Health Association, 1985). This procedure is based on detecting gas produced by coliforms growing in a medium containing lactose. Five tubes of the sterile medium are inoculated with 10 mL sample portions and then the set of tubes is incubated for 24-48 hrs at 35°C to permit any coliforms present to ferment the lactose with release of gas into the medium and entrappment in a fermentation vial. If growth and gas are produced, those individual positive cultures are verified by growth transfers into a second, more restrictive lactose medium for further evidence of gas production. This confirmation provides assurance the sample contained coliform bacteria, not some other bacteria causing a false positive reaction. The density of coliform bacteria per 100 mL is then estimated from the number of tubes positive by using a most probable number statistical table. The FT procedure requires at least 48 to 96 hours to determine a test result and is more labor intensive than the MF procedure.

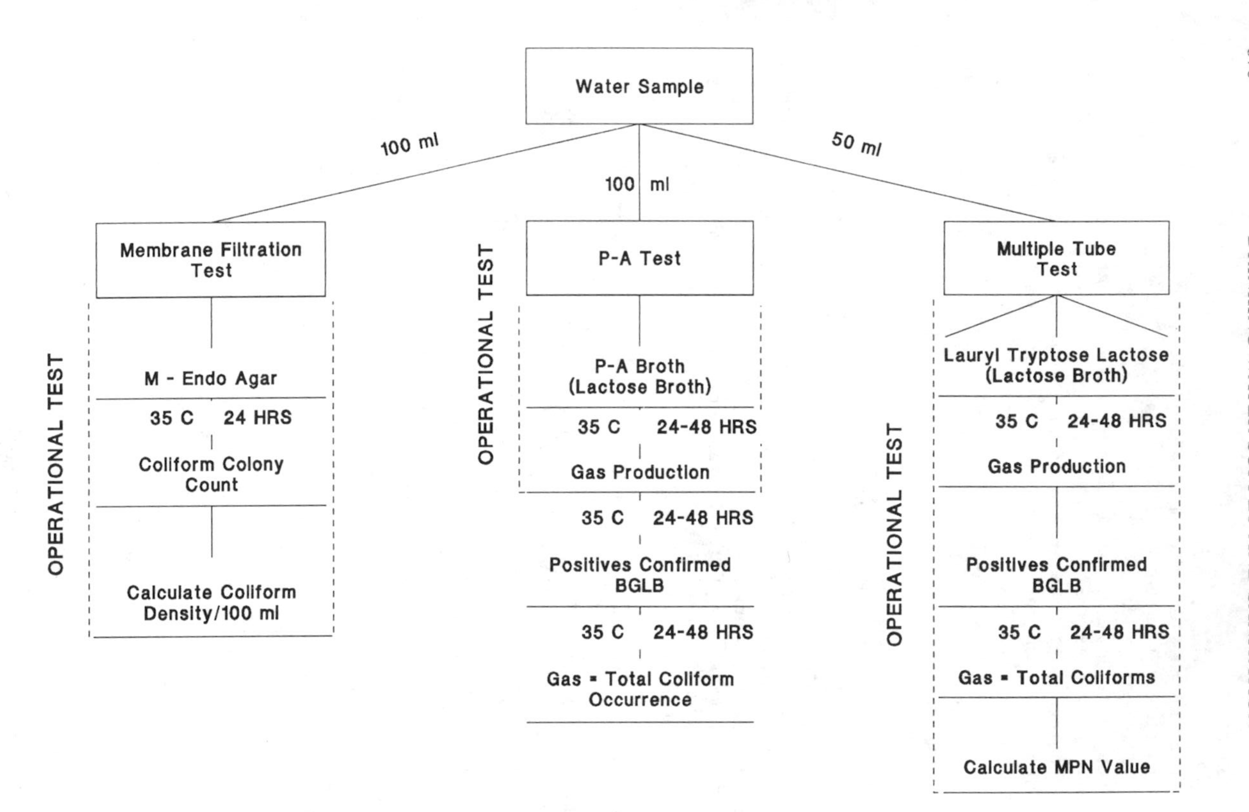

FIGURE 1. PATHWAYS FOR TOTAL COLIFORM DETECTION IN WATER SUPPLY

A more simplified test (Figure 1) and one that can be a very attractive operational test for small water plant operators is the presence-absence (P-A) test (American Public Health Association, 1985). While this procedure appears to be somewhat similar to a multiple tube fermentation test, the involvement in laboratory work is much less complicated. Basically, the procedure consists of inoculating 100 mL of sample into a bottle containing the appropriate concentration of a lactose type medium and a fermentation tube for gas entrappment. The bottle, with 100 mL sample, is incubated for 24 to 48 hours at 35°C and inspected for growth and gas production. If gas is noted in the fermentation tube or a color change (acid reaction) is observed, a small inoculumn of the culture is transferred to a tube of brilliant green lactose broth for verification that gas production again occurs and was related to coliform occurrence. Results of this test are completed within 96 hrs and reported as coliform present or absent. The only equipment needed is a 35°C incubator, sterile sample containers, bottles of P-A medium and a supply of sterile brilliant green lactose broth in culture tubes containing fermentation vials for gas entrappment. Media could be prepared and sterilized in a small autoclave at the water plant or obtained prepared for use from a central laboratory. If the P-A culture bottles are carefully premarked at 150 mL capacity (100 mL for sample plus 50 mL for medium with adjustment made for fermentation tube displacement in liquid) the samples could be added directly into the P-A bottle at the site of sample collection. With this accomplished, sterile sample containers would not be a necessary item in the test. Any chlorine residual in the water would be immediately neutralized by the medium constituents.

P-A Concept vs P-A Test

It is important to separate two different aspects associated with presence-absence information. The P-A concept is concerned with the frequency of coliform occurrence in a water supply over a specified time span. Such data can be obtained from conventional bacteriological tests using either the membrane filter or multiple tube procedure, simply by translating any coliform count or positive tube results into a coliform occurrence. This concept places equal emphasis on all positive samples, regardless of density, with a limit defined by a specific percentage of positive coliform occurrences permitted. For example, the presence-absence record of compliance could be based on degree of treatment: 5% positive occurrences permitted if the system uses complete treatment, 3% if treatment consists only of filtration and disinfection of surface waters and 1% for water supplies using disinfection as the only treatment control barriers. For the public health authority, information on how often there are coliform occurrences, over the long term, is an important indication of treatment effectiveness and operator skill in providing a continuous supply of safe drinking water.

From the operators viewpoint, it is desirable to have the capability for a simple bacteriological test which provides frequent checks on water quality produced. Before acceptance of the P-A test, two questions need to be addressed: (1), what are the technical considerations in performing the test and (2), how do the results of the P-A test compare with the more exacting laboratory procedures using either the membrane filter or multiple tube test? In a review of technical considerations, the P-A test ranks high for ease of examination. The sample can be added directly to the culture bottle at the field site, thus dispensing with the sterile sample bottle and dechlorinating agent. Preparation of medium in sterile culture bottles could be delegated to a cooperating public health laboratory and sufficient supplies stored in a dark, cool storage area for no more than four weeks. As another option, at least one commercial venture is exploring the potential for manufacturing prepared medium in a disposable culture bottle so that no medium preparation need be done in the water plant. The answers to the second question can be found in a statisitcal review of several evaluation studies now completed (Jacobs, et al., 1986; Pipes, et al., 1986; Caldwell and Seidler, 1987).

How Valid is the P-A Test

Operational tests used by the small water plant operator should not only be inexpensive and easy to perform but also acknowledged to produce data that is equivalent or better in precision to that of the standard laboratory procedures. It therefore becomes important to understand the validity of P-A test results obtained from a variety of water supplies in different geographical areas. This consideration is critical to both the operator proposing to use the test and the public health authority who needs to evaluate the water supply quality. Test sensitivity to coliform detection and parallel examinations by the P-A test and either the MF or FT procedure (or both) were done in three widely divergent geographical areas. A data base of 1,483 samples in Vermont (Jacobs, et al., 1986), 1,560 samples in Oregon (Caldwell and Seidler, 1987) and 2,601 samples in eastern Pennsylvania (Pipes, et al., 1986) obtained from small water systems, were analyzed statistically to determine if significant differences could be established between any of the three testing procedures. Comparative data in the Pennsylvania study did not include parallel examination by the FT test. The McNemars statistical test (Fleiss, 1981) was selected as the most appropriate way to compare the different coliform detection methods, since each water sample was examined in parallel by more than one method.

As noted in Figures 2 and 3, the P-A test significantly outperformed both the multiple tube and membrane filter tests for coliform detection in water samples from supplies in Vermont and Oregon. Details of the statistical analyses are given in Table 2. Further analysis of these data also revealed that the multiple tube fermentation test detected more samples containing coliforms than the conventional membrane filter procedure using M-Endo medium. The statistical analysis of the eastern Pennsylvania data (Figure 4) indicated that there was no significant difference in coliform detection by the P-A test and the MF method. The Chi-square value was only 1.27, signifying that comparative results between the P-A and MF test results were essentially equivalent. Why there was no clear cut superiority to the P-A test as the best method for coliform detection in all instances may be related to the state of vigor for coliforms in the water supply. For the eastern Pennsylvania study, many of the positive results were obtained from samples of a small water system that did not chlorinate or other-

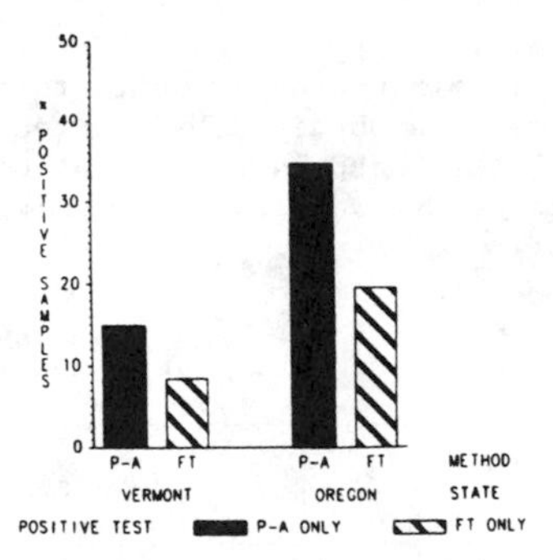

FIGURE 2. P-A VS FT METHOD COMPARISONS

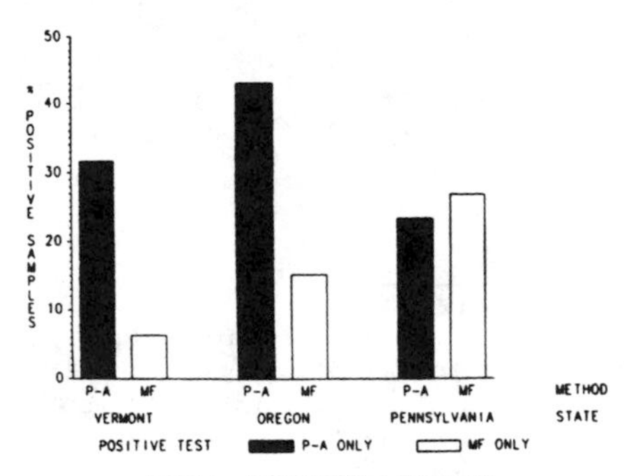

FIGURE 3. P-A VS MF METHOD COMPARISONS

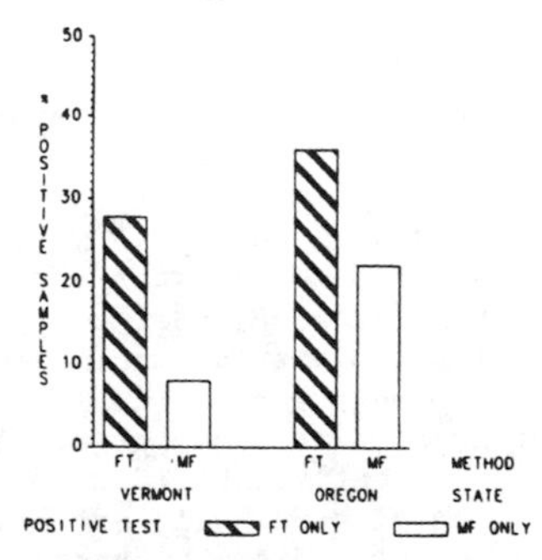

FIGURE 4. FT VS MF METHOD COMPARISONS

wise treat the groundwater supply. Regardless of this fact, these field studies on test performance do demonstrate that the P-A method equals or exceeds the results obtained by the recognized conventional procedures. Therefore, there should be no reason not to use the P-A test as either an operational tool in the small water plant or as part of the official monitoring data.

Table 2. McNemar's Test Results

Method Comparison	Study	McNemar's x^2	p-value
P-A vs FT	Vermont	5.3	0.02
	Oregon	12.7	<0.01
P-A vs MF	Vermont	52.0	<0.01
	Oregon	39.9	<0.01
	Pennsylvania	1.3	0.26
FT vs MF	Vermont	31.4	<0.01
	Oregon	8.3	<0.01

Operator Tests for Water Quality

Lacking information on plant effluent quality seriously restricts the rural water plant operator's ability to promptly respond to unsatisfactory water quality conditions through treatment adjustments. Perhaps this position leads to blind faith that water treatment can run automatically and produce a satisfactory water supply of uniform quality. This may be true for protected groundwater supplies but in many small water systems, surface and ground raw water quality does fluctuate because of poorly designed wells that do not protect water quality from contamination by improperly operated wastewater treatment systems, stormwater runoff and animal activity in the area adjacent to water supply intakes.

Monitoring finished water quality for several key characterisitcs is not beyond the realm of possibility in small water systems. There are available several basic techniques that water plant operators can learn to use with minimal training that will give them a measure of water quality being released as public water supply. Much critical information of immediate value can be obtained by frequent measurements for free chlorine residual, turbidity and coliform bacteria in the plant effluent. These three measurements will provide information respectively on the continued maintenance of a treatment control process, interference and protective shielding of microorganims and verification that the treatment process is effectively removing coliform bacteria. Chlorine and turbidity measurements can be made at frequent intervals in the day and provide immediate information using basic test kits that are readily available and easy to use. While coliform testing requires a minimum of 24 hours of processing time it is an important record of treatment effectiveness for controlling the microbial quality of water produced and should be done at least once per week.

Bacteriological testing of public water supplies in rural or remote areas is not beyond the reach of operator capability. Both the MF and FT procedures have been packaged into commercially available field kits that may be used. For sheer simplicity, however, the presence-absence (P-A) coliform test may be the method of choice. Adapting the P-A test into a simplified operational test would be ideal for on-site monitoring although it does not quantify the extent of contamination events that might suggest treatment breakdown, loss of distribution integrity or biofilm development. The procedure tests a single 100 mL sample, not 50 mL which must be divided into five replicate portions of 10 mL required in the multiple tube test. Materials required for the P-A test (media and associated glassware items) are more readily available and cost less in many third world countries than membrane filters and associated filtration equipment. While the P-A concept was originally developed with a modified lactose broth (P-A broth), there is no reason why the test could not be applied to other coliform broth formulations (lactose broth, lauryl tryptose broth or MacConkey broth)to which bromcresol purple is added to indicate acid production. To further simplify the P-A test application to small water systems use, it should be permissable to utilize positive test information obtained from culture bottles after 24 to 48 hours incubation at 35°C without any further positive result confirmation. While these positive results might occasionally contain a false positive reaction (gas produced by a non-coliform organism) the error would be on the conservative side for safety considerations. These options would provide the opportunity to use the P-A test in remote locations in the world, where regional suppliers of laboratory materials carry limited choices of bacteriological media. The only critical consideration would be in the preparation of triple strength medium to be dispensed to the culture bottles in 50 mL volume. Medium dilution by the addition of 100 mL sample would create normal strength medium.

P-A Test For Remote Monitoring

Regional or national public health authorities need to monitor the water quality of small systems in all areas of the District or Nation. This essential program should define a minimal number of water samples to be collected (on a monthly basis) from all public water supplies, regardless of their remote locations and submitted to the central laboratory for examination. Unfortunately, some samples may be in transit for several days before they reach the laboratory. During this time, the microbial flora in the water changes, often leading to adverse, uncharacteristic test results that suggest the water sample meets national drinking water standards when indeed it does not (McDaniels and Bordner, 1983).

Since preservation of bacteriological samples is difficult to achieve without loss of low density coliform occurrences through nutrient depletion, extended contact time with chlorine, toxicity of heavy metal impurities or microbial flora antagonism, the only other approach is to permit controlled growth during transit.

The P-A test could be a solution to this problem. The sample is added directly to the medium, then the culture bottles transported (maximum of 5 days) back to a central laboratory. The procedure is uncomplicated and requires very minimal effort by the water plant operator to perform (Pipes, et al., 1986). During transit, ambient air temperatures will influence the magnitude of sample culture growth, being slow during winter temperatures below 10°C and accelerated at summer or tropical temperatures that may reach normal (35°C) incubation temperature. Upon arrival in the laboratory, observation for gas production is made to determine what processing will be required. If turbidity and gas production is evident, the culture is confirmed in brilliant green lactose broth for evidence of coliform occurrence. If no gas production is noted, the culture bottle is incubated for 24 to 48 hrs at 35°C; then confirmed, if positive. Those culture bottles with no acid production or no visible turbidity but gas in the inner tube indicate air was shaken into tube during shipment, so these must be inverted to release the entrapped air and incubated 24 to 48 hrs as for the other negative cultures received for processing. The procedure is not labor intensive and would provide information on any coliform occurrences that were related to the original water sample.

Summary

Development of a practical microbiological monitoring program for small water systems is urgently needed at the local level for prompt detection of contamination followed by appropriate remedial actions. Three operational measurements are within the capability of the water plant operator: chlorine residual, turbidity and a basic test for coliform occurences using a presence or absence (P-A) concept. Evaluation of the P-A test in three different geographical areas demonstrated the procedure to outperform both the membrane filter and fermentation tube (MPN) tests for coliform detection. The use of a P-A test approach is recommended for small water system personnel because of its simplicity and adaptability to minimal resources available to the public system sector in rural and remote regions world-wide. Further adaptation of the test can be made to initiate field inoculation of the sample to be sent to a central laboratory for final processing and data gathering on water quality nationwide.

References

American Public Health Association (1985) Standard Methods for the Examination of Water and Wastewater. 16th ed., 1268 pp., Washington, D.C.

Caldwell, B. A. and Seidler, R. J. (1987) Comparison of Bacteriological Assays and Sampling Regimes for Increased Coliform Detection in Small Public Water Supplies. Appl. Environ. Microbiol. (In Press).

Craun, G. (1985) An Overview of Statistics on Acute and Chronic Water Contamination Problems, p 5-15. In Fourth Domestic Water Quality

Symposium: Point-of-Use Treatment and Its Implications. Water Quality Assoc., Lisle, Ill.

Fleiss, J. L. (1981) Statistical Methods for Rates and Proportions. 2nd ed. John Wiley and Sons, New York.

Geldreich, E. E. (1978). Bacterial Populations and Indicator Concepts in Feces, Sewage, Stormwater and Solid Wastes. In Indicators of Viruses in Water and Food, G. Berg ed., 422 pp., Ann Arbor Science Publishers Inc., Ann Arbor, MI.

Jacobs, N. J., Zeigler, W. L., Reed, F. C. Stukel, T. A. and Rice, E. W. (1986) Comparison of Membrane Filter, Multiple-Fermentation-Tube, and Presence-Absence Techniques for Detecting Total Coliforms in Small Community Water Systems. Appl. Environ. Microbiol., 51:1007-1012.

Kowal, N. E. (1982) Health Effects of Land Treatment: Microbiological. U.S. Environmental Protection Agency, EPA-600/1-82-007, Health Effects Research Laboratory, Cincinnati, Ohio.

Lippy, E. and Waltrip, S. (1984) Waterborne Disease Outbreaks, 1946 - 1980: A Thirty-Five-Year Perspective. Jour. Amer. Water Works Assoc., 76:60-67.

McDaniels, A. E. and Bordner, R. H. (1983) Effects of Holding Time and Temperature on Coliform Numbers in Drinking Water. Jour. Amer. Water Works Assoc., 75:458-463.

Pipes, W. O., Minnigh, H. A., Moyer, B. and Trog, M. A. (1986) Comparison of Clark's Presence-Absence Test and the Membrane Filter Method for Coliform Detection in Potable Water Samples. Appl. Environ. Microbiol. 52:439-443.

Pipes, W. O., Minnigh, H. A. and Troy, M. (In press). Field Incubation Of Clark's P-A Test for Coliform Detection. Proc. Water Quality Technology Confr., Portland, OR.

World Health Organization (1984) Guidelines for Drinking Water Quality, Vols. 1-3. Geneva, Switzerland.

APPLICATION OF ULTRAVIOLET DISINFECTION IN SMALL SYSTEMS

R. R. CARDENAS, JR.[1], B. A. BELL[2] and M. HATIM[3]

ABSTRACT

Testing and design of UV treatment units for small systems to assure adequate disinfection may be achieved by the bioassay method. This requires selection and growth of a specific organism for the desired UV range, preparation of a calibration curve, and injecting the test organism through the UV test unit over expected flow ranges and range of transmittance values for UV at 254 nM. Samples are collected for the influent and effluent and the survivors counted. The UV dose delivered by the unit is then compared to a standard curve. The required evaluation and comparative data are presented for two small [(5-40 gpm), (19-151 l/min)] UV disinfection units.

INTRODUCTION

The use of Ultraviolet (UV) disinfection for small potable water supplies has become appealing for developing nations. While the idea would not have invited much interest a generation ago, social changes in developing nations, especially latin america, as well as major technical advances changes that have taken place in the application of UV technology in the industrial nations suggests this may be an idea whose time has come. Advantages of UV disinfection in developing nations are numerous and include: operating requirements are usually limited to power and replacement lamps; little, or no, operator skill is required; no transport, handling,

---

1. Dept. of Civil & Environmental Engineering, Polytechnic University, Brooklyn, N.Y. 11201. Presently on Sabbatical Leave, Employed by Carpenter Environmental Associates, Inc. 406 Paulding Avenue, Northvale, N.J. 07647.

2. Vice President, Carpenter Environmental Associates, Inc. 406 Paulding Ave., Northvale, NJ 07647.

3. Graduate Student, Dept. of Civil & Environmental Engineering, Polytechnic University, Brooklyn, N.Y. 11201.

shipping, or storage of hazardous materials is required; and, no adverse effects result from overdosing.

BACKGROUND

Bacteria are killed by UV by the disruption of genetic structural units resulting in fatal damage. While cells may have the capability to "repair" this damage at low UV dosages, at high, or lethal dosages, repair is not usually possible.

When a specific number of cells (N_o) are exposed to a UV intensity (I), over a period of time (t), the survivors (N), will fit Equation 1, where Q, (a constant), is a function of the species of bacteria present:

$$N = N_o \exp[-It/Q] \qquad \text{(Eq. 1)}$$

The UV dose (D) required to kill bacteria is a function of the UV intensity (I) and exposure time (t) (Eq. 2):

$$D = It \qquad \text{(Eq. 2)}$$

At a constant wave length the intensity of the UV light (I) can be measured by the Beer-Lambert relationship (Eq. 3) which requires knowledge of the intensity of the light entering the water (I_o), and the UV absorption characteristics of the water being treated (a), and length of the light path (d):

$$I = I_o \exp[-ad] \qquad \text{(Eq. 3)}$$

UV Dosimetry

Until recently measurement of UV dose in a full scale disinfection unit was difficult. This was because UV is limited in its penetration of water and by water quality, specifically UV Transmittance at 254 nM. Moreover, for any disinfection unit there is a variable UV intensity throughout the unit because of reflection, refraction, and diffraction, complicated by complex flow patterns which may be variable with flow.

Measurement of UV dose can be carried out a number of ways including direct measurements, and chemical and biological methods. This paper will focus on biological methods using bacterial indicators, or bioassays.

Bioassays

The UV dosage required to kill most microorganisms is predictable and linear when collimated light is used. UV dosage can be readily measured using static, as well as continuous flow conditions (Mundlak, 1984). The use of a bioassay with a defined UV kill curve is appealing since bacteria in water physically model the movement of microorganisms through the UV unit. More importantly, a bioassay offers a promising tool in the design of a UV disinfection unit for a specific application. This is of special relevance for small applications where the water may be of variable quality and flow.

Bioassays are now commonly used in testing UV disinfection units for treated wastewaters. A number of large scale wastewater plants have employed bioassays to define the UV dosage under different conditions of flow.

At least four stages are necessary to carry out a bioassay, (1) Selection and growth of the test organism, (2) Determination of the calibration curve, (3) Testing of the disinfection unit and (4) Determination of the UV dosage delivered by the disinfection unit.

Approximate UV dosages necessary to kill specific microorganisms have been defined (Nagy, 1964). This information may be used to estimate the maximum UV dosage required to achieve a specific kill for a specific microorganism. Bacteria, for example, range in susceptibility from about 3,500 uW-sec/cm^2 to almost 9,000 uW-sec/cm^2. Coliform bacteria and most pathogenic bacteria can be killed in pure cultures at about 6,500 uW-sec/cm^2. Spore forming bacteria can require higher dosages, approaching 70,000 uW-sec/cm^2. Mold and yeast spores, can require considerably higher dosages, up to 330,000 uW-Sec/cm^2. Thus, an organism of appropriate UV resistance may be selected to measure the desired dose range.

Once the test organism is selected, a pure culture stock must be obtained, and mass grown for both calibration and testing. A calibration curve is prepared by exposing a known quantity of the test microbe to varying dosages of collimated UV light and determining the survivors. The results of such a test, using the bacteria M. lutea are presented as Figure 1.

In testing UV units the conditions of flow and water quality must be defined. The flow range to be tested is determined by the specific application. Water quality must be determined by sampling and laboratory analysis. The most important water quality parameters are UV transmittance (%T) at 254 nM, and suspended solids or turbidity.

The approaches that are used in testing full scale units include (1) Batch mixing of organisms and flow through the UV unit, or (2) Pulse or continuous injection of organisms.

During the test, about 100,000 organisms/mL are passed through the test unit and samples are removed immediately prior to entering the UV unit (N_o) and after exposure (N). The organisms are passed through the unit under varying conditions of flow, % T measured at 254 nM and lamps. To determine UV dosage the survivors (N/No) are compared to the standard curve.

DESIGN OF SMALL UNITS

The successful use of UV for disinfection depends greatly on a relatively high water quality, especially with respect to UV transmittance. Testing of the UV disinfection unit to define required disinfection dosage is basic to selecting the unit. The bioassay lends itself readily to this application.

The design approach can be organized into six distinct operational steps:

(1). Establishing Design Conditions.
(2). Collecting Background Data.
(3). Defining Test Conditions or Specifications.
(4). Selection of Test Units.
(5). Testing.
(6). Evaluation.

Design Conditions

Not all communities, households or situations will have the same needs. Under the most commonly employed conditions, UV disinfection is not used to kill parasitic cysts, spores or viruses. Nor does UV provide the persistence of chlorine. However, UV has the ability to kill most pathogenic bacteria and can be used to reduce coliform counts to acceptable levels.

While coliforms and most pathogenic bacteria can be killed at about 6,500 uW-sec/cm^2, coliform bacteria from waters containing sewage have been found to require greater UV dosages (Qualls, 1985). Based on experience with a full scale unit, we have found a UV target dosage of 15,000 uW-sec/cm^2 to be adequate to consistently reduce coliform levels sufficiently low to meet the permit requirements of 200 Fecal Coliform/100 mL (Cardenas, et al 1986). The wastewater, in this case was of relatively high quality.

More specifically the needs will be based on use (flow) and water quality, including variations in Percent Transmittance (%T) of UV at 254 nM, and suspended solids or turbidity. Thus, background water quality data must be obtained for the application contemplated.

Collecting Background Data

To be successful a UV unit should process only clear water that is relatively free of turbidity. However, low turbidity or solids data per se is not enough to guarantee success.

As a minimum it is advisable to gather background data on flow, especially during peak demand periods. Water quality information must be gathered to give the range of percent Transmittance (% T) at 254 nM. Additional support data should include total suspended solids, turbidity, and coliforms. These data should be collected over a period of time to give the range of values which may be expected.

Define Test Conditions

Based on the data, the test conditions can be defined, or, the performance specifications for the unit can be written. The most important parameters to be defined will be the required percentage of kill, the range of values for % T, suspended solids, raw water coliform concentrations, and the allowable reduction in UV lamp intensity.

Screening Test Units

Available test units may be sought from commercial sources. At this point data regarding cleansing, lamp monitoring, controls, maintenance, service backup, periodic testing, etc. as well as costs may be obtained.

Testing

Testing the UV unit is the heart of the design process. Testing must be carried out to simulate the required needs with respect to percent kill, range of flow, percent transmittance at 254 nM, and reduction of UV lamp intensity. Test runs need not be need not be lengthy. Usually only a few minutes at each flow rate is sufficient. Often a small, temporary, backyard plastic pool of about 1,000 gallons (378 L) can be used.

The following example will illustrate the required conditions.

Example:

For this example, it is assumed that there are two similar commercially available UV units with different lamp spacings, shown as Figure 1, that might treat potable water at an average flow of 20 gpm (76 L/min) with a range of from 5 to 40 gpm (18 to 156 l/min). Because of the lack of reliability of the water, we would also like to know the effects of lower water quality, measured as % T, for this water. A standard curve using M. lutea, selected on the basis of UV tolerance was carried out as shown on Figure 2.

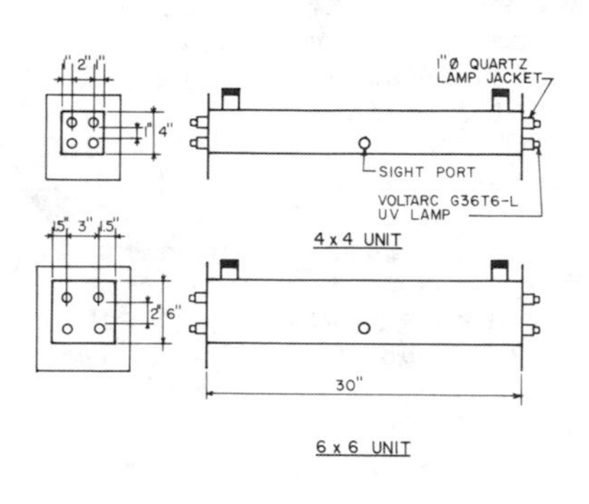

FIGURE 1. UV TEST UNITS, 4x4 AND 6x6.

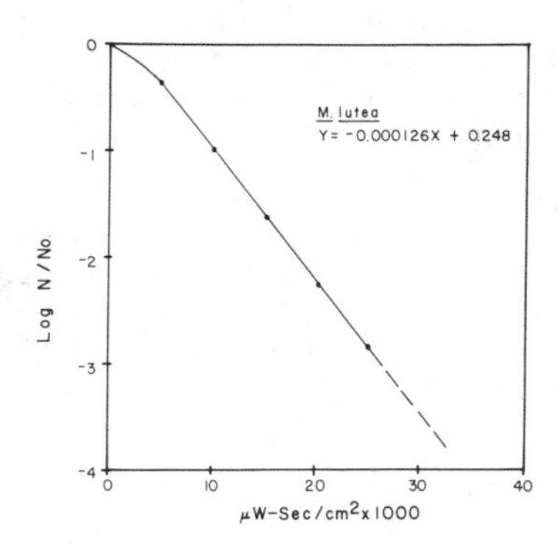

FIGURE 2. CALIBRATION CURVE USING COLLIMATED UV LIGHT.

The units were field tested by pumping water containing about 100,000 bacteria (M. lutea) through the unit at four different flow rates, 5, 10, 20 and 40 gpm (19, 38, 76, and 151 l/min), and at three different levels of percent transmittance values, 30, 50, and 70 % T, using 100 and 50% of the UV lamps. Influent and effluent samples were collected, in duplicate, for each run, and immediately removed to the laboratory for plating and counting.

The bacteria were counted, the results expressed in terms of survivors (N/N_o), and compared with the standard calibration curve for the test bacteria, M. lutea, (Figure 2). The results of the evaluation for the 6x6 unit are shown as Figures 3 and 4.

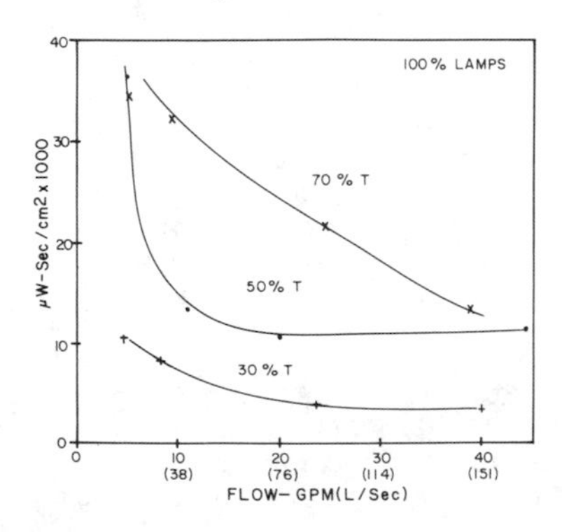

FIGURE 3. UV TESTING, 6X6 UNIT AND 100% LAMPS.

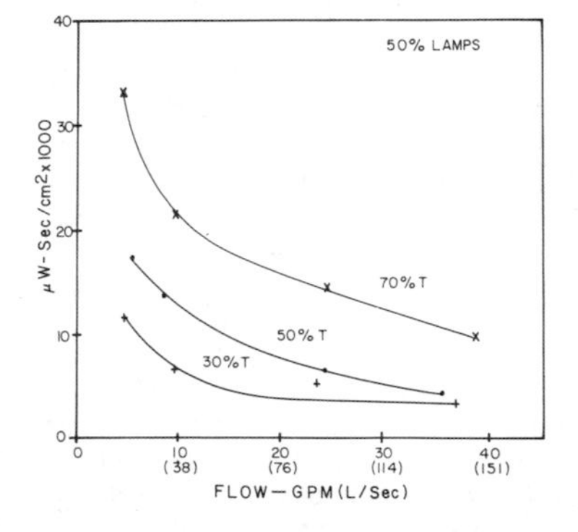

FIGURE 4. UV TESTING, 6x6 UNIT AND 50 % LAMPS.

From Figure 3, using 100% lamps with water quality at 70 % T, and at a flow rate of 20 gpm (76 l/min), it may be seen that the unit will deliver over 20,000 uW-sec/cm^2, well in excess of that required to kill coliform bacteria. However, UV kill for this unit quickly decreases with decreasing UV transmittance (%T). For example, only 11,000 and 4,000 uW-sec/cm^2 were measured at 50 and 30 % T. Also, at the higher flow rates of 30 and 40 gpm (114 and 151 l/min) poor coliform kills can be expected.

When 50 % of the lamps are used, as shown in Figure 4, at 70 % T, the 6x6 unit will still deliver the required UV dosage. However, at reduced UV transmittance and higher flow rates, performance is relatively poor for this lamp spacing.

The results of testing the 4x4 units are shown as Figures 5 and 6.

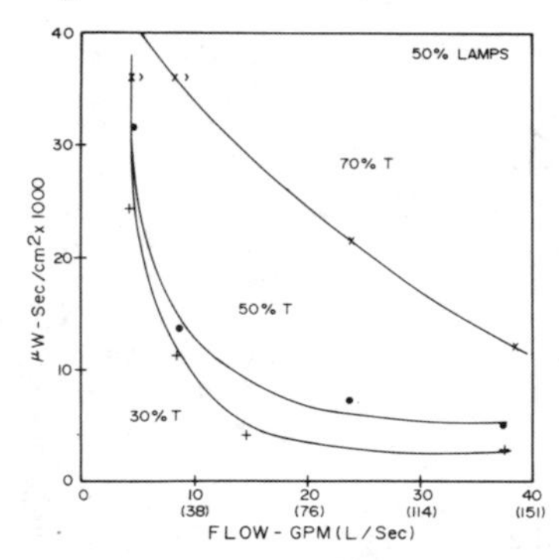

FIGURE 5. UV TESTING, 4X4 UNIT AND 100 % LAMPS.

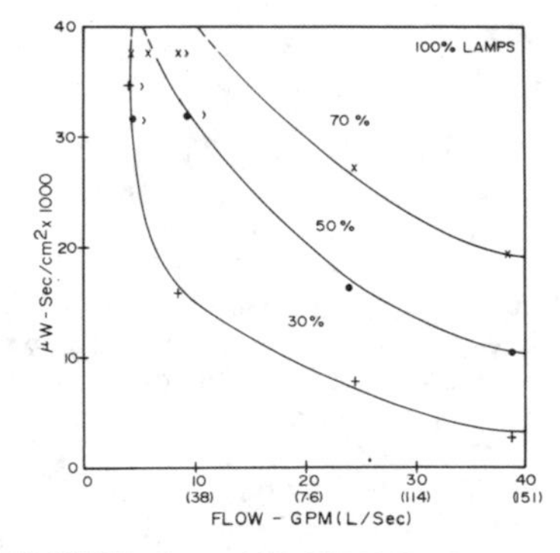

FIGURE 6. UV TESTING, 4X4 UNIT AND 50 % LAMPS.

As shown on Figures 5 and 6, the effect of closer lamp spacing greatly increases the UV dosage. In Figure 5, using 100% lamps at 20 gpm (76 l/min, at both 70 and 50%T this unit, can deliver UV dosages in excess of the required 15,000 uW-sec/cm^2. At higher flow rates of up to 40 gpm at 70%T, UV kill is still adequate. However, a combination of high flow and low % T does not give a good UV dosage, for example, at higher flow rates, say at 30-40 gpm, when the %T is below 70%, a UV doasage of greater than 15,000 uW-sec/cm^2 cannot be assured.

The same unit tested using 50% lamp output (Figure 6) shows that at 20 or 30 gpm (76-113 l/min) a UV dosage of 15,000 uW-sec/cm^2 can be achieved. At this reduced lamp output, however, below 70 % T, an acceptable UV dosage might not be achieved.

A comparison of these two units at 20 gpm (76 l/min) is shown as Figures 7 and 8.

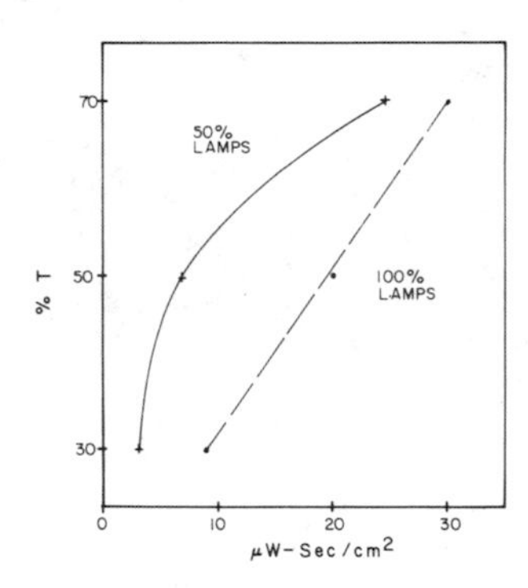

FIGURE 7. UV DOSAGE, 20 GPM (76 L/MIN), 6X6 UNIT.

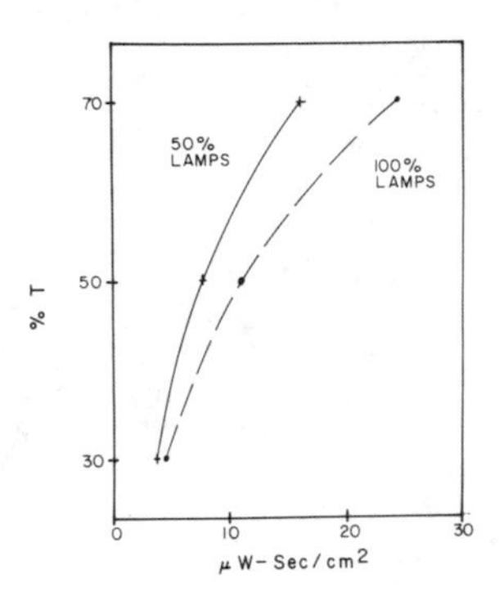

FIGURE 8. UV DOSAGE, 20 GPM (76 L/Min), 4X4 UNIT.

The advantages of the smaller lamp spacing provided by the 4x4 unit can be seen. When 100 % of the lamps are operating, even at reduced % T levels, the UV unit can deliver the required UV dosage. When lamps have been reduced to 50 %, the unit can still deliver 15,000 uW-sec/cm^2 at the design rate of 20 gpm (76 l/min). Performance for this unit, of course, can be compared for any flow rate.

REFERENCES

Cardenas, R. R., Jr., Ravina, L. and Lindsey, D., 1986, "Pilot to Full Scale Ultraviolet Disinfection at the Suffern Wastewater Treatment Plant," J. Environ. Systems 16 (1) 25.

Qualls, R., Ossoff, S., Chang, J., Dorfman, M., Dumais, C., Lobe, D., and Johnson, D. J., 1985, "Factors Controlling Sensitivity in Ultraviolet Disinfection of Secondary Effluents", Water Pollution Control Federation 57 (10) 1006.

Mundlak, E., 1984, "Biological Dosimetry Studies of Ultraviolet Disinfection Under Continuous Flow Operating Conditions," Ph. D. Thesis, Dept. of Civil and Environmental Engineering, Polytechnic Institute of New York, Brooklyn, N.Y.

Nagy, R. 1964, "Application and Measurement of Ultraviolet Radiation," Amer. Indust. Hyg. Assoc. J. 25 274.

Ultraviolet Sterilizes Drinking Water With Renewable Energy Systems

Christopher Anand Scott
ASCE Associate Member

ABSTRACT

High operating costs for conventional water sterilization techniques preclude their application in numerous water supply systems in developing countries. Ultraviolet (UV) disinfection, using 254 nm radiation, appears to be a cost-effective and appropriate alternative to chlorination, reverse osmosis, or ozone treatment for certain projects. Problems associated with the design and operation of UV sterilizers are presented. Their flexibility facilitates integration with an overall treatment system. Low power consumption, portability, and increasing availability of low cost, long life, 12 volt direct current UV lamps lend themselves to renewable energy or hybrid power supplies.

INTRODUCTION

Pathogens proliferate in the surface and groundwater supplies of most communities in developing countries. Although several techniques of sterilization of drinking water have been used in water supply projects, chiefly chlorination, the vast majority of the world's rural populace continues to consume contaminated water. Reluctance to disinfect water by conventional means may be attributed to economic and aesthetic considerations. Boiling water is simply too expensive, given the price of wood or charcoal in Asia, Africa, and South America. Reverse osmosis is not cost-effective except for highly saline raw water. Chlorine dosages required for eradication of certain micro-organisms, for example _streptococcus lactis_, result in foul tasting water, to say nothing of the high operating costs of dosing either toxic chlorine or ozone. For small scale, decentralized water treatment projects, a sterilization system is sought that requires low maintenance, is portable for application in remote areas, and that is flexible enough to be incorporated into an integrated water supply system.

Disinfection of drinking water by irradiation (with 254 nanometer wavelength UV) may be feasible for certain applications, primarily where either raw water quality (pH and temperature), or the strong taste of treated water disallow chlorination. However, chlorination provides a distinct

----

The author is a water resources consultant, currently at:
Development Alternatives
22 Palam Marg, Vasant Vihar
New Delhi 57, INDIA

advantage over UV treatment in the establishment of free chlorine residual to combat recontamination in the distribution network (Oliver and Carey, 1976, p. 2619). However, as mentioned above, the use of UV for small projects (without extensive distribution systems) obviates the need for a residual agent.

The manufacture of commercially available UV lamps with low power consumption and increased life (Zhang and Mee, 1986) allows a wide range of applications. Of particular interest is the use of UV with solar photovoltaic, wind, or microhydropower generators.

FUNDAMENTALS

Ultraviolet radiation may be subdivided into three types: UV-A, or longwave (400-330 nm); UV-B, or mediumwave (330-270 nm); and UV-C, or shortwave (270-200 nm). The lethal effect of UV varies with wavelength, peaking between 250 and 260 nm for all organisms. Micro-organisms, in particular, are susceptible to 254 nm UV-C radiation.

Fortunately, the earth's ozone layer attenuates virtually completely this band of the spectrum. Window glass, even cloth, block localized UV-C. The intensity of radiation in a given medium falls away exponentially with distance from the source, according to Beer's law:

$$I(r) = I_o e^{-kr}$$

where I_o is the incident radiation intensity, k the extinction coefficient, and r the penetration distance. Typical values of k (in cm^{-1}) range from 0.007-0.01 for distilled water to 0.3-0.4 for strained sewage effluent, with k for drinking water from 0.02-0.1 (Masschelein, 1983, p. 211). Factors affecting k include turbidity, color and ion concentrations. Evidently, pretreatment of water greatly increases penetration of UV.

The germicidal action of 254 nm radiation results from inactivation of the micro-organism's DNA. Bacteria and certain viruses may be eradicated through brief exposure to UV, though higher organisms (Daphniae, Euglena, etc.) and algae require significantly higher doses. Table 1 lists the UV dosage required for 90 percent and 99.99 percent extinction of various micro-organisms (Groocock, 1984, p. 168; Angehrn, 1984, p. 111; Masschelein, 1983, p. 218).

Organism	254 nm dosage (mJ/cm^2) for: 90% extinction	99.99%
Salmonella enteritidis	4.0	25
Salmonella typhimurium	8.0	25
Shigella paradysenteriae	1.7	20
Streptococcus lactis	6.2	25
Vibrio cholerae	3.4	25
Polio (virus)	3.2	25

Table 1:
UV dosage required for extinction of commonly occurring micro-organisms (WHO, 1984, p. 3).

DESIGN

Typical exposure time ranges from 3 seconds to several minutes, depending on sterilizer design and proximity of water to the UV source. The most common source of UV-C radiation is the uncoated mercury vapor lamp, which releases a high intensity spike at 253.7 nm. Hot cathode and cold cathode types are available; of the two, hot cathode lamps are preferable, due to higher optical efficiency, optimal lamp temperature of 40 degrees C (compared with 500 degrees C for the cold cathode variety), and longer overall life. Both types are available commercially as tube lamps. In 1980, the antimony vapor lamp was developed with higher efficiency, though costs are high and availability limited.

Design of the irradiation chamber is based on one of three flow regimes: turbulent, laminar, or well-mixed batch processing. With the aid of impellers, and sleeves (to encase the lamp and allow immersion in a pressurized conduit), maximum exposure of pathogens to radiation is achieved under turbulent flow conditions. Generally the encased UV source fits longitudinally inside the chamber of circular cross section, with room for one or more impellers. It is assumed under such conditions, that all micro-organisms pass against the sleeve, minimizing penetration distance. However, determining exposure time at that distance may be problematic.

Laminar flow irradiation allows for more precise calculation of exposure time, for minimum intensity occurs at the maximum distance. This may lead to simpler design and less expensive fabrication, and facilitates maintenance. Several designs are conceivable. A shallow, rectangular channel conveys water under a longitudinally placed lamp set at the focus of a parabolic reflector of the same length as the lamp. Baffles ensure mixing. Or, an encased, sinuous lamp (requiring custom fabrication) is immersed in a shallow rectangular channel. Finally, the standard tube lamp may be placed perpendicular to flow to act as a round-crested weir. In this last alternative, exposure time will be very brief; yet penetration distance will be no greater than critical depth.

A batch processing irradiation chamber may be considered. Variations could incorporate any of the design alternatives mentioned above. However, automated valves allow inflow and outflow, and would certainly increase power consumption, and consequently, cost.

Safety factors must be considered for several aspects of lamp operation. UV output is a function of: 1) time after startup, 2) applied voltage, 3) lamp temperature, and 4) lamp age. Characteristically, a warmup period of several minutes should be allowed before the UV lamp reaches 100 per cent of its rated output. For hot cathode lamps, UV output is roughly proportional to applied voltage, for the 90 to 110 percent of nominal applied voltage range (Masschelein, 1983, p. 209). Output may drop to 70 percent of the rated maximum at operating temperatures below 5 degrees C (Masschelein, 1983, p. 210). Optimal operating temperature is

40 degrees C. Output drops with age. After several thousand hours of operation, the lamp may be producing only 75 percent of the maximum rated output (Angehrn, 1984, p. 112). It is recommended that lamps operating eight hours per day be replaced at least once a year.

Control circuitry should be considered to detect lamp failure. As the mercury vapor lamp emits visible radiation in addition to UV-C, inexpensive photo cells, or calcium sulphide elements whose electrical resistance varies with incident light intensity, may serve this purpose. In a prototype sterilizer designed and tested by the author, such photo cells were coupled with a relay switch that closed a solenoid valve at the inlet to the laminar flow irradiation chanber. Additionally, a simple timer opened the valve several minutes after startup to allow the lamp to warm up.

OPERATION

Several auxiliary systems may be applied to augment the operation of the UV source. In addition to the sleeves and reflectors discussed above, chemical "sensitizers" can be added to the water to enhance the action of UV on organics in the untreated water. Chief among these are hydrogen peroxide (Malaiyandi, *et al*., 1980, p. 1135), and TiO_2 in the anastase phase (Carey and Oliver, 1980, p. 157). H_2O_2 is consumed during treatment, though TiO_2 can be recovered. Application of sensitizers necessitates dosing and recovery systems whcih increase both the complexity of design and the total cost.

The use of sleeves, reflectors, and sensitizers has implications for the fabrication and operation of UV sterilizers. Since window glass does not transmit UV-C, sleeves must be made of quartz. Optical quartz with high transmittance at 254 nm is prohibitively expensive--a single sleeve may run several tens of times the cost of the lamp alone. Additionally, immersed lamps must then be periodically cleaned of organic buildup, which may pose particular problems in the case of enclosed, turbulent flow, pressurized devices.

Lamps suspended above the water should make use of reflectors in order to maximize the available radiation. However, reflectance varies with wavelength. It has been found that highly polished Alzak reflectors have high reflectance at 254 nm. Aluminized mylar has poorer reflectance, to say nothing of degradation of mylar films under intense UV radiation.

To improve the efficiency of pathogen eradication, designers may wish to consider two types of pretreatment: filtration and ion-exchange. It is noted above that turbidity, color, and the presence of ions reduce the penetration of UV-C in water. Pre-filtration would lower the indices of turbidity and color, and may be a necessary process in the overall treatment. Ion-exchange resins reduce the concentration of dissolved ions, thus improving the efficiency of UV treatment.

The application of UV sterilizers in developing

countries raises several hardware problems. Lamps commercially available in industrialized countries may have to be imported. Transportation of delicate components (lamps and quartz sleeves) is likely to be costly. Finally, periodic replacement of various components requires some technical expertise.

CONCLUSION

UV sterilization of drinking water offers several unique advantages over conventional techniques (chlorination, reverse osmosis, and ozone treatment) and may actually be more appropriate for application under certain conditions. Compared to chlorine and ozone, UV requires less maintenance, is not toxic, is considerable faster, and is less dependent on water quality (considering suspended matter, temperature, and pH). Installation is less elaborate for UV systems than for any of the three mentioned above, and for a small scale project, operating costs are lower.

Mercury vapor lamps are inexpensive and consume 4 to 64 watts (GE, p. 2). Control circuitry requires an additional 10 to 50 percent of power. The prototype constructed and tested by the author used a 15 W lamp (General Electric G15 T8) to treat 2.0 liters per minute (of unfiltered water), achieving over 99 percent eradication of micro-organisms. The estimated operating cost was U.S.$ 0.045 per 1000 liters (1985 energy prices).

Currently, both AC and DC lamps are available. Without the use of DC/AC inverters, a DC lamp could be run directly from battery storage, powered by solar photovoltaic, wind, or microhydropower generators, possibly in conjunction with a diesel hybrid system. For example, a photovoltaic array with batteries and voltage regulator, sized for a 15 W system operating 8 hours per day, would cost less than U.S.$ 1000 for Cameroon, West Africa or India (1986 prices, which will fall). The initial outlay plus operating costs for a complete treatment system using UV sterilization for small scale applications is significantly less that for a comparable conventional system.

BIBLIOGRAPHICAL REFERENCES

Angehrn, M., "Ultraviolet Disinfection of Water," Aqua, no. 2, pp. 109-115, 1984.

Carey, John H. and Barry G. Oliver, "The Photochemical Treatment of Wastewater by Ultraviolet Irradiation of Semiconductors," Water Pollution Research Journal of Canada, vol. 15, no. 8, pp. 157-185, 1980.

General Electric, "Germicidal Lamps," Large Lamp Department Publication TP-122, undated.

Groocock, N.H., "Disinfection of Drinking Water by Ultraviolet Light," Journal of the Institution of Water Engineers and Scientists, vol. 38, no. 2, pp. 163-172, April, 1984.

Malaiyandi, Murugan and M. Husain Sadar, Pauline Lee, and Ron O'Grady, "Removal of Organics in Water Using Hydrogen Peroxide in Presence of Ultraviolet Light," Water Research, vol. 14, no. 8, pp. 1131-1135, 1980.

Masschelein, W.J., "Scope and Limitations of Disinfection of Water by Ultra-violet Radiation," Water Supply, vol. 1, no. 4, pp. 205-229, 1983.

Oliver, Barry G. and John H. Carey, "Ultraviolet Disinfection: An Alternative to Chlorination," Water Pollution Control Federation Journal, vol. 48, no. 11, pp. 2619-2624, November 1976.

World Health Organization, Guidelines for Drinking Water Quality, Volume 2, "Health Criteria and Other Supporting Information." Geneva: WHO, 1984.

Zhang, , and Mee, 'Commercial Mercury Vapor Lamps Adapted for Long Life,' Journal of Physics (E), April, 1986.

Prospects for the International Drinking Water Supply and Sanitation Decade

Frank Hartvelt, Senior Programme Officer,
United Nations Development Programme

Ladies and Gentlemen,

At the beginning of this conference we had the privilege to listen to Dr. Abel Wolman, who assured us that despite many constraints the Decade Programme is still alive. We also listened to Alexander Rotival, UNDP/WHO Water Decade Co-ordinator, who gave us a comprehensive overview on the Status of the Decade Today. We also heard the voice of optimism of Ambassador McDonald and words of encouragement from our ASCE and Puerto Rico hosts. Now that we are at the end of this conference I am facing the unenviable task of discussing the future of the Decade with you.

My speech today is dedicated to the thousands of people I met over the past years in the international community, governmental and non-governmental agencies, and local communities across the world who turned me into a true Decade crusader.

The past few days were fascinating in that the presentations and discussions oscillated between the unavoidable extremes of conventional large facilities with such buzz words as utility, rates of return, engineering, and small systems with their own, more gentle jargon such as community and women's participation, village caretakers, low-cost options and hygiene education.

But I perceived a very encouraging trend among the conference participants towards accepting more realistic and sensible Decade goals and activities taking into account recent experience, and prevailing conditions in terms of limitations and opportunities, as a point of departure for the expansion and improvement of water supply and sanitation programming during the remainder of the Decade and beyond.

It is now widely agreed that Decade projects must not only be technically, financially and economically sound but also socially and culturally viable. It is also agreed that we are all part of a Decade process whether we are engineers, doctors, managers, administrators, technicians, health workers, or social scientists, interacting with people who need clean drinking water and sanitation.

We should always keep in mind the extremely important positive consequences of Decade activities as expressed in lives saved, less sick or disabled people, less suffering, and more socially and economically productive people.

I should like to draw your attention to three events which took place around mid-Decade (in 1985-86) which, in our view, are not only a point of reference but also will shape Decade activities in the years to come. Firstly, the 1985 meeting of the Development Assistance Committee of the Organization for Economic Co-operation and Development, OECD/DAC, emphasized the need for an intensified and effective process of country-level aid co-ordination in the sector. To be effective, sectoral consultations and co-ordination should be a joint donor-recipient process, analysing and improving sector policies, investment programmes and institutional frameworks. Secondly, the mid-Decade report of the Secretary General of the United Nations on progress in the attainment of the goals of the Decade stressed that organizations of the United Nations system, as well as other international, bilateral and non-governmental organizations, should continue to enhance co-ordination of their development assistance activities at the global and national level and support the role of the UNDP Resident Representatives as focal points for the Decade at the country level. Both the Development Assistance Committee of the OECD and the UN Secretary General committed themselves and urged other agencies and governments to focus their attention and programmes increasingly on the following key areas: human resources development; community participation; health and hygiene education; institutional development; linkages between health, water and sanitation agencies; documentation and information. The UN agencies, members of the Decade Steering Committee (chaired by UNDP), are already active parties in this process. There is no doubt in my mind that voluntary agencies (the NGOs) active at the grassroot levels in thousands of small-scale projects all over the world must have been instrumental in making the big countries and major international agencies accept the above areas of focus.

Thirdly, an important international seminar on low-cost rural and urban-fringe water supply took place in October 1986 in Côte d'Ivoire which resulted, *inter alia,* in "The Abidjan Statement" outlining a five-point strategy (a copy of which is attached.) The outcome of the seminar is well summarized in the preamble which reads as follows:

> "Lasting health and economic benefits for the rural and urban-fringe populations of Africa can be achieved through increased community management of water supply and sanitation systems based on proven low-cost technologies. African governments and donors are urged to identify and commit adequate resources and provide all necessary support for the direct involvement of communities in choosing, managing and paying for their water and sanitation systems."

The main elements of the five-point strategy as developed during the Abidjan Seminar are as follows:

(1) the role of governments and donors, policies to standardize technology and socio-economic approaches, sustainability and replicability, inter-agency co-ordination;

(2) the involvement of communities - especially women - in decision-making and management; affordability of water supply systems;

(3) community water supply as an integral part of primary health care;

(4) choice of appropriate technology, in-country manufacture and distribution of handpumps and spare parts;

(5) community-based maintenance, supported by a national strategy of standardization of spare parts.

Although the seminar focussed on Africa we believe that its outcome is of great interest to all countries in the world which are committed to make substantial progress in the second half of the International Drinking Water Supply and Sanitation Decade (IDWSSD) and beyond. As you will appreciate, the importance of the Abidjan Statement does not only lie in its substance, but also in its endorsement by 100 participants representing both developing and developed countries. UNDP, UNICEF, the World Bank and WHO were among the international agencies which endorsed the statement.

In essence the Abidjan Statement records the conviction of national and international policy makers, administrators, engineers and other specialists, that a 180-degree turn from centrally-managed to community-based water supply programmes is now becoming more feasible in technological, financial and social terms. Especially, it gives participants from developing countries ammunition for discussions with their own policy makers in terms of future programmes.

Consequently, the five-point strategy adopted in Abidjan should provide the necessary guidance in the planning and implementation of low-cost rural and urban-fringe water supply programmes.

The worldwide economic and financial constraints experienced in recent years have severely limited government budgets and exacerbated problems of governments in implementing primary health care programmes including water and sanitation. Despite the divergent trends between diminishing resources and advances in health, water and sanitation technologies, products, measures and approaches there are ways to disseminate progress. These include (a) a greater reliance on families and communities in promoting the use of their own ingenuity, experience and resources, and (b) the improvement or rehabilitation by local authorities, with external support, of health, educational, agricultural and other essential infrastructure.

It is estimated that at present 1.3 billion people still need potable water and 1.6 billion people need sanitation. By the year 2000 another 1 billion people will be unserved unless public and private forces shift into a higher gear.

The UN system is committing both human and financial resources to expand and improve water and sanitation programmes. In recent years major donor countries have joined our Decade programmes: the Federal Republic of Germany, Switzerland, Canada, Denmark, Norway, Sweden, Finland and Holland. We are optimistic that other donors will join us as well.

- 4 -

UNDP, the World Bank, UNICEF and WHO recently agreed to strengthen co-operation and co-ordination among themselves and with other agencies. These agencies called "The Four", will more and more pool their human resources and, wherever possible, jointly undertake projects using the strengths of each agency. As outlined earlier, based on experience of the first half of the Decade the main elements to expand and improve water supply and sanitation programmes are now known, and perhaps more importantly, agreed upon by virtually all governments of developing countries, UN system agencies, the OECD/DAC countries and other organizations including non-governmental agencies.

In a spirit of full participation of all interested parties in both developing and developed countries UNDP, the World Bank, UNICEF, WHO and other UN agencies have agreed to focus their efforts on the following areas during the remainder of the Decade.

(1) Reforms of outdated policies and regulations, sector planning and programme preparation

(2) design, implementation and evaluation of low-cost demonstration and investment projects or components in other projects. Incorporation of women's participation, community management, hygiene education and other primary health care components wherever possible; inclusion of income generating activities and socio-cultural research, and allocation of resources for these purposes.

(3) human resources development including training to strengthen institutions and communities.

(4) continued operational research and feedback into the above programmes in such areas as local manufacturing geared towards village level operation and maintenance pumps (VLOM) for drinking water and small-scale irrigation, drilling (cost reduction), borehole design and construction, water quality, sanitation in urban-fringe areas (e.g. small-bore sewers), cost recovery, liquid and solid waste management focussing on recycling (tens of millions of dollars can be saved or generated through recycling); and, more generally speaking in management, operation and maintenance.

(5) dissemination of general and technical information.

With the exception of waste management the above objectives are tilted towards rural and urban fringe areas which clearly deserve priority attention at present.

However, the international community may also be forced by circumstances to direct more attention and resources to urban water and sanitation problems in the future as the populations of cities are increasing at an alarming rate.

In conclusion, I should like to extend an invitation to all engineering and other associations and societies represented at this conference to adopt Decade guidelines and promote Decade activities through conferences and seminars in both developed and developing countries, publications, fund raising, and, wherever possible to become co-sponsors and partners in water and sanitation activities in the Third World.

Congratulations to ASCE for organizing this most successful conference. Thank you.

Puerto Rico
29/5/1987

CONCLUSIONS AND RECOMMENDATIONS: HUMAN RESOURCES DEVELOPMENT

Horst Otterstetter and Michael Potashnik***

Introduction

In the Human Resources Development Track (HRD) presentations and discussion dealt with human resources in the larger sense. Included are aspects of institutional and cultural environment; management, educational and training practices; as well as dissemination of technological information and community involvement in water supply and sanitation.

Conclusions and Recommendations

The main conclusions are as follows:

1. Overall Conclusion

The overall conclusion was that the ultimate goal of institutional and human resource development is to enable countries to manage their own resources with their own people satisfying their own needs.

2. Institutional performances and practices

Conclusion:

It was recognized that frequently managers of water and sanitation institutions lack the necessary skills in order to, among others, institute adequate management practices based upon well developed processes of strategic, tactic and operational planning. Management information systems are needed to support sound planning and decision making.

Recommendation:

Human resources development in the large sense, meaning recruitment, selection, employment training and motivation, must be a permanent function of any water and sanitation institution and in order to make a real contribution to the achieve-

*Regional Engineer Advisor, Pan American Health Organization; Washington, D.C.
**Union Chief, International Training Network for Water and Waste Management, World Bank; Washington, D.C.

ment of the goals and objectives, HRD planning must be integrated into the overall planning process of the institution.

3. Training as an institutional function

Conclusion:

Training requires time to establish itself as an institutional function. Additionally, the moral and financial full support from top management and participation of middle management is necessary for the success of this training function. The training needs are determined whenever there are performance problems due to the lack of knowledge and skills.

Recommendation:

Activities destined to satisfy training needs should be designed based upon the gap between the job profile and the personal profile of the occupant of the post. Twinning agreements between institutions provide valid alternatives for the acquisition of knowledge and skills.

4. Engineering education

Conclusion:

At present, engineering education does not fully respond to the real needs of water supply and sanitation institutions and the health sector as a whole. Aspects and concepts of management; public health and appropriate technology to serve low-income rural and urban communities are missing from most sanitary engineering curricula. Insufficient time is allocated for college engineering students to acquire practical field experience. These conclusions are valid for school, both in developing as well as in developed countries. The International Training Network for Water and Waste Management is an important Decade initiative to help promote education and training on appropriate low-cost water supply and sanitary techniques and approaches.

5. Community involvement

Conclusion:

In order to achieve overall development, the total community—men, women and children—should be involved in the process. Some of the motivation necessary for this participation can be obtained by allowing the community to make contributions at all stages: from needs assessment through operation and maintenance of water and sanitation systems. The conclusion was reached that there are, by now, plenty of experiences of active participation of women in the conceptualization, construction, operation and maintenance of improved community water supply systems, to prove the benefit of their participation.

Based upon the above conclusions, the following are recommended to the specific audiences indicated:

Recommendation. To Practicing Sanitary Engineers

To expand their personal profile by seeking the knowledge and skills necessary for improving their management capabilities which will allow them to perform well when they reach management positions.

Recommendation. To Water And Sanitation Institutions

To institute the human resources development function and integrate it with their strategic, tactic and operational planning. To seek twinning arrangements with other institutions as a strategy for human resources and institutional development.

Recommendation. To Sanitary and Environmental Schools In Developing And Developed Countries

To add a management and a public health dimension to their courses, as well as to incorporate the concepts of appropriate and low cost technology for water supply and sanitation.

To seek twinning arrangements as a strategy for curricula enrichment as well as for student and teacher development.

Recommendation. To International And Bilateral Development Agencies

To encourage the involvement of the total community—men, women and children—in all phases of implementation of water supply and sanitation systems they sponsor or finance.

To support continuously for at least 5 years any human resource development and training projects they may sponsor in order to allow them to be incorporated permanently as an organizational function and to make a substantial contribution to the level of knowledge and the technological culture of the water and sanitation institutions.

Recommendation. To The Practicing Engineer And Engineering Educators In The United States

Through individual efforts and through the professional associations to increase knowledge and understanding of water supply, sanitation and the environmental health situations in the developing world. This will enable them to be better educators and more effective collaborators with their professional colleagues in the developing countries.

CONCLUSIONS AND RECOMMENDATIONS: FINANCE AND ECONOMICS

Harold Shipman and John Kalbermatten***

Introduction

In planning this part of the program, an effort was made to devote a portion of the time allotted to a review of past practices and policies applied to the financing of projects in the water and sanitation sector by international funding sources over the past ten to fifteen years. The program which finally evolved, reflected this effort only partially since a number of papers were included which, although relevant to the topic, did not address specifically the question of past and future policies in the international arena. Papers presented, therefore, ranged from the highly technical, such as how to optimally design a water distribution system, to the economic justification of water supply investments. Discussions were equally wide ranging, from how to do suggestions to philosophy; they were also very spirited, leading at least one participant to suggest that interdisciplinary debates should stress learning from experience rather than parceling out of blame for past mistakes.

Although the presentations and discussions were as wide ranging as the background of the participants were varied, a motif, if not a consensus developed quickly: users are able and willing to contribute to, and participate in the development of water supply and sanitation to a far greater extent than planners and functionaries realize. This then became one of the major points of discussions, with two other topics, economic and financial justification and institutional issues, also receiving a great deal of attention.

Context

There are about three and a half years left in the International Drinking Water Supply and Sanitation Decade. Participants in the ASCE conference on Resource Mobilization, therefore, had an opportunity both to assess progress made so far and to look beyond the Decade. Clearly, resource generation for the Decade has been insufficient. However, progress has been made in the use of low cost technical alternatives. Equally important, officials and users alike are not willing to use these alternatives. As a consequence, the resource gap is not the major impediment to progress it once was thought to be and the stage is set for significant progress during the remaining years of the Decade and the years beyond. Conclusions drawn from participants' presentations and discussions indicate clearly that substantial increases

*Professional Engineer; Chevy Chase, Maryland
**Kalbermatten Associates; Washington, D.C.

in service coverage can be achieved if low cost alternatives are consistently used and if user participation plays a major role in project development, implementation and operation.

Willingness and Ability to Pay

The willingness of the users to pay for water supply and sanitation services and their ability to do so was an issue raised and remained a principle discussion topic thereafter. The main conclusions of this interchange can be summarized as follows:

Conclusion 1.

Both willingness and ability to pay are usually greater than assumed by project developers.

Conclusion 2.

Project proponents-financiers, governments and their water agencies—must make greater efforts to determine willingness and ability to pay; they must then reflect these findings in the design, financial and institutional arrangements covering the project.

Conclusion 3.

Dialogue with users must be established to prevent the design and implementation of projects which often result in decreasing, even destroying, the users willingness to pay and thus leading to project failures.

Conclusion 4.

Tariff structures should be designed to satisfy economic, financial and social considerations including internal subsidies from the richer to the poorer users.

Conclusion 5.

There is need for broad dissemination of reliable information on matters of willingness and ability to pay. All project developers and decision makers should be exposed to the facts surrounding these issues. More attention should be given to conveying this information to the NGO's.

Economic and Financial Justification

Conclusion 1.

International agencies involved in the financing of water supply projects in the past have usually required that a sound economic justification for investment be established. They have found that because of the difficulty in measuring the benefits of water in the reduction of disease, in reducing fire losses, and in its contribution to other community amenities, the only precise measurement is the financial benefit as measured by the revenue generated from water sales. It has been common therefore to note that in the justification of water projects financed by, among others, the World Bank, a statement is carried which says that the economic return which will include benefits to health, will be over and above the financial return. In the discussion of this issue, it was recognized that clear distinctions do exist between the economic and financial justification of projects. It was also recognized that there is much that needs to be done to bring more substance to the means for measuring the economic benefits which water can bring to a community. Among the areas needing exploration are those of water's contribution to production of raw materials, its benefits to commerce and industry, particularly small industry; and its ability to attract greater resources of other types needed for national development.

Conclusion 2.

The discussions also brought out the need for improving technical and administrative arrangements through reductions of costs of operations and the improvement of efficiency. Better use of manpower, better retention of capable people, establishing causes for, and acting to reduce the high amounts of non-revenue (unaccounted-for) water, better billing and collection practices, and better metering policies were just a few of the areas needing attention.

Conclusion 3.

There was agreement that grants and subsidies for water supply should not be used for payment of operation and maintenance costs. Where government policy provides financial assistance to poor areas and to special areas confronted with unusual problems, it is possible to apply such contributions to capital costs without seriously disrupting efforts to stimulate self support by the water agency.

Background and Experience related to WHO publications on rural water and excreta disposal, and to recent work on latrine and handpump design being carried out under UNDP funding by the World Bank

Some of the broad observations made covering the overall presentations on this topic can be summarized as follows:

Conclusion 1.

The basic public health concepts incorporated in the latrine designs shown and described in the original WHO publication have not changed. Recent work on improving designs have been directed primarily at giving these facilities a greater appeal and attraction to encourage better acceptance by the people. The major problem confronting everyone working in this subsector is that of changing habit patterns and stimulating use and acceptance to build and use latrines. This remains as it has been over the years the number one problem. In this regard, it appears that the support which the World Bank has given to latrine programs, added to WHO's continuing promotion of safe excreta disposal practices, are showing some encouraging signs. It seems likely, however, that the urban fringe and unsewered central areas of cities will feel the need and be the places where greatest activity is likely to be focused for the foreseeable future.

Conclusion 2.

The WHO publication on rural water supply, like its counterpart on excreta disposal, has not been revised since publication in 1959. It remains as a basically sound treatise on the public health justification for safe water, and on the various types of well and spring improvement suited to protecting them against contamination. Recent work on hand pump improvements have been directed primarily at the operation and maintenance aspects of well water supplies and have not altered to any extent the basics of well protection presented in the WHO publication.

Conclusion 3.

One area of progress brought out in the discussions was that of the improvements realized in recent years in the work of water exploration and in well drilling. Current methods have literally revolutionized the rate at which wells can now be put down.

Hopefully the results of the very extensive work done on hand pump design will begin to show up in longer lasting pumps and fewer breakdowns. However, there is no evidence that the well or the hand pump have yet been designed which do not require maintenance and repair. The means by which these are to be done is being concentrated at the village level and the hope is that emphasis on the village capability for maintenance, will bring a major improvement to the task of keeping these facilities in operation.

Funding the Decade Water/Sanitation Efforts

Conclusion:

Annexes 1 to 3 present an overview of the sources and amounts of funds estimated to have come from the various organizations, agencies and institutions in support of projects, technical assistance, and coordination work to date. There is also presented some estimates of needs for the future.

Overall Recommendations

Given the pervasive governmental control of the sector, it is tempting to address recommendations to governments. It is equally tempting to recommend actions on all of the problems identified. Although the following recommendations can be implemented by governments, it is more important that they be acted upon without delay by conference participants and by those active in the sector without governmental actions and policy changes. The number of recommendations is kept to a minimum in the hope that this will lead to a concentration of efforts and thus lead to visible results within a reasonable time.

Recommendation 1.

Project design should be based on willingness and ability to pay. This requires better studies in urban areas and greater dialogue with rural user communities than has been customary up to now. Results of this approach should be disseminated widely permitting everyone active in the sector to learn from the experience.

Recommendation 2.

A greater effort should be made to establish the economic justification of water and sanitation projects. Among other approaches, attempts should be made to quantify the value of services provided to commerce and industry, and to ensure that proposed investments represent the most efficient solution through a mix of service standards and technologies suited to the situation.

Recommendation 3.

There should be strict adherence to the doctrine that water systems charge for water the full cost. The means by which sewer service charges should be charged and collected remains a major problem where more experience and more information is needed and must be collected.

A Final Conclusion And Recommendation

The Water and Sanitation Decade when it closes in 1990 will find few countries with targets and goals fully reached. ASCE should join with all the other national and international agencies in advocating the continuation of a follow-up program which will support and encourage all countries to renew their work in this sector by establishing new targets, new financial needs, and new manpower training plans in the fields of water supply, sewerage, drainage, and waste disposal. Whether the target date is set at 1995 or the year 2000 is not important. It is important to support the developing nations of the world in their continuation of an effort which must be ongoing and essential to the health and welfare of their peoples.

CONCLUSIONS AND RECOMMENDATIONS: TECHNOLOGY AND ENGINEERING

Dennis A. Warner and Charles Morse***

Introduction

The papers presented in the Technology and Engineering Track were wide ranging and covered a variety of water supply and sanitation issues. The speakers put forth recommendations and views to which the audience was generally supportive. Time for questions and comments was limited because of the number of papers, even with parallel sessions.

Nevertheless, a number of relevant points were made and most of them are contained in the conclusions and recommendations presented below. These have been grouped into the following three area: Levels of Service, Role of Technology, and Institutional Aspects.

Levels of Service

Conclusion 1.

People often pay more for traditional forms of low quality water supply and sanitation services then they would be required to pay for properly-engineered, improved levels of service.

Financially viable water and sanitation projects are possible, even in low income areas. The opportunities for developing such projects are probably greater than is currently perceived. A key task of the engineer is to identify where and how such projects can be implemented.

Conclusion 2.

A two-tiered water system, whereby a small amount of potable water is provided to the users and a larger amount of non-potable water is drawn from traditional sources, is a useful option. This system is applicable in areas where resources are limited and traditional water sources are both abundant and convenient.

Recommendation:

Greater consideration should be given to the implementation of two-tiered water systems as a means of reducing costs and expanding population coverage.

*Deputy Project Director, Water and Sanitation for Health Project, Arlington, Virginia
**Consultant; Arnold, Maryland

Conclusion 3.

The denial of essential public services, such as water and sanitation, is frequently used as a means of controlling, or slowing, rural-to-urban population migrations. This is done in both the developed and developing worlds.

It is not clear whether this is at all effective in the developing countries. In fact, considerable evidence indicates that the presence (or absence) of water and santitation has little effect upon rural drift to the cities, as can be seen in the favellas of Latin America, the Bidonvilles of West Africa, and the squatter shantytowns in many other parts of the world.

Conclusion 4.

Few countries have water quality standards which are applicable to rural conditions. Since rural water supplies rarely can meet official standards, rural water quality is largely ignored.

Recommendation:

Water quality standards should be realistic and achievable and should serve to encourage progressive improvement in water quality over time. National governments should be encouraged to establish water quality standards appropriate to their rural areas. Where necessary, such standards should allow for the presence of various levels of contaminants, such as fecal coliforms, before additional water treatment actions are required. As national capabilities for meeting these standards improve over time, the standards should be raised to new levels appropriate to the country in question.

Conclusion 5.

The increased health and convenience benefits of changing from public taps to private house connections are likely to be considerably greater than the increased costs. Unfortunately, these increased benefits are difficult to assess in monetary terms and, therefore, difficult to compare directly with costs.

There is need for better methods of assessing the incremental benefits and costs of different levels of service. Willingness to pay is a partial method.

Conclusion 6.

Appropriate service levels are related to available resources and the perceptions and expectations of the users.

Local inputs from the user community are essential to the identification of appropriate service levels.

Conclusion 7.

Oral rehydration therapy (ORT) is receiving a great deal of attention as a means of combating severe dehydrating diarrheas. Some development agencies have expanded their ORT programs at the cost of reduced water and sanitation programs.

There is a tendency to put undue reliance on the long-term benefits of ORT programs.

Funding the Decade*

There is a need to provide an idea of the world-wide dimensions of:

a. the *total funding needs* in order to reach the goals of water supply and sanitation for all;
b. the *present level of funding* from all sources: domestic (national government and communities) *plus* external (loans and grants); and
c. the *sources of funding.*

The following figures may give an idea at least of the order of magnitude.

They represent only a very personal guesstimate, but based on WHO statistics with adjustments following field experience.

Projections for the next 5-10 years are based on the recent major policy shifts of the majority of governments in the developing countries jointly with the international donor community.

These projections, however, are for *new* installations only, *not* for rehabilitation or replacement of old systems, which would add considerably to the below figures.

An informal UNICEF working paper is under preparation with more details on global funding needs. In spite of the source, any figures still remain rough guesswork. The realistic possibilities hinge on many other concerns.

A. TOTAL FUNDING NEEDS

For all developing nations:
(US dollars in 1985 values)

$150 billion of which for	
conventional urban systems	$ 80 billion
poor urban and rural	70 billion
Total	$150 billion

For a ten year period, this would mean $15 billion per year. This is not much above present funding levels *but* implies wholesale use of lowest-cost technologies and massive community inputs in cash, kind and labor (and decision-making/planning).

B. PRESENT LEVEL OF FUNDING

of water and sanitation installations
in all developing countries:

*Martin Beyer, UNICEF

Total $10-12 billion per year

Assuming $10 billion per year, this investment distributed as follows for first half of water/sanitation Decade; the years 1981–1985:

Number of beneficiaries (new users) added (million persons):

	Water	Sanitation
URBAN	200M	150M
RURAL	200M	100M
TOTAL	400M	250M
Investment $ billion	25	25
per capita $	62.5	100

C. SOURCES OF FUNDING PER YEAR
present level (1985)

1. Total	$ billion	Percentage
Domestic	7.5	75%
External	2.5	25%
Total	10	100%

2. External inputs per year; by sources:

(US$ millions)

Input	Source	Trad. HC	Low-cost	Total
LOANS	World Bank	500	100	600
	Regional Development banks	400	-	400
	Intergovt. (Arabs, etc.)	130	20	150
	Bilateral [3]	200	50	250
	Commercial	300?	-	300?
	Subtotal loans	1530	170	1700
GRANTS [1]	UN system [2]	20	130	150
	Bilateral [3]	50	450	500
	NGO/PVO's [4]	-	150	150
	Subtotal grants	70	730	800

Notes: 1 - Equipment, materials, "non-supply" (cast, etc.)

2 - See table 3. for breakdown

3 - E.g., USAID and other government donor agencies

4 - Non-governmental organization/Private voluntary organizations.

LOANS + GRANTS TOTAL	1600	900	2500

3. United Nations system inputs per year
(grants in US$ million)

Source	$ million	Work Scope
WHO	30	Methods, health research, standards,
ILO, UNESCO, FAO	3	human resources development, training,
Reg. Econ. Comm.	2	education, policy promotion, situation monitoring and evaluation.
UNDP	5	As above, global development and coordination
UNCDF	5	Implementation (financing)
UNTCD (Wat. Res. Sec.)	10	Implementation/W.Africa & water res.
UNICEF	65	Implementation/global
Subtotal	120	
Emergency inputs	30	Draughts, earthquakes, conflicts, etc.
Total	150	

4. United Nations system 1987:

Number of water/sanitation/environmental specialists employed (guesstimate):

Organization	Headquarters		Field	Total	Remarks
UN Secretariat (DIESA)	New York	1	-	1	Coordination
Regional Economic	Santiago	1	-	1	)
Commissions	Geneva	1	-	1	) also work
	Addis Ababa	1	-	1	(with water
	Baghdad	1	-	1	) resources
	Bangkok	1	-	1	)
UNEP (Env. Proj.)	Nairobi	1	-	1	
WHO (Health)	Geneva	6	70	76	
UNDP/World Bank (Dev.)	Washington/ New York	20	60	80	
UNTCD (Water Resources)	New York	6	30	36	
UNICEF (Children)	New York	3	150	153	
ILO (Labour)	Geneva	1	-	1	HRD/Training
FAO (Agriculture)	Rome	1	-	1	Micro-img.
UNESCO (Educ. & Sci.)	Paris	1	-	1	Education

INSTRAW (Women)	Santo Domingo	1	-	1	Social Research
TOTAL (approx.):		46	310	356	

There is a need to better inform the development community and policymakers that, while ORT is a temporary solution, water is a permanent solution. ORT is, without question, a valuable life-saving too, but water supply is even more important as a life-maintaining intervention.

Role of Technology

Conclusion 1.

Engineering and technology are often viewed as the sole input to solutions for development problems. A narrow engineering approach can only work in areas with well-developed supporting infrastructure.

Recommendation:

Solutions to development problems must include consideration of many issues besides engineering. Because of weak physical and human infrastructure in most developing countries, engineers should be even more sensitive to non-technical issues in international development than they must be in domestic (U.S.) work.

Conclusion 2.

Engineering curricula in American universities tend to give excessive emphasis to technical subjects. Little attention is given to broader, non-technical subjects.

Recommendation:

Universities should incorporate into their engineering curricula additional courses in social sciences, economics, and public health.

Conclusion 3.

Engineers are better at understanding technologies than they are at understanding people. When technology-based solutions do not work, there is a tendency among some engineers to place primary blame upon the people.

Recommendation:

Engineers should remember that it is easier to change technologies to suit people rather than to change people to suit technologies. The engineer needs to understand the people's point of view in order to select and apply appropriate technologies.

Conclusion 4.

Successful development projects require the involvement of the user community. Engineers tend to design community water and sanitation projects from a technical standpoint with little direct input from the community.

Recommendation:

Engineers should learn how to communicate with the community in order to better understand their needs. Special efforts should be given to obtaining the support of some, especially if projects directly effect their welfare.

Conclusion 5.

The engineering profession often is too concerned with high technology and state-of-the-art solutions. This bias is especially pronounced in the universities and professional societies.

Recommendation:

Redefine the concept of state-of-the-art, changing it from a narrow, high technology focus to a broader multidisciplinary approach. Give the broad approach professional status and acceptance equal to that of the narrow approach. Define the concept to include state-of-the-art applications of appropriate technologies to water and sanitation development.

Institutional Aspects

Conclusion 1.

Institutions tend to clone themselves. This prevents the institution from growing and expanding beyond its original disciplinary bounds.

Recommendation:

Engineering institutions must find ways to remain open to innovation and receptive to ideas from other disciplines. In particular, development-related ideas from the social sciences and public health must be an integral part of engineers working in water supply and sanitation development.

Conclusion 2.

Engineering societies in the industrialized world have much to offer to, and much to learn from, the developing world. At present, communication between the two worlds is not well developed.

Recommendation:

Engineering societies in America should increase communication on a professional level and provide outreach assistance to engineering societies in the developing world.

Conclusion 3.

Engineering societies in America give little attention to the problems and needs of the third world.

Recommendation:

Engineering societies in America should be as interested in international development issues as they are in domestic issues. The societies should be willing to take positions on issues of U.S. development policies and funding levels. Membership awareness within the societies could be heightened through the establishment of international development committees, or, better yet, through the creation of a Development Division.

Conclusion 4.

There is no formal mechanism or organization in the U.S. with responsibilities to coordinate information and activities related to the Water Decade.

Recommendation:

Create a Coordinating Committee for the Decade composed of representatives of professional societies having interest in water and sanitation development in the third world. The responsibilities of this committee should include informing its members of Decade activities, marshaling public opinion for Decade initiatives, and supporting innovative approaches to Decade programs.

Conclusion 5.

The origins of the Water Decade are now almost 30 years old. After three decades of attention to water supply and sanitation issues, much has been done but much still remains.

Recommendations:

1. Because of the magnitude of problems remaining, the Water Decade should be continued through the 1990's and on into the 21st century.

2. Using the experience and lessons gained from the Decade to date, the new Decade should be continued, strengthened, and broadened to include several related new areas, such as (a) vector control and the elimination of diseases associated with water and sanitation (e.g., guinea worm, schistosomiasis, etc.) and (b) solid wastes in urban areas.

Conclusion 6.

Engineers and institutions in the developed countries often appear to consider research and development only from Europe and North America without appreciation for the knowledge of work going on in developing countries.

Recommendation:

The engineering societies should encourage research work in developing countries and attendance at foreign conferences. They also should encourage proficiency in foreign languages.

CONCLUSIONS AND RECOMMENDATIONS: OPERATIONS AND MAINTENANCE

James Jordan and Carl Bartone***

General

The single most important conclusion that resulted from the conference with respect to Operation and Maintenance (O&M) is the realization that with the need for quality water in lesser developed countries being so enormous as was noted by several speakers at the conference, the need for effective maintenance programs is, if anything, even greater.

A general review of the O&M track reveals two other important observations can be made with respect to O&M. The first is that recognition of the necessity of effective operations and maintenance programs in order to sustain the investments in water and sanitation that have made or will be made in LDCs is growing. This is demonstrated quite effectively by the fact that O&M is afforded equal status at the conference with the other three tracks. In addition, several speakers gave examples of bi- or multi-lateral agencies that have included a significant O&M component in their project. The World Bank, among others, was cited as an example of this trend. One speaker concluded that "lack of maintenance *is* the problem" (with respect to providing water for all people in LDCs). The conference verified that poor O&M is a serious problem, but that steps are being taken to change this situation.

The second observation is that successful water supply and sanitation programs require that a multi-disciplinary approach to O&M be taken. Good O&M programs take in account human resource development, appropriate technology, willingness and capability of the recipients of the water supply and/or sanitation system to financially support the system as well as logistics and the availability of spare parts and supplies. O&M cannot be considered as simply an engineering exercise or a funding question. Effective O&M is a planned program incorporating a number of different disciplines.

Specific Conclusions and Recommendations

Conclusion 1.

Inadequate attention is being given to operations and maintenance when water supply and sanitation (ws&s) systems are in the planning stage.

*Operations and Maintenance Specialist, Water and Sanitation for Health Project; Arlington, Virginia

**Project Officer, World Bank; Washington, D.C.

Recommendation:

External support agencies (ESA) should determine before funds are committed to a project if adequate resources and plans are available to operate and maintain the system.

Conclusion 2.

For projects in rural areas, communities must be actively involved in the construction, operation and maintenance of their ws&s systems if the systems are to continue to function. Local government will not be able to afford to take care of all of the ws&s systems that will be constructed.

Recommendation:

Project planners must incorporate community participation into any project being undertaken in rural communities.

Conclusion 3.

Most small communities provide part-time operators for their water system. This does not provide sufficient time for all of the maintenance tasks, particularly preventive, that should be done for water supply systems. In addition, operators (caretakers) are not given training until the water system is under construction.

Recommendation:

Training of operators should be initiated earlier perhaps, during the design phase.

Conclusion 4.

Poor and inadequate communication is a major cause of ineffective operations and maintenance. In large urban systems, managers are often not given proper direction from upper management with respect to acceptable operating parameters, and mid and lower level managers do not keep their management informed of system conditions and potential problems. In rural settings, water committees are not given proper instructions on how to report problems to regional maintenance crews.

Recommendation:

Project planners need to insure that proper information and data will flow among those agencies, governmental departments, consumers and others involved in ws&s system O&M.

Conclusion 5.

Ineffective record-keeping and lack of analysis of existing data can lead to numerous problems in operating and maintaining ws&s systems. In particular, proper equipment and network maintenance is difficult to achieve without good O&M manuals and system drawings.

Conclusion 6.

We should be collecting data on the number of systems that are still working in addition to data on those that have been built. This is the only way that an accurate picture of the status of O&M in LDCs can be determined.

Conclusion 7.

The training and development of operators is much less expensive than that for engineers, and the former are more likely to remain in the country.

Recommendation:

Project planners should take special care to insure that funds allocated for training are directed so as to maximize the return on investment. Both technical and management training for O&M personnel is needed.

Conclusion 8.

On the urban side, the need for appropriate technology was noted. In particular, complicated automatic control systems need to be eliminated.

Recommendation:

It is recommended that the establishment of certification programs could help to raise the level of competence of ws&s system operators, particularly for small community systems.

Conclusion 9.

Local manufacturing of equipment is extremely important for effective operations and maintenance.

Conclusion 10.

The cost of O&M when it is carried out at the village level is much less than the cost of centralized operations and maintenance.

Recommendation:

Project planners need to emphasize the importance of VLOM (village level operation and maintenance) equipment when designing a ws&s project. This concept goes hand-in-hand with the need for community participation.

Conclusion 11.

Several alternatives to chlorine based disinfection systems were presented. The conclusion is that they are promising, but require further field work.

Recommendation:

It was suggested in one session that the measurement of total coliform is of little value in untreated rural ws systems. Testing should be for *E. Coli* as an indicator of fecal contamination.

Conclusion 12.

It was emphasized that in addition to the need to protect against microbiological and chemical disease causing agents, it is necessary to assure acceptable aesthetic quality of drinking water.

Without acceptable taste, odor and appearance, some consumers may resort to unregulated water even though the latter may be unsafe from a health perspective.

Presentation made by Ing. Luis U. Jáuregui at the Closing Session of the International Conference on Resource Mobilization for Drinking Water Supply and Sanitation in Developing Nations.
San Juan, Puerto Rico - May 29, 1987

Ladies and Gentlemen,

I feel very pleased and honored with this opportunity of addressing you at this important and challenging Conference in my capacity as President of the Mar del Plata United Nations World Water Conference.

I would like to start with some reflections and comments on the Conference itself and, particularly, on the International Water Supply and Sanitation Decade.

In the first place, I must again emphasize that the Decade was a political decision taken by all governments of both developed and developing nations, who brought to the international arena the issue of the existing critical situation in the area of Water Supply and Sanitation, and undertook the commitment to find and implement solutions in a cooperative effort. This commitment has since been renewed and the launching of the International Water Supply and Sanitation Decade in 1980 is a good example of it.

Secondly, -and allow me to personalize in Professor Wolman and others present here today- I would like to mention that the role many of us played and the personal satisfaction we derived from our involvement in this important project, concurrently entailed a challenge and a commitment to assure implementation, of which we were well aware and which we seriously assumed.

In third place, let me say that although I have some doubts about the degree of success already achieved, or that may reasonably be expected, from the Mar del Plata Action Plan that covers the overall spectrum of topics analyzed at the Conference, I feel strongly positive and confident about what we have done so far and, better still, about what we are presently doing, beyond statistics, in the specific area of Water Supply and Sanitation.

The so-called Mar del Plata Action Plan approved by the Conference, includes the notion and efforts in Water Supply and Sanitation to be accomplished in the 80's on a global basis, defining some strategies, objectives and implementation approaches that by no means represent a "radical change" from the manner in which we had been performing at that very time, and from the methods and criteria we had been using.

In fact, and beyond an excessive idealism -if the willingness to supply drinking water and sanitation to all the people in the world, if possible, by 1990, can be termed that way- I strongly believe that the fundamental idea of change was full of pragmatism, that it recognized the need for realism, tried to break away from paternalism, and made people, governments and society as a whole, responsible for reversing the problem.

Actually, the Decade approaches called for:

- increased coverage of services - levels to be set up by countries, prioritizing underserved rural and periurban areas.

- complementarity in developing water supply and sanitation.

- programs promoting self-reliant systems that people can afford.
  - we started managing the concept of cost-effective techniques adapted to each specific social, cultural and economic environment.

- mobilization of local resources (existing/available/to be developed, consumption, tariffs, productivity, efficiency, redistribution of government expenditures).

- the need to prioritize community involvement at all stages of project implementation, starting with the identification of the need for these types of services.

- association of water supply with relevant programs in other sectors, particularly with primary health care, concentrating for example on health education, human resources development and the strengthening of institutional performance.

- the intensification of international technical cooperation, and encouragement of technical cooperation between developing countries (TCDC) - calling for determination and, essentially, effort coordination from those countries and institutions (governmental and non-governmental organizations) that are in a position to cooperate.

Here, I would like to stress again that the Decade's main action in the area of Water Supply and Sanitation should be carried out at country level or, better said, at developing country level, since the latter were the principal recipients of the Decade's efforts and final results will depend on what countries are doing as well as on our ability to organize the necessary technical and financial assistance to enhance the countries' capabilities. This means shared responsibility for success or failure.

Let me tell you, since I heard some people say the Decade is failing, that the Decade's torch is still blazing in most of the countries.

The Decade was launched at the most unfavorable time for developing countries; they were affected by a deep global economic recession and most member nations were overwhelmed with the problem of a huge foreign debt -probably the worst in their history- depriving them of the availability of more local and foreign funds.

This led us to re-emphasize all the other areas mentioned before, i.e. the Decade approaches; plus permanent motivation of political levels in each country; plus the need to optimize (rehabilitation, operation and maintenance) investments made in existing facilities.

And really here, in these areas, is where we found the most critical and sectorial weaknesses and constraints. In effect, in global terms, we found:

- Insufficient water supply and sanitation and financially weak sector institutions;

- a widespread neglect of sanitation development as compared with water supply;

- insufficient attention to operation, maintenance and rehabilitation of existing water supply and sanitation systems;

- inadequate technology, level of services, design and performance criteria used;

- a lack of organized community participation and hygiene education, and

- insufficient coordination among donors and external support agencies (technical and financial) and among them and recipient countries, in identifying what is needed, how and when the problem could be solved, and who would participate in the solution.

However, we started to work knowing we had traced the problem to its deepest roots; that strengthening the institutions, developing people, getting people involved and educated, motivating policy-makers, among other activities, requires patience, perseverance and time; that time should not extend beyond 1990, although realizing our work and efforts would not end at that very moment because changing and adapting personnel, institutions and social values and cultures must per force be gradual to allow people to cope with these changes.

Colleagues, to show you we are already on the move and producing results, I'll just mention the amount of knowledge, experience, effort and money spent in the area of technology, looking for simplicity, affordability and performance. And this was the result of a really cooperative effort that paved the way for other efforts.

Trying to avoid comparisons, have you noticed the large number of people attending the Institutional Development and Human Resources sessions in this Conference? This is a clear indicator that we are making inroads and that results will come.

Continuity of actions and programs in the identified critical areas, no doubt, will possibilitate achieving the necessary operating changes in a permanent and positive manner.

At present, we are working on the foundation of a very tall and solid building, but we are still working very close to ground level and there is nothing too spectacular for people to notice from a distance. It is common knowledge that a soldier in a battle sees only the immediate action -and confusion- without knowing how the battle as a whole is progressing.

I consider this Conference an example of the necessary continuity I mentioned earlier on. New ideas, discussions, proposals, conclusions and recommendations in the 4-track areas covered, that coincide with the main sectorial constraints, no doubt will supply new inputs for actions.

I would like to insist, also, in the specific area of people motivation. This motivation can be exercised upwards or downwards, i.e. governmental authorities, politicians on one hand, community and users, on the other hand.

I shall refer to the first one. I do not want to act as the politicians' champion and I am not a politician myself, but I feel frustrated with the approach used by many of us in our professional capacity when dealing with policy-makers and decision-takers. As in football or basketball, attack is the best defence; likewise, we tend to blame them for our problems to hide our own shortcomings.

The change needed here is that we must be able to reach them with clear-cut proposals and projects that are well defined, well prepared and have sound justifications; projects they can easily "buy" if we also manage to sell them the idea of the benefits he or his political group or party will reap from them.

I heard a concept at this Conference with which I readily concur. This is that "the largest support votes a politician can get come through well-managed public utilities that serve the people's needs". Colleagues, this is "music" for me; I leave it to you.

There is another comment I wish to make. We have been talking about changes. For changes to be positive they must be led by the top management in a Water Supply and Sanitation agency. Now, top management is a broad concept; actually, it is integrated by "managers", and this is precisely what we desperately need in the sector; not bureaucrats or frustrated professional engineers in managerial positions.

We must work on a short and medium term basis in this area. We need to incorporate and develop managers capable of interacting with people, leading people, planning and organizing with sound economical concepts the use of all the available resources. People with firm beliefs in concepts such as productivity, efficiency, cost recovery, among others.

Developing countries simultaneously experience a large number of high priority problems to resolve and cannot afford to continue acting without adhering to these principles if they are to avoid postponing any longer the solution of their population's basic needs. However, the education and development of these managers takes time.

Allow me now to revert to one of the areas to which I assign fundamental importance, i.e. Human Resources and Institutional Development, and to emphasize again that the ultimate objective of Institutional Development and Human Resources is to enable the countries to manage their own resources with their own people, according with their own cultures, satisfying their own identified needs. We must avoid the temptation to create permanent dependencies in this area.

With no intention to compete with our admired Professor Abel Wolman, who three years ago gave us his 3-M's rule as identification of the Decade's major constraints to obtain its objectives (Motivation - Manpower and Money, in this order), I would like to give you my 5-C's vision of what is needed to achieve the Decade's objectives in the future, beyond what we shall do by 1990.

CRISIS. This will be the characteristic of the environment in which we'll have to work. Its nature may be economical, population growth or another one. This crisis will call for pragmatism which in turn will require, as a constant, the optimization of available resources. The approach will be "Do more with what we have".

CONTINUITY: This concept or idea is basic and fundamental. Many of the actions already implemented or to be implemented have to deal with cultural and behavioral changes of people, institutions and/or societies. They take time but produce better and more permanent results. For instance, switching over from centrally managed to community-based projects will not happen overnight. Continuity is synonimous of stability of people involved in the projects and is of significant importance.

CREATIVITY: The rational change of a critical situation calls for the permanent search of new alternatives to solve the problems. The search for innovative alternatives in technological, organizational and educational aspects as well as on the level of services supplied, among others, is fundamental.

And, last but not by any means least,

COMMITMENT AND COURAGE: To do and modify all that needs to be done and changed in order to achieve the Decade objectives and to increase community participation -and this includes both rural as well as urban populations- in the development and operation of the projects.

Colleagues, on the occasion of a pre-Conference panel, it was very rewarding for me to listen to prestigious engineering societies of the U.S.A. expressing their interest in participating more actively than up to the present in the development of the Decade, and undertaking the commitment to intensify communications among them and to provide outreach assistance to professional associations of developing countries.

The words used by U.S. Ambassador J. Mac Donald were of the same positive nature when referring to his country's resumed support to the International Water Supply and Sanitation Decade objectives.

The Decade is not failing. The millions and millions of people whose basic water and sanitation requirements were provided during this period, perhaps would still be lacking them where it not for the tremendous global effort the launching of the International Water Supply and Sanitation Decade entailed. But the magnitude of the effort to be made and the road that was chosen, i.e. structural change, require more time than the years in a decade.

I am persuaded that the world community as a whole should start a New International Decade, that of the 90's, to assure the necessary continuity to the actions already initiated.

It was said at this Conference that for millions of people Water Supply and Sanitation could be a commodity product, but they are still now an absolute necessity for millions and millions of them. We must work hard to give them the right answer.

WATER AND SANITATION FOR ALL — FROM UTOPIA TO REALITY

Martin G. Beyer *

(Speech held at the banquet of the American Society of Civil Engineers during their International Conference on Water Supply and Sanitation in San Juan, Puerto Rico, 28 May 1987)

Antes que nada quisiera rendir homenaje a nuestros huespedes sobre todo al Colegio de Ingenieros y Agrimensores y el Instituto para el Fomento de Puerto Rico. Quisiera bien ofrecer mis agradecimientos para toda hospitalidad y la recepción tan amable y calurosa que se le nos ofrecieron.

Before beginning my speech, I must explain to you that I am presenting this in my capacity of Official Orator of the International Drinking Water Supply and Sanitation Decade (1981–1990).

It came that way during the first consultative meeting of donor countries to the Decade, organised by the World Health Organisation jointly with the Federal German Technical Co-operation Agency (GTZ). This was held in Königswinter, a small town on the Rhine near Bonn in October 1984 and consequently called the "Oktoberfest".

*UNICEF

The concluding act was held in a wine cellar late one evening. It was preceded by rather thorough testing of the local wines. As you know, those of us who deal professionally with water, know how dangerous that can be, so we never touch the stuff anyway. Therefore we went in for the wine with our usual professional thoroughness.

When this spirited meeting drew to an end, the then UNDP Co-ordinator for the Water Decade, Dr. Peter Lowes, Alexander Rotival's predecessor and a distinguished Canadian lawyer and political scientist, rose. He told us that at Oxford University by tradition there always is an official orator for such occasions. Furthermore, that eminent official was supposed always to speak in Latin. Then Peter Lowes proceeded by anointing me as the official Orator for the Water Decade, exhorting me to likewise thank our hosts in Königswinter – in Latin. Which I did – Kitchen Latin, of course.

With this preamble I take the pleasure now at long last to begin my speech by addressing you in the appropriate manner:

- PROFESSOR ABEL WOLMAN, EXCELLENTISSIME DOMINE, PRAEFECTUS HONORIS CAUSA (Most Distinguished Honorary Chairman) SOCIETATIS AMERICANAE FABRUM TOGATORUM VEL CIVILIUM (of the American Society of Civil Engineers);

- DOMINA ET DOMINE HOMO FABER HENRICE RUIZ (Ingeniero and Mrs. Enrique Ruiz), PRAEFECTUS COLEGII HOMINUM FABRUM ET AGRIMENSORUM (President of the Colegio de Ingenieros y Agrimensores de Puerto Rico);

- DOMINE HOMO FABER MONTANILLE MONTANARI, PRAEFECTUS ILLAE CONFERENTIAE (Engineer Monty Montanari, Chairman of this Conference);

- EXCELLENTISSIME DOMINE IOHANNES MACDONALD VEL FILIUS DONALDI (His Excellency John MacDonald), LEGATUS (Ambassador) CIVITATIUM CONSOCIATARUM AMERICAE (of the United States);

- DOMINAE ET HOMINES INGENUI (Ladies and Gentlemen):

The subject of my speech to you tonight is "Water and Sanitation for All - From Utopia to Reality, or in Latin: DE DECENNIO GENTIUM AQUAE ET SANITATIONIS AMBIENTALIS - AB UTOPIA AD REBUS VERIS. In other words, it will give you an idea in a nutshell, how to get from nice resolutions at major international meetings to satisfy the needs for basic life-giving services to an entire planet.

During the present conference we had a good look at the Water Decade. We had quite a deal of statistics. They indicate that up to now, we have just been able to keep up with the population increase. We are still left with the same figures for the overall needs to build new installations for water supply to 1.3 billion people and sanitation for 2 billion, as where we were seven years ago.

In reality, I believe no one of us at the beginning of the Decade in 1980-81 thought that the noble goal of water and sanitation might be reached by 1990. It would have meant providing new water supply and sanitation installations to 500,000 people each day during the ten years of the Decade.

Yet, it was the only goal that should and could be aimed at. It would take longer than the ten years, but there is no reason to despair.
A lot has been achieved.
We have learned a lot.
After all, some 400 million new water users and sanitary excreta disposal facilities were added in the first half of the Decade.

Our work, not only for water and sanitation but for development as a whole, is like warfare. It is a war on want, a war on hunger and thirst, a war on disease and poverty. As in a war or in military service for that sake, don't ask too many questions - go ahead and do it!

Work for development also is like criminality: You have to do it on a sufficiently large scale. Only then it pays - and you are less apt to get punished. Thus there is hope for the Water Decade, if we apply these principles properly.

The objectives: Health and well-being for children, mothers and others

The objectives for our work on water supply and sanitation are founded on those of the United Nations - working towards a better and peaceful world for ourselves and our children.

The workings of the United Nations are not easy to understand for those, who are not directly engaged in the day-to-day work within the system. Yet, you all are part of the United Nations by implication, through the partnership of your countries in the main task, which is peace-keeping in this world, underpinned by the second major task - that of promoting and supporting development.

The importance of the efforts towards development is illustrated by the fact that overwhelming part of the about 50,000 people directly employed by the United Nations and its different organisations, work with very concrete matters of technical, economic and social development.

The work of UNICEF, jointly with other organisations, notably the other major partners, now known as "The Four" - UNDP (United Nations Development Programme), WHO and the World Bank, aims at the survival, better health and well-being of children, mothers and others.

The results are anticipated to be a higher quality of life for the poorer populations. In its turn, this also on a longer term basis would contribute to the lowering of the population growth rate.

Some results: The Southern Sudan experience

On the short-term level, improvements in water supply and sanitation rapidly result in tangible improvements in peoples' lives. A good example is Southern Sudan, an area as remote as any in this world. The distances to the nearest ports, either Port Sudan on the Red Sea or Mombasa in Kenya, are 2,500 kilometres (1,500 miles). The landscape is a vast wooded savanna, half of the year very hot and dry, the other half of the year, flooded by the White Nile and converted into the world's largest swamp area, the "Sudd".

UNICEF helped start up a water-well drilling cum handpump installation programme there in 1974. We had to start from scratch, following the end of seventeen years of a literally bloody civil war that had devastated large tracts of land.

The inhabitants belong to several tribal groups, such as the Dinka and the Nuer, living in villages scattered all over the vast plains. They are half-nomadic cattle-herders and derive the rest of their living from subsistence farming.

Water used to be fetched in buckets from faraway miserable water holes and contaminated at that. A survey carried out ten years ago, showed that women, invariably the beasts of burden for water and firewood in the developing countries, the women in Southern Sudan on the average spent six hours a day, just to fetch a few daily gallons of water for their homes from miles away.

Another part of the same survey showed, what happened to women in villages that had been blessed with handpumps close to their homes. Of the many hours saved, the women used some hour and a half to work in the fields for food production and one hour for brewing of beer. The local beer in those parts of Africa in moderate quantities is a safe and nutritious drink. The beer-brewing households also sell their beer to others and in this way contribute to the family economy.

Last but not least, two hours were recorded as spent on rest. The restoration of physical and mental energy is as good a health factor as any.

In a way, this particular handpump programme in Southern Sudan also contributed to the peace-keeping efforts of the United Nations. All men, old and young, walk around, each armed with three or four short but nasty spears. They use them for hunting but also for fighting. In fights over cattle, women and water, they used to inflict terrible spearwounds on one another.

I remember at an early stage in our work in Southern Sudan, visiting a government hospital in a town, called Gogriel. This is way out in the boondocks, 100 kilometres north of Wau. That hospital had one hundred beds and was staffed by a nice young medical doctor from Uganda.

Of the hundred beds, however, ninety were filled with men and boys, bandaged all over and treated for spearwounds. A few years and several hundred new drilled wells and handpumps later, the same hospital in Gogriel had none of the spearwound patients and could now serve the more "legitimate" cases. The advent of the handpump to the villages had eliminated a major cause for conflict, the scarcity of water...

The major problem - population pressure

Returning to the world-wide scene, the constraints and problems encountered, are legio. The main problem we all are facing, is that overshadowing one of over-population and the continuing high rates of population growth. This has led to an over-exploitation of soils and a deterioration of the environment with dire consequences for the water resources.

The lowering of water tables and the increased contamination and pollution of surface and groundwaters are part of this problem. This touches on all aspects of planning, design and implementation of any type of water supply and sanitation programmes.

The impact of the population increase stands out clearly, when looking at the figures for the total world population:

In Roman times, 2,000 years ago, around the birth of Christ, our forebears were only	100 million
A hundred years ago, when our grandparents were teenagers, 1887	1,300 million
Sixty years ago, in 1927, when Charles Lindbergh flew to Paris	2,000 million
Right now in 1987 we are	5,000 million
By the year 2000 we'll be	6,000 million
and, at least according to forecasts by the Club of Rome and others, by the year 2150 the population curve would have reached a plateau of some	10,500 million people.

Many of the other problems and constraints in trying to reach the goals of the Water and Sanitation Decade, technical, economic, social and political, have been analysed and discussed in our conference here in Puerto Rico during the last few days. They are described abundantly in many publications.

Therefore there is no need for me to bore you with a repetition of them. Rather, I would like to talk to you about the experiences, the action and the solutions.

The past is the key to the future

First the experiences.

To quote one of the fathers of the geosciences, the great British geologist Charles Lyell from the beginning of last century, who said about the rocks and the soils of the world: "The past is the key to the present".

For modern development, not the least for the development of appropriate infrastructures, including water and sanitation, this insight could be expanded to include the future.

Much of what the ancients did, can still show the way. Vide the works of the ancient Chinese, who drilled 3000 feet deep wells two thousand years ago, the great water works of the Khmer around Angkor Wat or of King Anuruddha/Anuwratha in Sri Lanka and Burma or of the Mayas in Yucatan.

Even though their services were incomplete, the ancient Romans showed the way. They had their famous *aquaeducti* managed and cared for by the *curatores aquarum*. The public latrines, *foricae*, in the Roman Empire were kept up – by special *conductores foricarum*. These levied small fees from the individual users, much as is done in the popular public latrines *cum* bathhouses in Patna or Ahmedabad in India. They gave in a similar way to the individual citizens a chance to relieve him – or herself and then to wash themselves properly afterwards.

In all these activities to improve people's lives, the important role of the non-governmental or private voluntary organisations in introducing innovations and stimulate development should be highlighted. Without their pioneering we would not be where we are today. We can thank them for many of the low-cost but efficient technologies and approaches we now use all over the world, the use of rapid drill rigs or VIP latrines (not necessarily for VIPs from the United Nations but Ventilated Improved Pit latrines). Give them a hand.

Highlights among the experienes

Among the many experiences that at long last are being brought into play under the Water Decade and its vigorous continuation beyond 1990, I would only highlight a few:

* system thinking: water and sanitation as entire social systems, based on sound and - in the widest sense of the word - appropriate technologies.

* Basis of the works in the communities.

* Balance and complete interplay between the people of the communities, the government systems and the external donors.

* Integration of water supply and sanitation with the other sectors of development.

* Make sanitation with excreta disposal palatable to the public. This is a matter of communication. (A prime reference is the famous book by Charles H. Sale, "The Specialist", about the experiences of a latrine builder in the Appalachians in the 1920s.)

* Acceptance by the users-to-be and willingness to pay.

As an example I remember a visit to a valley in the Hindukush mountains of Afghanistan many years ago. With UNICEF's assistance, the government just had installed a gravity feed scheme to one village, the first one to benefit from this kind of service in that particular area. Water from a protected natural spring up on a mountain side was led through a 4 kilometre pipeline to public standposts or taps around the mosque in the middle of the village.

When we arrived, there were at least two hundred elders from twenty villages around, serious, bearded gentlemen in turbans. They handed over petitions to us for more systems like this first one. A few had signed the petitions in writing. Most had just provided their thumb prints - the literacy rate being very low.

Then they explained to us that when some months earlier, the government engineers and the UNICEF water project officer had come up to the first village in order to plan for the new system, everybody in that village had run away into the hills. They thought it was the tax collector...

Slowly and reluctantly they returned and had had themselves half-heartedly convinced to build this system. Most of them, all men, of course, thought, why bother, since it anyway was the women who would carry the water.

When at long last, the water began to flow by itself into the village through the new system, even the men saw the benefits to themselves and to the women. In this case, the practical demonstration won them over, and also generated a demand from a large area around with a clear will to pay.

* Motivation and education, by some called "social marketing", is an indispensable part of the whole system.

This is nothing new. From two thousand years ago there is an inscription on a house-wall in Pompeii: QUISQUIS HIC MINXERIT AUT CACAVERIT IOVIS OPTIMI MAXIMI IRAM PROVOCABIT - "Whosoever would pee or defecate here, will provoke the wrath of Jupiter, the Biggest and the Best". Apparently, the warning was not heeded, since, as you know, Pompeii and its inhabitants were destroyed by the eruption of Vesuvius in the year 79 A.D.

* Proper follow-up of the activities. There is the need to secure continuous operation and maintenance of the facilities. We have to continuously learn from the activities, and to use monitoring and evaluation as managing tools.

* Patience. The work to get water supply and sanitation facilities to everybody also means changing and improving the ingrained hygienic habits of at least half the world's population. Rome was not built in one day. The same applies to our world of the future. No programmes or projects can be expected to have lasting effects, unless the time allowed for their individual implementation would be seen in terms of at least half or a full generation - somewhere between fifteen and thirty years.

The action and the ingredients for success

Now to the action. With the experiences accounted for, we have the ingredients for success.

* The technologies are reasonably well known. That is a question of adaptation and the appropriate proportions in the mix between different levels of technologies.

* Financing and funding. Together with the technologies this is a lesser problem. When there is a good project, the funds normally can be found. With the new orientation by the international donor community, spearheaded by the World Bank, for more emphasis on the low-cost options for water and sanitation, a much greater impact on country-level policies, planning and implementation can be expected.

* Again, the proper balance between users, government organisation, the private sector and external agencies.

* People constitute the major factor. We have first the users. It is for their benefit that the work is carried out. The health and well-being especially of the poorest two billion people on this globe hinges on the quality of their water supply and sanitation. The users also are or should be the major partners in planning for and carrying out and in the upkeep of the installations.

* There are secondly the policy- and decision-makers. There are the politicians and the administrators, who need the motivation and the political will.

* Thirdly the teachers in a wide sense, for spreading the sanitation and health message. They can be school teachers, health workers, engineers or craftsmen. Rain, sleet or snow, the message must get through, then to be followed by training for proper use, operation and maintenance.

* Fourth, there are the planners and implementers. They include the engineers, not just as technicians but as citizens, members of the community, members of a multidisciplinary team. They, if any, also share the responsibility as teachers.

* Other ingredients, more specifically for us who participate in this conference, and for our confratres and sorores in the professions, will follow in the next chapter, which also is the last.

Aim at the stars

Reaching to the end, my famous last words will constitute a SALVE ATQUE VALE (Welcome and Goodbye). It can also be summarized with the words, WHERE IS THE BEEF?
Or as we say on the Forum Romanum:
UBI CARO BUBULA?

Thus the question is: What is in it for all of us?
Even more so – what is in it for the ASCE members?

I would answer this question with three items:

1) Understanding
2) Opportunities
3) Communications

Number One: UNDERSTANDING

This concerns the understanding of development and the place of water and sanitation in it. It is a matter of understanding the dimensions and implications of the problems and their solution from all angles: Humanitarian, social, cultural, political and economical.

Number Two: OPPORTUNITIES

This is a two-way street. On the one hand there is the chance to improve the lot of millions and billions of human beings and to help diminish the great iniquities between the Third World and the others, and within the communities to give a lift to the poorest populations.

On the other hand there is the possibility for sound business, in the form of supply of equipment and materials, or as transfer of knowledge. As always in this world, one good thing does not exclude the other.

Number Three: COMMUNICATIONS

Learn more: For the ASCE members and their associates – more information of the kind that was provided during the present conference.

The main organisations in the United Nations system, involved in this work, are willing to help spread the knowledge of development work and the mechanism.

Include more information in the professional journals and other publications. A good example was the article on the Water Decade some years ago in ASCE's journal Civil Engineering.

Arrange for this topic to be presented at regional or local ASCE meetings around the United States.

Associate with other professional fields: With the systems approach to water supply and sanitation, the inputs into the actual programmes and projects gain more weight, if done jointly with the health and social sciences. A closer joining of forces, as ASCE has it in the fields of health with APHA, the American Public Health Association through the American Academy of Environmental Engineers, and with the corresponding organisations dealing with the social sciences might be highly productive. This would add weight and relevance to ASCE, not only as a professional but also a civic body, helpful in guiding United States policy-makers in such domestic and foreign issues, as are based on technologies.

Engineers are often too modest in relation to their role and responsibility in the community. A broadening of their scope would be to the advantage of programmes, projects and the people who benefit therefrom.

Major issues need closer links between engineering and other fields: With the issues concerned with the more immediate goals of the Water Decade come others, closely connected. They may be better safeguarded in the industrialised countries but are of equal importance in the Third World.

The concern for water resources, their conservation and management is one. There is an urgent need to revive the ideas and promotion behind the Mar del Plata Action Plan for the water resources of the world, as it came out of the United Nations Water Conference in 1977.

There is also the question of saving what we can of the rapidly deteriorating environment around the world. It is symptomatic and encouraging to learn e.g. of the World Bank in the course of its present re-organisation setting up a separate environmental department. ASCE again could have a major role in furthering an optimal adjustment of engineering planning to the environmental needs, especially when its members get involved in work in the developing countries.

These outlooks into wider realms bring us to the close of this most pleasant conviviality, a feat of linkage in itself on a personal basis. Here the old adage - and that will be the last Latin words you'll hear from me tonight - is well in place:

MENS SANA IN CORPORE SANO

A Sound Mind in a Healthy Body.

As to the Water Decade, it sometimes makes me think of a very rural Swedish saying: The old farmer takes his son outdoors one night, points upwards and says: "Aim at the stars, my son, at least it can land you on top of the cowshed!".

Aim at the stars - it can land us on top of the water and sanitation problem. With some perseverance, Utopia can yet become reality!

ASCE INTERNATIONAL CONFERENCE ON
RESOURCE MOBILIZATION FOR DRINKING
WATER SUPPLY AND SANITATION IN DEVELOPING NATIONS
CARIBE HILTON HOTEL - MAY 26-29, 1987

EXHIBITORS AND SPONSORS

AMERON PROTECTIVE LININGS DIVISION

ATLANTIC PIPE CORPORATION

CAMP, DRESSER & MCKEE, INC.

CH2M HILL, INC.

HAZEN & SAWYER PC

INDUSTRIAS VASSALLO, INC.

INSITUFORM DE PUERTO RICO

KROFTA ENGINEERING CORPORATION

MALCOLM PIRNIE, INC.

METCALF & EDDY, INC.

THE PITOMETER ASSOCIATES, INC.

ROY F. WESTON, INC.

RURAL HOUSING IMPROVEMENT, INC.

SUBJECT INDEX

Page number refers to first page of paper.

AUTHOR INDEX

Page number refers to first page of paper.